HIGH CONDUCTIVITY SOLID IONIC CONDUCTORS

RECENT TRENDS AND APPLICATIONS

HIGH CONDUCTIVITY SOLID IONIC CONDUCTORS

RECENT TRENDS AND APPLICATIONS

Editor

Takehiko Takahashi

Nagoya University

World Scientific

Singapore • New Jersey • London • Hong Kong

Published by

World Scientific Publishing Co. Pte. Ltd.
5 Toh Tuck Link, Singapore 596224
USA office: 27 Warren Street, Suite 401-402, Hackensack, NJ 07601
UK office: 57 Shelton Street, Covent Garden, London WC2H 9HE

British Library Cataloguing-in-Publication Data
A catalogue record for this book is available from the British Library.

HIGH CONDUCTIVITY SOLID IONIC CONDUCTORS
Recent Trends and Applications

ISBN 978-9971-5-0750-3 (hardcover)
ISBN 978-981-4434-29-4 (ebook for institutions)
ISBN 978-981-4551-88-5 (ebook for individuals)

For any available supplementary material, please visit
https://www.worldscientific.com/worldscibooks/10.1142/0729#t=suppl

PREFACE

Many books on "High Conductivity Solid Ionic Conductors" have been published up to the present. Some authors have referred to these high conductivity solid ionic conductors, as "Superionic Conductors" since the early 1970s. There are also many authors, however, who regard this terminology as inappropriate, since a connection is implied with electronic "Superconductors". As the ionic conductivity in solids can vary over a wide range of values with temperature and the ionic conductivity values in solids are in the range of those in "Semiconductors" at the highest, some authors have referred to these high conductivity solid ionic conductors, as "High Conductivity Solid Electrolytes" or "Optimized Ionic Conductors". Also a term "Fast Ionic Conductors" has been used for high conductivity solid ionic conductors since the early 1970s. However, as the electrical conductivity is expressed by the product of the concentration of conducting species and its mobility, there are some authors who believe that this term is not appropriate to apply to high conductivity solid ionic conductors because this is sometimes misunderstood as if merely the mobility of ions is considered in discussing high ionic conductivities in solids.

However, a special terminology may be used at least for high conductivity solid ionic conductors like α-AgI and β-alumina, as the conduction for which there is a need for explanation by some special mechanisms.

In 1986, a term "Hyperionic Conductors" was proposed in place of "Superionic Conductors" at the "Gordon Research Conference on Solid State Ionics". The term "Solid State Ionics" which was the theme of this conference, was named originally by T. Takahashi, the editor of this book, about a quarter of century ago. It means a field involving all kinds of ionic phenomena in solids, and "The International Conference on Solid State Ionics(SSI)" is to be held every other year. The 6th Conference was held in 1987 in the Federal Republic of Germany, and the papers shown in the Figure were presented at this conference. The "Solid State Ionics" journal has been published bi-monthly since 1980. In 1987, "The International Society for Solid State Ionics" was founded. This book is planned partly in recognition of that Society.

The editor would like to thank all of the authors of this book for their excellent contributions and to express his thanks to Ms. P. H. Tham and Ms. Kim Tan, the editors of World Scientific Publishing Co. Ltd. for their encouragement to publish this book.

Nagoya, Japan
December, 1988

T. Takahashi

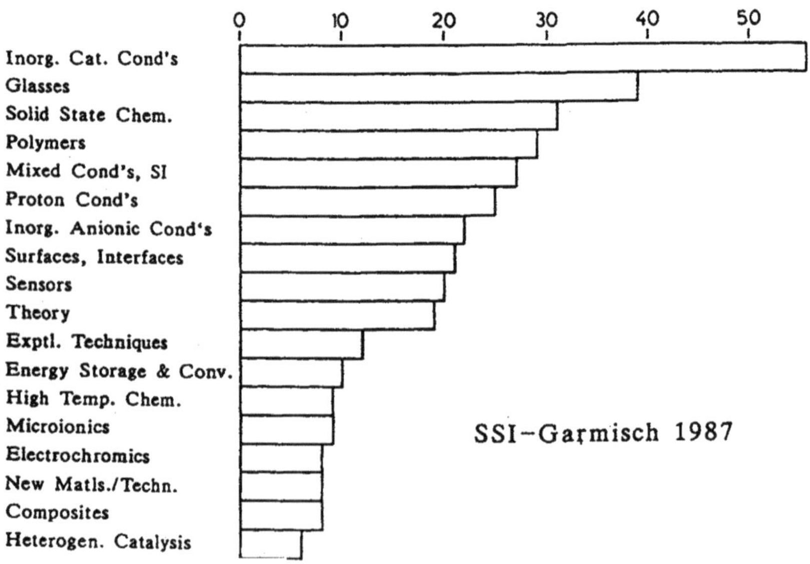

CONTENTS

HIGH CONDUCTIVITY SOLID IONIC CONDUCTORS

RECENT TRENDS AND APPLICATIONS

HIGH CONDUCTIVITY SOLID IONIC CONDUCTORS: THE PAST AND THE PRESENT

Takehiko TAKAHASHI

Professor Emeritus of Nagoya University
Present Address:2-106 Umemorizaka-Nishi,Meito-Ku,Nagoya
465 Japan

ABSTRACT

Though the phenomena of the electrical conduction due to the movement of ions in solids have been recognized since the mid-nineteenth century,high conductivity solid ionc conductors have been studied widely only after the 1960s. Here, by high conductivity solid ionc conductors, we mean solids exhibiting high ionic conductivities at fairly low temperatures below their melting points. In this chapter, the history of solid ionc conductors is described briefly.

1. INTRODUCTION

Ideal solid ionc conductors are unique in the respect that they block the electronic current in spite of their exhibiting electrical conduction in the solid state which is accompanied with mass transfer.

Before the early 1960s, only stabilized zirconia at high temperatures and silver iodide above 150°C were known to exibit high ionic conductivities in the solid state. In the early 1960s, however, Ag_3SI was found to show a high silver ion conductivity at room temperature.

Since then, various kinds of solid cation conductors exhibiting high conductivities at room temperature have been found. These are stoichiometric compounds having cation disordered phases. Then in the middle 1960s, sodium β-alumina was proposed as a high conductivity sodium ion conductor which is a nonstoichiometric compound possessing a high intrinsic vacancy concentration.

These discoveries of high conductivity solid ionic conductors in the 1960s will indicate that there was drastic progress in the field of solid ionc conductors in the 1960s. So, in this chapter, the history of solid ionic conductors is described dividing it into before and after the 1960s, respectively.

2. STUDIES ON SOLID IONIC CONDUCTORS BEFORE THE 1960S

2.1 Cationic Conductors

In 1854, Buff[1)] studied the electrical conductivity of heated glasses and in 1882, Kohlrausch[2)] described the electrical conductivities of solid silver halides. In 1884, Warburg[3)] electrolyzed solid glasses, and Lehmann[4)] electrolyzed solid silver iodide in 1885. In 1888, Warburg and Tegetmeier[5)] carried out transference measurements, and in 1889, Lehmann[6)] studied on the movement of ions in silver iodide in the solid state. In 1893-94, Wiedemann[7)] published a series of text books on "Elektricität", in Volume 1 and 2 of which some descriptions of solid electrolytes are seen. In 1904-06, Haber and Tolloczko[8)], and in 1913, Bruni and Scarpa[9)] demonstrated that Faraday's law could be applied for solid ionic conductors. In 1914, Tubandt and Lorenz found that the conductivity of solid silver iodide near its melting point was higher than that of the molten material[10)]. Early in the 1920s, Tubandt and his co-workers[11)] carried out systematic transference number measurements to show that silver halides are cationic conductors. In 1923, Joffé[12)] introduced the concept of interstitial ions and vacancies by which the theory of defects in solids was started. Frenkel[13)] was the first to apply the law of mass action to interstitials and vacancies in 1926. In 1930,

Wagner and Schottky[14] found the so-called Schottky disorder in alkali halides and reported a thermodynamics of disorder in solids. These were the starting points for the theory of ionic conduction in solids. And in 1931, Tubandt et al.[15] and Jost and Schweitzer[16] in 1933 measured the cationic transference numbers of sodium and potassium halides at relatively high temperatures to find that in sodium halides, the cationic ion transference number was near unity whereas that in potassium halides was not.

In 1934, Strock[17] studied on the structure of α-AgI to postulate that the silver ions in α-AgI are distributed statistically over a large number of sites which are nearly energetically equivalent, whereas the iodide ions form a well-ordered structure. This structure was once called average structure which is an extreme case of Frenkel disorder and it was studied also by Rahlf[18]. Though this picture has been modified somewhat since then, it was late in the 1970s that the structure of α-AgI was made fairly clear. This α-AgI is stable between the transition point of β-AgI to α-AgI, 147°C, and 557°C, the melting point of α-AgI which means that silver iodide shows a high silver ion conductivity in the solid state only at relatively high temperatures. β-AgI exhibits only a low silver ion conductivity below 147°C.

Thus, no solid ionic conductors had been found which showed a high cationic conductivity in the solid state at room temperature before 1960.

2.2 Anionic Conductors

The history of anionic condutors in the solid state began when Faraday[19] noted the remarkable ionic properties of PbF_2 in the solid state in 1834 which was followed by Nernst[20] when he produced his "glower" in 1900. He proposed an anionic conductor such as ZrO_2 with small amount of added Y_2O_3 or CeO_2 to use as a light source of the electric lamp. The anionic conductors which Nernst finally used contained 85 w/o ZrO_2 and 15 w/o Y_2O_3 and was called the "Nernst mass". The Nernst glower could be operated for a long period of time on direct current though electrolysis was found to occur. It was explained that the Nernst mass was an oxide ion conductor at high temperature and any loss of oxygen liberated at the anode was balanced by an equal amount of oxygen taken into the Nernst glower at the cathode.

This kind of oxide ion conductor is called the stabilized zirconia and as the stabilizers, CaO[21,22,23,24], MgO[25], and Y_2O_3[20,24,26] had been used to utilize them practically at high temperatures around 1000°C. Extensive use of them has been found since Kiukkola and Wagner[27] carried out electromotive force measurements in 1957 to determine free enthalpies for chemical reactions at high temperatures.

Transference measurements were carried out by Tubandt et al.[28] on barium and lead halides to find that their anionic transference numbers are almost unity at relatively high temperatures.

3. SOLID IONIC CONDUCTOR STUDIES AFTER THE 1960S

3.1 Cationic Conductors

3.1.1 Silver and Copper(I) Ion Conductors

In the picture of the α-AgI structure, which was developed from the observed intensities of the diffraction lines in the Debye x-ray pattern, the silver ions had been considered to be almost free to move three dimensionally suggesting that they are in a "quasi-molten" or "liquid-like" state. This was also suggested by the anomalously low entropy of fusion for silver iodide (11.3 J/degree·mol) which is in contrast to the values for the alkali halides that are close to a mean value of 24 J/degree·mol. The value of the transition entropy of AgI, 14.5 J/degree·mol, at 147°C also suggested that the silver ion sub-lattice might be considered to melt at 147°C whereas the remaining iodide sub-lattice collapsed at 557°C[29,30,31]. In such a structure, a simple jump-diffusion model was applied to calculate the silver ion conductivity using microwaves, far-infrared and quasielastic neutron scattering data for powdered AgI in the early 1970s by Funke[32].

However, there was a difference between the calculated and measured conductivity at 250°C which suggested that the silver ions do not move isotropically in α-AgI.

Neutron diffraction analysis[33] and EXAFS measurements[34] in the late 1970s have indicated that there is a large anharmonic contribution to the thermal vibrations of the silver ions in α-AgI in the directions <100>[35] or <110>[36] suggesting that either of these directions should be regarded as channel-like diffusion paths for the silver ions in α-AgI.

Late in the 1950s, Reuter and Hardel, and Takahashi and Yamamoto found Ag_3SI which has a silver ion conductivity of 10^{-2} S/cm at 25°C, and reported their results early in the 1960s[37,38]. Since then, stimulated by this discovery of a room temperature high conductivity silver ion conductor being synthesized by combining AgI and foreign ions, many inorganic and organic anions and cations have been introduced into the lattice of silver iodide to obtain high conductivity silver ion conductors at room temperature in the crystalline state as well as in the glassy state[39,40,41].

Because of the similarities in the properties of the silver ion and the copper(I) ion, high conductivity copper(I) ion conductors at room temperature have been synthesized by Takahashi et al. in 1973 using a similar method as in the case of silver ion conductors. Since then, many copper(I) ion conductors exhibiting high conductivities at room temperature have been found[39,42,43].

3.1.2 Alkali Ion Conductors

There are a few solid alkali ion conductors which exhibit high conductivities at room temperature, whereas many high conductivity solid alkali ion conductors have been found at higher temperatures.

Lithium iodide mixed with a high surface area alumina, which is called a composite solid electrolyte has been reported by Liang in 1973 to have an unexpectedly high lithium ion conductivity of the order of 10^{-4} S/cm at room temperature[44,45]. Moreover, polar polymers such as polyethylene oxide, polypropylene oxide, polyether-substituted polyphosphazene and other polymers or copolymers dissolved with inorganic lithium salts have been discovered by Fenton et al. in 1973 to have high lithium ion conductivities above 70°C[45,46,47]. At higher temperature, Li_2O based compounds and solid solutions with other metal oxides, or $LiAlCl_4$, Li_3N, and Li_3N-LiX(X=Cl, Br, I) have been found from the late 1970s to early in the 1980s as high conductivity lithium ion conductors[45,46,48]. Many glassy and amorphous lithium ion conductors have also been found in the middle 1980s[45,46,49].

In 1967, Yao and Kummer[50] reported that β-alumina exhibited high sodium ion conductivity, though the history of β-alumina goes back at least fifty years when Rankin and Merwin[51] reported on this material misconceiving it as a crystallographic modification of alumina. Generally, $A_2O \cdot 11M_2O_3$(A=mono-valent metal, M=tri-valent metal) is called A-β-alumina. Na-β-alumina is the most famous sodium ion conductors, when A is Na and M is Al. The two most important members of this family are the hexagonal β-alumina and the rhombohedral β"-alumina. In β-alumina, the interstitialcy mechanism accounts for the high two-dimensional ionic conductivity, while in β"-alumina, conduction occurs through the movement of cation vacancies[52].

There are also sodium ion conductors with framework structures which are formed by tetrahedra MO_4, octahedra MO_6, or both. Of these networks,the second and the third types are known to exhibit three dimensional high sodium ion conductivities at relatively high temperatures after late in the 1970s[53]. Glassy and polymeric sodium ion conductors have also been found in the early 1980s[54].

One dimensional potassium ion conductors which have a hollandite type structure and two dimensional potassium-β"-alumina type conductors have been known. Further, three dimensional potassium ion conductors have been found in the K_2O and other metal oxide systems[55]. These potassium ion conductors have been known since early in the 1970s.

3.1.3 Poly-valent Ion Conductors

It has been known that the sodium ions in both β- and β"-alumina can be exchanged not only by mono-valent cations such as Li^+, K^+ and Ag^+ but also by poly-valent cations such as Ca^{2+}, Ba^{2+}, Cd^{2+}, Nd^{3+}, Eu^{3+}, Er^{3+}, and many other di- or tri-valent cations. Most of them have been found by Farrington et al. in the early 1980s. They have the ionic conductivities of the order of 10^{-6} S/cm at 40°C and 10^{-1} S/cm at 300-400°C[56].

3.1.4 Protonic Conductors

Because of the extremely small size of protons and their lack of an electronic charge cloud, protons are not subject to the usual ion-repulsion effects that make other ions spatially separate individuals. It seems more likely that protons in solids would be transported with an associated ion such as H_3O^+, OH^-, or NH_4^+ than by themselves. Thus, three different mechanisms are considered. These are liquid-like transport of H^+, Vehicle mechanism, and H_3O^+/H_2O jump. The third is the most important and is named Grotthuss mechanism in which the proton jumps are followed by reorientation of donating and/or accepting groups[57].

Further, in 1980, Takahashi and Iwahara found that some solid solutions having a perovskite structure show high protonic conductivities in hydrogen atmosphere at high temperatures[58].

3.2 Anionic Conductors

3.2.1 Fluoride Ion Conductors

Of halide ion conductors, fluoride ion conductors have been studied mainly since the middle 1960s. There are three types of fluoride ion conductors which have the crystal structures of the fluorite, tysonite, and YF_3 type, in which the YF_3 type conductors have relatively low fluoride ion conductivities even at high temperatures[59].

Some high conductivity solid chloride ion conductors have been found at relatively high temperatures.

3.2.2 Oxide Ion Conductors

Stabilized zirconias show the oxide ion conductivities based on the oxide ion vacancy mechanism. As the stabilizers, CaO, Y_2O_3, Yb_2O_3, Sm_2O_3, Sc_2O_3, Gd_2O_3, and Nd_2O_3 are known[60,61]. Takahashi and Suzuki found that the oxide ion conductivities usually show a time dependence when the amount of stabilizer is smaller than that for giving the highest conductivity[62]. This is attributed to the oxide ion vacancy ordering in this range.

Thoria and ceria solid solutions exhibit high oxide ion conductivities at high temperature when di- or tri-valent metal oxide such as CaO, Y_2O_3, or La_2O_3 is added to the hosts[60,63]. The thoria solid solution shows the p type semiconduction under a high partial pressure of oxygen and pure oxide ion conduction at relatively low oxygen partial pressure. In ceria solid solutions, tetra-valent cerium ion is unstable in a reducing atmosphere.

In 1971, Takahashi and Iwahara found that by substituting a part of A or B sites in the perovskite type oxide, ABO_3, with other metals whose valency are lower than A or B, respectively, oxide ion vacancies appear to exhibit both the oxide ion and hole mixed conduction in air at high temperature. Under a low partial pressure of oxygen, however, it shows high oxide ion conductivity, the value of which is comparable to those of stabilized zirconias[60,64].

Above 730°C, Bi_2O_3 shows a fluorite type face centered cubic structure named δ type which exhibits the oxide ion conductivity of 10^0 S/cm. Below 730°C, it transforms to the monoclinic structure of α type which shows a p type conduction. In order to obtain the high oxide ion conductivity phase at lower temperatures, di-, tri-, penta-, or hexa-valent metal ion has successfuly been introduced into the lattice of Bi_2O_3 by Takahashi and Iwahara in 1970[60,65]. The obtained high conductivity oxide ion conductors have a face centered cubic or rhombohedral single phase, or both face centered cubic and rhombohedral phases according to the ionic radius of the added cation.

4. SUMMARY

In this chapter, the development of high conductivity solid ionic conductors is described mainly around the 1960s. The brief descriptions after the 1960s are to avoid the overlaps with those in the other chapters.

In any case,the further developments of high conductivity solid ionic conductors can be expected to provide new materials which will support innovative new technologies.

REFERENCES

The following references after the 1960s are restricted mainly to the reviews and books in order to reduce overlaps with the references in the other chapters.

1) Buff, H., Ann. Chem. Pharm. 90, 257 (1854).

2) Kohlrausch, W., Ann. Physik u. Chem. N.F. 17, 642 (1882).

3) Warburg, E., ibid. 21, 662 (1884).

4) Lehmann, O., ibid. 24, 1 (1885).

5) Warburg, E. and Tegetmeier, F., ibid. 32, 455 (1888).
Tegetmeier, F., ibid. 41, 18 (1890).

6) Lehmann, O., ibid. 38, 396 (1889).

7) Wiedemann, G., "Die Lehre von der Elekticität", Vrelag F. Vieweg u. Sohn, Braunschweig, 1, 553, 565, 815; 2, 491 (1893/1894).

8) Haber, F. and Tolloczko, A., Z. anorg. Chem. 41, 407 (1904); 51, 245, 289 (1906).

9) Bruni, G. and Scarpa, G., Rend. reale accad. naz. Lincei 22, 483 (1913).
10) Tubandt, C. and Lorenz, E., Z. Physik. Chem. 87, 513 (1914).
11) Tubandt, C., Z. anorg. allgem. Chem. 115, 105 (1921).
Tubandt, C. and Reinhold, H., Z. Elektrochem. 29, 313 (1923).
12) Joffé, A., Ann. Phys. (Leipzig) 72, 461 (1923).
13) Frenkel, J., Z. Phys. 35, 652 (1926).
14) Wagner, C. and Schottky, W., Z. phys. Chem. B11, 163 (1930).
Schottky, W., ibid. B29, 335 (1935).
15) Tubandt, C., Reinhold, H. and Liebold, G., Z. anorg. allgem. Chem. 197, 225 (1931).
16) Jost, W. and Schweitzer, H., Z. phys. Chem. B20, 118 (1933).
17) Strock, L. W., Z. phys. Chem. B25, 441 (1934); B31, 132 (1935).
18) Rahlf, P., ibid. B31, 156 (1936).
19) Faraday, M.,"Experimental Researches in Electricity", Taylor and Francis, London, Art.1339 (1839).
20) Nernst, W., Z. Elektrochem. 6, 41 (1900).
21) Duwez, P, Odell, F. and Brown Jr., F. H., J. Amer. Ceram. Soc. 35, 107 (1952).
22) Trombe, F. and Foex, M. Compt. Rend. 236, 1783 (1953).
23) Dietzel, A. and Tober, H., Ber. Dtsch. Keram. Ges. 30, 47, 71 (1953).
24) Kingery, W. D., Pappis, J., Doty, M. E. and Hill, D. C., J. Amer. Ceram. Soc. 42, 393 (1959).
25) Baur, E. and Preis, H., Z. Elektrochem. 43, 727 (1937).
26) Hund, F., Z. phys. Chem. 199, 142 (1952).
27) Kiukkola, K. and Wagner, C., J. Electrochem. Soc. 10, 308, 379 (1957).
28) Tubandt, C., "Handbuch der Experimentalphysik" (eds. W. Wien and F. Harms), Akademische Verlagsgesellschaft, Leipzig, 12, Part 1, 383 (1932).
29) O'Keeffe, M., "Fast Ion Transport in Solids" (ed. W. van Gool), North-Holland Publ. Co., Amsterdam, p.233 (1973).
30) van Gool, W., "Solid Electrolytes" (eds. P. Hagenmuller and W. van Gool), Academic Press, New York, p.9 (1974).

31) O'Keeffe, M. and Hyde, B. G., Phylos. Mag. 33, 219 (1976).

32) Funke, K., Kalus, J. and Lechner, R. E., Solid State Commn. 14, 1012 (1974).
Eckold, G., Funke, K., Kalus, J. and Lechner, R. E., Phys. Lett., 55A, 125 (1975): J. Phys. Chem. Solids 37, 1097 (1975).
Funke, K., "Superionic Conductors" (eds. G. D. Mahan and W. L. Roth), Plenum Publ. Corp., New York, p.183 (1976): "Solid Electrolytes" (eds. P. Hagenmuller and W. van Gool), Academic Press, New York, p.77 (1978).

33) Cava, R. J., Reidinger, F. and Wuensch, B. J., Solid State Commn., 24, 411 (1977).

34) Boyce, J. B., Hayes, T. M., Stutis, W. and Mikkelsen, J. C., Phys. Rev. Lett. 38, 1362 (1977).
Boyce, J. B. and Hayes, T. M., "Superionic Conductors" (ed. M. B. Salamon), Springer-Verlag, Berlin Heidelberg, p.5 (1979).

35) Funke, K., Höch, A. and Lechner, R. E., J. de Phys.41,C6 (1980).

36) Hoshino, S., Sakuma, T., and Fujii, Y., Solid State Commn. 22, 763 (1977).

37) Reuter, B., and Hardel, K., Angew. Chem. 72, 138 (1960): Naturwiss. 48 61 (1961): Z. anorg. allgem. Chem. 340, 158, 168 (1965): Ber. Bunsenges. phys. Chem. 70, 82 (1966).

38) Takahashi, T. and Yamamoto, O., Denki Kagaku 32, 610 (1964); 33, 346 (1965); 5, 181 (1967): Electrochim. Acta 11, 779 (1966).

39) Takahashi, T., J. Appl. Electrochem. 3, 79 (1973): "Progress in Solid Electrolytes" (eds. T. A. Wheat, A. Ahmad and A. K. Kuriakose), Energy, Mines and Resources, Canada, Ottawa, p.221 (1983).

40) Geller, S.,"Solid Electrolytes" (ed. S. Geller), Springer-Verlag, Berlin Heidelberg, p.41 (1977).

41) Kasper, J.S., "Solid Electrolytes" (eds. P. Hagenmuller and W. van Gool), Academic Press, New York, p.217 (1978).

42) Takahashi, T.,"ibid.", p.201 (1978).

43) Matsui, T. and Wagner,Jr., J. B., "ibid.", p.237 (1978).

44) Liang, C., J. Electrochem. Soc. 120, 1289 (1973).
Wagner,Jr., J. B., "Solid State Batteries" (eds. C. A. C.

Sequeira and A. Hooper), Martinus Nijhoff Publishers, Dordrecht, p.77 (1985).

Poulsen, F. W., "Transport-Structure Relations in Fast Ion and Mixed Conductors" (eds.F. W. Poulsen, N. H. Andersen, K. Clausen, S. Skaarup and O. T. Sørensen), Risø National Laboratory, Roskilde, p.67 (1985): Risø-M-2540, 94pp. (1986).

45) Shahi, K., Wagner, J. B. and Owens, B. B., "Lithium Batteries" (ed. J. P. Gabano), Academic Press, London, p.407 (1983).

46) Raistrick, I. D. and Huggins, R. A., "Lithium Battery Technology" (ed. H. V. Venkatasetty), John Wiley & Sons, New York, p.205 (1984).

47) Armand, M. B., J. Power Sources 14, 11 (1985): "Solid State Batteries" (eds. C. A. C. Sequeira and A. Hooper), Martinus Nijhoff Publishers, Dordrecht, p.63 (1985), Ann. Rev. Mater. 16, 245 (1986).

Bonino, F. and Scrosati, B., "Materials for Solid State Batteries" (eds. B. V. R. Chowdari and S. Radhakrishna), World Scientific Publishing Co., Singapore, p.41 (1986).

Vincent, C. A., Progress Solid State Chem. 17, 145 (1987).

Papers Presented at the 1st International Symposium on Polymer Electrolytes which was held in St.Andrews, Scotland on 17-19, June, 1987.

48) Huggins, R. A., Electrochim. Acta 22, 773 (1977).

Shannon, R. D., Taylor, B. E., English, A. D. and Berzins, T. ibid. 22, 783 (1977).

Schoonman, J. and Bottelberghs, P. H., "Solid Electrolytes" (eds. P. Hagenmuller and W. van Gool) Academic Press, New York, p.335 (1978).

49) Singh, K. and Rokade, S., J. Power Sources 13, 151 (1984).

Tuller, H. L. and Button, D. P., "Transport-Structure Relations in Fast Ion and Mixed Conductors" (eds. F. W. Poulsen, N. H. Andersen, K. Clausen, S. Skaarup and O. T. Sørensen) Risø National Laboratory, Roskilde, p.119 (1985).

Angell, C. A., Solid State Ionics 18/19, 72 (1986).

Levasseur, A., "Materials for Solid State Batteries" (eds. B. V.

R. Chowdari and S. Radhakrishna) World Scientific Publishing Co., Singapore, p.97, p.119 (1986).

50) Yao, Y. F. T. and Kummer, J. T., J. Inorg. Nucl. Chem. 29, 2453 (1967).

51) Rankin, G. A. and Merwin, H. E., J. Amer. Chem. Soc. 38, 568 (1916).

52) Kummer, J. T., "Progress in Solid State Chemistry" (eds. H. Reiss and J. O. NcCaldin) Pergamon Press, Oxford, 7, 141 (1972).
Kennedy, J. H., "Solid Electrolytes" (ed. S. Geller) Springer-Verlag, Berlin Heidelberg, p.105 (1977).
Collongues, R., Théry J. and Boilot, P., "Solid Electrolytes" (eds. P. Hagenmuller and W. van Gool) Academic Press, New York, p.253 (1978).
Whittingham, M. S., "Materials Science in Energy Technology" (eds. G. G. Libowitz and M. S. Whittingham) Academic Press, New York, p.455 (1979).
Pelton, A. D., Dubreuil, A. and Malenfant, M. "Progress in Solid Electrolytes" (eds. T. A. Wheat, A. Ahmad and A. K. Kuriakose) Energy, Mines and Resources Canada, Ottawa, p.503 (1983).
Moseley, P. T., "The Sodium Sulfur Battery" (eds. J. L. Sudworth and A. R. Tilley), Chapman and Hall, London, p.19 (1985).
"Electrolyte Characterization in Proc. of the Symposium on Sodium-Sulfur Batteries" (eds. A. R. Landgrebe, R. D. Weaver and R. K. Sen) Electrochem. Soc., Pennington, p.65-175 (1987).

53) Hong, H. Y-P., "Solid State Chemistry of Energy Conversion and Storage" (eds. J. B. Goodenough and M. S. Whittingham), Amer. Chem. Soc., Washington, D. C., p.179 (1977).
Palanisamy, T., Kerr, R. L. and Maloy, J. T. "Fast Ion Transport in Solids" (eds. P. Vashishta, J. N. Mundy and G. K. Shenoy) North-Holland, New York, p.427 (1979).
Goodenough, J. B., "Progress in Solid Electrolytes" (eds. T. A. Wheat, A. Ahmad and A. K. Kuriakose) Energy, Mines and Resources Canada, Ottawa, p.53 (1983).
Boilot, J. P., Collin G. and Colomban, Ph., ibid., p.91 (1983).
Nygren, M., "Transport-Structure Relations in Fast Ion and Mixed

Conductors" (eds. F. W. Poulsen, N. H. Andersen, K. Clausen, S. Skaarup and O. T. Sørensen) Risø National Laboratory, Roskilde, p.51 (1985).

Clearfield, A., Subramanian, M. A., Rudolf, P. R. and Moini, A., Solid State Ionics, 18/19, 13 (1985).

54) Susman, S., Delbecq, C. J., McMillan, J. A. and Roche, M. F., Solid State Ionics 9/10, 667 (1983).

Andersen, N. H., Kjems, J. K. and Poulsen, F. W., Risø-M-2489, Risø National Laboratory, Roskilde, p.49 (1985).

55) Réau, J.-M., Delmas, C. and Hagenmuller, P., "Solid Electrolytes" (eds. P. Hagenmuller and W. van Gool) Academic Press, New York, p.381 (1978).

Watelet, H., Besse, J.-P., Baud, G. and Chevalier, R., Mat. Res. Bull. 17, 863 (1982).

Hong, H. Y-P., "Solid State Chemistry of Energy Conversion and Storage" (eds.J. B. Goodenough and M. S. Whittingham) Amer. Chem. Soc. Washington, D. C., p.179 (1977).

Delmas, C., Maazaz, A., Fouassier, C., Réau, J. M. and Hagenmuller, P., "Fast Ion Transport in Solids" (eds. P. Vashishta, J.N. Mundy, and G.K.Shenoy) North-Holland, New York, p.451 (1979).

56) Dunn, B. and Farrington, G. C., Solid State Ionics 18/19, 31 (1985).

57) Poulsen, F. W., Risø-M-2244, Risø National Laboratory, Roskilde, 23pp. (1980).

"Solid State Protonic Conductors I" (eds. J. Jensen and M. Kleitz) Odense University Press, Odense, 339pp. (1982).

"Solid State Protonic Conductors II" (eds. J. B. Goodenough, J. Jensen and M. Kleitz) Odense University Press, Odense, 417pp. (1983).

"Solid State Protonic Conductors III" (eds. J. B. Goodenough, J. Jensen and A. Potier) Odense University Press, Odense, 284pp. (1985).

Chandra, S., "Transport-Structure Relations in Fast Ion and Mixed Coductors" (eds. F. W. Poulsen, N. H. Andersen, K. Clausen, S.

Skaarup and O. T. Sørensen) Risø National Laboratory, Roskilde, p.365 (1985).

Wheat, T. A., Ahmad, A., Kuriakose, A. K. and Canaday, J. D., J. Can. Ceram. Soc., 54, 32 (1985).

Goodenough, J. B., Methode Enzymol., 127, 263 (1986).

58) Takahashi, T. and Iwahara, H., Rev. de Chim. minerale, 17, 243 (1980).

Iwahara, H, Esaka, T., Uchida, H., Yamaguchi,T. and Ogaki, K., Solid State Ionics, 18/19, 1003 (1986).

Mitsui, A., Miyayama, M. and Yanagida, H., ibid. 22, 213 (1987).

59) Réau, J.-M. and Portier, J., "Solid Electrolytes" (eds. P. Hagenmuller and W. van Gool) Academic Press, New York, p.313 (1978).

Hagenmuller, P., Z. Chem. 23, 1 (1983).

60) Takahashi, T., "Physics of Electrolytes" (ed. J. Hladik) Academic Press, London, 2, 989 (1972).

Obayashi, H. and Kudo, T. "Solid State Chemistry of Energy Conversion and Storage" (eds.J. B. Goodenough and M. S. Whittingham) Amer. Chem. Soc., Washington D. C., p.316 (1977).

Worrell,W. L., "Solid Electrolytes" (ed. S. Geller) Springer-Verlag, Berlin Heidelberg, p.143 (1977).

Dell, R. M. and Hooper, A. "Solid Electrolytes" (eds. P. Hagenmuller and W. van Gool) Academic Press, New York, p.291 (1978).

Subbarao, E. C. and Maiti, H. S. "Progress in Solid Electrolytes" (eds. T. A. Wheat, A. Ahmad and A. K. Kuriakose) Energy, Mines and Resources Canada, Ottawa, p.281 (1983).

Sørensen, O. T., Johannesen, Ø. and Clausen, K., "Transport-Structure Relations in Fast Ion and Mixed Conductors" (eds. F. W. Poulsen, N. H. Andersen, K. Clausen, S. Skaarup and O. T. Sørensen) Risø National Laboratory, Roskilde, p.93 (1985).

61) The Proceedings of the 1st (Cleveland in 1980), 2nd (Stuttgart in 1983) and 3rd (Tokyo in 1986) International Conferences on the Science and Technology of Zirconia (Published in the volumes of "Advances in Ceramics").

62) Takahashi, T. and Suzuki, Y., Proc. J. Internat'l. Etude Piles Combust. 2, 378 (1967).

63) El Adham, K. and Hammou, K. "Progess in Solid Electrolytes" (eds. T. A. Wheat, A. Ahmad and A. K. Kuriakose) Energy, Mines and Resources Canada, Ottawa, p. 313 (1983).
Kilner, J. A. and Faktor, J. D., "ibid." p.347 (1983).

64) Takahashi, T. and Iwahara, H., Energy Conversion 11, 105 (1971).

65) Takahashi, T. Iwahara, H., Mater. Res. Bull. 13, 1447 (1978).
Cahen, H. T. "Thesis-Bi_2O_3-M_2O_3 Systems", Univ. Utrecht, 107pp. (1980).

TRANSPORT IN IONIC SOLIDS: THEORETICAL ASPECTS

W. Dieterich

Fakultät für Physik, Universität Konstanz, D-7750 Konstanz 1, F.R.G.

ABSTRACT

Some general theoretical concepts applicable to transport phenomena in different types of solid ionic conductors are reviewed. We start by discussing the notion of the effective single-particle potential and its relevance to crystalline framework conductors. In the major part of this article we consider stochastic lattice gas models and random electrical networks as general schemes to treat strongly correlated and disordered systems. Applications to experiments demonstrating some general behaviour are indicated, and directions for future research are pointed out.

1. INTRODUCTION

Within the past decade considerable progress has been achieved in our basic theoretical understanding of the phenomenon of fast ionic transport in solids. In particular, many of the physical properties of optimized ionic conductors of the α-AgI type are well understood by now. Current basic research in solid state ionics is stimulated by the properties of a wide spectrum of new materials, including crytalline structures, mixed conductors, glasses, polymers or composites. Important themes are the identification and the consequences of specific transport mechanisms in these systems. Moreover, the investigation of interface processes is playing an increasing role.

The developments in these different areas bear a variety of challenging theoretical problems. Clearly, an important aim is to calculate material properties in a quantitative and predictive way from microscopic parameters. Most advances in this domain come from molecular-dynamics (MD) simulations. Structural and dynamic properties of simple materials, including structural phase changes, and thermodynamic properties have been treated. Major trends are to extend the accessible time scales and to apply the method to more complex structures. We shall not discuss this subject further, but refer the reader to the recent reviews by Vashishta[1)], Gillan[2)] and Jacucci[3)].

Another direction of theoretical research is aiming at a more qualitative understanding of the characteristic features of special classes of solid electrolytes. This aspect is covered by the excellent review of Ratner and Nitzan[4)], which contains an extensive discussion of ionic transport in glassy and polymeric electrolytes and in framework crytals. In this article we limit our discussion to a few general concepts and demonstrate their utility in specific applications. Thereby we try to keep the overlap with Ref.[4)] as small as possible. In Section 2 we discuss classical density functional theory as a relatively simple analytic tool to compute the density distribution of mobile ions in crystalline framework conductors. Comparison with previous MD-studies shows that this method can yield quantitative information on the density of Ag^+ - ions in α-AgI-type conductors. In addition it provides valuable insight into the quantitative role of ion-ion correlations in these systems. We also comment briefly on a transport-structure relationship based on the effective single-particle potential.

To treat the correlated ion dynamics at frequencies much below atomic vibrational frequencies, one usually has to resort to a coarse-grained description in terms of hopping models. In fact, many-particle hopping models provide a rather general and powerful approach in statistical mechanics for studying classical transport processes both under equilibrium and non-equilibrium conditions. Most of our knowledge in this area is based on Monte Carlo simulations[5)]. Regarding solid electrolytes, the application of hopping models is clearly limited to systems with a simple defect structure. Section 3 starts out with some general remarks on the formulation of many-particle hopping and on specific effects of the Coulomb-interaction. Backward correlations resulting from many-particle interactions and effects of disorder are then discussed as possible mechanisms which lead to a pronounced dispersive regime in the dynamic conductivity. We also address some non-equilibrium phenomena, in particular the properties of diffusion fronts.

In Section 4 we turn to a completely macroscopic level. It is well-known that systems containing disorder on macroscopic length scales can show a rich behaviour in their transport properties. The interpretation naturally leads to the ideas of percolation and fractal geometries. We illustrate this by two examples in solid state ionics, namely transport in dispersed ionic conductors and the ac-response of non-ideal electrodes which are regarded as surface fractals.

2. EFFECTIVE SINGLE-PARTICLE POTENTIAL

The equilibrium density $\rho(r)$ of diffusing ions in fast ion conductors is an interesting quantity in several respects. Experiments e.g. on α-AgI - type compounds[6] show a substantial delocalization of diffusing ions around the preferred interstitial sites, which reflects the structure of the dominant conduction pathways. Moreover, the variation of the density along the conduction paths is closely related to transport properties. To discuss this, let us introduce the effective single-particle potential $V_{eff}(\underline{r})$ via the relation

$$\rho(\underline{r}) = \text{const} \exp(-\beta V_{eff}(\underline{r})) \quad . \tag{2.1}$$

From a comparison of structural and transport data of different materials, Schulz and Zucker concluded that the activation energy determined from the static conductivity is close to the saddle-point energy of the effective potential[7,8]. This observation is supported theoretically by extensive Langevin-dynamics simulations[9,10] and a recently proposed dynamic mean-field theory[11]. Finally, comparing $V_{eff}(\underline{r})$ with the bare single-particle potential $V(\underline{r})$ originating from host ions only, one can deduce quantitative information on the role of commensurability effects, i.e. on the degree of competition between short-range correlations among the mobile ions and the ion-host interaction.

The problem, which arises, is to calculate the density $\rho(\underline{r})$ from interionic potentials. Apart from the standard approach of MD-simulations, it seems that classical density functional theory[12] can yield reasonably accurate densities in a simple way, as shown by Roman et al.[13] and Billi et al.[14]. Under the assumption of a rigid framework lattice, the effective potential can be decomposed into an external part $V(\underline{r})$ and a correlation potential $V_{corr}(\underline{r})$. The external potential is determined by the positions of host ions and their interaction with the mobile ions. The idea is to regard $V_{corr}(\underline{r})$ as a functional of $\rho(\underline{r})$ and to perform a functional expansion. To lowest order one obtains

$$V_{eff}(\underline{r}) \simeq V(\underline{r}) + \int d^3r' \; c(|r-r'|,\bar{\rho})(\rho(\underline{r}')-\bar{\rho}) \qquad (2.2)$$

where $c(r,\bar{\rho})$ denotes the Ornstein-Zernike direct correlation function of a homogenous fluid of mobile ions with average density $\bar{\rho}$. For reasonable densities $\bar{\rho}$ one can approximate $c(r,\bar{\rho})$ by the well-known result for the one-component plasma[15]. The density profile $\rho(\underline{r})$ is then calculated self-consistently from (2.1) and (2.2).

Results obtained for α-AgI and β-Ag_2S are shown in Fig. 1 together with MD-data[16,17]. Clearly, the simple approximation (2.2) correctly reproduces the main trends which are also observed in experiments[6]. The most interesting result is a change in the structure of conduction paths as a consequence of the different strength of correlations in the two materials. In α-AgI, the correlation potential is rather weak and Ag^+ - ions diffuse according to the valley-structure of the external potential. In β-Ag_2S, on the other hand, the repulsion among Ag^+ - ions is substantially stronger, and the correlation potential becomes essential. This leads to conduction paths t - C - t involving the octahedral sites C, as seen from Fig. 1.

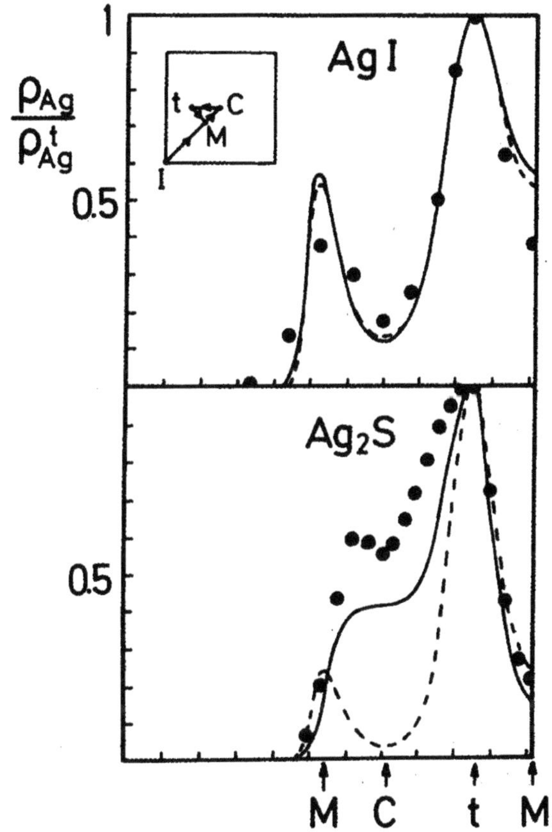

Fig. 1 Density profiles of Ag^+ - ions in α-AgI and β-Ag_2S in the face of the cubic unit cell along paths indicated in the inset (after Ref.[14])
---- density following from the bare potential V($\underline{r}$)
—— self-consistent solution including correlations.
Data points represent MD-results according to Ref.[16] for α-AgI and Ref.[17] for β-Ag_2S.

Despite of its success, the simple version of density functional theory described so far needs extensions in different directions. For example, the problem of deformations of the framework lattice[18] or, more generally, of a two-component formulation[19] and the role of higher-order terms in the expansion (2.2) remain to be investigated. In principle, densitiy-functional theory can also be used to calculate the pair-correlation function and the static structure factor. This is done by adding to $V(\underline{r})$ the potential of one mobile ion held at a fixed position $\underline{r}_0$. The conditional density $\rho(\underline{r}|\underline{r}_0)$ obtained in this situation directly yields the pair-correlation function $g(\underline{r},\underline{r}_0)$.

A more general problem is how to extend the theory to the dynamic response of framework electrolytes. Dynamic models in this area are mostly based on a one-component description in terms of interacting Brownian particles subjected to the external host potential $V(\underline{r})$. The collective dynamics of this model has been studied extensively by different techniques. Details and references are found in Ref.[4]. Here we only comment on a transport-structure relationship arising from that model. Its mean-field treatment in Ref.[11] contains the effective potential $V_{eff}(\underline{r})$ and the inhomogeneous Ornstein-Zernike function $c(\underline{r},\underline{r}')$ as structural input quantities. It turns out that $c(\underline{r},\underline{r}')$ does not explicitely enter the static conductivity. Hence, within mean-field theory, the static conductivity $\sigma(\omega = o)$ is given by the conductivity σ_0 of non-interacting particles in the presence of the effective potential,

$$\sigma(\omega = o) = \sigma_0[V_{eff}] \quad . \tag{2.3}$$

Evaluating the r.h.s. of this equation in terms of a prefactor and a thermally activated factor, one is led to the interpretation of the effective potential barrier as the activation energy measured in a static conductivity experiment. This conclusion is supported by extensive Langevin dynamics simulations[9,10]. These calculations also show that,apart from the high-friction regime, the prediction of (2.3) regarding the prefactor is only of qualitative validity. Nevertheless, equation (2.3) represents a simple relationship between structural and transport properties, which should be useful under rather general conditions.

The above-mentioned mean-field theory also suggests that the effect of ion-ion interactions in the two quantities $V_{eff}(\underline{r})$ and $c(\underline{r},\underline{r}')$ largely cancel with respect to the vibrationel resonance in the dynamic conductivity. The position of the resonance peak is therefore essentially determined by the bare potential $V(\underline{r})$.

3. HOPPING MODELS

3.1 Stochastic Lattice Gas

Correlation effects in the ionic motion beyond the validity of mean-field theories represent a particularly interesting subject in the physics of ionic conduction. The most important theoretical tool in that area is the master-equation approach of many-particle hopping. In this approach, the state of mobile ions at a given instant of time is specified by a set of occupation numbers $n_{\underline{\ell}} = \pm 1$ for sites $\underline{\ell}$ of some prescribed lattice. The configurations of the system, $\underline{n} = \{n_{\underline{\ell}}\}$, evolve in time due to particle hopping. Assuming hops only between nearest neighbours $\underline{\ell}$ and $\underline{\ell} + \underline{\delta}$, the probability in configuration space $p(\underline{n},t)$ satisfies the master equation

$$\frac{dp(\underline{n},t)}{dt} = \frac{1}{2} \sum_{\underline{\ell},\underline{\delta}} \left[w_{\underline{\ell},\underline{\ell}+\underline{\delta}}(\underline{n}') \, p(\underline{n}',t) - w_{\underline{\ell},\underline{\ell}+\underline{\delta}}(\underline{n}) \, p(\underline{n},t) \right] \tag{3.1}$$

where $\underline{n}'$ denotes the configuration which results from $\underline{n}$ by interchanging the occupations $n_{\underline{\ell}}$ and $n_{\underline{\ell}+\underline{\delta}}$. For an isolated system the transition rates $w_{\underline{\ell},\underline{\ell}+\underline{\delta}}(\underline{n})$ obey the detailed balance condition

$$w_{\underline{\ell},\underline{\ell}+\underline{\delta}}(\underline{n}) \, e^{-\beta H(\underline{n})} = w_{\underline{\ell},\underline{\ell}+\underline{\delta}}(\underline{n}') \, e^{-\beta H(\underline{n}')} \tag{3.2}$$

Here $\beta = (k_B T)^{-1}$ and

$$H(\underline{n}) = -\frac{1}{2} \sum_{\underline{\ell} \neq \underline{\ell}'} V_{\underline{\ell}-\underline{\ell}'} \, n_{\underline{\ell}} \, n_{\underline{\ell}'} - \sum_{\underline{\ell}} \varepsilon_{\underline{\ell}} \, n_{\underline{\ell}} \tag{3.3}$$

denotes the lattice gas Hamiltonian involving site energies $\varepsilon_{\underline{\ell}}$ and pairwise interactions $V_{\underline{\ell}-\underline{\ell}'}$.

The main limitation of the stochastic lattice gas model formulated in this way lies in the fact that the underlying lattice is assumed to be rigid and effects of the particle inertia are neglected[20]. Still the general model (3.1) is extremely useful in describing correlated low-frequency atomic motions in many different areas, including ionic conduction[21,6], hydrogen-diffusion in metals, two-dimensional diffusion of adsorbed atoms or kinetics of clustering near first-order phase-transitions. Much of the progress in these areas is due to the increasing power of Monte Carlo simulations[5].

For systems at equilibrium the following dynamic correlation-functions are of particular interest. One is the coherent dynamic structure factor

$$S_{coh}(\underline{q},\omega) = \frac{1}{2\pi N} \int_{-\infty}^{\infty} dt \sum_{\underline{\ell},\underline{\ell}'} \langle n_{\underline{\ell}}(t)\; n_{\underline{\ell}'}(0) \rangle \; e^{-i\underline{q}(\underline{\ell}-\underline{\ell}')} \; e^{-i\omega t} \tag{3.4}$$

which determines the cross-section for inelastic neutron-scattering. Let us first consider the case of short-range interactions, which pertains to diffusion of neutral atoms or to ions in mixed ionic-electronic conductors, where the Coulomb interactions among the ions are screened by the conduction electrons. In that case the behaviour of the coherent structure factor at long wave-lengths ($q \to 0$) is dominated by diffusion. This means that the expression (3.4) shows a quasielastic Lorentzian line with a half-width $D_{coll}\; q^2$. In the simplest case of a lattice with equivalent sites, the mean-field result for the collective diffusion constant D_{coll} can be written as

$$D_{coll} = \langle \alpha \rangle \; a^2/S(q \to 0) \tag{3.5}$$

where a is the jump distance and S(q) the static structure factor. The average hopping rate $\langle \alpha \rangle = \langle w_{\underline{\ell},\underline{\ell}+\underline{\delta}} \rangle /2c$ (c is the particle concentration) is usually written in a factorized form[21),

$$\langle \alpha \rangle = \alpha_o \; V W \tag{3.6}$$

where α_o is the hopping rate for infinite dilution. The vacancy availability factor V is defined in terms of the pair-correlation-function for nearest neighbours. The change of the local potential barrier due to interactions is contained in the effective frequency-factor W. Let us remark that for systems with attractive interactions the appearance of the static structure factor in the denominator of (3.5) leads to the well-known slowing down of diffusion, as the critical temperature is approached.

The single-particle motion is described by the incoherent dynamic structure factor $S_{inc}(\underline{q},\omega)$, which is defined in terms of the correlation function of the tracer occupation $n^*_{\underline{\ell}}$. A convenient representation is[22,23)]

$$S_{inc}(\underline{q},\omega) = \frac{1}{\pi} \mathrm{Re}(-i\omega + \langle \alpha \rangle \gamma(\underline{q})\ f(\underline{q},\omega))^{-1} \tag{3.7}$$

Here $\gamma(\underline{q})$ denotes the lattice structure factor known from the Chudley-Elliott model[24)], and $f(\underline{q},\omega)$ is a wave-vector and frequency-dependent correlation factor. The conventional tracer correlation factor f_t is obtained from $f(q,\omega)$ in the limit $q \to 0$ and $\omega \to 0$.

Because of its relative simplicity, the case of hard-core particles on a regular lattice with $H(\underline{n}) \equiv 0$ and transition rates

$$w^o_{\underline{\ell},\underline{\ell}+\underline{\delta}}(\underline{n}) = \alpha_o\ (n_{\underline{\ell}} - n_{\underline{\ell}+\underline{\delta}})^2 \tag{3.8}$$

deserves special attention. The collective dynamics of this 'ideal lattice gas' can in principle be worked out completely. The density-fluctuations behave as in the case of single-particle hopping[6)], a result which can be generalized to disordered lattices containing randomly blocked sites[25)]. Exact results have also been derived for four-body correlation functions on ordered lattices[26)]. Such quantities pertain to nuclear magnetic resonance and also to light scattering. In fact, depolarized light scattering spectra in CuI [27)] and ZrO_2 [28)] indicate the presence of a pair-mechanism, where the simultaneous occupation of nearest-neigbour sites leads to an anisotropic contribution to the polarizability. As a consequence, depolarized scattering is determined by the pair-density correlation function[26].

On the other hand, no exact solution is available for the single-particle dynamics in the ideal lattice gas. From Monte Carlo simulations[21], Green-function techniques[29] and continued fraction expansions[30], however, accurate results for the correlation factor $f(\underline{q},\omega)$ have been obtained, and our information on the single-particle dynamics in that model is rather complete. A discussion of the broad literature on transport properties in the presence of nearest-neighbour interactions is beyond the scope of this article. Instead we refer the interested reader to some recent publications[21,5,31,32].

3.2 Coulomb Interactions

A characteristic feature of solid electrolytes are the unscreened (long-range) Coulomb-forces among the moving defects. One would therefore like to study the properties of a Coulomb lattice gas, where

$$V_{\underline{\ell}-\underline{\ell}'} = \frac{e^2}{|\ell-\ell'|} \quad . \tag{3.9}$$

Static correlations among charged defects on ordered lattices with $\varepsilon_{\underline{\ell}}$ = const. (see equation (3.3)) have been obtained within the mean-spherical approximation[33]. To define the dynamics of the lattice gas, it is natural to assume that the effective activation energy, which enters the transition rate $w_{\underline{\ell},\underline{\ell}+\underline{\delta}}$, is determined by the local electric field[23]. This is equivalent to the symmetric expression

$$w_{\underline{\ell},\underline{\ell}+\underline{\delta}}(\underline{n}) = w^{o}_{\underline{\ell},\underline{\ell}+\underline{\delta}}(\underline{n}) \exp[\beta(H(\underline{n}) - H(\underline{n}'))/2] \tag{3.10}$$

Let us address some special properties of a Coulomb lattice gas, which differ qualitatively from their counterparts in a lattice gas with short-range interactions. First, in the long wave-length limit, the static structure factor of a charged system behaves as $S(q) \sim q^2/q_D^2$, where $q_D = (4\pi e^2 \beta\rho)^{1/2}$ denotes the Debye-Hückel wave-vector and ρ the particle density. The dynamic structure factor takes the form

$$S_{coh}(\underline{q},\omega) = \left(\frac{q^2}{\pi\, q_D^2} \right) \mathrm{Re}(-i\omega + 4\pi\, \sigma(\omega))^{-1} \quad ; \quad q \to 0 \quad . \tag{3.11}$$

This shows that the charge-relaxation spectrum becomes independent of q as $q \to 0$, as opposed to the above-mentioned diffusive quasielastic peak in the case of short-range interactions[34]. Secondly, the tracer correlation factor f_t shows a $\rho^{1/2}$ - dependence as $\rho \to 0$ [23],

$$1 - f_t = \text{const.}\; e^3 (k_B T)^{-3/2} \rho^{1/2} \tag{3.12}$$

in contrast to the conventional linear dependence on ρ. Note that the remaining factors in (3.12) are independent of any microscopic material parameters. In some alkali-oxid glasses a cusp-like decrease of the correlation factor with increasing alkali-oxid content has been observed recently[35], and a quantitative comparison of these findings with (3.12) would be interesting.

Theoretical attempts to extend (3.12) to the regime of higher densities have been based on a mode-coupling analysis of the corresponding memory-function $M_{inc}(q,\omega)$ [23], which represents the essential non-trivial part in the correlation factor $f(q,\omega)$. In this theory a particle is coupled to a dynamic local potential barrier, induced by the Coulomb-interaction with the surrounding particles. Calculating the time-evolution of the local barrier, however, is a complex problem, and explicit results are limited to moderate coupling strengths.

At this stage it is natural to seek for a more phenomenological approach. The relaxation theory by Funke[36,37] is precisely based on the idea of a dynamic local potential. After an initial forward hop of a particle, this potential relaxes and readjusts to the new particle position. During this process the barrier $\delta(t)$ against backward hops increases until it reaches its maximum in the final (relaxed) configuration. Therefore, the probability $W(t)$ that a backward hop occurs after time t, satisfies an equation containing a time-dependent rate constant $\Gamma(t)$. Schematically, $\dot{W}(t) = -\Gamma(t)\, W(t)$ with $\Gamma(t) \propto \exp(-\beta\delta(t))$. From the complete set of equations derived from this physical picture,

Funke obtains a consistent description of nuclear magnetic resonance, quasielastic neutron scattering and conductivity data. It turns out that the ac-conductivity is well approximated by the expression

$$\sigma(\omega) = \sigma(\infty) - \frac{\sigma(\omega)-\sigma(o)}{1+(i\omega\tau)^s} \quad . \tag{3.13}$$

In the limit $\omega\tau \ll 1$, one obtains a power-law dependence, $\sigma(\omega)-\sigma(0)\sim(i\omega)^s$, with an exponent $s \lesssim 1$. This is consistent with the so-called 'universal dynamic response' observed in many different low-conductivity materials[38,39].

3.3 Hopping in Random Media

In a Coulomb lattice gas model with unscreened interactions one expects a certain distribution of the local electric field and, consequently, a rather wide distribution of local hopping rates. This suggests that part of the many-body effects in an ionic conductor might be accounted for by a single-particle description employing a statistical distribution of hopping rates. While such analogies have not been worked out in detail, the problem of hopping in random media is a broad subject in itself, and excellent reviews emphasising different aspects have been published[40-42]. Direct examples of hopping transport in random atomic structures are provided by ionically conducting glasses, mixed-alkali systems and polymeric materials. Other applications are in the field of macroscopically disordered structures, whose transport properties can be modelled by random electrical networks (see Section 4).

Generally one starts from standard rate equations for the site occupation numbers $n_i(t)$,

$$\frac{dn_i}{dt} = \sum_j \left(W_{j\to i}\, n_j(t) - W_{i\to j}\, n_i(t)\right) \tag{3.14}$$

where the hopping rates $W_{i\to j}$ are random variables. For a three-dimensional system away from a possible percolation threshold one may apply effective medium theory to obtain the transport properties. In connection with the problem of hopping conduction in amorphous semiconductors, Movaghar et al. have formulated an effective medium theory suitable for topologically disordered lattices[43,44]. Schirmacher and Schirmer have carried out explicit calculations of the dynamic structure factor (with application to hydrogen-diffusion in amorphous metals), the conductivity $\sigma(\omega)$ [45] and the nuclear spin-lattice relaxation rate T_1 [46]. The general assumption is that

$$W_{i\to j} = \nu_o(r_{ij}) \exp(-\beta E_{ij}) \tag{3.15}$$

with random jump distances r_{ij} and random energy barriers $E_{ij} > 0$ related to the local energies ε_i. For fixed-range hopping, $\nu(r_{ij}) \equiv \nu_o$, and under the assumption of a probability distribution $p(E_{ij})$ = const., the conductivity is obtained as shown in Fig. 2. At low temperatures and frequencies $\omega \ll \nu_o$ it is consistent with (3.13) with an exponent s close to unity. At zero frequency, the conductivity is thermally activated with an activation energy proportional to $(p(E_{ij}))^{-1}$. The transition from dc-behaviour to the dispersive regime occurs at frequencies of the order of the hopping frequency in the dc-limit. Experimental data of the temperature- and frequency-dependent conductivity of $Na_2O \cdot SiO_2$ glasses[47] are well reproduced by this model.

Transport properties in the vicinity of a percolation threshold are generally characterized by scaling relations and critical transport exponents. Such a situation might be realized in certain mixed-alkali β-alumina compounds[48,25]. As observed by Bruce et al.[49], the potassium ions in Na_{1-x}/K_x - β-aluminas preferentially replace the sodium ions on the Beever-Ross (BR) - sites in the conducting planes. A pronounced minimum in conductivity occurs when all the BR-sites are occupied by K^+ - ions. This suggests that the diffusion paths get blocked as the ion exchange x ($x \lesssim 0.5$) is increased. The associated drop in conductivity should therefore reflect a percolation transition. The broadening of electric modulus spectra observed by Bruce et al.[50] seems to confirm this picture. Clearly, an explanation of mixed-alkali effects on the basis of interionic forces is still lacking. Nevertheless, relying on the above-mentioned structural information on the site preferances of the different alkali ions, it is interesting to

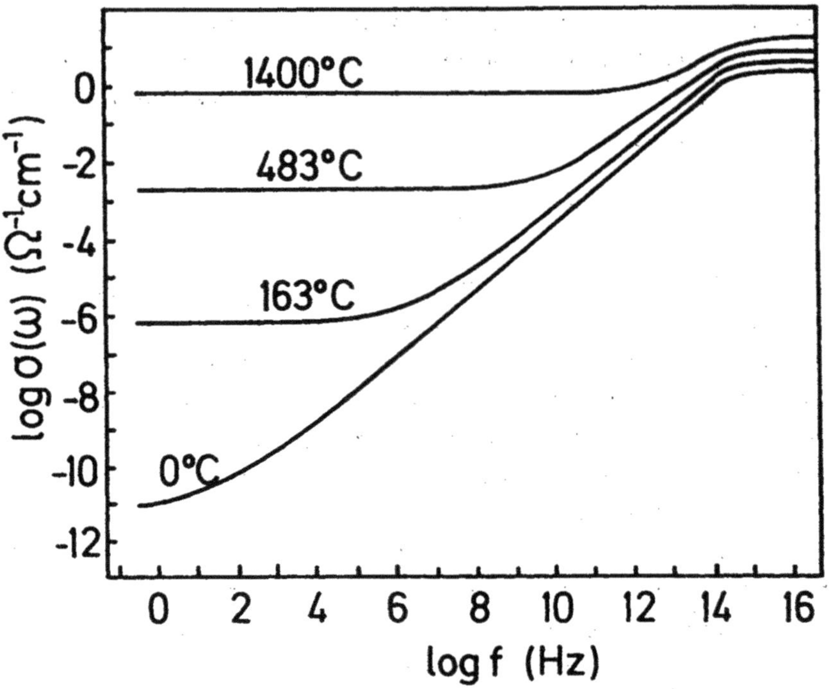

Fig. 2 Dynamic conductivity of a random hopping model with a constant distribution of barriers, which is regarded as a fit-parameter (after Ref.[45]). The curves are in excellent agreement with the temperature- and frequency-dependent conductivity of Na_2O - SiO_2 glasses[47].

study further consequences of the percolation idea. In the present context, the simplest model for correlated transport in a percolating system is a hard-core lattice gas containing a random distribution of certain blocking complexes. The static conductivity σ obtained from

Monte Carlo simulations of a model of this type is in good overall agreement with the observations[25]. If a non-vanishing mobility of the blocking particles is neglected, a sharp threshold concentration x_c exists, where the conductivity goes to zero according to

$$\sigma \sim (x_c - x)^{\mu} \quad ; \qquad x < x_c \quad . \tag{3.16}$$

Here μ is the conductivity exponent, given by $\mu \simeq 1.3$ for diffusion in a two-dimensional plane. Further Monte Carlo simulations of tracer diffusion in this model have shown that the scaling properties of the time-dependent mean-square displacement $< r^2(t) >$ close to x_c remain essentially unaffected by the hard-core interactions. At x_c one should therefore expect "anomalous diffusion" of the form

$$< r^2(t) > \sim t^k \tag{3.17}$$

where k can be expressed in terms of μ and static percolation exponents[51]. In two dimensions, $k \simeq 0.66$. An extension of this model to the case of more realistic interactions seems to be a difficult, but important task.

So far we considered hopping on networks with a given static disorder. In many cases, however, particles are diffusing in a matrix which undergoes configurational changes on time scales comparable with the experimental observation time. For example, the segment motion in polymeric solid electrolytes can become sufficiently rapid to have a significant influence on the dynamic conductivity of ions moving through the polymer network. To deal with such cases of dynamic disorder, Druger et. al.[52] have recently developed a generalized hopping model, where the hopping probabilities $W_{i\to j}$ are randomly reassigned after a certain renewal time λ^{-1}. For uncorrelated renewals following a Poisson process, they obtain the remarkable result that the dynamic conductivity $\sigma(\omega)$ is obtained by analytic continuation from the conductivity σ_o in the absence of renewals,

$$\sigma(\omega) = \sigma_o\,(\omega - i\lambda) \ . \tag{3.18}$$

This relation provides a convenient frame to analyse e.g. the temperature- and pressure-dependence of the conductivity in cases where λ is a known function of intrinsic properties of the polymer network[53].

3.4 Non-Equilibrium Properties

Stochastic lattice-gas models are particularly useful to study non-equilibrium processes. There are many important processes in solid state ionics, e.g. ion-exchange reactions, diffusion along and across interfaces or the formation of intercalated compounds, where a system is driven by an inhomogeneous particle concentration. A few special cases of this type are briefly reviewed in the following.

i) Steady state correlations.
An interacting lattice gas in a non-equilibrium steady state generally develops long-range correlations. The pair-correlations in a hard-core lattice gas induced by a constant gradient in particle density have been obtained analytically by Spohn[54]. Such correlations become significantly enhanced, if the local energy levels ε_i are not uniform[55]. We only mention that the the existence of such microscopic correlations will also affect the time-evolution of the macroscopic concentration profile in a non-stationary situation. This has been shown explicitely in a simple model for diffusion across an interface represented by a step in the site energies, where the (non-equilibrium) correlations lead to an increased interface resistance[56].

ii) Diffusion fronts.
Let us look more closely to a situation typical of many solid-state diffusion processes, where particles are released at a planar source at $x = 0$ and diffuse into the initially empty space $x > 0$. On a macroscopic level one is primarily interested in the concentration profile $p(x,t)$. Without explicit interactions, the profile follows from the ordinary diffusion equation, with a diffusion constant D_0. The result is

$$p(x,t) = \mathrm{erfc}(x/\ell_D(t)) \qquad (3.19)$$

In this purely macroscopic description the penetration depth $\ell_D(t) = 2(D_0t)^{1/2}$ is the only length scale. Simulating this process in terms of a discrete lattice gas, Rosso et al.[57,58] have recognized that intriguing geometrical properties arise on smaller length-scales. In fact, during the diffusion process a front is generated which has a fractal structure. On a two-dimensional (d=2) square lattice the diffusion front is defined as the limit of the cluster of occupied sites connected via first neighbours to the source, see Fig. 3. It turns out that for

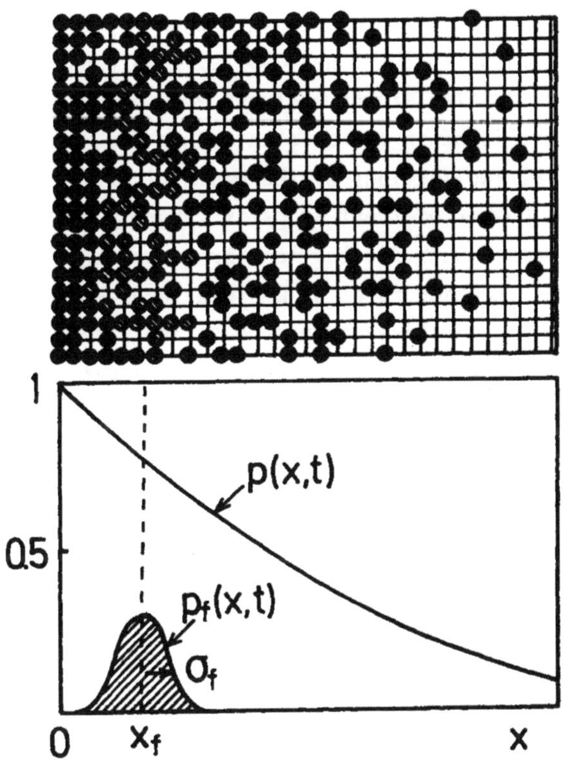

Fig. 3 Realization of a 2-d diffusion front in the presence of the concentration profile p(x,t). The distribution of front atoms $p_f(x,t)$ is indicated (after Ref.[62]).

sufficiently long times t the geometrical properties of the front are intimately related to percolation. The mean position x_f of the front is determined by $p(x_f,t) = p_c$, where $p_c = 0.59$ is the threshold for site percolation. Furthermore, the front has a typical width σ_f, which scales as $\sigma_f \sim \ell_D^{\alpha}$. The exponent α is given in terms of the correlation length exponent $\nu = 4/3$ according to $\alpha = \nu/(1+\nu)$. On length scales $r < \sigma_f$ the front atoms represent a self-similar object with fractal dimension $d_H = 7/4$. Considering the same process on a three-dimensional (d=3) simple cubic lattice, the front extends within a region determined by $1 - p_c' > p(x,t) > p_c$. Here $p_c = 0.31$ is the usual (first neighbour) site percolation threshold in d=3 and $p_c' = 0.097$ the threshhold for connections up to third neighbours. This front region in d=3 can be regarded as a material where almost each atom is a surface atom. Further details are found in the original paper[58]. Here let us mention only a few physical implications of the geometrical structure of diffusion fronts. The number of front atoms is strongly fluctuating during the diffusion process. Such fluctuations are a source of low-frequency noise[59]. Chemical reaction rates at an interface created in this way may show a specially sensitive dependence on the size of the adsorbed atoms[60]. In d=3, the front atoms represent an 'ideal porous medium'[58] with interesting physical properties, which remain to be investigated.

Diffusion of particles with attractive interactions have recently been studied by Kolb et al.[61]. At sufficiently high temperatures a non-linear diffusion equation with a concentration-dependent diffusion constant can be derived. More generally, for temperatures above the critical temperature T_c the concentration profile satisfies a scaling form, $p(x,t) = \pi(x/\langle x(t)\rangle)$, where the penetration depth scales as $\langle x(t)\rangle \sim t^{1/z}$. On approaching T_c, the dynamic exponent z becomes larger than two, indicating a critical slowing down of diffusion. The study of the geometrical structure of the front is related to the general problem of percolation of interacting particles[62].

4. INHOMOGENEOUS SYSTEMS AND RANDOM NETWORKS

As mentioned in the Introduction, we discuss here some special ionic transport phenomena which arise from macroscopic disorder. The general approach is to construct an appropriate electrical network model and to treat the corresponding network equations.

4.1 Dispersed Ionic Conductors

Random mixtures of an ionic conductor and a dispersed insulating phase can show a dramatic increase in conductivity as compared to the pure conductor. This effect, originally discovered by Liang[63)], is due to an enhanced conductivity along the internal interface between the two phases. For details and the possible microscopic mechanisms we refer to Maier[64)] and the article by Wagner[65)].

For understanding the macroscopic aspects of ionic conduction in these materials, the role of interface percolation is essential. Bunde et al.[66,67] have recently proposed an effectively three-component electrical network model, which emphasises the distinct interface properties. An extension of this model to include ac-effects is illustrated in Fig. 4 [68)]. The insulating phase is represented by the shaded

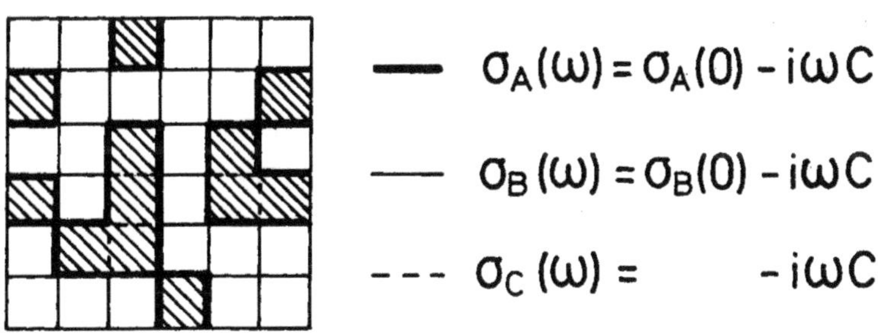

Fig. 4 Two-dimensional illustration of the random network model for dispersed ionic conductors. Shaded region: insulator; Non-shaded region: ionic conductor (after Ref.[66,68)]).

clusters originating from a random occupation of squares with probability p on a 2-d square lattice. Three types of bonds can be distinguished, as indicated in the Figure, and conductances corresponding to the interface (A), the ionic conductor (B) and the insulator (C) are assigned to theses bonds. The conductivity enhancement at the interface is expressed by the paramter $\tau = \sigma_A(0)/\sigma_B(0) \gg 1$. Ac-effects are taken into account by a capacitive element C in all the bonds.

The electrical network problem defined in this way displays a first threshold concentration p_c' where interface percolation (percolation of A-bonds) sets in and a second threshold $p_c'' = 1 - p_c'$, where the network undergoes a conductor-insulator transition. In three dimensions, $p_c' = 0.097$, which is the threshold for a site-percolation problem defined via third-neighbour-connections, see also Section 3.4.

In order to obtain the complex dynamic conductivity $\Sigma(\omega)$ or the dielectric constant

$$\varepsilon(\omega) = 1 + 4\pi i \; \Sigma(\omega)/\omega \;, \tag{4.1}$$

one has to evaluate the corresponding Kirchoff - equations. Qualitatively, disregarding any detailed critical effects associated with the thresholds p_c' or p_c'', one may apply effective medium theory[69,70]. More accurate results and a non-trivial critical behaviour can in principle be obtained by numerical simulation or by position-space renormalization methods. At zero frequency it is convenient to use the well-known mathematical equivalence of Kirchoff's equations with a corresponding random walk problem which is easily accessible by simulation. Then $\Sigma(0)$ is obtained from the time-dependent mean-square displacement $\langle r^2(t)\rangle$ of the walk in the limit $t\to\infty$,

$$\Sigma(0) \propto \lim_{t\to\infty} \langle r^2(t)\rangle / t \quad . \tag{4.2}$$

For compositions p in the range between p_c' and p_c'', the static conductivity (4.2) found in this way passes through a broad maximum[67] in good agreement with experiments by Poulsen on the LiI/Al_2O_3 - system[71]. An explicit example for $\tau = 10^3$ is displayed in Fig. 5 by the full curve.

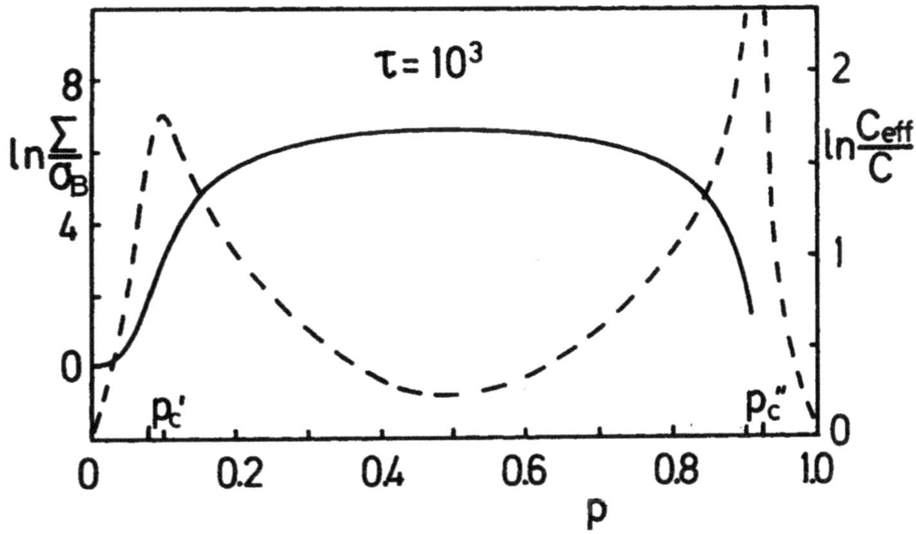

Fig. 5 Static conductivity (full line) and effective capacitance (dashed line) as a function of composition for a 3-d network model with a conductivity enhancement $\tau = 10^3$ at the interface (after Ref.[68]).

Experimentally, the detailed variation of the conductivity with composition and the maximum enhancement depend on the size of dispersed particles[65]. Such effects are reproduced successfully by a generalized network model, as shown by Roman and Yussouff[72]. In this model the shape of particles and their size relative to the thickness of the highly conducting layer can be varied. We also remark that in general there is no simple symmetry among the two thresholds.

For finite frequencies, the bond conductances are complex and the analogy between Kichoff's equations and a random walk does not hold. Then one has to rely on more elaborate numerical techniques like the transfer-matrix technique or the relaxation method. However, as shown by Blender et al.[58], a simple position-space renormalization method leads to rather accurate results, which we summarize briefly. Let us focus our attention to the vicinity of p_c', a region well accessible by experiment. Near p_c' and under the conditions $\tau \gg 1$, $\omega \ll \tau\, \omega_B = \tau\, \sigma_B/C$,the conductivity satisfies the well-known scaling form[41],

$$\Sigma(\omega) \sim \sigma_A(\omega)\ \Delta p^{\mu}\ \varphi_{\pm}\ \left(\frac{\sigma_B(\omega)/\sigma_A(\omega)}{\Delta p^{s+\mu}} \right) \tag{4.3}$$

where $\Delta p = |p-p_c'|$. The exponents μ and s and the scaling functions $\varphi_{\pm}$ (the two signs refer to the regions $p \gtrless p_c'$) are universal quantities, independent of geometrical details of the system, but depending on the dimensionality. For d=3 one has $\mu \simeq 1.9$ and $s \simeq 0.7$.

Let us discuss some of the main consequences of equation (4.3). At p_c' it follows that

$$\Sigma(\omega) \sim \sigma_B(0)(1 - i\omega/\omega_B)^{u}\ (\tau - i\omega/\omega_B)^{1-u} \tag{4.4}$$

where $u = \mu/(\mu+s) \simeq 0.75$. In the frequency-range $\omega_B \ll \omega \ll \tau\,\omega_B$, a power-law dependence $\Sigma(\omega) \sim (i\omega)^u$ is obtained, which reflects itself in a constant phase angle behaviour in the complex impedance diagram. In Fig. 5 we also show the effective capacitance $C_{eff} = (\varepsilon(0)-1)/4\pi$ as a function of composition. In the vicinity of p_c' it develops a sharp maximum which scales as $C_{eff} \sim \tau^{1-u}$. This maximum is reminiscent of the divergence $C_{eff} \sim |\Delta p|^{-s}$ expected to hold near a transition from a conductor to a perfect insulator. Measuring the effective capacitance should therefore yield direct information on the critical concentration p_c' in an actual material. It would be interesting in turn to study experimentally the scaling properties (4.3) which are a direct consequence of the fractal structure of the network of σ_A - bonds.

4.2 Fractal Electrodes

So far we considered a system with macroscopic disorder in the bulk. Another important subject of practical interest is concerned with transport across a disordered interface between two media. Considerable attention is currently devoted to the problem of the dynamic impedance of rough or porous electrodes in contanct with an electrolyte[73]. Within certain length-scales such electrodes may be modelled by surface fractals. Typically one observes a power-law behaviour of the dynamic impedance for frequencies of the order $\omega/2\pi \sim 10^3$ Hz,

$$Z(\omega) \sim (i\omega)^{-\eta} \quad . \tag{4.5}$$

The question is how to relate the exponent η $(0 < \eta < 1)$ to the fractal dimension or to some more general geometric characterization of the interface. To answer this, one has to solve the Laplace equation under appropriate boundary conditions at the interface[74] or a corresponding discrete network problem. Several proposals for model-studies of the impedance in terms of hierarchical networks[75] or in terms of processes similar to diffusion-limited aggregation[76] are in the literature. Here we limit ourselves to recall a simple model, proposed by Sapoval[77], which illustrates some of the conditions for obtaining an electrical response of the type (4.5). The model consists of cylindrical pores of length L, which act in parallel. The pores contain an electrolyte with bulk resistivity ρ, and the interface is characterized by a (generally complex) interface resistance r. On the opposite side of the interface a constant potential is maintained. The diameter a of pores is supposed to be sufficiently small such that the pores display a one-dimensional response with an admittance

$$Y(a) \simeq (a^2/\rho\ \lambda(a))\ \tanh(L/\lambda(a)) \tag{4.6}$$

where $\lambda(a) = (ra/\rho)^{1/2}$ is the penetration depth. If the pore diameters are distributed according to $dN/da \sim a^{-D}$ (between limits $0 < a < a_0$), then the electrode embedded in three-dimensional space can be regarded as a surface fractal with fractal dimension D. The total admittance Y_{tot} is now obtained by summing (4.6) over all a. Assuming $2 < D < 5/2$, the result is

$$Y_{tot} = \int_0^{a_0} Y(a)\, \frac{dN}{da}\, da \sim \begin{cases} r^{-1/2} & \text{if} \quad \lambda(a_0) \ll L \\ r^{D-3} & \phantom{\text{if}} \quad \lambda(a_0) \gg L \end{cases} \tag{4.7}$$

As r increases, the penetration depth $\lambda(a_0)$ for the largest pore increases and (4.7) shows a cross-over from the 'one-dimensional' regime to the fractal regime. Setting $r \sim (i\omega C)^{-1}$, which corresponds to a blocking electrode with an interface capacitance C, equation (4.5) is recovered together with the previously suggested relation $\eta = 3-D$ [75,73]. In general, however, for electrodes with a realistic geometry no simple relation between η and D can be expected to hold, and the detailed geometric and physical conditions for observing a response of the form of (4.5) remain to be clarified.

ACKNOWLEDGEMENTS

The author would like to thank R. Blender, A. Bunde, J.F. Gouyet, J. Petersen, H.E. Roman, B. Sapoval and M. Yussouff for fruitful discussions. This work was supported in part by the Deutsche Forschungsgemeinschaft, Sonderforschungsbereich 306.

REFERENCES

1) Vashishta, P., Solid State Ionics 18 & 19, 3 (1986)
2) Gillan, M., "Dynamical simulations of fast ion conduction", in "Ionic solids at high temperatures", edited by A. Stoneham (World Scientific Publ. Co., Singapore, 1987)
3) Jacucci, G., Solid State Ionics, in press
4) Ratner, M.A. and Nitzan, A., Solid State Ionics, in press.
5) For reviews see "Applications of the Monte Carlo Method in Statistical Physics", edited by K. Binder (Springer-Verlag, Berlin, 1984)
6) Dieterich, W., Fulde, P. and Peschel, I., Adv. Phys. 29, 527 (1980)
7) Cava, R.J., Reidinger, R. and Wuensch, B., J. Solid State Chem. 31, 69 (1980)
8) Schulz, H. and Zucker, U.H., Solid State Ionics 5, 67 (1981); Schulz, H., Ann. Rev. Mat. Sci. 12, 351 (1982)

9) Rosenberg, R.O., Boughaleb, Y., Nitzan, A. and Ratner, M.A., Solid State Ionics 18 & 19, 127 (1986)

10) Boughaleb, Y., Ratner, M.A. and Nitzan, A., in "Superionic Solids and Solid Electrolytes", edited by A. Lasker and S. Chandra (Academic Press, in press)

11) Thomas, U. and Dieterich, W., Z. Physik B 62, 287 (1986)

12) For a review see Evans, R., Adv. Phys. 28, 143 (1979)

13) Roman, H.E. and Dieterich, W., J. Phys, C: Solid State Phys. 19, L801 (1986)

14) Billi, F., Roman, H.E. and Dieterich, W., Solid State Ionics, in press

15) Chaturvedi, D.K., Senatore, G. and Tosi, M.P., Nuovo Cimento 62, 375 (1981)

16) Vashishta, P. and Rahman, A., Phys. Rev. Letters 40, 1337 (1978)

17) Vashishta, P. Ebbsjö, I., Dejus, R. and Sköld, K., J. Phys. C: Solid State Physics 18, L291 (1985)

18) Gillan, M., Roman, H.E. and Dieterich, W. unpublished

19) March, N.H. and Tosi, M.P., Phys. Chem. Liquids 11, 89 (1981)

20) An extension of the lattice gas model to include inertia effects has been proposed by Gobron, T. and Gouyet, J.F., preprint

21) Murch, G.E., in "Diffusion in Crystalline Solids", edited by G.E. Murch and A.S. Nowick (Academic Press, 1984) p. 379

22) Mori, H., Progr. Theor. Phys. 34, 423 (1969)

23) Bunde, A. and Dieterich, W.,Phys. Rev. B 31, 6012 (1985)

24) Chudley, C.T. and Elliott, R.J., Proc. Phys. Soc. London 77, 353 (1961)

25) Harder, H., Bunde, A. and Dieterich, W., J. Chem. Phys. 85, 4123 (1986)

26) Dieterich, W. and Peschel, I., J. Phys. C: Solid State Phys. 16, 3841 (1983)

27) Nemanich, R.J., Martin, R.M. and Mikkelsen, J.C., Jr., in "Fast Ion Transport in Solids, Electrodes and Electrolytes", edited by P. Vashishta, J.N. Mundy and G.K. Shenoy (North-Holland, New York, 1979) p. 547

28) Suemoto, T. and Ishigame, M., Phys. Rev. B 33, 2757 (1986)
29) Tahir-Kheli, R.A. and Elliott, R.J., Phys. Rev. B 27, 844 (1983)
30) Chaturvedi, D.K., Phys. Rev. B 28, 6868 (1983)
31) Ishii, T., Sato, H. and Kikuchi, R., Phys. Rev. B 34, 8335 (1986)
32) Ertl, W., Froböse, K. and Jäckle, J., preprint
33) Bunde, A., Chaturvedi, D.K. and Dieterich, W., Z. Physik B 47, 209 (1982)
34) March, N.H. and Tosi, M.P., Coulomb Liquids" (Academic Press, London, 1984) ch. 5
35) Kelly III, J.E., Cordaro, J.F. and Tomozawa, M., J. Non-Cryst. Solids 41, 47 (1980)
36) Funke, K., Solid State Ionics 18 & 19, 183 (1986)
37) Funke, K., Zeitschrift für Physikalische Chemie Neue Folge 154, 251 (1987)
38) Jonscher, A.K., Nature 267, 673 (1977)
39) Ngai, K.L., Comments Solid State Phys. 9, 141 (1980)
40) Haus, J.W. and Kehr, K.W., Physics Reports 150, 264 (1987)
41) Havlin, S. and Ben-Avraham, D., Adv. Phys.,in press
42) Bunde, A., Solid State Ionics, in press
43) Movaghar, B. and Schirmacher, W., J. Phys. C: Solid State Phys. 14, 859 (1980)
44) See Böttger, H. and Bryksin, V.V., "Hopping Conduction in Solids" (VCH Verlagsgesellschaft, Weinheim, 1985) and references therein
45) Schirmacher, W., Solid State Ionics, in press
46) Schirmacher, W. and Schirmer, A., Solid State Ionics, in press
47) Wong, J. and Angell, C.A., "Glass, Structure by Spectroscopy" (Marcel Dekker, New York 1976)
48) Sato, H., Wada, K., Suzuki, A. and Abkar, S.A., Solid State Ionics 18 & 19, 178 (1986)
49) Bruce, J.A., Howie, R.A. and Ingram, M.D., Solid State Ionics 18 & 19, 1129 (1986)
50) Bruce, J.A., Hunter, C.C. and Ingram, M.D., Solid State Ionics, in press

51) See Stauffer, D., "Introduction to Percolation Theory" (Taylor & Francis, London 1985)

52) Druger, S.A., Ratner, M.A. and Nitzan, A., Phys. Rev. B 31, 3939 (1985)

53) Granek, R., Nitzan, A., Druger, S. . and Ratner, M.A., Solid State Ionics, in press

54) Spohn, H., J. Phys. A 16, 4275 (1983)

55) Blender, R. and Dieterich, W., Solid State Ionics 18 & 19, 240 (1986)

56) Billi, F., Blender, R., Dieterich, W. and Frisch, H.L., Z. Physik B 64, 35 (1986)

57) Rosso, M., Gouyet, J.F. and Sapoval, B., Phys. Rev. B 32, 6053 (1985)

58) Rosso, M., Gouyet, J.F. and Sapoval, B., Phys. Rev. Letters 57, 3195 (1986)

59) Sapoval, B., Rosso, M., Gouyet, J.F. and Colonna, J.F., Solid State Ionics 18 & 19, 21 (1986)

60) Bunde, A., Gouyet, J.F. and Rosso, M., J. Phys. A: Math. and Gen. 20, 6127 (1987)

61) Kolb, M., Gouyet, J.F. and Sapoval, B., Europhysics Letters 3, 33 (1987)

62) See Gouyet, J.F., Solid State Ionics, in press and references therein

63) Liang, C.C., J. Electrochem. Soc. 120, 1289 (1973)

64) Maier, J., "Composite Electrolytes", in "Superionic Solids and Solid Electrolytes", edited by A. Lasker and S. Chandra (Academic Press, in press)

65) Wagner, Jr., J.B., this volume

66) Bunde, A., Dieterich, W. and Roman, E., Phys. Rev. Letters 55, 5 (1985)

67) Roman, H.E., Bunde, A. and Dieterich, W., Phys. Rev. B 34, 3439 (1986)

68) Blender, R. and Dieterich, W., J. Phys. C: Solid State Phys. 20, 6113 (1987)

69) Brailsford, A.D., Solid State Ionics 21, 159 (1986)

70) Rojo, A.G. and Roman, H.E., Phys, Rev. B 37, in press

71) Poulsen, F.W., Hessel Andersen, N., Kinde, B. and Schoonman, J., Solid State Ionics 9 & 10, 119 (1983)

72) Roman, H.E. and Yussouff, M., Phys. Rev. B 36, 7285 (1987)

73) See Sapoval, B., Rosso, M. and Gouyet, J.F., "Fractal Physics and Superionic Conductors", in "Superionic Conductors and Solid Electrolytes", edited by A. Lasker and S. Chandra (Academic Press, in press) and references therein

74) Halsey, T.C., Phys. Rev. A 35, 3512 (1987)

75) Liu, S.H., Phys. Rev. Letters 55, 529 (1985)

76) Chu, Y.T. and Bates, J.B., preprint

77) Sapoval, B., Solid State Ionics 23, 253 (1987)

THE PADDLE-WHEEL MODEL FOR ION CONDUCTION IN SOME SOLID PHASES

Arnold Lundén
Department of Physics, Chalmers University of Technology
S-412 96 Göteborg
SWEDEN

John O. Thomas
Department of Chemistry, University of Uppsala
Box 531, S-751 21 Uppsala
SWEDEN

ABSTRACT

The high-temperature phases of Li_2SO_4, $LiNaSO_4$, $LiAgSO_4$ and $Li_4Zn(SO_4)_3$ differ from other high-temperature phases in terms of their transition enthalpies, cation diffusion properties, crystal structures, light scattering characteristics, etc. It has been concluded from this that the cation mobility is enhanced by a strong coupling to the rotation of the otherwise translationally immobile sulphate ions. The term "paddle-wheel" or "cogwheel" mechanism has been coined to depict this general phenomenon.

1. INTRODUCTION

It was reported in 1921 that solid lithium sulphate showed a drastic increase in its electrical conductivity as the temperature approached 580°C[1]. This attracted little attention until the discovery that the structure of the high temperature phase was fcc for Li_2SO_4 and bcc for $LiNaSO_4$[2,3], and that the electrical conductivity of both phases was of the same order of magnitude as that of the fused state[3]. It is quite natural that electrical conductivity is the property which attracts greatest interest in a solid electrolyte material. Many other types of studies have been made, however, for the group of four high-temperature sulphate phases which also includes bcc $LiAgSO_4$ and non-cubic $Li_4Zn(SO_4)_3$, e. g. cation diffusion, latent heat of phase transitions, X-ray diffraction, elastic and inelastic neutron scattering, light scattering (Brillouin and Raman), electromigration, thermomigration (thermal diffusion), rheology, thermal expansion, zone refining and computer simulations (molecular dynamics). Isotope effects have been detected for the electrical conductivity, electromigration, thermomigration and phase transitions. The results of some of these studies have been summarized in a recent review[4].

The results from all these investigations have been compared with relevant studies of other types of solids. It is concluded that the four sulphate phases can be considered as both plastic crystals (rotator phases) and solid electrolytes. Moreover, it is suggested that the strongly coupled rotational motion of the translationally static sulphate ions enhances the mobility of the cations. The term cogwheel[5] or paddle-wheel[6] mechanism has been coined to depict this effect. None of the various experiments have given results which conflict with this present interpretation, and no alternative model has so far been proposed to explain all properties of the four sulphate phases; see a recent series of discussions[7-11].

The strongest evidence for the proposed transport mechanism comes from studies of cation diffusion, transition enthalpies, and neutron and X-ray scattering, while the results of other studies provide indirect support.

2. SOLID SOLUBILITIES AND PHASE DIAGRAMS FOR SULPHATE SYSTEMS

Phase diagrams have been determined for a number of binary

systems with Li_2SO_4 as one of the components[4,12]. For the fcc phase of Li_2SO_4, solubilities of at least 4 mol% were found for Na^+, Ag^+, K^+, Be^{2+}, Mg^{2+}, Ca^{2+}, Zn^{2+}, Cd^{2+} and Mn^{2+} ions, while the solubility is of the order of 1 mol% or less for Rb^+, Cs^+, Sr^{2+} and Ba^{2+}. Carbonate ions are soluble to about 10 mol% in fcc Li_2SO_4, and iodide ions to at least 10 % in bcc $LiAgSO_4$[13]. On the other hand, the solubility is very small for chloride and bromide ions in fcc Li_2SO_4.

3. MOBILITIES FOR MONO- AND DIVALENT CATIONS

Measurement of the electrical conductivity for a pure salt usually gives sufficient information on ion mobility, while it can be necessary to use other experimental techniques in the presence of several ion-types.

It is well-known that, for materials in which the ionic conductivity can be attributed to a lattice defect mechanism, this mechanism is strongly affected by the presence of aliovalent ions. To take an example: the conductivity of the hexagonal phase of Na_2SO_4 at 550°C is $7x10^{-4}$ S/cm for reagent grade salt but 0.25 S/cm for heavily doped $Li_{0.33}Na_{1.33}Zn_{0.22}SO_4$[14]. In contrast, doping has little influence on the conductivity of the four above-mentioned sulphate phases. It makes little difference whether homo- or aliovalent ions are added[10]; typically, the conductivity of the fcc phase of $Li_{1.5}Zn_{0.25}SO_4$ is only 30 % higher than that of pure Li_2SO_4 at the same temperature.

It is a general characteristic of other types of solid electrolyte phases that the size of the mobile ion must be neither much smaller nor much larger than the size of the bottlenecks in the host structure[15,16].

The diffusion of mono- and divalent cations has been studied in all four sulphate phases[17-22], and the results are summarized in Table 1. All experiments were performed using the conventional method of placing a thin layer of salt containing the diffusing ion at the top end of a long salt column. Chemical analysis was used to detect the diffusion of impurity ions, while radioactive tracers were used for

Table 1. Cation diffusion in four sulphate phases.
N.B. The diffusion coefficients of Li_2SO_4 are quoted for 1023K, while those of the other three salts are for 823K.

	Li_2SO_4			$LiNaSO_4$		$LiAgSO_4$		$Li_4Zn(SO_4)_3$	
Ion	radius	$D\times10^5$	Q(D)	$D\times10^5$	Q(D)	$D\times10^5$	Q(D)	$D\times10^5$	Q(D)
	Å	cm^2/s	eV	cm^2/s	eV	cm^2/s	eV	cm^2/s	eV
		(1023K)		(823K)		(823K)		(823K)	
Li^+	0.68	4.16	0.34	1.00	0.63	1.03	0.52	1.30	0.19
Na^+	0.97	3.73	0.31	0.93	0.64	0.77	0.38		
Ag^+	1.26	2.78	0.30	0.67	0.59	1.00	0.52		
K^+	1.33	2.08	0.19	0.44	0.90				
Tl^+	1.47	1.78	0.19	0.37	0.68				
Rb^+	1.47	1.14	0.075						
Mg^{2+}	0.66	0.22	0.60	0.13	0.72				
Ca^{2+}	0.99	0.51	0.46	0.083	0.54				
Ba^{2+}	1.34	0.8*	(0.4)						
Zn^{2+}	0.74	0.22	0.54					0.14	0.28
Cd^{2+}	0.97	0.54	0.43	0.17	0.34				
Mn^{2+}	0.80	0.6*	(0.4)						
Pb^{2+}	1.20	0.87	0.32	0.20	0.17				

* Estimated from a single measurement at another temperature.

Na^+, Ag^+ and Zn^{2+} ions when these were abundant in the phase studied. Li^+ diffusion was studied by using enriched ^6Li. In a typical experiment, measurements were made at four or five different temperatures. While all the cations listed in Table 1 are highly mobile, the diffusion coefficients for two trivalent cations were found to be of the order of 10^{-8} cm^2/s at 750°C[23], and 3×10^{-9} cm^2/s for the sulphate ions of the fcc phase at 750°C[5].

The activation energies for diffusion in the fcc phase, Q(D), are about the same for the Li^+, Na^+ and Ag^+ ions, while they are significantly smaller for the larger K^+, Tl^+ and Rb^+ ions. Their high mobility is very difficult to understand on the basis of any conventional hopping mechanism. As for the divalent cations, Q(D) is always larger than for monovalent ions with approximately the same radius and, in an early interpretation[20], it was possible to estimate formation and migration energies. This approach fails, however, for bcc $LiNaSO_4$, where the activation energies of the divalent ions are smaller than those of the corresponding monovalent ions (the only exception being the Li^+-Mg^{2+} pair). For the two bcc phases, the mobility is concentration dependent for the Na^+ and Ag^+ ions. Thus, when the ion in question is abundant, both diffusion coefficient and activation energy are about the same as for Li^+, while they are considerably smaller if the ion is present in small concentrations.

Table 2 lists the activation energies of electrical conductivity, $Q(\sigma T)$, lithium ion diffusion, Q(Li), and sulphate ion reorientation, $Q(\tau)$. The reorientation time τ is estimated from the Raman spectra[24). While Q(Li) is about 75 % of $Q(\tau)$ for the three cubic phases, $Q(\sigma T)$ is of the order of 0.4 eV for all four phases. Table 2 also gives the electrical conductivity σ at 550°C (i. e. the conductivity of the fcc

Table 2. Activation energies of electrical conductivity, $Q(\sigma T)$, diffusion of lithium ions, Q(Li) and of sulphate ion reorientation, $Q(\tau)$. The conductivities and the Li^+ diffusion coefficients at 823K are also given, i. e. the data of fcc Li_2SO_4 are extrapolated beyond the phase transition at 850 K.
H_R = D(Li)/D(cond.) = Haven Ratio. D(cond) = calculated conductivity diffusion coefficient

Salt		Li_2SO_4	$LiNaSO_4$	$LiAgSO_4$	$Li_4Zn(SO_4)_3$
Structure		fcc	bcc	bcc	non-cubic
$Q(\sigma T)$	eV	0.43	0.41	0.40	0.36
Q(Li)	eV	0.34	0.63	0.52	0.19
$Q(\tau)$	eV	0.43	0.86	0.72	
Conduct.	S/cm	0.86	0.93	1.17	1.00
$D(Li)\times10^5$	cm^2/s	1.59	1.00	1.03	1.30
H_R		0.95	0.41	0.50	0.64

phase is extrapolated below the transition temperature), and the Haven Ratio $H_R = D^*/D(\text{cond})$, where D^* is the measured (tracer) diffusion coefficient (elsewhere in this paper denoted by D), and D(cond) is the calculated conductivity diffusion coefficient. It should be remembered that the assumptions underlying the definition of the Haven Ratio are not valid for solid electrolytes[25,26], but in the present case its value can be taken as an indication that the structural differences between the fcc phase and the two bcc phases affect the cation transport mechanism.

Electromigration experiments confirm that K^+ ions have a high mobility in fcc Li_2SO_4[27]. High cation mobilities have also been verified by means of two different types of experiments in which temperature gradients have been established along a salt column. Thus, the rate at which a thermoelectric force is built up in fcc Li_2SO_4 is in satisfactory agreement with the measured diffusion coefficient of Li^+ ions[28]; thermomigration of various cations is also a fast process[29].

4. SOME PROPERTIES OF PLASTIC CRYSTALS

Molecular crystals are formed by the great majority of organic compounds and by many inorganic compounds. Such substances are often polymorphic, and it was pointed out by Timmermans in 1935 that the high temperature phase I often has a special character for which he proposed the name "plastic crystal"[30-34]. This type of material can be considered as intermediate between an ideal crystal and a liquid. (It should be noted that there are similarities, but also clear differences, between liquid crystals and plastic crystals[32].) Of the properties typical of plastic crystals, none is entirely unique. It is common that the entropy of melting (ΔS_m) is much lower than the 14 J/molK characteristic of argon, krypton and xenon (see Table 3). Furthermore, the entropy ΔS_t of the II-I phase transition is usually larger than that of melting, cf. Table 3. The molecules of plastic crystals are often globular, and the crystallographic symmetry for phase I is high; usually, but not always, cubic. It is also common that

Table 3. A comparison between some molecular crystals, three lithium-rich sulphate phases and two other ionic phases.

T_t = temperature of II-I phase transition

T_m = melting point of phase I

ΔS_t = entropy of II-I phase transition

ΔS_m = entropy of melting

The structure of phase I is also listed; the properties of some molecular crystals are very complex[34].

	T_t K	T_m K	ΔS_t J/molK	ΔS_m J/molK	structure
Ar		84.0		14.03	fcc
N_2	35.6	63.3	6.45	11.43	hex.
CH_4	20.5	90.7	3.18	10.38	fcc
SiH_4	63.75	88.5	10.93	7.54	tetrag.?
CF_4	76.2	89.5	19.43	7.83	non-cubic
CCl_4	225.4	250.3	20.35	10.05	fcc
Li_2SO_4	850	1133	29.18	7.94	fcc
$LiNaSO_4$	791	888	31.23	small	bcc
$Li_4Zn(SO_4)_3$	755	852	33.38	small	non-cubic
AgI	420	828	14.87	5.97	bcc
Na_2SO_4	520	1157	12.50	18.15	hexagonal

diffraction patterns from these materials have only a few reflections.

The distance between the centres of neighbouring polyatomic molecules (or in our case SO_4^{2-} ions) is always found to be smaller than the diameter of the sphere which would circumscribe a freely rotating molecule[30]. This fact is occasionally overlooked, and some authors[35] claim that each molecule must have sufficient free volume for it to reorient within its cage. Heat capacity analyses[30] indicate that C_{rot} is much closer to the characteristic 6R for three-dimensional torsional oscillations than to 3R, corresponding to free rotation. To summarize: plastic molecular crystals are characterized as possessing long-range order but with local orientational disorder of the structural units.

The rotation-like motion of these units can be interpreted as stepwise jumps of the molecule between instantaneous equilibrium states.

In Table 3, the transition entropies of three lithium-rich sulphate phases are compared with those for some molecular crystals and for two other ionic compounds (α-AgI and phase I of Na_2SO_4). The ratio $\Delta S_t/\Delta S_m$ is 3.7 for fcc Li_2SO_4. This is significantly higher than the value of 2.5 for bcc AgI, in which both ions are monatomic. For Na_2SO_4, ΔS_m is larger than ΔS_t; this is the normal situation for ionic compounds.

Elastic constants have recently been determined by Brillouin scattering studies of single crystals for fcc lithium sulphate[36] and bcc silver iodide[37]. The results are shown in Table 4, along with data for two molecular crystals and two ionic compounds for comparison. The elastic constants C_{11}, C_{12} and C_{44} are used for calculating the bulk modulus, $B = (C_{11} + 2C_{12})/3$, and the elastic anisotropy $A = 2C_{44}/(C_{11}-C_{12})$. The bulk modulus is the reciprocal of the volume compressibility. A detailed interpretation of the elastic constants of these two solid electrolyte materials has been presented recently[36,37].

Raman spectra[24,38] of Li_2SO_4, $LiNaSO_4$ and $LiAgSO_4$ give evidence of sulphate ion reorientation, while no such motion is observed in Na_2SO_4. The reorientation time of the sulphate ion is found to correspond well with the 2 ps reported from a molecular dynamics study[39] of fcc Li_2SO_4. Activation energies $Q(\tau)$ have been calculated for the

Table 4. Elastic constants, C, bulk modulus, B, and elastic anisotropy, A, of some materials. Unit 10^9 N/m^2.

Material	T(K)	C_{11}	C_{12}	C_{44}	B	A
γ-O_2	54	2.60	2.06	0.28	2.24	1.02
CH_4	90	1.96	1.45	0.92	1.62	3.6
Li_2SO_4	920	23.0	18.0	12.0	19.7	4.8
AgI	613	14.8	12.9	6.3	13.5	6.63
AgBr	623	36.3	25.5	8	29.1	1.48
AgCl	295	58.6	35.8	6.2	43.4	0.54

reorientation of the sulphate ions[24], see Table 2 and the discussion in Sect. 3.

Similarities between some solid electrolyte materials and molecular plastic crystals have also been observed in stress-relaxation studies[40].

5. ISOTOPE EFFECTS

Isotope effect studies can often give information on transport mechanisms. For thermomigration (thermal diffusion), the heat of transport seems to be higher for the bcc phases $LiNaSO_4$ and $LiAgSO_4$[41] than for fcc Li_2SO_4[42]. For the latter phase, isotope effects have also been found for electromigration[43], the electrical conductivity[44] and the diffusion of sodium ions[45]: the electrical conductivity is 4.7 % higher for $^6Li_2SO_4$ than for $^7Li_2SO_4$, and ^{22}Na diffuses 2.8% faster than ^{24}Na. For diffusing isotopes, it is common to use the relation:

$$b_1/b_2 - 1 = -f\,\Delta K((m_1/m_2)^{1/2} - 1)$$

where m_1 and m_2 are the ionic masses, b_1 and b_2 are the ionic mobilities, f is a correlation factor, and ΔK denotes the fraction of the kinetic energy at the saddle point associated with the diffusing ion[46]. For small differences between the two isotopes, the equation can be transformed to

$$\Delta K = 2(\Delta D/D)/(\Delta m/m)$$

where f is given the value unity. For the conductivity study, one obtains $\Delta K_{Li} = 0.61$, while $\Delta K_{Na} = 0.64$ from diffusion measurements.

6. OTHER ROTATOR PHASES.

$Na_2S_2O_3$ has been studied by von Benda and von Benda[47]. Phase I has a bcc structure with an X-ray diffraction pattern similar to that of bcc $LiNaSO_4$; the electrical conductivity is 0.3 S/cm at 670 K and the activation energy $Q(\sigma T)$ is 0.33 eV. It is suggested that, not only phase I, but also the trigonal phase II is a rotator phase. In fact, several molecular crystal materials exist which have more than one plastic phase[30]. Phase II has an electrical conductivity of the order of 0.1 S/cm at 620 K, with an activation energy of 0.7 eV. The number of X-ray diffraction lines is greater than that commonly found in plastic crystals. On the whole, phase II of sodium thiosulphate appears to show little similarity with the four sulphate phases discussed here. A stronger parallel may, in fact, be drawn with the suggestion that sulphate orientational disorder is of importance for the relatively high electrical conductivity of phase I and II of $LiKSO_4$[48].

A conductivity study[49] was earlier interpreted as evidence that phase I of Li_2WO_4 is a plastic crystal. This has been seriously challenged, however, by subsequent conductivity and X-ray diffraction work[50]. There are some striking similarities between alkali metal selenates and the corresponding sulphates[51], from which it would seem likely that several ionic plastic crystals occur among the selenates. Considerable experimental difficulties might be expected, however, since Li_2SeO_4 is volatile at elevated temperatures[52], and it is also known that selenates are carcinogenic.

7. THE STRUCTURE OF FCC-Li_2SO_4

As mentioned in Sect. 1, the concept of a "paddle-wheel" type effect was first introduced in connection with structural studies of the high temperature phase of Li_2SO_4[6]. It can be of interest, therefore, to review the diffraction work which has been carried out on this material, to clarify the exact nature of the experimental evidence for the existence of such a mechanism. It must immediately be stated that conventional X-ray or neutron diffraction studies alone are intrinsically incapable of producing definitive evidence to support a correlated coupling between SO_4^{2-} rotation and (in this case) Li^+ ion diffusion. Diffraction studies provide only a space- and time-averaged structural picture, and say nothing directly about local correlation effects.

Early X-ray powder diffraction work[2] clearly indicated an fcc structure above the transition from the low-temperature monoclinic

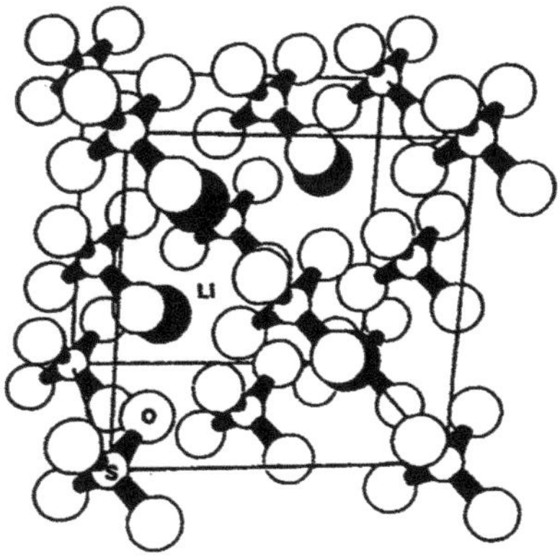

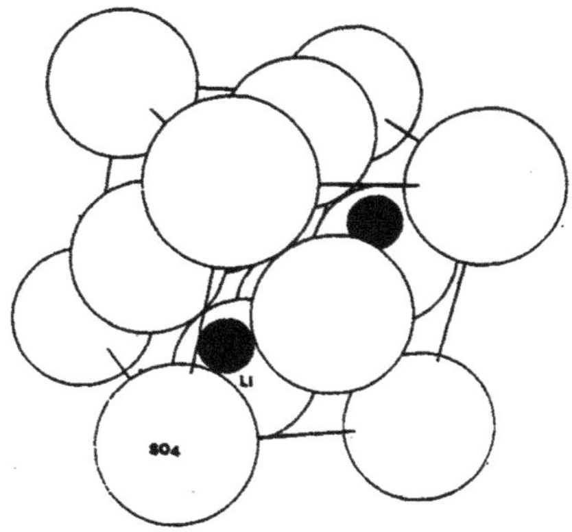

Figure 1a. The ordered fcc ($F\bar{4}3m$) model of Li_2SO_4. Isotropic temperature factors used in the plot: 5, 10 and 17 $Å^2$ for S, O and Li^+, respectively.

Figure 1b. The spherically disordered SO_4^{2-} fcc model (Fm3m) for Li_2SO_4. The SO_4^{2-} ion is drawn as an open circle of radius 1.49 Å with a 15 $Å^2$ isotropic temperature factor. The Li^+ ion is again drawn as a solid circle with a 17 $Å^2$ isotropic temperature factor.

phase at 850 K (see Fig. 1a), but was unable to provide a model which agreed well with the experimental intensity data. Nilsson, Tofield and Thomas[6)] achieved more success using neutron powder diffraction data. Although very little Bragg intensity could be observed (only two "strong" reflections), good agreement between experiment and model was obtained for a structure in which spherically symmetric SO_4^{2-} ions are sited at the origin, and Li^+ ions are at the non-centrosymmetric site (1/4, 1/4, 1/4), with exceptionally large temperature factors B ($8\pi^2\overline{u^2}$) ≈ 30 $Å^2$, where $\overline{u^2}$ is the m. s. vibrational amplitude. This structure is depicted schematically in Fig. 1b.

It was on the basis of this model that the intuitive suggestion was made for the existence of some form of correlation between anion rotation and cation diffusion, whereby the translation of Li^+ ions between equivalent sites is facilitated – indeed, stimulated – by a coupled rotation of SO_4^{2-} ions. The rotational movement of the oxygen atoms, as it were, opens and closes the available pathways between the Li^+ rest sites. The distance of closest approach between Li and O atoms consistent with this simple model is indeed 1.57 Å, so that some suggestion of a local correlation whereby these atoms could avoid each other would not seem too implausible! The sum of radii for Li^+ and O is about 1.9 Å.

It therefore became of ultimate interest to examine this question more closely when a large (roughly centimeter size) single crystal was grown successfully in situ in a furnace adapted for the collection of neutron diffraction data[53]. A much more detailed structural model could be refined using the roughly 20 observed reflections. The best fit is obtained for a model in which the SO_4^{2-} anion is still allowed to be spherically symmetric, but is now associated with a vibrational model involving a combination of isotropic and (highly significant) fourth-rank vibrational tensor terms (δ_{ijkl}'s) (see[54]). In practice, this merely implies a centrosymmetric distortion of the spherical space- and time-averaged distribution for the SO_4^{2-} group (see Fig. 2).

The refined model for the Li^+ is even more complex: 90% of the ions are refined at their "old" (1/4, 1/4, 1/4) sites, along with isotropic and (again) highly significant non-centrosymmetric third-rank vibrational tensor terms (γ_{ijk}'s). The remaining 10 % are spread as a spherical "shell" around the SO_4^{2-} ions at 2.8 Å from the S-site at (0,0,0), and are constrained to share the same thermal model as the SO_4^{2-} anion. The interpretation of this model is that the Li^+ ions have a highly distorted distribution (again see Fig. 2), in which the majority occupy the vicinity of the (1/4, 1/4, 1/4) site, while a smaller number are in a more mobile mode in a "conduction pathway" around the rotating anion. This is essentially the structural implication of the proposed paddle-wheel type effect, and can thus be taken as indirect evidence for the existence of such a mechanism. Possible Li^+ diffusion routes are indicated in the nuclear distribution in the (110)

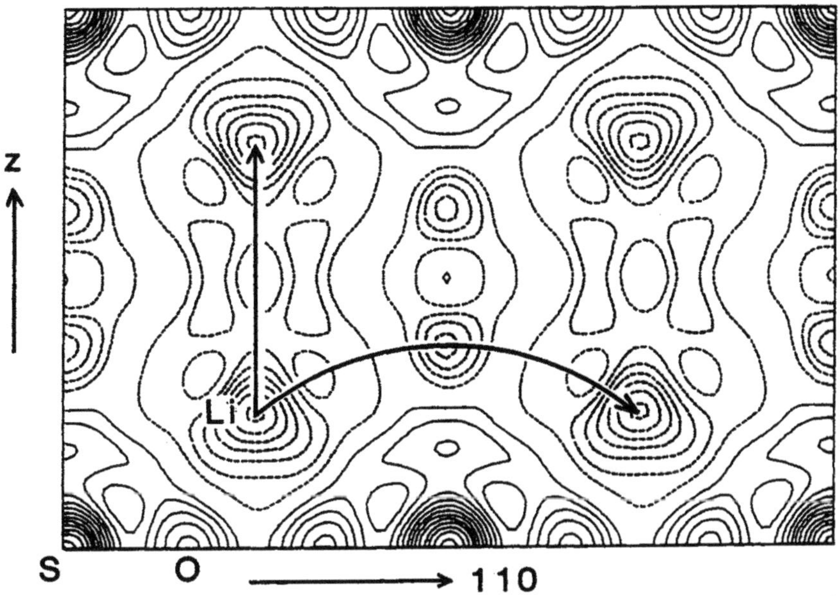

Figure 2. An observed nucleus density (F_o) synthesis in the (110) plane obtained from a recent single-crystal neutron diffraction experiment[53]; see text. Note that the negative contours for Li^+ are a result of the negative neutron scattering length of natural Li. The arrows indicate two possible Li^+ conduction pathways.

plane plotted in Fig. 2. Clearly, this in no way confirms a paddle-wheel type mechanism and, particularly, says nothing about any possible enhancement effect on the ionic conductivity. It is encouraging, however, that this considerably more sensitive source of structural information no more than adds subtle detail to the earlier, more simplistic, freely rotating anion model.

8. POSITIONS OF SILVER AND SODIUM IONS

As stated in Sect. 2, there is a large solid solubility of silver and sodium ions in fcc Li_2SO_4. X-ray and neutron powder diffraction studies of $Li_{1.6}Ag_{0.4}SO_4$[55] have shown that the Ag^+ ions are distributed between tetrahedral and octahedral sites in the rough proportion 7:3 at 838 K. This redistribution of cations compared with pure fcc Li_2SO_4 might be the main cause of the conductivity increase discussed in Sect. 3.

The bcc phases of $LiAgSO_4$ and $LiNaSO_4$ have also been studied by means of X-ray and neutron powder diffraction[56]. It is found in both cases that the Li^+ ions are in tetrahedral positions, while the Ag^+ and Na^+ ions are in octahedral positions.

9. GENERAL CONCLUSION

A paddle-wheel model is consistent with the results of the many different types of experiments that have been carried out on the high-temperature phases of Li_2SO_4, $LiNaSO_4$, $LiAgSO_4$ and $Li_4Zn(SO_4)$, and presumably also $Na_2S_2O_3$. It is still not clear, however, just how general this type of behaviour will prove to be in high temperature ionic transport mechanisms.

ACKNOWLEDGEMENTS: We are extremely indebted to all the people who have contributed to our knowledge of high-temperature lithium-rich sulphate phases over the last 30 years or more; we hope sincerely that their work has been accurately cited herein.

REFERENCES

1. Benrath, A. and Drehkopf, K., "Über die elektrische Leitfähigkeit von Salzen und Salzgemischen", Z. Phys. Chem. 99, 57-70 (1921)
2. Förland, T. and Krogh-Moe, J., "The structure of the high temperature modification of litium sulfate", Acta Chem. Scand. 11, 565 (1957).
3. Förland, T. and Krogh-Moe, J., "The structure of the high-temperature modification of sodium lithium sulfate", Acta. Cryst. 11, 224 (1958)
4. A. Lundén, "Sulphate based solid electrolytes and their applica-

tions in batteries, sensors and heat storage", in: Materials for Solid State Batteries, eds B. V. R. Chowdari and S. Radhakrishna (World Scientific Publ. Co., Singapore 1986) p. 149-160.

5. Kvist, A. and Bengtzelius, A., "Tracer diffusion and electrical conductivity of cubic lithium sulphate and a transport model for cubic sulphates", in: Fast Ion Transport in Solids, ed. W. van Gool (North-Holland, Amsterdam 1973) p. 193-199.

6. Nilsson, L., Tofield, B. C. and Thomas, J. O., "The structure of the high-temperature solid electrolyte lithium sulphate at 908 K", J. Phys. C: Solid State Phys. 13, 6441-51 (1980).

7. Gundusharma, U. M., MacLean, C. and, Secco, E. A., "Rotating sulfate contribution to electrical conductivity in Li_2SO_4 and $LiNaSO_4$", Solid State Commun. 57, 479-481 (1986).

8. Lundén, A., "Evidence for and against the paddle-wheel mechanism of ion transport in superionic sulphate phases", Solid State Commun. 65, 1237-40 (1988).

9. Secco, E. A., "Ion transport in sulfates: percolation mechanism versus paddle-wheel mechanism", Solid State Commun. 66, 921-3 (1988)

10. Lundén, A., "Enhancement of cation mobility in some sulphate phases due to a paddle-wheel mechanism", Solid State Ionics. In press.

11. Secco, E. A., Solid State Ionics. In press.

12. Lundén, A., Schroeder, K. and Ljungmark, H., "Phase diagrams of binary Li_2SO_4-$MeSO_4$ (Me=Be, Mg, Ca, Sr, Ba, Zn, Cd, Mn)", Solid State Ionics, In press.

13. Lundén, A., Kvist, A. and Ljungmark, H., "Solid electrolytes in the system Li_2SO_4-Ag_2SO_4-AgI", Proc. Symp. on Electro-Ceramics and Solid State Ionics, Honolulu Oct. 1987. Eds. H. L. Tuller and D. M. Smyth, The Electrochemical Society, Proceed. 88-3, 139-148.

14. Careem, M. A., Heed, B., Lundén, A. and Mellander, B.-E., unpubl.

15. Skaarup, S., "Solid electrolytes", J. Materials Education 6, 667-730 (1984).

16. Kennedy, J. H., "The beta-aluminas", in Solid Electrolytes, ed. S. Geller. Springer-Verlag, Berlin 1977, p. 105-141.
17. Kvist, A. and Trolle, U., "Self-diffusion and transport mechanism in cubic lithium sulphate", Z. Naturforsch. 22a (1967) 213-214.
18. Kvist, A. and Bengtzelius, A., "Diffusion in cubic sulphate. I Univalent cations in pure lithium sulphate", Z. Naturforsch. 23a, 679-682 (1968).
19. Bengtzelius, A., Kvist, A. and Trolle, U., "II. Li^+ and Ag^+ ions in bcc $(Li,Ag)_2SO_4$", Z. Naturforsch. 23a, 2040-42 (1968).
20. Kvist, A., Bengtzelius A. and Trolle, U. "III. Divalent cations in pure lithium sulphate", Z. Naturforsch. 23a, 2042-2045 (1968).
21. Kvist, A., Bengtzelius, A. and Schiraldi, A., "Diffusion measurements in the cubic high temperature modification of $LiNaSO_4$ and $LiAgSO_4$" in: Diffusion Processes, eds. J. N. Sherwood, A. V. Chadwick, W. M. Muir and F. L. Swinton (Gordon and Breach, London 1971) Vol 2, p. 523-529.
22. Lundén, A., Bengtzelius, A, Kaber, R., Nilsson, L., Schroeder, K. and Tärneberg, R, "Phase diagram, electrical conductivity and cation diffusion of the system lithium sulphate - zinc sulphate", Solid State Ionics 9&10, 89-94 (1983).
23. Bengtzelius, A., Josefson, A.-M., Kvist, A. and Schroeder, K. "Phase diagram and transport properties of cubic lithium sulphate with small quantities of thulium sulphate and aluminium sulphate", Z. Naturforsch. 25a, 1921-1924 (1970).
24. Börjesson, L. and Torell, L. M. "Contrasting behavior of the SO_4^{2-} symmetric Raman mode in high and low conducting sulfates", Proc. Electrochem. Soc. (Molten Salts) 86-1, 21-28 (1986).
25. Mundy, J. N., "Diffusion and ionic conductivity in solid electrolytes", in Fast Ion Transport in Solids, edited by P. Vashishta, J. N. Mundy and G. K. Shenoy (North-Holland, Amsterdam, 1979) p. 159-164.
26. Murch, G. E., "The Haven Ratio in fast ionic conductors", Solid State Ionics 7, 177-198 (1982).

27. Lundén, A., Floberg, A. and Mattsson, R., "Electromigration in fcc Li_2SO_4(K) and in hcp $Li_{0.4}Na_{1.6}SO_4$", Z. Naturforsch. 28a, 1857-1861 (1973).

28. Kvist, A. and Lundén, A. "Thermoelectric power and thermal diffusion in cubic lithium sulphate", Z. Naturforsch. 19a, 1058-1064 (1964).

29. Lundén, A. and Olsson, J.-E. "Thermal diffusion and related phenomena in cubic (fcc and bcc) sulphate lattices", Z. Naturforsch. 23a, 2045-2052 (1968).

30. Staveley, L. A. K., "Phase transitions in plastic crystals", Ann. Rev. Phys. Chem. 13, 351-367 (1962).

31. Schmid, D., and Wannagat, U., "Plastische Kristalle I, Kristallisationsverhalten und Thermodynamik", Chemiker-Ztg 98, 575-582 (1974).

32. Schmid, D., "Molekulare Rotation im festen Zustand", Chemiker-Ztg 99, 12-18 (1975).

33. Parsonage, N. G. and Staveley, L. A. K., Disorder in Crystals. Ch. 9 & 10. Clarendon Press, Oxford 1978.

34. Sherwood, J. N. (Ed.) The Plastically Crystalline State. J. Wiley, New York 1979.

35. Bartoli, F. J. and Litovitz, T. A., "Raman scattering: Orientational motions in liquids", J. Chem. Phys. 56, 413-425 (1972).

36. Aronsson, R. and Torell, L. M., "Elastic constants of a solid-electrolyte fcc Li_2SO_4 single crystal obtained by Brillouin scattering", Phys. Rev. B, 36, 4926-4934 (1987).

37. Börjesson, L. and Torell, L. M., "Elastic constants of a superionic alpha-AgI single crystal determined by Brillouin scattering", Phys. Rev. B 36, 4915-25 (1987).

38. Börjesson, L. and Torell, L. M., "Reorientational motion in superionic sulphates: A Raman linewidth study", Phys. Rev. B 32, 2471-77 (1985).

39. Impey, R. W., Klein, M. L. and MacDonald, I. R., "Structure of the fast-ion conducting phase of solid lithium sulphate", J. Phys. C:

Solid State Physics 17, 3941-44 (1984) and J. Chem. Phys. 82, 4690-98 (1985).

40. Jansson, B. and Sjöblom, C.-A., "The rheological properties of AgI and Li_2SO_4 solid electrolytes", Rheol. Acta 16, 628-634 (1977) and 20, 360-369 (1981).

41. Lundén, A., "Thermal diffusion in molten and solid (bcc) sulfate mixtures, mainly isotope effects", Z. Naturforsch. 24a, 1673-74 (1969).

42. Lindqvist, K. and Lundén, A., "Thermodiffusion der Li-Isotope in festem Lithiumsulfat", Z. Naturforsch. 16a, 626-627 (1961).

43. Lundén, A., "Der Isotopieeffekt bei der elektrolytischen Überführung der Lithium-Ionen in festem Lithiumsulfat", Z. Naturforsch. 15a, 365 (1960) & 17a, 142-146 (1962).

44. Kvist, A., "The electrical conductivity of solid and molten $^6Li_2SO_4$ and $^7Li_2SO_4$", Z. Naturforsch. 21a, 487 (1966); Schroeder, K. and Kvist, A., "The dependence on the ^{6}Li-content of the electrical conductivity of solid and molten lithium sulphate", Z. Naturforsch. 25a, 1365-1366 (1970).

45. Bengtzelius, A. and Kvist, A., "Diffusion in cubic sulphates IV ^{22}Na and ^{24}Na in fcc Li_2SO_4", Z. Naturforsch. 24a, 468 (1969).

46. Mullen, J. G., "Isotope effect in intermetallic diffusion", Phys. Rev. 121 1649-58 (1961); LeClaire, A. D., "Some comments on the mass effect in diffusion", Phil. Mag. 14, 1271-84 (1966).

47. von Benda, H. and von Benda, K., "Zur Polymorphie des $Na_2S_2O_3$", Z. Naturforsch. 34b (1979) 957.

48. Pimenta, M. A., Abelard, P., Echegut, P. and Gervais, F., Solid State Ionics, In press.

49. Kvist, A. and Lundén, A., "The electrical conductivity of solid and molten lithium tungstate", Z. Naturforsch. 21a, 1509 (1966).

50. Kaber, R. and Poulsen, F. W., unpublished.

51. Mehrotra, B. N., "Phase diagram and crystallography of the system Na_2SeO_4-K_2SeO_4", N. Jb. Miner. Mh. 1977, 398-407 and 1978, 485-490.

52. Shekhtman, I. G. and Burmakin, E. I., "Solid electrolytes in the Li_4SiO_4-Li_2SeO_4 system", Izv. Akad. Nauk SSR, Neorg. Mater. 21, 91-93 (1983); Inorg. Mater. 21, 78-80 (1985).

53. Andersen, N. H., Kaber, R., Nilsson, L., Thomas, J. O. and Lundén, A., unpublished.

54. International Tables for X-ray Crystallography, Vol.IV. Kynoch Press, Birmingham 1974.

55. Nilsson, L., Andersen, N. H. and Kjems, J., "The structure of the solid electrolyte $Li_{1.6}Ag_{0.4}SO_4$ at 565°C", Solid State Ionics 6, 209-214 (1982).

56. Nilsson, L., Andersen, N. H. and Lundén, A., unpublished

INERTIA-EMF IN SOLID IONIC CONDUCTORS
- MEASUREMENTS ON α-AgI AND $RbAg_4I_5$

Wolfgang Koch and Hans Rickert
Lehrstuhl für Physikalische Chemie I der Universität Dortmund
D 4600 Dortmund 50

Abstract

Investigations on the inertia-EMF in solid ionic conductors have been carried out in recent years [1-3]. The inertia-EMF equals the product of the ratio mass to charge of the mobile ions, the length and the acceleration of the solid ionic conductor. The development of sensors to measure accelerations by the use of these materials seems to be possible.

1. INTRODUCTION

The effect of forces of inertia is known to everybody from everyday life. If a railway coach is decelerated each passenger perceives the forces of inertia. A standing passenger has to ensure adherence in order to compensate these forces. If a solid material contains mobile particles, for example electrons in metals or ions in solid ionic

conductors, these particles are also exposed to forces of inertia if the material is accelerated.
Investigations on the forces of inertia in metallic conductors have already been carried out by Tolman et al. [4] in 1914. If a metallic conductor is decelerated a voltage pulse can be measured between its ends. The resulting voltage pulse is proportional to the ratio of mass to charge of the electrons and is caused by the inertia of the mobile electrons during acceleration or deceleration. An electrical field is set up within the metallic conductor. This inertia-effect of the electrons in accelerated metals is generally named after its discoverer i.e. "Tolman-effect". Tolman et al.[5] tried to investigate the forces of inertia in liquid electrolytes as well, especially in alkali iodide solutions, as Colley [6] has done before in 1882. In his experiments Colley abruptly decelerated tubes containing aqueous solutions of cadmium iodide. The results of these measurements could not be explained in a simple way. Besides the opposite and often different charges of cations and anions, which both are mobile in the electrolyte solution, thus causing an intricate coupled transport, relaxation processes may occur. Due to these difficulties investigations on inertia effects of liquid solutions have not been continued at that time.

In contrast to the investigations on liquid electrolytes measurements on solid ionic conductors [1-3] can be carried out under very simple conditions. In solid ionic conductors typically only one ionic species is mobile while the other ions constitute the rigid partial crystal. EMF-measurements on accelerated or decelerated solid electrolytes have been performed in recent years by Betsch, Rickert and Wagner [1,2]. In these experiments the solid silver ion conductors α-AgI at 170°C and $RbAg_4I_5$ at 25°C and 120°C have been decelerated by dropping rods of these materials from different heights onto a plastic material placed on the ground. The authors could show that the voltage pulses occuring between the electrodes at the ends of the rod-like solid ionic conductors are equal to the product of the length of the conductors,the ratio mass to charge of the silver ions and to the velocity of the sample just before striking the ground. Further investigations have been carried out by Koch and Rickert [3]. They exposed rods of $RbAg_4I_5$ to a sinusoidal motion parallel to the main axis of the rod. The applied frequencies ranged from 50 Hz to 5 kHz. In these experiments the acceleration at any time could be calculated and compared with the measured EMF as a function of time.

Since the mass of the silver ions is about $2 \cdot 10^5$ times as large as that of the electrons ("Tolman-effect") the inertia-effect obtained for a decelerated ionic conductor produces an electrical field which is about $2 \cdot 10^5$ times higher than that for a metallic conductor under similar conditions. Due to this fact the development of accelerometers based upon the use of solid ionic conductors seems to be possible. A further advantage of these accelerometers results from the independence of the inertia-EMF on temperature as will be shown later.

An observation related to the inertia-EMF was reported by Kompan [7]. He showed that the generation of a gravity-induced EMF occurs in the gravitational field of the earth. A further paper by Ukse et al. [8] deals with the inertia of mobile ions in $RbAg_4I_5$- and $Cu_4RbCl_3I_2$-samples being exposed to a radial acceleration in a centrifuge with a simultaneous transport of ions in the conductor.

Ogorelec et al. [9] discussed the sedimentation in mixed conductors that occurs in a centrifugal field.

2. THEORETICAL CONSIDERATIONS

We consider a rod of the solid ionic conductor $RbAg_4I_5$ in contact with silver electrodes at both ends at constant temperature as shown in fig. 1. Such an arrangement represents an isothermal symmetrical galvanic cell which is schematically notated as follows

$$Ag(1) \quad | \quad RbAg_4I_5 \quad | \quad Ag(2) \qquad (I)$$

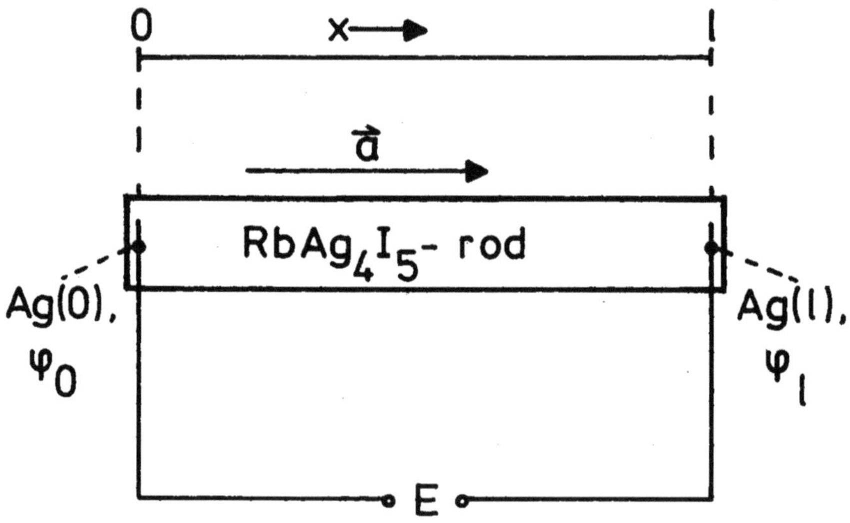

Figure 1: Solid state galvanic cell for the measurement of the inertia-EMF.

In $RbAg_4I_5$ the silver ions have a high mobility already at room temperature, whereas the other ions of the compound form a rigid partial crystal. The rod of the solid ionic conductor ($RbAg_4I_5$) with the two silver electrodes at its ends thus constitutes the solid state galvanic cell (I) which may be exposed to a definite acceleration in the longitudinal direction at a certain instant of time, whereby the cell is under open circuit conditions.

At this point we do not discuss how a constant acceleration with respect to time can be achieved experimentally. Due to the acceleration the mobile ions try to move relative to the rigid partial lattice of the non-mobile ions. But, in a very short time an electrical field is built up. The time constant for the generation of this electrical field is of the order of 10^{-11} s as will be discussed in part 2.3. In this short time interval a stationary state is set up where the mobile ions are accelerated in the same way as the non-mobile ones. In the following we regard this stationary state with zero current flow in the solid electrolyte.

The description of this state depends on the chosen reference frame. In all reference frames Newton's second axiom reads:

$$\vec{F} = m\,\vec{a} \qquad (1)$$

This law states that the force $\vec{F}$ acting on a mass m is equal to the product of this mass and the acceleration a the mass experiences. First, we choose the non-accelerated laboratory system as a reference frame and the rod of the solid ionic conductor is accelerated or decelerated with respect to this frame. In the steady state, as discussed above, mobile and non-mobile ions of the solid ionic conductor will be accelerated in the same way. The force acting on the mobile ions is the electrical force $\vec{F}_e$:

$$\vec{F}_e = e\,\vec{E} \tag{2}$$

Here e denotes the charge of the silver ions, equal to the elementary charge and $\vec{E}$ denotes the electrical field, a vector quantity which will be distinguished notationally from the EMF E by using the arrow symbol. It is assumed that no other forces are present and, especially, that no pressure gradients occur. Pressure gradients may be practically eliminated by a suitable experimental set-up.

In the case discussed i.e. that $\vec{F}_e$ is the only contribution to the total force $\vec{F}$ it follows from eqns (1) and (2):

$$e\,\vec{E} = m\,\vec{a} \tag{3}$$

Eqn. (3) expresses the fact that the electrical charge e multiplied by the electrical field $\vec{E}$ is equal to the mass m multiplied by the acceleration $\vec{a}$. This equation is the basis for all considerations to follow.

Eqn. (3) may also be derived using another frame of reference. A convenient one for the problem discussed is the accelerated reference frame of the rigid partial crystal. Relative to this system the mobile and the non-mobile ions are at rest, that means they do not move with respect to this reference frame as far as the stationary conditions apply. This is equivalent to the statement that no current is flowing in the solid ionic conductor.

Then the sum of all forces now acting on the mobile ions is equal to zero. In the accelerated frame of reference forces of inertia $\vec{F}_{in}$ have to be attributed to all masses m. These forces are given as follows:

$$\vec{F}_{in} = - m \vec{a} \tag{4}$$

In addition the electrical force $\vec{F}_e$ is acting on the mobile ions according to eqn. (2). As the sum of all forces on the mobile ions is zero in the accelerated frame of reference one obtains for the stationary state of zero current

$$- m \vec{a} + e \vec{E} = \vec{0} \tag{5}$$

Eqn. (5) is in fact identical with eqn. (3). This equation as already stated above will be the basis for the following considerations.

We assume that the rod of length l in fig. 1 is uniformly

accelerated at all positions, i.e. for all values of the coordinate x the same acceleration is present. Therefore, the field strength is independent of the position and constant along the rod for a given acceleration. It is now easy to integrate the electrical field strength E over the length of the rod from x = 0 to x = 1:

$$\int_{0}^{1} \vec{E}\cdot d\vec{x} = \vec{E}.\vec{1} = -(\varphi_1 - \varphi_o) \qquad (6)$$

Here φ_1 and φ_o denote the electrical potentials at both ends of the rod. $(\varphi_1 - \varphi_o)$, i.e. the difference between the electrical potentials at the ends of the accelerated conductor, equals the inertia-EMF E:

$$\varphi_1 - \varphi_o = E \qquad (7)$$

Combining eqns. (3), (6) and (7) we obtain the following expression for the inertia-EMF of cell (I):

$$E = -\frac{m}{e}\, l\, a \qquad (8)$$

This EMF is given by the product of the ratio mass to charge (m/e) of the mobile charge carriers in the solid silver ion conductor $RbAg_4I_5$, the length l of the conductor and the instantaneous acceleration $\vec{a}$ applied to the rod.

2.1 Dropping Experiments

We consider a rod of the solid ionic conductor, as described before. This rod is dropped along the direction of its main axis from different heights to a plastic material on the ground (fig. 2) where it is decelerated from the velocity v at time t_o just before striking to the velocity zero i.e. to rest at time t_1.

Integrating the inertia-EMF and corresponding by the r.h.s. of eqn. (8) over the time interval from t_o to t_1 one obtains the following expression for the resulting voltage pulse:

$$\int_{t_o}^{t_1} E dt = -\frac{m}{e} l \int_{t_o}^{t_1} a\, dt \tag{9}$$

Assuming that the rod is decelerated from its velocity $v(t_o)$ just before striking to the velocity $v(t_1) = 0$, i.e. to rest, the integration of the acceleration gives:

$$\int_{t_o}^{t_1} a\, dt = -v\,(t_o) \tag{10}$$

Combining eqns. (9) and (10) the following equation results:

$$\int E dt = \frac{m}{e} l\, v\,(t_o) \tag{11}$$

As can be seen the resulting voltage pulse for a conductor being decelerated should be equal to the product of the ratio mass to charge (m/e) of the silver-ions, the length l of the conductor and its velocity $v\,(t_o)$ just before striking the ground. Eqn. (11) may be verified by carrying out dropping experiments with solid ionic conductors, whereby the EMF-time-curve is recorded with a memoryscope.

Figure 2: Arrangement for the dropping experiments (schematic).

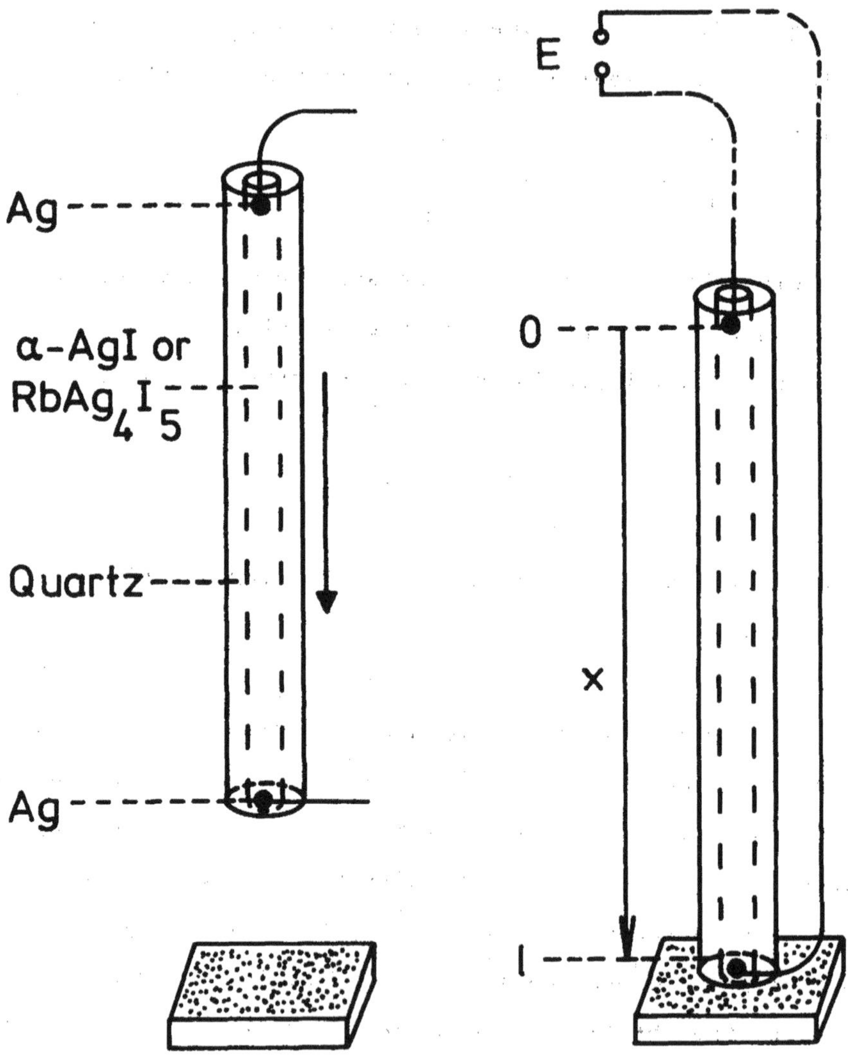

We should point out that in the case of the discussed dropping experiments the inertia-EMF is integrated over time to yield the voltage pulse. It is desirable to measure the inertia-EMF and compare it with the instantaneous acceleration in a direct way.
This is possible by exposing the solid ionic conductor to a periodical motion, for instance to a sinusoidal motion.

2.2 Simple Harmonic Motion

We now consider the rod of the solid ionic conductor $RbAg_4I_5$ with the silver electrodes, i.e. the galvanic cell (I), in a situation where it is exposed to a simple harmonic motion in the longitudinal direction. For this motion the displacement y can be described by:

$$y = y_o \sin 2 \pi \nu t, \qquad (12)$$

where y_o denotes the amplitude of displacement, ν the frequency of the oscillatory motion and t the time. The second derivative of eqn. (12) with respect to time yields the following expression for the acceleration a:

$$a = \ddot{y} = - 4\pi^2 \nu^2 y_o \sin 2\pi\nu t \qquad (13)$$

From eqns. (8) and (13) the time dependent inertia-EMF results:

$$E(t) = \frac{m}{e} 4\pi^2 l \nu^2 y_o \sin 2\pi\nu t \qquad (14)$$

From this expression we can see that a sinusoidally varying voltage with frequency ν is expected to occur between the electrodes. Its amplitude E_o should be equal to the product of the ratio mass to charge, the length l of the ionic conductor, the maximum displacement y_o, the square of the frequency ν^2 and $4\pi^2$:

$$E_o = \frac{m}{e} 4\pi^2 l \nu^2 y_o \qquad (15)$$

Fig. 3 shows the displacement y, the acceleration a and the inertia-EMF E, according to eqns. (12), (13) and (14) for a given sinusoidal motion of frequency ν and displacement amplitude y_o.

2.3 Stationary State Condition

Now we discuss the electrical time constant τ_{el} which is identical with the time interval during which the electrical field strength has reached a value of $(1-1/e)\vec{E}$, after applying a constant acceleration to the rod. $\vec{E}$ is the electrical field strength under steady state conditions for the case of constant acceleration. The electrical time constant has been calculated [1)] by regarding the rod of $RbAg_4I_5$ as a series connection of a resistor and a

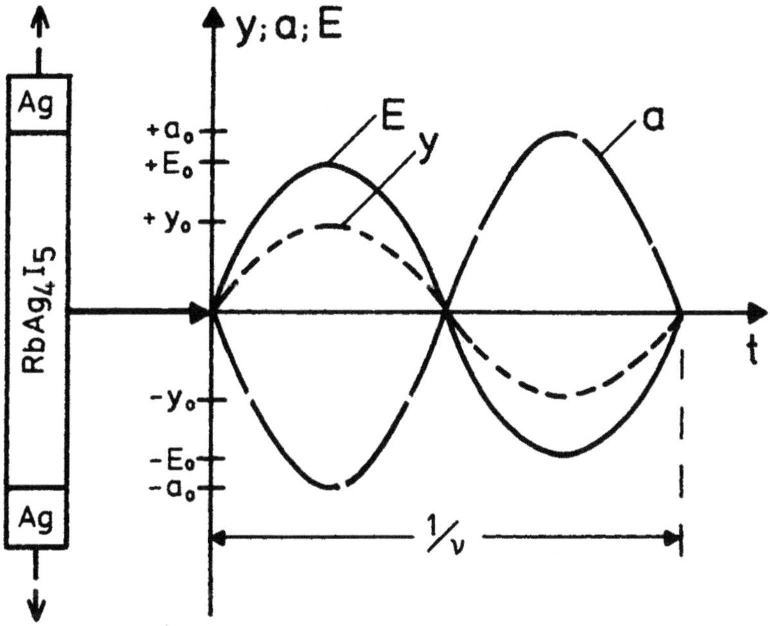

Figure 3: Sinusoidal displacement "y", acceleration "a" and the expected inertia-EMF "E" vs. time t.

capacitor. The time constant for such a RC-circuit is given by:

$$\tau_{el} = R\,C \tag{16}$$

where R denotes the electrical resistance and C the capacitance. In the case of the rod it follows for the resistance:

$$R = \rho\, l\, /\, A \tag{17}$$

Here ρ denotes the specific resistance of $RbAg_4I_5$, l the length of the rod and A its cross-section. The capacitance of the rod is given by

$$C = \varepsilon\, \varepsilon_o\, A/l \tag{18}$$

where ε_o is the vacuum permitivity and ε the relative permitivity. Combining eqns.(16), (17) and (18) one obtains for τ_{el}:

$$\tau_{el} = \rho\, \varepsilon\, \varepsilon_o \tag{19}$$

It can be seen that τ_{el} is independent of the values for l and A. Using the data [10, 11] for the specific resistance ρ (ρ = 3.70 Ω cm) and the relative permitivity ε (ε = 10) the electrical time constant has been calculated [2] to 10^{-11}s for $RbAg_4I_5$. This can be interpreted as follows: if the solid ionic conductor is exposed to an acceleration, already after some 10^{-11} s a stationary currentless state is obtained. Due to this a steady state is built up in the ionic conductor for processes in which the acceleration varies in time

intervals larger than 10^{-11}s.
But there is another reason why a rod of an ionic conductor cannot be accelerated in an infinite short time.
This reason is of mechanical nature. If a force $\vec{F}$ is applied to one frontal face of a rod at time t_o, this force "appears" at the opposite face with a certain time delay τ_{mech}. This time is given by

$$\tau_{mech} = \frac{l}{v} \qquad (20)$$

where l is the length of the rod and v the velocity of sound propagation in the rod. τ_{mech} could be termed mechanical time constant. For a solid ionic conductor which is surrounded by a support material it is reasonable to use the velocity of sound of the support material. It follows from these considerations that oscillatory motions of the rod with frequencies higher than $1/\tau_{mech}$ can not be described using a straightforward approach.
For a rod with the length $l = 4$ cm and the velocity of sound $v = 9.6 \cdot 10^3$ ms $^{-1}$ for corundum (sintered α- Al_2O_3), used as supporting material, one obtains a mechanical time constant of about 4 µs.
Due to the very small electrical time constant of about 10^{-11} s it can be deduced from the foregoing that the mechanical time constant τ_{mech} is the quantity which determines the response time for the generation of

the inertia-EMF along a $RbAg_4I_5$-rod after a force has been applied to the front side of the solid electrolyte.

3. RESULTS AND DISCUSSION

3.1. Dropping Experiments

Dropping experiments with α-AgI-rods [1)] at 170°C and $RbAg_4I_5$-rods [2)] at 25°C and 120°C have been carried out with rods of different lengths. A typical result for the voltage pulse of a $RbAg_4I_5$-sample, that occurs during the deceleration of a $RbAg_4I_5$-rod (l = 7.4 cm), is shown in fig. 4 for a velocity $v\ (t_o)$ of 1.44 m s^{-1} i.e. the velocity just before striking the ground. The velocity $v\ (t_o)$ was measured using a light barrier. One can see that the inertia-EMF rapidly increases and passes through a maximum of about 67 μV which corresponds to an acceleration of 82 g (g = 9,81 m s^{-2}). The period of braking takes about 3 ms. The instantaneous acceleration of the rod can be calculated in principle from the value of the inertia-EMF at the corresponding instant. In order to check eqn.(11) the EMF was graphically integrated to get the value for $\int Edt$. This value was in good agreement with the one calculated from eqn. (11) using the measured impact velocity of the rod. One obtains $\int Edt = 1.2 \cdot 10^{-7}$ Vs for the given conditions as shown in fig. 4.

Such measurements have been carried out for different ve-

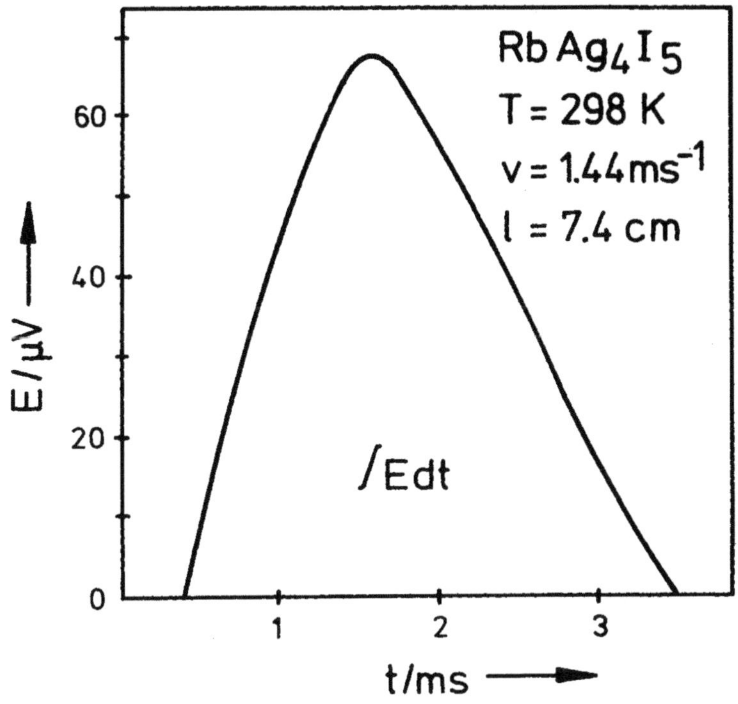

Figure 4: Example of a voltage-time-curve resulting from a dropping experiment (voltage pulse, schematic).

locities.

Fig. 5 shows the voltage pulse $\int Edt$ as a function of the velocity for an α-AgI-rod with a length of 38.5 cm at 170°C. In agreement with eqn. (11) the voltage pulse is equal to the product of the velocity $v(t_o)$ of the α-AgI-rod, the ratio mass to charge of the silver ions and the length of the rod. Measurements [2] on $RbAg_4I_5$-samples at various temperatures have shown that the voltage pulse $\int Edt$ is independent of the temperature as it is predicted by eqn. (11). The resistance of the solid ionic conductor is a function of temperature but not the value of the voltage pulse itself. The shape of the voltage pulse depends on some specific properties of the material used for the deceleration, for example on its stiffness, but the value of the integrated EMF-time-curve, i.e. the voltage pulse is not affected by using different materials for a given impact velocity and a given length of the rod.

3.2. Sinusoidal Motion

Rods of $RbAg_4I_5$ have been exposed to a sinusoidal motion [3] which is described by eqn. (12). For this situation it can be seen from eqns. (12) and (13) that the rod experiences a sinusoidal acceleration, too. In these experiments the length l of the solid ionic conductor, the frequency ν of

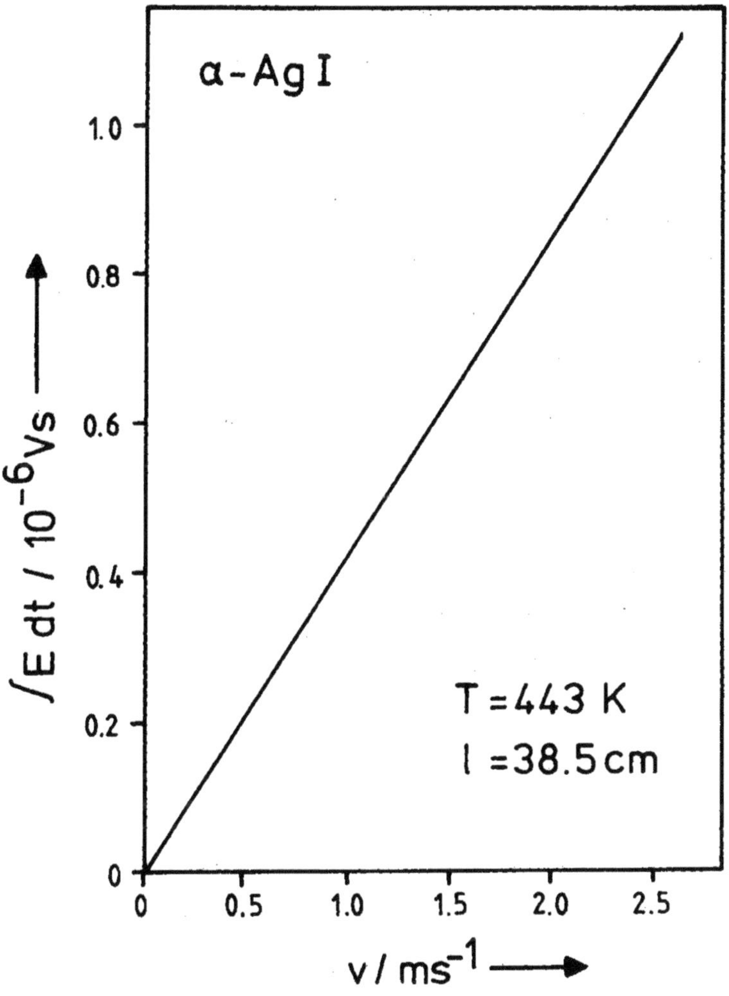

Figure 5: Voltage pulse ∫ Edt as a function of the velocity of the α - AgI sample just before striking the ground (T = 443 K).

the oscillatory motion and the maximum displacement y_o have been varied. For frequencies ranging from 50 Hz to 5 kHz an electrodynamic shaker was used, whereas for frequencies lower than 50 Hz a mechanical device with an eccentric drive was applied. An example of an EMF-time curve is given in fig. 6 for an oscillatory motion of a $RbAg_4I_5$-rod of length $l = 4$ cm at a frequency $\nu = 50$ Hz and a maximum displacement $y_o = 1.3$ cm. One obtains an EMF sinusoidally varying with time, where the frequency of the inertia-EMF is the same as the one of the driving oscillatory motion, these findings being in agreement with the theoretical considerations (eqns. (12) and (14)). For the amplitude of the inertia-EMF a value of 57 µV has been obtained, which coincides with the amplitude E_o of the inertia-EMF calculated from eqn. (15).

In fig. 7 the amplitude E_o of the inertia-EMF is depicted as a function of the sqare of frequency ν^2 of the oscillatory motion. The amplitude E_o is proportional to the square of the frequency ν^2. Further investigations [3] have shown that E_o is also proportional to the maximum displacement y_o and the length l of the $RbAg_4I_5$-rod, the factor of proportionality being $4\pi^2 \frac{m}{e}$. These results are in agreement with eqn. (15). Up to 5 kHz a phase shift between the displacement y and the EMF E has never been observed. This zero phase delay has tacitly been assumed for the schematic

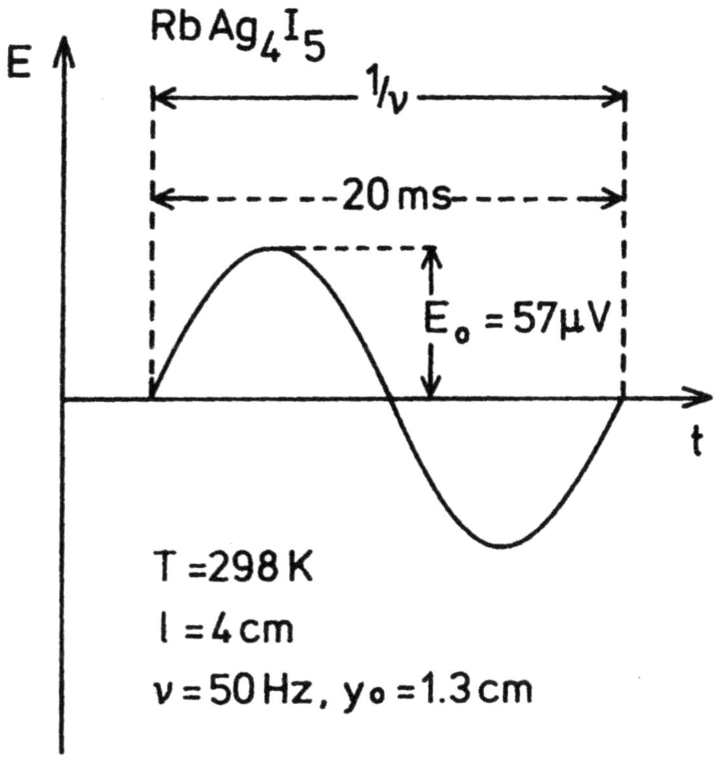

Figure 6: Sinusoidal inertia-EMF at a frequency of the sinusoidal motion of 50 Hz and a displacement amplitude y_o of 1.3 cm for a $RbAg_4I_5$-rod.

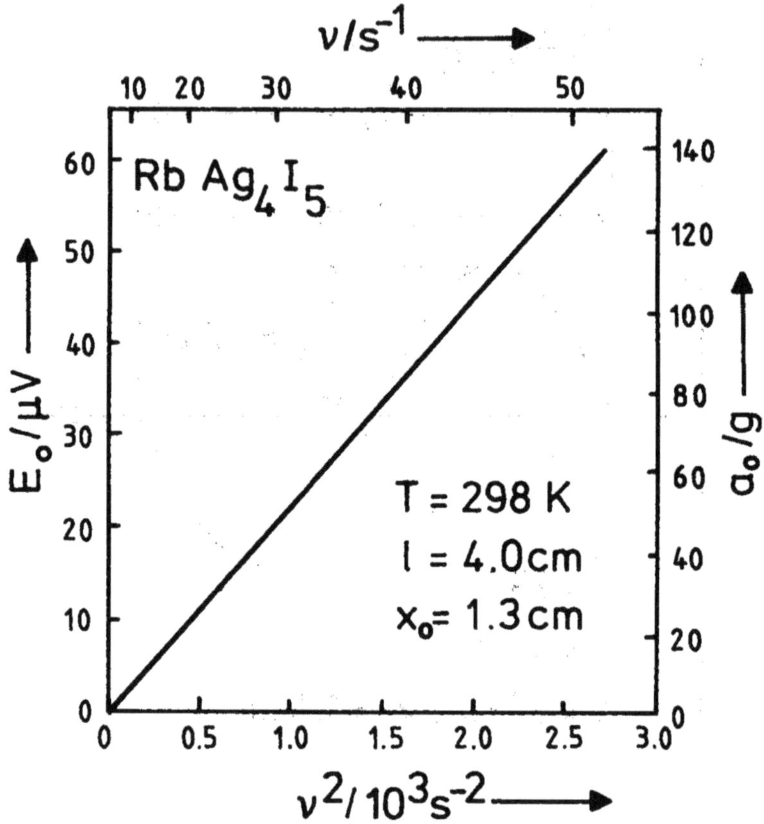

Figure 7: Amplitude of the inertia-EMF E_o as a function of the square of the oscillation frequency ν^2 for a displacement amplitude y_o of 13 mm.

representation of the theoretical E- and y-curves shown in fig. 3. The investigated range of frequencies was only limited by the mechanical performance of the shaker. One can expect that up to higher fequencies no phase shift will occur.

3.3. Conclusion

The experiments described above show that accelerations can be measured by the use of solid ionic conductors. Thus, the development of sensors with solid electrolytes for acceleration measurements seems to be possible.

REFERENCES

1) Betsch, M., Rickert, H. and Wagner, R., Ber. Bunsenges. Phys. Chem. 89, 113 (1985)

2) Betsch, M., Rickert, H. and Wagner, R., Solid State Ionics 18/19, 1193 (1986)

3) Koch, W. and Rickert, H., Solid State Ionics 1988

4) Tolman, R.C., Osgerby, E.W., Stewart, T.D., J. Am. Chem. Soc. 36, 466 (1914)

5) Tolman, R.C. and Stewart, T.D., Phys. Rev. 8, 97 (1916)

6) Colley, Wied. Ann. 17, 55 (1882)

7) Kompan, M.E., JETP let. 37, 327 (1983)

8) Vershinin, N.N., Malov, Y.I., Ukse, E.A., Elektrokhimiya 19, 102 (1983)

9) Ogorelec, Z., Solid State Ionics 15, 199 (1985)

10) Owens, B.B., Argue, G.R., J. Electrochem. Soc. 117, 898 (1970)

11) Roemer, H., Schwarz, D., Unruh, H.G., Luther, G. and Funke, K., Solid State Ionics 11, 253 (1983)

PHENOMENOLOGY OF FAST ION CONDUCTING GLASSES: FACTS AND CONFUSIONS

C.A. Angell
Department of Chemistry
Purdue University
West Lafayette, IN 47907

ABSTRACT

The distinction between fast ion *glass* and salt-in-*polymer* electrolytes is made in terms of the energetics of conductance and viscous flow processes in the *fluid* states of the two solid media. The difference in the case of glassy systems leads to a separation in time scales (decoupling) of conductance and viscous relaxation processes on cooling, the extent of which is revealed by plotting conductivities for rival glasses on a reduced temperature scale, T/T_g. The chemical factors which lead to large decoupling for cation-conducting and anion-conducting glasses are discussed. The article then focuses attention on clarifying aspects of the subject which are frequently misunderstood, or which are subject to conflicting experimental reports. Examples discussed are the relation between a.c. conductivity and single frequency NMR "activation" energies, high pressure effects on fast ion glass conductivities, and the conflict between thermodynamic and other evidence for cluster pathway models of fast diffusion and conductivity processes.

1. INTRODUCTION

For often-stated reasons, principally to do with the ease of thin film fabrication and the continuous variability of composition, amorphous solid electrolytes offer certain outstanding advantages for applications purposes over other types of fast ion conducting solids.[1–3] Amongst the amorphous solid electrolytes there exist two distinct classes,[4] the polymer salt electrolytes and the fast ion conducting glassy electrolytes of which only the latter will be of concern in this chapter. It is, however, helpful by way of introduction to the subject to distinguish between the two classes in some direct manner and thus we give an initial section to this question.

The author has provided a perspective on this field on three recent occasions,[4,5] and another would risk being repetitive and dull. Therefore in this article we will only briefly review the principal phenomenology and will devote our effort instead to dealing in detail with certain problems which are a source of some confusion, if not controversy, in the literature and which we consider to be of primary importance.

2. POLYMER SALT VS. VITREOUS SALT ELECTROLYTES

The distinction between vitreous salt electrolytes and polymer salt electrolytes is very much the same as the distinction between vitreous salt electrolytes and aqueous electrolyte solutions. The difference is only that the solvating species in the latter case is chosen to be part of a macromolecule, and this leads the solution to have a macroscopic viscosity which is determined by factors other than the local rearrangement probability which governs the macroscopic viscosity of normal liquids. Thus the microscopic rearrangements which permit the exchange of ions between coordination shells, hence permit the electrical conductivity, proceed independently of the chain disentanglement phenomena which must occur in order for macroscopic flow to occur in these systems. This has the consequence that the polymer salt system, while microscopically liquid, and conducting like a solution, can be macroscopically *solid*,[4–8] i.e. can have a shear rigidity, due to the macroscopic viscosity and its relaxation time having become greater than 10^{13} poise (and 10^2 seconds) respectively.

Since it is the microscopic rearrangements which carry most of the thermodynamic strength, the liquid-like properties of the polymer are little affected by the acquisition of a shear rigidity and only fall to values typical of the glassy state at much lower temperatures, T_g. The glass transition observed is comparable in most respects to that observed in concentrated aqueous or nonaqueous electrolyte solutions. Indeed the conductivities of aqueous and polymer solutions, compared at equal multiples of their glass transition temperatures, differ by only 1-2 orders of magnitude.[9]

A useful way of pointing out the distinction between the glassy and polymer-salt classes of amorphous ionic conductors is to compare the energetics of their electrical and viscous flow characteristics at temperatures where both are exhibiting fluid-like properties, i.e. above their glass transition temperatures. This is most conveniently done using low molecular weight polymers where the fluid state is accessed at lower temperatures.[9]

In Fig. 1 we compare the Arrhenius activation energies defined in the conventional way as Rdlog(η or σ)/d(1/T), for the fast ion glassformer 35AgCl·45AgI·20CsCl, and the salt in polymer solution $NaCF_3SO_3$ + PPO (4000 molecular weight, composition, [-O-]:[Na^+] = 16) as they vary in temperature normalized to the temperature of the glass transition, T_g. Included in Fig. 1 are data for one composition in the well-studied aqueous electrolyte system LiCl–H_2O.[10,11]

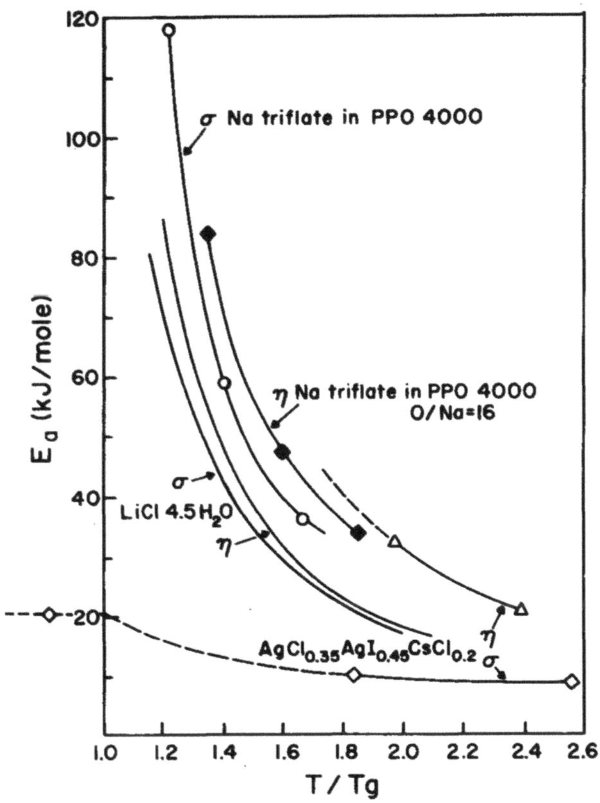

Figure 1. Comparison of the T_g-scaled temperature dependences of activation energies for conductivity and viscosity in three liquids: a salt-in-polymer solution, a salt-in-water solution, and molten salt mixture. All viscosities, and the solution conductivities, behave the same way. The low, and weakly temperature dependent, barrier for the molten salt conductivity results in a superionic conducting glassy state.

Fig. 1 shows that, from the point of view of energetics of mass transport there is very little distinction between the electrical and viscous transport processes in the polymer salt and aqueous solution systems. In the fast ion glass case, however, the energetics of the electrical conduction process are quite distinct from those of fluid flow. It is this energetic independence of the electrical conduction process which makes the fast ion glass phenomenon possible, and also makes it academically interesting. At high temperatures in the fluid state, the activation energies are comparable and the two processes may be considered to occur cooperatively or in a coupled manner. As temperature decreases, however, the electrical process becomes decoupled from the viscous process and, if normalized to the viscosity, occurs on a shorter and shorter timescale. The extent to which the electrical processes decouples from the flow process during cooling largely determines how good a solid state ionic conductor the glass will be. Of course, to be useful as a solid electrolyte, T_g must be above the temperature of application, whereas, as pointed out many times,[4–8)] for a polymer-salt rubber to be useful, T_g should be very far below the temperature of the application.

A useful way of quantifying the conductivity/viscosity decoupling is through the direct comparison, in dimensions of time, of the responses to electrical stresses on the one hand and shear mechanical stresses on the other. The ratio of these response times, τ_η/τ_σ, which is dimensionless, has been called the *decoupling index*[4,5)], and has been utilized as a figure of merit for the different fast ion glass systems. Its assessment requires a little discussion.

3. THE DECOUPLING INDEX

The decoupling index is defined as the ratio of the structural relaxation time to the electrical relaxation time. Each relaxation time is obtained either as a most probable time directly from a.c. experiments discussed below, or as a Maxwell relaxation time, the ratio of the ordinary linear response coefficient, viscosity η or specific resistivity ρ, respectively, to the appropriate modulus:

$$\tau_s = \eta_s/G_\infty\,, \quad \tau_\sigma = \rho/M_\infty \tag{1}$$

where G_∞ is the (infinite frequency) shear modulus, and M_∞ is the (infinite frequency) electrical modulus, $M_\infty = (\varepsilon_\infty e_0)^{-1}$. In a few cases the data necessary to define both relaxation times over a considerable range of temperature are available, hence the behavior of their ratio can be depicted (see below).

These time scales can be determined directly over a range of temperatures by measuring the frequencies of maximum energy dissipation ("loss") from alternating electrical and mechanical stress fields of different frequency. Using ultrasonic techniques, both shear and bulk viscosity relaxation times can be obtained and comparisons can be made with electrical relaxation times measured using conventional admittance bridges (near T_g for normal ionic liquids, and below T_g for superionic cases). Typical electrical loss spectra for a fast-ion conducting *glass* ($6AgI{\cdot}4Ag_2B_4O_7$) and for a weakly conducting ionic *liquid* ($3KNO_3{\cdot}2Ca(NO_3)_2$) are shown in Fig. 2(a) and longitudinal mechanical loss spectra for the same weakly conducting ionic liquid at somewhat higher temperatures (due to the higher frequencies of the ultrasonic and hypersonic techniques applied) are shown in Fig. 2(b).

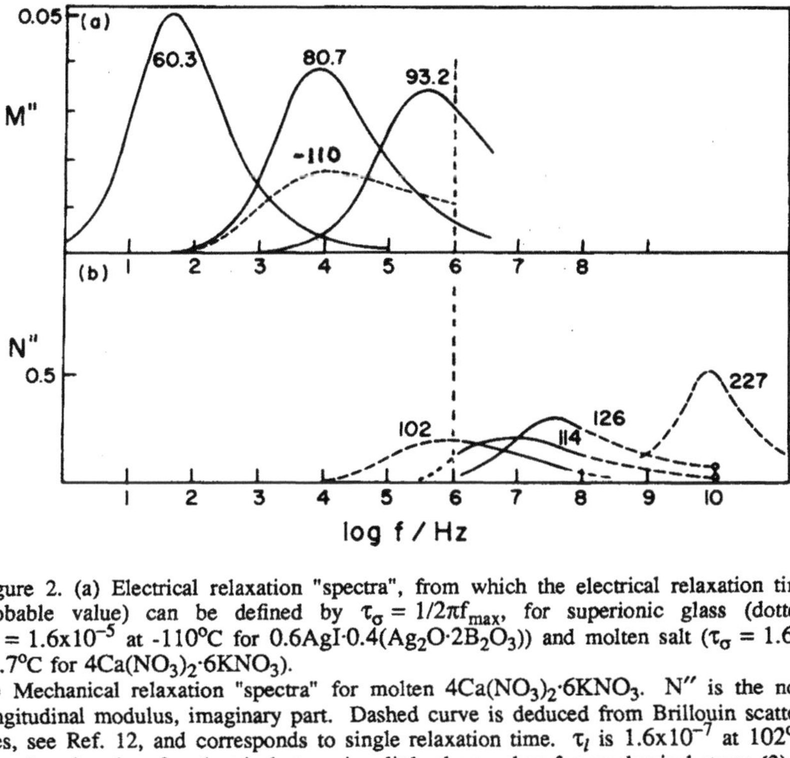

Figure 2. (a) Electrical relaxation "spectra", from which the electrical relaxation time (most probable value) can be defined by $\tau_\sigma = 1/2\pi f_{max}$, for superionic glass (dotted curve $\tau_\sigma = 1.6\times10^{-5}$ at -110°C for $0.6AgI{\cdot}0.4(Ag_2O{\cdot}2B_2O_3)$) and molten salt ($\tau_\sigma = 1.6\times10^{-5}$ at 80.7°C for $4Ca(NO_3)_2{\cdot}6KNO_3$).
(b) Mechanical relaxation "spectra" for molten $4Ca(NO_3)_2{\cdot}6KNO_3$. N″ is the normalized longitudinal modulus, imaginary part. Dashed curve is deduced from Brillouin scattering studies, see Ref. 12, and corresponds to single relaxation time. τ_l is 1.6×10^{-7} at 102°C. Note (1) relaxation time for electrical stress is a little shorter than for mechanical stress (2) electrical relaxation narrows with decreasing temperature while mechanical relaxation narrows with increasing temperature.

Relaxation times τ_σ and τ_l obtained from loss peak of Fig. 2 by writing $\tau_\sigma = 1/2\pi f_{max(\sigma)}$ $\tau_l = 1/2\pi f_{max(l)}$, are plotted as closed and open triangles in Fig. 3 for the case of $3KNO_3 2Ca(NO_3)_2$. Their behavior is contrasted with that of some all-halide superionic glasses and their liquids, using Maxwell relaxation times to augment the relaxation spectra and extend the temperature range. Fig. 3b shows the variation of the decoupling index, $R_\tau = \tau_\eta/\tau_\sigma$ with temperature down to the glass transition. At T_g equilibrium is lost and the liquid shear relaxation time can no longer be determined. Fig. 3b shows clearly the enormous differences in conductivity/viscosity decoupling which can occur amongst different inorganic glassforming liquid systems. The substances of greatest interest as far as fast ion conduction in the amorphous state is concerned are, of course, those with the largest decoupling indexes at T_g. However, a large decoupling index is not all that is needed to provide for a high *ambient* temperature conductivity.[4b,5] Data for several glasses of different decoupling index (judging by reported conductivities extrapolated to T_g) are collected in Fig. 4a and it is seen

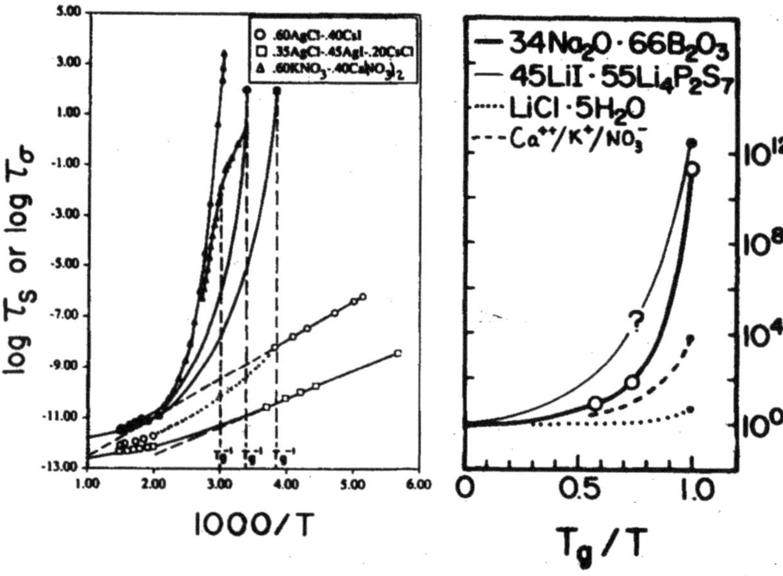

Figure 3. (a) Variation of mechanical and electrical relaxation times with temperature (Arrhenius plot) for liquid and glassy states of $4Ca(NO_3)_2{\cdot}6KNO_3$ from Fig. 2, and two superionic AgCl-AgI-CsCl cases. The more AgX-rich of the two shows 10^{13} decoupling in the liquid before T_g terminates equilibrium measurements. (b) Variation of decoupling index $R_\tau = \tau_l/\tau_\sigma$ with temperature (Arrhenius plot) for several systems.

that systems which have very high T_g values may have relatively poor ambient temperature conductivities despite very favorable decoupling ratios. The utility of scaling temperature dependent transport data by the glass transition temperature is shown by the T_g-scaled plotting of Fig. 4a data in Fig. 4b.

We now consider briefly the chemical conditions which should be met to insure large values of the decoupling index for cationic and, alternatively anionic motions and, in so doing, we will review the best fast ion conducting glassy systems which have been discovered. We will then consider certain problems encountered in studying and interpreting the conductance mechanism, and the structures underlying the fast ion conducting glass phenomenon.

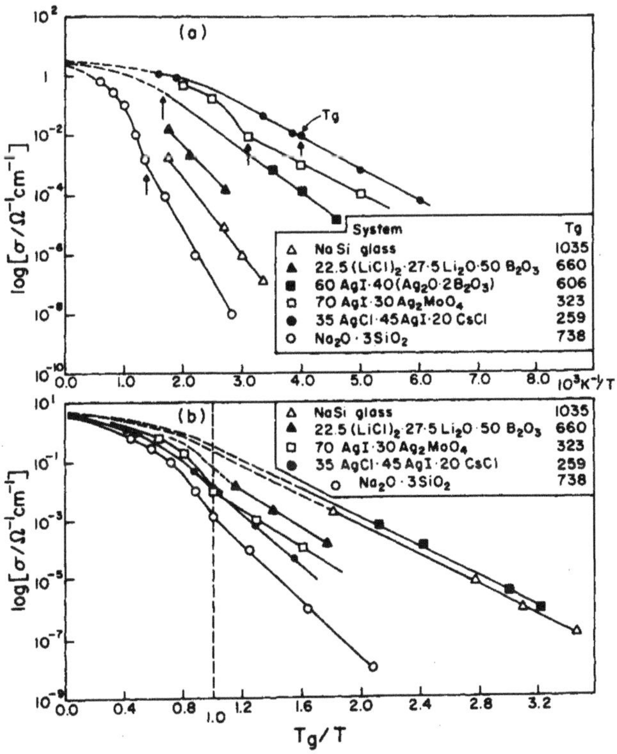

Figure 4. Arrhenius plots of conductivity data for glasses of different T_g and different decoupling indices showing how both factors play a role in determining the ambient temperature conductivity (part a). Part (b) shows how the relative decoupling in different cases can be displayed by a T_g-scaled Arrhenius plot.

4. SYSTEMS YIELDING FAST CATION AND FAST ANION CONDUCTING GLASSES

4.1 Cationic Conductors

The most studied of the cationic fast conducting glasses have been those in which the component AgI is dissolved in an oxide glass consisting of Ag_2O and one of the common Lewis acid glassformers. [13] The two most widely studied cases are those in which the Lewis acid is P_2O_5 or B_2O_3, the latter system being particularly interesting not only because of the apparently high decoupling indexes but also because of the high glass transition temperatures which it exhibits even at a mole fraction of silver iodide as high as 0.7[14]. Not only have the glassforming regions and the conductivities in these regions been determined for this system, but a range of mechanical relaxation studies extending over 10 orders of magnitude in frequency have been conducted. Some of these we will discuss in more detail later on.

Glasses with high conductivities can also be formed with the Lewis acids GeO_2[1] and no doubt Bi_2O_3, TeO_2, etc., but these have been less studied. The thio-analogs of such systems have also been broadly investigated and, although more difficult to work with because of their tendency to oxidize and to absorb water, they bestow the advantage of lower glass transition temperatures and thus higher conductivities, see Fig. 4. The highest ambient temperature conductivity yet reported is found in the system $AgI–Ag_2S–B_2S_3$.[16,17].

Slightly less complex in character are the related glasses in which silver iodide is mixed with an oxysalt of silver, such as Ag_2MoO_4 or Ag_2CrO_4. These are special cases of the first group in which the Lewis acid base reaction has been taken to the last stage when all $L^{n+}–O–L^{n+}$ bridges (L^{n+} is the Lewis acid cation e.g. B^{3+}, P^{5+}, Cr^{6+}) have been broken. They provide outstanding examples of the breakdown of the Zacchariesen rules for glassforming systems. Presumably, they owe their existence to the relatively low lattice energies of the compounds which form in the systems with the consequence that the liquidus temperatures are low relative to the forces between ions in the liquids, which implies high viscosities and slow crystallization kinetics. An exception to the latter condition, however, is the system $AgI–Ag_2SO_4$ which, despite exceptionally low liquidus temperatures, can only be vitrified by extremely fast quenching. There appear to be anomalously high nucleation rates in this system, perhaps associated with the nucleation of superionic anion-disordered crystals of entropy almost as high as those of the liquid from which they form.[18] Some interesting behavior is found in ternary systems in which the oxyanion shape is maintained constant, but the size is varied by changing the atom at the tetrahedron center from a small Se(VI) to the larger W(VI).[18] The (AgI-$Ag_2M[VI]O_4$) systems are characterized by low glass transition temperatures which, according to Fig. 4, favors high ambient temperature conductivity, even when the decoupling indices are not exceptionally high.

The thio-analogs of these systems have so far not been studied to the best of the writer's knowledge.

More recent and interesting amongst the cation-conducting superionic glass discoveries are those in which the maximum conductivity is *not* conditioned by the maximum content of α-AgI or its analog. There are two cases, each based on Ag^+ conductivity, in which the best conducting composition is a glass close in composition, and probably therefore similar in microstructure, to a superionic crystal of structure related to AgI. The crystal structures in question are $RbAg_4I_5$ and Ag_3SI. A vitreous phase analogous to the first of these was reported by Liu, et. al.[19] in the ternary system AgI-AgCl-CsCl, and presumably owes its existence to the disorder associated with the mixing of anions. The glass related to Ag_3SI is a composition almost identical to the crystal compound itself, but one which is stabilized in the glassy state (i.e., the crystallization kinetics slowed down in the liquid) by addition of a relatively small amount of a third component.[20] Sun et. al.[20] found that As_2S_3 and Sb_2S_3 were satisfactory but a number of others will probably also serve. Using twin roller quenching, the glassforming region can be pushed almost to the AgI-Ag_2S boundary. The possibility that mixing halide ions at the Agx–Ag_2S stoichiometry might as in the first case, promote glass formation, has not yet been investigated. The glass transition temperatures in each of these cases are very low and the ambient temperature conductivity of 0.9Ag_3SI·0 is the highest yet recorded.

Most of the efforts to interpret the high conductivity of glasses in the AgI-Ag oxysalt systems have involved the assumption that some form of clustering of AgI is occurring with intercluster connections allowing for fast Ag^+ diffusion throughout the sample. Indeed, many of the properties in the pseudo-binary systems with AgI as one component tend to the behavior of α-AgI (the high temperature superionic phase of the pure compound) with increasing AgI content,[21] see Fig. 5. On the other hand, such a postulate flies in the face of one of the very few thermodynamic studies of solutions in this type of system[22] which showed clearly that the mixing of AgI with the second component (in this case $AgPO_3$) proceeded with the *release* of energy. Exothermic mixing implies a tendency to chemically order rather than to cluster, hence the AgI cluster hypothesis is not supported. We will discuss this important question in considerable detail in a later section with particular reference to the corresponding *alkali* halide + oxyanion systems.

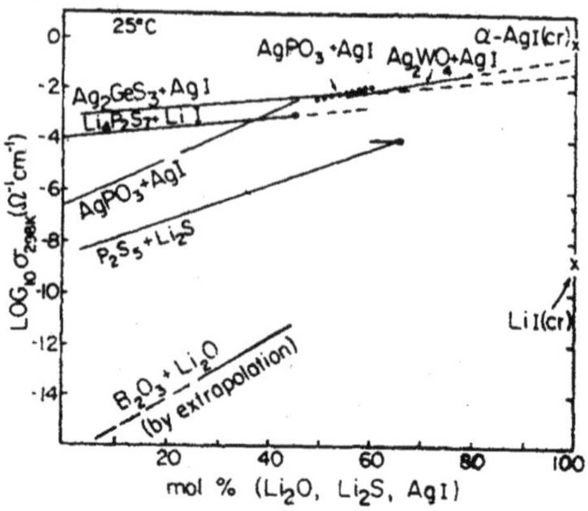

Figure 5. Variation of physical properties of AgI-containing glasses indication extrapolation to values for the fast conducting phase of crystalline silver iodide.

The superionic glasses with compositions essentially those of known superionic crystals are clearly disordered systems imitating the structures of the ordered phase. This implies that the decoupling which occurs during cooling is a reflection of the way in which decreasing thermal amplitudes allow a more and more integrated amorphous phase version of the open channel crystal structure to form, with the consequent increasing development of the intrinsic silver ion mobility. In general, it must then be expected that the conductivity of glasses of these compositions will fall below those of the parent crystals, (a relation which is by no means general among ion conducting systems: vitrification often produces a conducting phase from an insulating crystalline material).

All of the above systems presumably have analogs in which the fast ion is Cu^+, or in some cases Tl^+. We will not go into these here except to note that an attempt to obtain glasses with properties approaching the exceptional conductivity of the crystal $RbCu_4I_5$, was not successful.[23)]

More important for high energy solid state batteries are the glasses in which the mobile cation is an alkali metal. Academically, these are also interesting because of the greater fidelity with which their behavior can be simulated by the methods of molecular dynamics due to the simpler forms of the pair interaction potentials which are appropriate for the inert gas structure cations.

Most of the principles governing the formation of silver ion-conducting glasses apply to the alkali containing systems. Thus the best ambient temperature alkali conducting glass is a low T_g system containing an alkali iodide LiI in large mole fraction in a thio analog of an oxysalt viz. $Li_4P_2S_7$[24a)] or $Li_2S{\cdot}B_2S_3$.[24b)] On the other hand, the equivalent of the AgI-Ag oxysalt and the AgI-CsCl type glasses are not yet known.

The most studied of these systems are the cases LiI–$LiPO_3$ and LiI + LiP_2S_7. The glasses, however, are hygroscopic and relatively difficult to work with, and for many purposes the less hygroscopic chloroborate systems have been preferred. A system on which a great deal of work has recently been performed is the system LiCl–Li_2O–B_2O_3 described initially by Button, et. al.[25)]. Again, the highest conductivities are found at the edge of the glassforming range at the maximum LiCl content. The LiCl solubility in $Li_2O{\cdot}nB_2O_3$ (n=1,2,3) is significantly higher than that of LiBr or LiI so the LiCl-based system has the highest conductivity. We will discuss this system, with respect to structural insights from computer simulation studies, later on. For the moment we only note the existence of the naturally occurring compound boracite, the composition of which is $2LiCl{\cdot}2Li_2O{\cdot}3B_2O_3$, and the implication that a chemical ordering of LiCl in the glass is involved in its favorable conductivity.

Again, as in the case of the silver containing systems, there are similar systems based on other Lewis acid oxides, GeO_2, P_2O_5, TeO_2, etc., and also there are higher conducting thio-analogs. A notable feature of the latter systems is that the glassforming range always extends to higher alkali in the case of thio rather than oxy compounds, e.g. where the orthosilicate glasses can only be obtained under extreme quenching conditions,[26)] it is easy to obtain thio-based analog glasses at compositions even exceeding the ortho-thio-silicate composition.[27)] Oddly enough it does not appear that the conductivity is as strongly favored by high basicity in these systems as in the oxides.

Among the more practical alkali conducting glasses, although certainly not the highest conducting at ambient temperatures, are the "Nasiglass" type compositions described by Susman, et. al.[28)] These are glassy analogs of the "Nasicon" type fast ion conducting crystals, and owe their properties to the open network structure characteristic of the latter compound. These are systems with ***high*** glass transition temperatures and almost complete decoupling, see Fig. 4b. The decoupling index for Nasiglass at its T_g, for instance, is 10^{14}.

4.2 Anion-Conducting Glasses

Although fluoride anion conducting glasses such as the lead fluorosilicates[29)] have been known for a long time, it is only recently that concerted efforts to obtain really high fluoride ion-conducting glasses have been made. The best cases so far reported have been those with high PbF_2 contents. These glasses are not obtained in the system which first comes to mind (PbF_2+$Pb(PO_3)_2$) which only vitrifies at high $Pb(PO_3)_2$ contents[30)]. However, it turns out that either a replacement of $Pb(PO_3)_2$ by $Al(PO_3)_3$[31)] or addition of a third component such as MnF_2 to the all-lead cation system,[32)] gives easy vitrification at lead fluoride contents as high as 80 mole percent. While by no means rivalling the silver or alkali cation conducting glasses, these latter anion-conducting systems and some B_2O_3-based analogs[33)] reach conductivities of 10^{-4} $\Omega^{-1}cm^{-1}$ at 200°C, comparable with $Na_2O{\cdot}2B_2O_3$. Some data are shown in Fig. 6. The best *crystalline* fluoride ion conductor is the compound $PbSnF_4$ which shows a remarkable $10^{-1}\Omega^{-1}cm^{-1}$ at ambient temperature (see Fig. 6). Attempts to imitate this structure within the glassy state by addition of SnF_2 to the PbF_2+$Pb(PO_3)_2$ system, however, failed due to unsatisfactory glassforming ability. It now appears that extremely high fluoride ion conductivity may be obtained in a stannous fluoride-rich composition based on the binary SnF_2+$Sn(PO_3)_2$ described by Tick, et. al.[34)] Details have yet to be published.

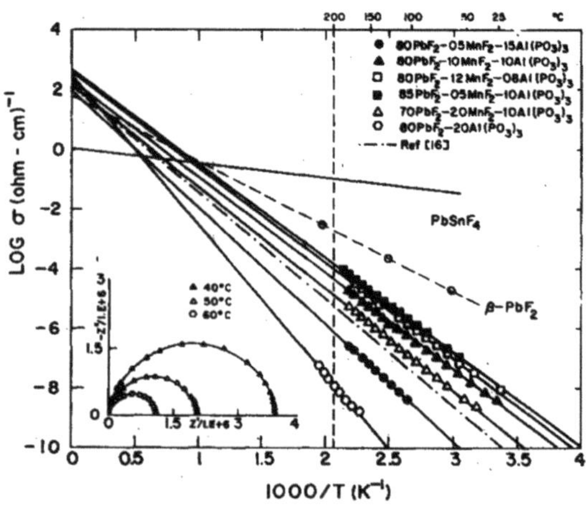

Figure 6. Arrhenius plots of data for high fluoride ion conductivity glasses. Comparisons are made with crystalline β–PbF_2 and $PbSnF_4$, the best fluoride ion solid electrolytes known. (from ref. 31, reproduced by permission).

4.3 Mixed Anion-Cation Conducting Glasses

Since the system PbF_2–$Al(PO_3)_3$ rich in PbF_2 is a good fluoride ion conducting system and since one of the more startling early discoveries of high alkali conducting glasses was the LiF–$Al(PO_3)_3$ glass rich in LiF, it is reasonable to inquire about the possibility of pseudo-binary glasses cutting across the PbF_2–LiF–$Al(PO_3)_3$ system in which there is a transition from anion conduction to cation conduction. A hint that interesting phenomenology in such systems would be found was given by the study of Reau, et. al.[35] on the effect of LiF additions to a fluorozirconate anion-conducting glass. These authors showed that a sharp minimum in the ionic conductivity occurred at 20% added LiF where they supposed the transition from anion to cation conduction occurred.

A detailed study of the more straight-forward ternary system, LiF–PbF_2–$Al(PO_3)_3$ at constant 20 mole % $Al(PO_3)_3$ has allowed some clarification of this interesting situation. In brief, Kulkarni and the author[32b] concluded that the conductivity minimum is *not* due to a mobility crossover but is instead just a reflection of a minimum in the glass transition temperature and, more important, in the decoupling index. Their best estimate was that the changeover from anionic to cationic conductivity occurred at a higher LiF content of about 50 mol % and was indicated by a maximum in the dispersion of the electrical modulus. A key diagram from this study is shown in Fig. 7. A resolution of this problem should be available through suitable pulsed-gradient diffusion measurements since both Li and F are very favorable nuclei for NMR studies.

5. CONDUCTIVITY IN RELATION TO THE KINETIC VARIABLES, COOLING RATE (HISTORY) AND MEASURING RATE (PROBE FREQUENCY)

5.1 History Dependence

Up to this we point we have described the conductivity of a glassy fast ion conductor as if it were a clear-cut quantity depending on only the composition of the glass in question. In fact this is not true since the conductivity, like any other property of the glassy material, is to a greater or lesser extent history-dependent. Long ago,[37] it was shown for the case of two Li^+-conducting glasses, one of very high and one of low conductivity, that the cooling rate had a very significant effect on the glass conductivity the effect being larger for the poorly-conducting glass. Indeed, the cooling rate dependence of glass conductivity in the case of classical oxide glasses has been very thoroughly studied.[38] The reason that the cooling rate effect is small for fast ion conducting systems and large for insulating glasses is discussed by the author in an earlier review.[4a] The relevant diagram and explanatory caption is reproduced herein as Fig. 8 in view of the fact that the problem has now been made the subject of a

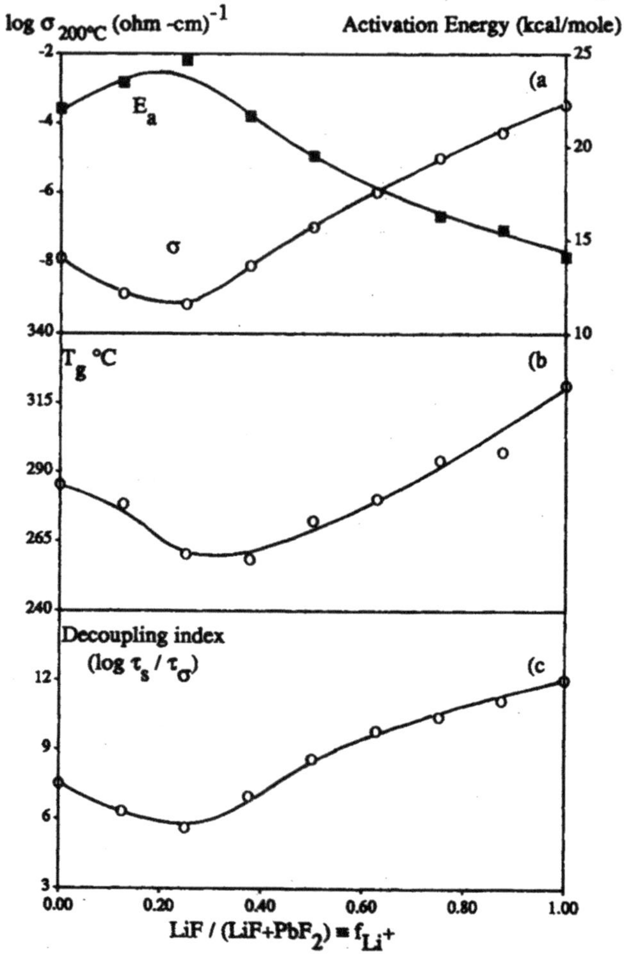

Figure 7. Variation of various physical properties with composition in the mixed anion - cation conducting system $LiF–PbF_2–Al(PO_3)_3$. The minimum appears to be due to a minimum in the decoupling of fast ion from viscous modes, rather than to a well-defined mobility cross-over. The latter may occur at higher LiF contents.

detailed study for a superionic glass by Kawamura and Shimoji[39)] for the case of $0.7AgI{\cdot}0.3Ag_2MoO_4$ glass. The latter authors also describe a model, based on an adaptation of the free volume model for liquid viscosities, which provides a quantitative analysis of the effects we have qualitatively described in Fig. 4(b). Essentially, the distinction between conductance and viscosity in a decoupled system can be

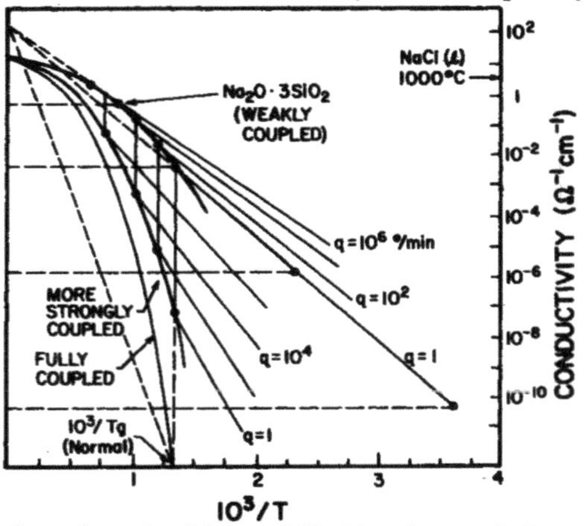

Figure 8. Representation of conductivities of liquid and quenched states of three different glasses all with the same "normal" (10^0/min) T_g, but differing in decoupling index (e.g. Na^+, K^+ and Ba^+ trisilicates). The heavy plot, upper set, is actual data for $Na_2O{\cdot}3SiO_2$. The effect of higher fictive temperature due to increasing quench rates on glass conductivity is demonstrated for Na^+ and 'K^+' cases by families of finer lines. Comparison of these at 100°C shows how "fast" high T_τ glasses benefit least from quenching.

parameterized using the long–recognized[40)] fact that the difference in temperature dependence of viscosity and conductivity can be assigned to a difference in the B parameters of the Vogel-Tammann-Fulcher equations for conductivity σ, and fluidity ϕ

$$\sigma,\phi = A_\sigma,A_\phi \exp -(B_\sigma,B_\phi/T-T_o) \quad , \quad (\phi = 1/\eta) \qquad (3)$$

where A, B, and T_o are constants. Since T_o is believed to be a thermodynamic characteristic of the liquid, hence independent of the transport process being studied, differences in transport mechanism must be reflected in the parameter B. This can be interpreted in terms of different critical free volume fluctuation needs v^* for conductivity and viscosity in the free volume model[41b)] or different fundamental free energy barriers $\Delta\mu$ for the "rearranging group" in conductivity and viscosity, in the Adam-Gibbs model.[41c)] In any case, the decoupling can be described by the ratio B_σ/B_η. The smaller the ratio the more weakly coupled the conductivity is to the viscosity. This index has the advantage of being single-valued for each system, rather than varying with temperature as in the case of R_τ. Kawamura and Shimoji show how different values of a quantity equivalent to this index lead to different glass state conductivity-temperature relations analogous to Fig. 4(b) and how different cooling rates led, at constant decoupling, to different glass state conductivities at any chosen temperature.

The evidence from both theory and experiment is that manipulation of the cooling rate is not a factor which is of great utility in improving the conductivity of a given fast ion conducting system. Other problems associated with the long term dimensional stability of glasses are exacerbated by quenching in high energy structures, hence this is not likely to be a variable of practical importance as far as fast ion conducting glasses are concerned. On the other hand, if insulating properties are of interest, then thorough annealing (equivalent to very slow cooling) can improve insulating properties by more than an order of magnitude.

5.2. Frequency Dependence: a.c. Conductivity and the Connection to NMR Investigations of Superionic Glasses

All of the above considerations have been made for the DC conductivity or frequency independent of vitreous materials. Both for fundamental reasons and for the purpose of relating conductivity measurements to other investigative methods utilizing oscillating fields such as NMR, it is of interest and importance to examine the conductivity under AC conditions.

We distinguish two important cases dependent on whether or not the conductivity relaxation time at the glass transition falls within the range of the measuring instruments. Most commercial admittance bridges function between 10 and 10^6 Hz. We show this frequency range as a shaded area on a diagram, Fig. 9, in which the temperature dependence of the conductivity relaxation frequency ($f_{max} = 1/2\pi\tau_\sigma$) is plotted in Arrhenius form for two relevant cases. The curve A shows a relatively strongly coupled system in which, the conductivity at the glass transition temperature is only ~ 10^{-9} Ω^{-1}cm^{-1} ($\tau_\sigma \cong 10^{-2}$ 5) while curve B shows a superionic conductor with a conductivity at the same glass transition temperature of 10^{-2} Ω^{-1}cm^{-1}. Each thick curve illustrates the behavior of the DC conductivity, showing the break at T_g where the system becomes nonergodic with respect to temperature change for a given cooling rate (of which three are shown). In the strongly coupled (poor conducting) case, this implies that the bridge frequencies can exceed the conductivity relaxation time within the *liquid* state. When the field oscillates too rapidly for the jumping ions to follow it, we obtain the maxima in electrical loss illustrated in Fig. 2. At this point in temperature for a given field frequency, unless the electrical relaxation is strictly exponential[0] the conductivity measured at that fixed frequency would depart from the DC conductivity curve (heavy line) and adopt an Arrhenius temperature dependence with a lower apparent activation energy. The activation energy exhibited will depend on the temperature at which the system becomes nonergodic with respect to the oscillating field. A practical case is illustrated in Fig. 10(a) for the proton-conducting aqueous solution $HZnCl_3 \cdot 5H_2O$.[41] This case is interesting because the departure from exponential relaxation which causes the frequency dependence of the conductivity, is *not* well described by the "universal response function"

$$\theta(t) = \exp -[(t/\tau)^{\beta}] \tag{4}$$

which seems to apply to most other liquid and glassy state relaxations.[11,42]

The second case is more interesting because the loss of equilibrium with respect to temperature comes at a conductivity relaxation frequency well above the capability of the measuring instrument. In this case, the dc conductivity Arrhenius plot, curve B in Fig. 9, will exhibit a *second* glass-like transition (break in $d\sigma/dT$) under fixed frequency measurements since it is *already* nonergodic with respect to temperature at the temperature where the system relaxation time comes within range of the measuring system. This is shown in Fig. 9 by the series of thin lines branching from the σ_{DC} plot, curve B. We give a laboratory example in Fig. 10(b) for the case of $0.6AgI \cdot 0.4Ag_2B_4O_7$. Here the apparent activation energies for conductivity are small with respect to the DC activation energy and depart from the DC curve at temperatures which depend on the measuring frequency held constant. The apparent activation energy is temperature-dependent near the point of departure from the d.c. plot and is frequency-dependent, being lower the higher the frequency. The ratio of the two activation energies in the span of data shown in Fig. 7(b) varies from 0.29 to 0.33 in the frequency range $10\text{-}10^7$ Hz. This is somewhat smaller than the value of β in the KWW function which fits these data, 0.48.

We pointed out in an earlier review[5] that exactly the same situation holds in the case of nuclear magnetic resonance studies with the additional fact that the high characteristic frequency of the oscillating magnetic field ($10\text{–}100 \times 10^6$ Hz) (which takes the place of the oscillating electric field in the cases we have just described) means that the NMR measurements are almost *always* being performed in the frequency-dependent, hence low activation energy, regime. This is illustrated by the two dotted curves included in Fig. 6. This connection between the two types of measurements has rarely been recognized in the literature, though Ngai[42] has pointed out that the departure from BPP behavior can be accounted for if fractional exponential, rather than exponential, relaxation is assumed. Rather, the low activation energy of the NMR results has often been interpreted in terms of some specific short range motion model.

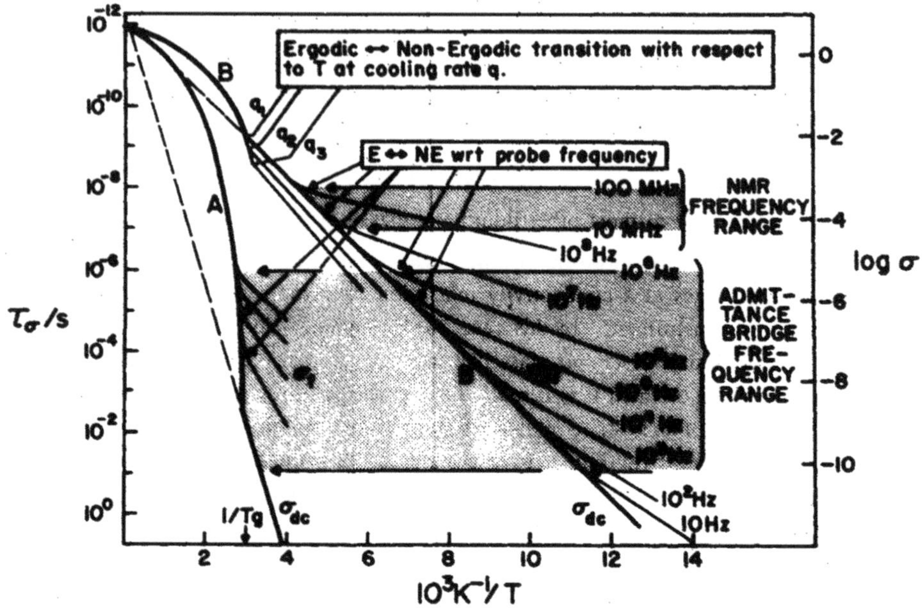

Figure 9. Variations of conductivity and conductivity relaxation time for two liquids, A and B, with the same glass transition temperature but very different decoupling indexes. The shaded areas show the characteristic frequency ranges of conductivity measurements and NMR relaxation measurements. The diagram indicates that NMR measurements (and also conductivity measurements in most of the frequency range) will show frequency-dependent behavior in the liquid state in case A, but only in the glassy state in case B. The latter will therefore show two breaks in conductivity-temperature behavior, the first at the glass transition temperature where the system becomes non-ergodic with respect to temperature, and the second where the system becomes non-ergodic with respect to the frequency of the applied stress. The temperature at which the second break occurs will depend on the frequency of the measurement as indicated by different fill lines for different frequencies. The diagram shows how it is only in the most highly decoupled vitreous conductors, or at the highest temperatures that NMR measurements will "see" the glass behaving ergodically with respect to probe frequency, and therefore will give the same activation energy as the d.c. conductivity.

The activation energy which is obtained, however, is determined entirely by the departure from exponential relaxation behavior and the value of the probe frequency relative to the temperature range of the data, hence has no more structural information than is contained in the Kohlrausch-Williams-Watts relaxation function. The connection of the latter to structure at this point has not been established. Most NMR measurements are conducted at a single frequency so the frequency dependence of the "activation energy" has escaped notice. If high temperature measurements are conducted on a highly decoupled system in which the high temperature frequency- independent branch can be explored then the activation energy found on the high temperature (ergodic wrt. probe frequency) is the same as found for conductivity,[43] as would be expected from Fig. 9.

Since the d.c. conductivity characteristic frequency Arrhenius plot extrapolates to the quasi-lattice vibration frequency it is obvious that relaxation times obtained by NMR or from lower temperature a.c. conductance measurements, will extrapolate to a

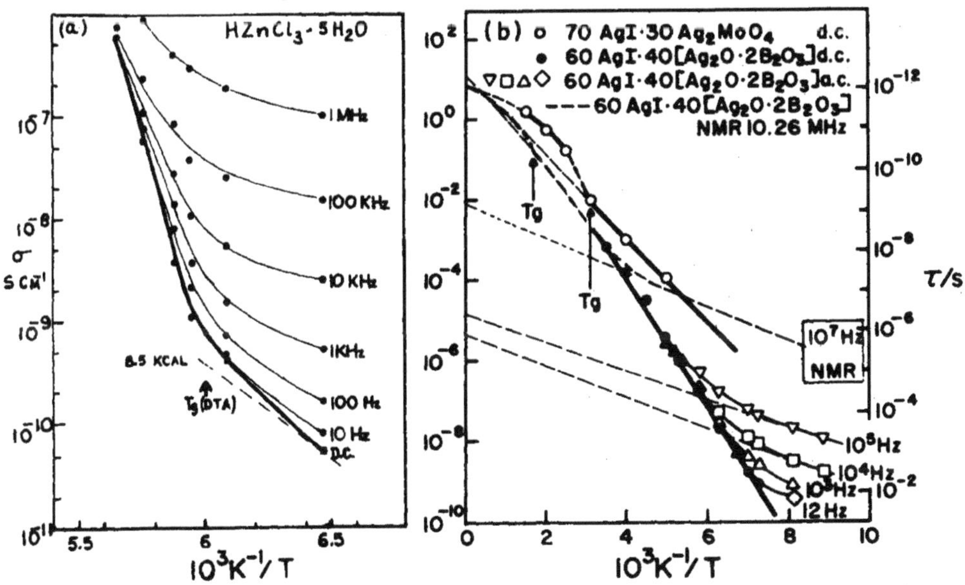

Figure 10. Experimental data for two laboratory examples of the cases A and B of Fig. 9. (a) the proton conducting system $HZnCl_3{\cdot}5H_2O$ in the frequency range 1MHz to DC. The glass transition is indicated by the break in the D.C. conductivity. (b) fixed frequency AC conductivity data and NMR relaxation time data for silver iodoborate glass showing parallels in phenomenology associated with the second non-ergodic regime.

pre-exponent well *below* the quasi-lattice vibration frequency. This has been a source of some consternation in interpreting NMR results which the present discussion elucidates. To refer to a laboratory system, we include in Fig. 10 some data from the study of Martin et. al.[44] on $xAgI\cdot(1-x)(Ag_2O\cdot 2B_2O_3)$, x = 0.55 and 0.65, averaging data for the two compositions to obtain data comparable with the conductivity data for x = 0.6.

The frequency dependence of NMR relaxation rates has been discussed in some detail by Funke in a recent paper[45] in which, however, the nonexponential nature of relaxation in fast ion conducting systems is not taken into account.

6. CONDUCTIVITY IN RELATION TO THERMODYNAMIC VARIABLES: PRESSURE

A potentially interesting and often quoted result in fast-ion conductor studies has been the enhancement of conductivity of a silver-ion conducting glass by as much as an order of magnitude observed with increase of pressure.[46] This report raises the question of whether or not such an effect may be general for Ag^+- conducting systems (the opposite effect is observed for alkali ion-conducting glasses including the FIC glass $LiCl–Li_2O\cdot 2BrO_3$).[25] Of particular interest, if the pressure enhancement effect is correct, is the possibility of achieving a permanent enhancement of conductivity by freezing in a configuration of high conductivity by carrying out a vitrification under pressure. This possibility has now been tested in a cyclic pressure-temperature study on the much investigated case of $AgI-AgPO_3$. Zhou and the author[47] have found that for $T \ll T_g$ the conductivity decreases with increasing pressure at the rate expected from the classical transition state theory, $(\partial \ln\sigma/\partial P)_T = -\Delta V/RT$, with ΔV equal to the molar volume of Ag^+ ions, as for alkali silicates[48] and borates.[49] Near and above T_g the rate is smaller because configurational compression can occur at a greater rate per unit pressure than in the glassy state. This brings ions into the conducting volume at a rate per unit pressure which offsets the decrease in their mobility. Cooling under pressure then results in a glass which, on return of pressure to ambient, has a *lower* conductivity than the original glass at the same temperature. Heating to T_g, where the frozen-in compression is released, then restores the original conductivity. These results, which were made in a simple cell at low (up to 1500 bar) but *reproducible* pressures raises some question about the reliability of the original higher pressure results, notwithstanding the different system involved.

7. CORRELATION OF CONDUCTIVITY (RESPONSE OF SYSTEM TO ELECTRIC STRESS) WITH MECHANICAL RELAXATION (RESPONSE TO MECHANICAL STRESS)

We earlier made comparisons of a.c. electrical field and magnetic field (NMR) responses. Another probe of fast ion motion can be made by submitting the system to mechanical stress and observing the time or frequency-dependent changes which reflect the fast ion relocation to new and more favorable sites. This matter has been the subject of a number of recent studies which together cover the frequency range from 0.4 Hz to 10^{10}Hz.[50)] The comparisons are interesting and informative but will not be discussed in detail here since the author has currently reviewed the available material in a separate article to which the interested reader may refer.[51)]

8. CONDUCTIVITY IN RELATION TO VITREOUS STRUCTURE

In an earlier section, mention was made of the fact that plots of various properties of superionic glasses as a function of increasing AgI content indicate an extrapolation to the properties of crystalline α-AgI. This type of behavior has been widely interpreted in terms of some sort of microdomain or cluster model, and it is the purpose of this part of the present article to call this matter into question.

Mention was made earlier of the fact that a clustering of components to give microregions of AgI seems incompatible with the thermodynamic information on heats of mixing of AgI in $AgPO_3$. The calorimetric observations in fact indicate the opposite, namely that the AgI component enters into intimate (more intimate than random) mixing with the $AgPO_3$ since heat is evolved during the solution, i.e. negative deviations from ideal solution laws are manifested. We recall that a direct connection can be made between fluctuations in composition and the free energy of mixing derived from thermodynamic experiments. This is the subject of the Bhatia-Thornton theory[52)] which allows one to obtain thermodynamic data from information on neutron scattering from composition fluctuations. Such measurements, which have been carried out on numerous metallic alloy and metal molten salt type systems,[53)] show quantitative agreement between theory and experiment and lead us to assert that the presence of composition fluctuations rich in AgI in these solutions is simply incompatible with the experimental calorimetric data. Clearly there is a need for more such data since there is one report,[54)] unconfirmed by more recent work,[55)] of microinheterogeneities, which were assumed to be AgI clusters in an AgI-$AgPO_3$ matrix.

We have earlier noted that there is much similarity between the type of solutions which yield high conductivity in AgI + Ag-oxysalt systems and those which lead to high conductivity in alkali halide + alkali oxysalt systems. In this connection we note an alternative interpretation of the structure of the high conducting solution by Button, et. al. in the case of LiCl + $Li_2O{\cdot}B_2O_3{\cdot}$ glasses. These authors suggested that the LiCl component "inserts itself interstitially" into the lithium borate matrix, a description which is a verbalization of the concept of chemical ordering[53)] as opposed to clustering. We have recently sought to support the interpretation by simulations which we have described in more detail elsewhere.[(57)]

Stimulated by the new observations of mechanical relaxation on the GHz timescale in 0.6 LiCl·0.4 [$Li_2O{\cdot}2B_2O_3$],[56)] Syed and Kieffer in this laboratory have simulated this solution and found conductivities close to, but above, the experimental values as would be expected for a glass with the high fictive temperature obtained in the ultra fast quenched simulated glass. Although the energy of mixing in these calculations is only obtained with large error bars it appears that deviation from ideal mixing is slightly positive, suggesting a weak tendency to cluster in this system. While this should be true at LiCl-rich compositions where laboratory experiments show two liquid phases coexisting, it is unexpected at low LiCl concentrations where the existence of the compound boracite ($2LiCl{\cdot}3Li_2O{\cdot}7B_2O_3$) would suggest the opposite state of affairs. Laboratory studies of the enthalpy and free energy of mixing, which should be easily performed, are needed to resolve this question.

9. CONCLUDING REMARKS

We have devoted most of this article to the problems of conceptualizing the nature and complexities of vitreous fast ion conducting glasses at the expense of detailed discussions of models developed for interpretation of the phenomena. In respect to the latter, we could do little to improve on the recent review of this aspect of the subject presented by M. D. Ingram[58)] and refer the reader interested in this aspect of the subject to this source.

10. REFERENCES

1. Souquet, J.L., Solid State Ionics *5*, 77 (1981).
2. Ravaine, D., J. Non-Cryst. Sol., *38 & 39*, 353 (1980).
3. Ingram, M.D., and Vincent, C.A., Ann. Rep. Chem. Soc., 23 (1977)

4. (a) Angell, C.A., Solid State Ionics, *9 & 10*, 3 (1983).
(b) Angell, C.A., in *Relaxations in Complex Systems*, ed. K. Ngai and G.B. Wright, National Technical Information Service, U.S. Department of Commerce, Springfield, VA 22161 (1985).

5. Angell, C.A., Solid State Ionics, *18 & 19*, 72 (1986).

6. Armand, M.B., Chabagno, J.M., and Duclot, M.J., in "Fast Ion Transport in Solids" Eds. P. Vashishta, J. N. Mundy, and G. K. Shenoy, North-Holland 1979.

7. Armand, M.B., Solid State Ionics, *9/10*, 745 (1983).

8. Papke, B.L., Ratner, M.A., and Shriver, D.F., J. Electrochem. Soc., *49*, 1694 (1982).

9. McLin, M. and Angell, C.A., J. Phys. Chem. (Letters), *92*, 2083 (1988).

10. (a) Bressel, R.D., Ph.D. Thesis, Purdue University, 1971.
(b) Moynihan, C.T., Balitactac, N., Boone, L. and Litovitz, T.A., J. Chem. Phys., *55*, 3013 (1971).

11. Howell, F.S., Bose, R.A., Macedo, P.B., Moynihan, C.T., J. Phys. Chem, *78*, 639, (1974).

12. (a) Torell, L.M., J. Chem. Phys., *76*, 3467 (1982).
(b) Angell, C.A. and Torell, L.M., J. Chem. Phys., *78*, 937 (1983).

13. Minami, T., J. Non-Cryst. Sol., *56*, 15 (1983); *73*, 293 (1985).

14. Chiodelli, A. and Magistris, A., Mat. Res. Bull., *17*, 1 (1982).

15. Souquet, J.L., Solid State Ionics, *5*, 77 (1981).

16. Carette, B., Robinel, E. and Ribes, M., Glass Tech. Ber: *24*, 157 (1983).

17. Burckhardt, W., Makta, M., Sevassen, A. and Hagenmuller, P., Mat. Res. Bull., *19*, 1083 (1984).

18. Senapati, H. and Angell, C.A. (to be published).

19. Liu, C., Sundar, H.G.K. and Angell, C.A., Mat. Res. Bull., *20*, 525 (1985).

20. Sun, H.W., Tanguy, B., Rean, J.M., Videau, J.J., Portier, J., and Hagenmuler, P., J. Solid. St. Chem. *70*, 141, (1987).

21. Martin, S.W. and Schiraldi, A., J. Phys. Chem., *89*, 2070 (1985).

22. Reggiani, J.C., Malugani, J.P. and Bernard, J., J. Chim. Phys., *75*, 846 (1978).

23. Liu, C. and Angell, C.A., Solid State Ionics, *13*, 105 (1984).

24. (a) Malugani, J.-P., Wasniewski, A., Dorean, M., Robert, G., and Al Rikabi, A., Mat. Res. Bull., *13*, 427 (1978).
(b) Burckhardt, W., Makta, M., Levasseur, A. and Hagenmuller, P., Mat. Res. Bull., *19*, 1083 (1984).

25. Button, D.P., Tandon, P.P., Tuller, H.L. and Uhlmann, D.R., Solid State Ionics, *5*, 655 (1981).

26. Tatsumisago, M., Minami, T. and Tanaka, M., J. Amer. Ceram. Soc., *64*, C97 (1984); *66*, 890 (1983).

27. Susman, S., Boehm, L., Volin, H.J. and Delbecq, C.T., Solid State Ionics, *5*, 667 (1981).

28. Susman, S., Delbecq, C.J., McMillan, J.A. and Roche, M.F., Solid State Ionics, *9 & 10*, 667 (1983).

29. Shultz, P.C. and Mizzoni, M.S., J. Amer. Ceram. Soc., *56*, 65 (1972).

30. Sundar, H.G.K., Martin, S.W. and Angell, C.A., Solid State Ionics, *18 & 19*, 437 (1986).

31. Kulkarni, A.R. and Angell, C.A., Mat. Res. Bull., *21*, 1115, (1986).

32. (a) Kulkarni, A.R. and Angell, C.A., Mat. Res. Bull., *21*, 1115 (1986).
(b) Kulkarni, A.R., Sundar, H.G.K. and Angell, C.A., Solid State Ionics, *24*, 253 (1987).
(c) Kulkarni, A.R. and Angell, C.A., J. Non-Cryst. Sol., *99*, 195 (1988).

33. Shelby, J.E., J. Am. Ceram. Soc., *68*, C177 and 551 (1985).

34. P. Tick (unpublished work)

35. Reau, J.M., Senegas, J., Aomi, H., Hagenmuller, P. and Poulain, M., J. Sol. St. Chem., *60*, 159 (1985).

36. Senegas, J., Reau, J.M., Aomi, G., Hagenmuller, P., Poulain, M., J. Non-Cryst. Sol., *85*, 315 (1986).

37. Boehm, L. and Angell, C.A., J. Non-Crystalline Solids, *40*, 83 (1980).

38. Boesch, L.P. and Moynihan, C.T., J. Non-Cryst. Sol., *17*, 144 (1975).

39. Kawamura, J. and Shimoji, J., J. Non-Cryst. Sol., *88*, 295, 286 (1986).

40. (a) "Transport in Low-melting Molten Salt Systems, in Molten Salts; Characterization and Analysis," C.A. Angell and C.T. Moynihan, Ed. G. Mamantov. Marcel Dekker, New York, June 1969, p. 315.
(b) Cohen, M.H. and Turnbull, D., J. Chem. Phys., *31*, 1163 (1959).

(c) Adam, G. and Gibbs, J.H., J. Chem. Phys., *43*, 139 (1965).

41. Hodge, I.M. and Angell, C.A., J. Chem. Phys., *67*, 4 (1977).
42. (a) Ngai, K.L. and Jain, H., Solid State Ionics, *18 & 19*, (1986).
    (b) Ngai, K.L., Ann. N.Y. Acad. Sci., *484*, 150, (1986).
43. Muller-Warmuth, W., "From Simple to Complex Liquids," Presented at a Conference at the University of Bielefeld, W. Germany, Sept. 1985; E. Gobel, W. Muller-Warmuth, and H. Olyschlager, J. Mag. Resonance, *36*, 371 (1979).
44. Martin, S.W., Bischof, H.J., Mali, M., Roos, J. and Brinckmann, D., Solid State Ionics, *18 & 19*, 421 (1986).
45. Funke, K., Zeit, Phys. Chem. NeueFolge, *154*, 5.251, (1987).
46. Hemlata, S., Parthasarathy, G., Lakshmikumar, S.T. and Rao, K. I., Phil. Mag. B, *47*, 291 (1983).
47. Angell, C.A. and Zhou, J. (to be published)
48. Hamann, S.D., Aust. J. Chem., *18*, 1, (1965).
49. Ryan, M. and Smedley, S., J. Non-Cryst. Sol. *65*, 29, (1984).
50. Borjesson, L., Torell, L.M., Martin, S.W., Liu, C. and Angell, C.A., Physics Letters, *125*, 330 (1987).
51. Angell, C. A. in "Materials Chemistry and Physics" Special issue, Ed. T. Minami, in press, 1988
52. (a) Bhatia, A.B. and Thornton, D.E., Phys. Rev. B 2, 30004, (1979)
    (b) Yokokawa, H., Kleppa, O.J., and Nachtrieb, N.H., J. Chem. Phys. *71*, 4099, (1979).
53. (a) Ruppersburg, H. and Egger, H., J. Chem. Phys., *63*, 4095 (1975).
    (b) Saboungi, M-L., Marr, J.J. and Blander, M., J. Chem. Phys., *68*, 1375 (1978).
54. Tachez, M., Mercier, R., Malugani, J.P., Solid State Ionics, *18 & 19*, 372, (1986).
55. Torell, L M. and Borjesson, L. (private communication)
56 Borjesson, L. and Torell, L.M., Solid State Ionics, *25*, 85 (1987).
57. Syed, R., Kieffer, J.A. and Angell, C.A., Mat. Res. Soc. Symp. (1988) (in press).
58. Ingram, M.D., Phys. Chem. Glass, *28*, 215 (1987).

POLYMERIC SOLID ION CONDUCTORS
The Immobile Solvent Concept

Michel Armand
Laboratoire d'Ionique du Solide
ENSEE-Grenoble B.P. 75 - 38402 Saint-Martin-d'Hères France
Michel Gauthier
Centre de Recherche d'Hydro-Québec (IREQ)
Montée Sainte-Julie VARENNES JOL 2PO Québec Canada

1 INTRODUCTION :

The motion of ions in polymeric matrices in the absence of a solvent is a relatively new phenomenon whose importance has been recognized in the past decade [1,2]. While polyelectrolytes, i.e. polymeric backbones with attached ionized groups are of considerable theoretical (chain conformation) and practical (ion exchange resins...) importance, they are conductive only in the presence of polar solvents. With the exception of Nafion®, cationic polyelectrolytes are only soluble or swollen by water or liquids of comparable hydrogen bonding ability (polyols, e.g. glycerol ; amides e.g. N-methyl acetamide), as the ionic moeties tend to cluster in large (20 - 100 Å) aggregates with high lattice energy. Thus, the electrochemical properties of such plasticized polymers are limited by the solvent and related to hydrogen lability in the case of protic media. Yet, an important difference stems from the immobility of one type of ion attached to the polymer backbone (transference number $t+ = 1$) while the solvent molecules migrate with the mobile ions (e.g. 4-20 molecule of H_20 per H^+ in ion exchange membranes). The conductivity takes place in interconnected "pockets" of solvent + ions, whose structure is quite similar to that of the free solutions.

Polymer electrolytes (PE) correspond to a different concept : the macromolecule itself acts as a solvent for a salt which becomes partially dissociated in the matrix, leading to ionic conductivity. No solvent is required for the conduction process though the preparation of the polymer-salt complex is routinely made out of a solvent of both constituents. It is interesting to note that the formation of complexes between ions and macromolecules in solution is quite frequently found and well studied ; in such case, the polar moeties on the macromolecule compete successfully with the solvent as a ligand. However, most of these complexes will not exhibit any conductivity in the dry state. It is remarkable that despite the intense work on polymers in general, the existence of an intrinsic conductivity either ionic or electronic (even if poorly characterized semiconductors were known) was only recognized in the 70's [1,2,3].

Following the early observation by Wright [1] of conductive complexes formed with poly(ethylene oxide), these materials have drawn the attention of the electrochemical community and have become an appreciable part of Solid State Ionics. The subject is now too vast to be covered completely in this chapter ; recent and extensive reviews [4,5,6], including a complete book [7] have been dedicated to the subject and are recommended for more detailed information.

We shall describe here the salient features of such polymeric electrolytes. As will be shown, the mechanisms leading to diffusion are quite unique when compared with classical solid electrolyte like β alumina or glasses and deserve special attention. They also differ from liquids even if the transport mechanisms show definite similarities since the ions diffuse unsolvated, thus suggesting the notion of the "immobile solvent".

2 THERMODYNAMICS OF COMPLEX FORMATION :

2.1 Site Control of Complex Formation :

In previous works, the formation of polymer electrolytes has been discussed from *ab initio* considerations. We shall take an approach where more comparisons are made with liquid solutions coordination chemistry which is well documented.

The existence of polar groups in polymers is a common feature, except for fully aliphatic macromolecules. Thus, the formation of stable complexes, both in solution and solid-state, is to be expected, and commonly observed (i.e. metal-containing proteins). There is a so-called "polymer effect", enhancement of the complex stability as compared to discrete molecules, due to the cooperative effect of neighbouring groups attached to a common backbone (i.e. poly(crown ethers); more rigorously, this effect is due to favourable configurational entropy.

The development of polymer-salt complexes parallels that of host-guest chemistry, one of the most active domain in this science during the past decade [8]. However in both cases, little or no effort has been devoted to anion-host interactions, probably because of the interest for kinetic effects in chemical reactions (nucleophilic substitutions) where dramatic increases are observed with solvated cations, and also considering the difficulty in stabilizing a negative charge with Lewis acids like ≡B◇ or ≡Al◇, which do not form easily bonds that are stable enough to construct host molecules or polymers.

Favourable free energies for dissolution are observed when the interactions between at least one of the ionic component and the solvent compensate for : i) the loss of lattice energy of the salt MX, ii) creation of vacancies in the solvent, in respective order of magnitude. The enthalpy effects are obtained from the energy of interaction of the ion with the dipoles of the solvent. In the close vicinity of the ion, the magnitude of the dipole moment is however less important than the quasi-chemical overlap of the orbitals of the ion and those of the ligand.

For non polarizable cations, ("hard ") Lewis acids including alkali and alkaline earth, the best donors are in the sequence :

–O– > –NH– >> –S–

For soft cations, i.e. those having partially filled "d" or to a lesser extent "f" orbitals, there is a net preference for solvation by secondary nitrogen (tertiary =N– showing usually steric restrictions) :

–NH– > –S– > –O–

The loss of entropy from the salt crystal is partially compensated by some ordering of the solution when the ions are present ; complexation is highly facilitated when the geometry of the ligand is pre-arranged to form cyclic structures where the ion is part of a 5-, more rarely 6-membered ring.

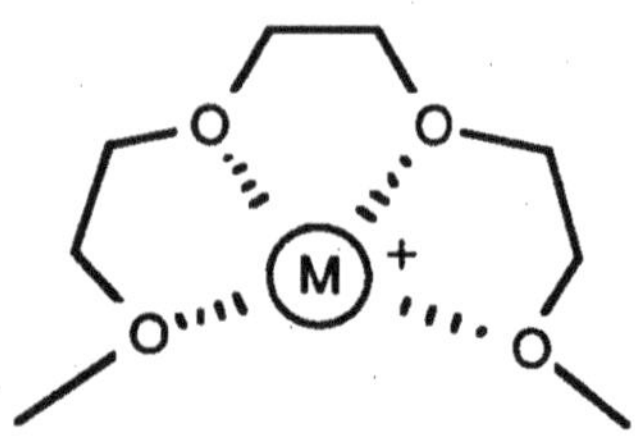

From the above mentioned considerations, the most favourable simple polymer structures for complex formation are :

$(\text{-CH}_2\text{CH}_2\text{O})_n\text{–}$ poly(ethylene oxide)

$(\text{-CH}_2\text{CH(CH}_3\text{)O})_n\text{–}$ poly(propylene oxide)

$(\text{-CH}_2\text{CH}_2\text{NH})_n\text{–}$ poly(ethylene imine)

Most polymer electrolytes studied today include these units with the few exceptions of polyesters and nitriles.

An interesting number is the concentration of electron pairs, for instance $3.15 \times 10^{22}/cm^3$ for PEO, to be compared to $6.7 \times 10^{22}/cm^3$ for water and $2.8 \times 10^{22}/cm^3$ for ammonia, reflecting the spatial distribution of the solvating units.

Anions play a key role in solvation thermodynamics. In water or methanol solutions where most of the host-guest chemistry have been studied, the hydrogen bonds exert a levelling effect on the stability of the negative charge carriers. In aprotic liquid or polymer solvents, dramatic differences can be observed that depend on the charge dispersion. Large delocalized anions, either "soft "(I^-) or "hard" bases (ClO_4^-), require little solvation. In fact the order :

$$Cl^- < Br^- < I^- \approx SCN^- < ClO_4^- \approx CF_3SO_3^- < AsF_6^-$$

parallels that of the lattice energy of their salts. For lithium derivatives, a well defined threshold at 850 kJ. for the solubility of lithium salts has been found [9].

Table 1 summarizes, under the presentation of a Mendeleiv chart, the formation of complexes for a large number of cations, with PEO as "solvent".

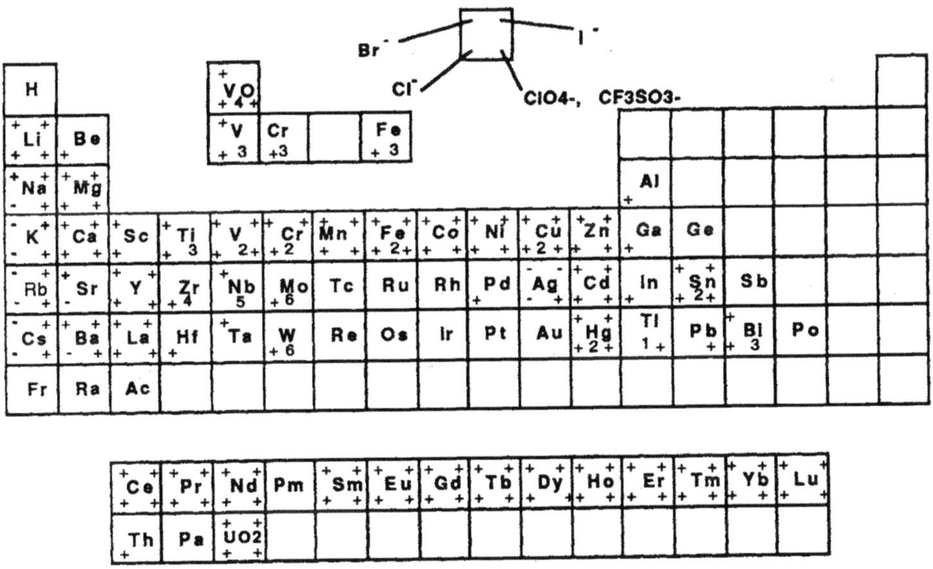

Table 1

Complex formation between PEO and various metal salts

+ : complex formed ; - no evidence of complex

As apparent from these results, PEO is able to solvate most metal derivatives, including, besides alkali metals, alkaline and rare earth, transition metals especially in their +2 state. In fact, *the solvating properties of PEO exactly parallels that of water* ; this result is to be expected considering the similar donicities of oxygen in H_2O and in ethers. The difference lies in the inability of ethers to solvate anions : Large solvation energies of all +2 and +3 elements suffice to induce the formation of complexes, even with the small Br^-, while the larger I^- is required for the heavier, less solvated +1 cations (K^+). The salts formed from both "soft" cations (Ag^+) and anions (I^-) show no solubility, while the weaker hard-soft interactions as in $AgCF_3SO_3$ allow competition of the ether oxygens and the anion. The free energy of the reaction :

$$<LiX> + P(EO)_n \leftrightarrow P(EO_nLiX)$$

has been determined [10] using the electrochemical cell :

$$- Li / P(EO_nLiX) / poly(viologen) +$$

the polyviologen :

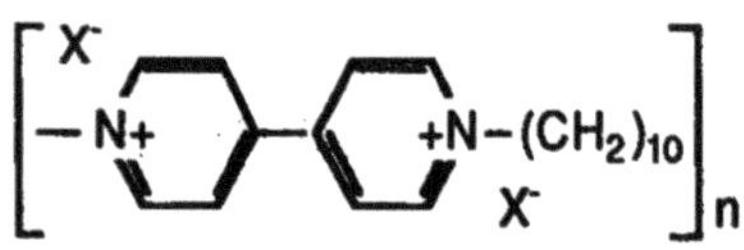

acting as an anion-reversible electrode through the couple V^+/V^{++}. For values of n ranging from 8 to 100, the stability constant of the complexes was found to vary from $\approx 10^3$ to $\approx 10^6$,assuming that the composition n= 2 contains some salt at unit activity . These numbers are quite similar to those calculated for crown ethers.

Poly(propylene oxide) on the other hand is far less solvating due to the steric hindrance introduced by the methyl group ; again, the decrease in stability constant parallels that of crown ethers made from the same PO units when compared to EO-based structures. No quantitative values are reported for the complexing power. For instance, it was reported that High Molecular Weight PPO does not dissolve appreciably any potassium salts, even those with low lattice energies [32]. On the other hand, Low MW PPO (≈2000) does dissolve alkali metal thiocyanates to some extent, including KSCN [11], however these salts precipitate out of the polymer phase at 80 to 150° C. In such case, the relatively high concentration of terminal OH groups for LMW PPO able to interact with the anion may explain these discrepancies. The IR spectra of the hydroxyl endings of these oligomers is strongly affected by the introduction of salts.

Poly(ethyleneimine), with a NH secondary amine group in place of the ether oxygen, represents an interesting class of complexing agents with "soft" nitrogen centers, to be compared with "cyclams" like :

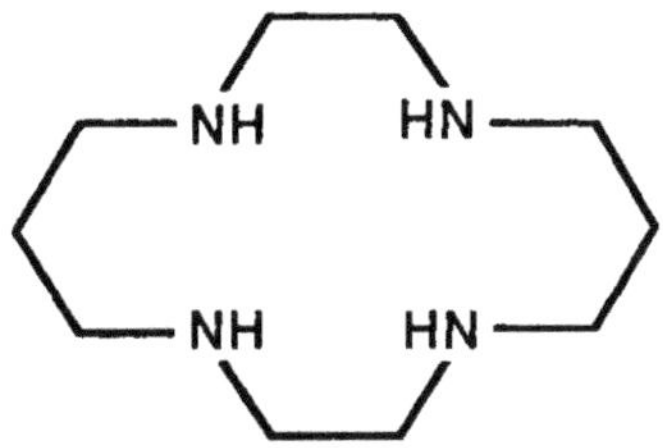

which show extraordinary complexation capabilitities for transition metal cations : Ag^+, Cu^{++}, $Co^{++/+++}$, Ni^{++}, Zn^{++}... ($k \approx 10^{22}$ for Cu^{++} from water) [8]. PEI, either linear or branched (see §3) is a good complexing agent for the same transition metal ions ($k > 10^{12}$), and shows in addition a solvating power towards alkali metal salts similar to that of PEO ; in this case the higher donicity of the nitrogen is partially compensated by a lower concentration of electron pairs and a smaller negative charge on the heteroatom, as compared to ethers (local electrostatic effect). The donicity effect prevails largely for transition ions.

As shown in § 3.1, other donor moeties, esters, amides, cyano... groups can be used for solvation. However, the principles remain similar.

2.2 Phase Diagrams :

The knowledge of the structure and morphology at the microscopic level is mandatory since, as discussed below (§ 4.2), polymer-salt complexes are conductive only when in the elastomeric state. The maximum change in solvation entropy is understandably obtained with a regular spacing of the solvating units in the polymer chain. This, in turn, favours crystallinity in the macromolecule, especially if the repeat unit is highly symmetrical, which is the case of the EO $-CH_2CH_2O-$ segments. While space lacks in this chapter to detail the complexity of phase diagrams for polymers in general, mention must be made that the melting point depends upon the crystal size, which increases with the annealing time, and the crystal type (folded chain, fibers). PEO has been studied in details [12, 13]. Thus, HMWPEO (>100,000) has a melting point of 66°C and is ≈ 85% crystalline. If we take this fraction as in quasi-equilibrium with the locked-in entropy due to chain entanglement and consider only the simplest crystal arrangement obtained by chain folding, the phase diagram is simply treated as that of discrete molecules ; this was first done by Sorensen et al.[14]. Three techniques have been used to characterize these complexes : i) DSC, a classically tool in macromolecular physics [15, 16], ii) optical microscopy, where the anisotropic polymer cystals are observed with a polarizing microscope[17, 18] , iii) NMR, which allows to determine the numbers of nuclei with a magnetic spins: ^{1}H ^{7}Li, ^{19}F, ^{23}Na in each phase, determined by the length of the T_2 decay time[19,20,21], shorter for the crystalline phase, iv) X ray diffraction [22,17,108].

All soluble Li and Na salts reported up to now form with PEO *at least one crystalline complex*, with melting points in the range 150-200°C with three monomer units/salt : $P(EO)_3MX$. Recently, Chatani and Okamura[108] have determined the structure of the $P(EO)_3NaI$ compound, showing an helical stucture as shown in Fig. 1.

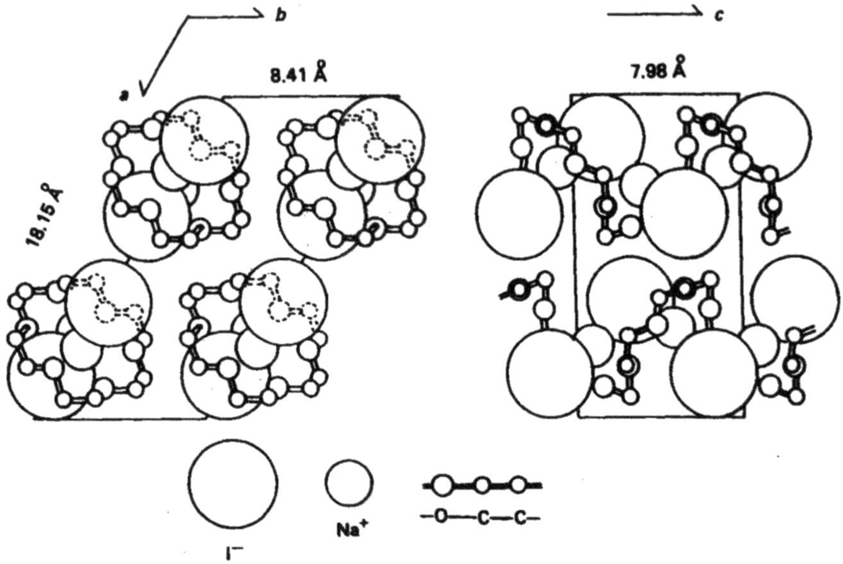

Fig. 1 : The structure of crystralline $P(EO)_3NaI$ showing the helical structure. From ref. 108) with permission.

In addition, some salts form a 6 :1 complex. Symmetry and/or ion size apparently are determinant for the formation of this latter phase. This is the case for instance of $LiClO_4$, $LiAsF_6$ (symmetry) ; also of $LiC_6F_{13}SO_3$ and $LiC_8F_{17}SO_3$, the longer chain homologues of $LiCF_3SO_3$ which itself only forms the 3 :1 complex. Between PEO and the leanest complex an eutectic exists. Its liquidus and composition are quite close to that of the pure polymer for the 3 : 1 systems, e.g. 60°C, n ≈ 100 for PEO -$LiCF_3SO_3$ (Fig. 2), and much more depressed and concentrated for 6 :1 systems : 40°C and n ≈ 9 for PEO- Li ClO_4 (Fig. 3). The kinetics of crystal formation are very slow for the latter. The existence of even richer complexes (2 :1, 1 : 1) is quite possible, but not yet thoroughly proven. For the heavier alkalies and NH_4^+, the ratios are 4:1 and 8:1 with the same criterions for size [23]. The stoichiometries are not known precisely for the multicharged cations (Mg, Ca^{++}, La^{+++}...) but a 3:1 phase seems to exist also.

PPO has an optically active carbon and can be polymerized into syndiotactic, highly

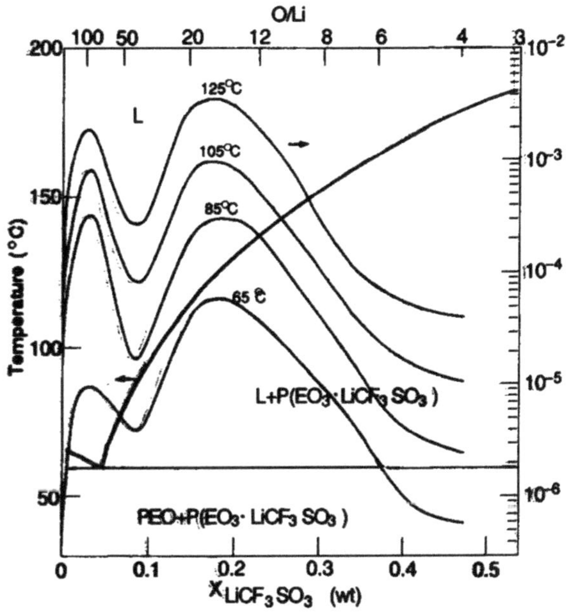

Fig. 2

Phase diagram (left scale) and conductivity isotherms for the PEO-$LiCF_3SO_3$ system. From ref. [6] with permission.

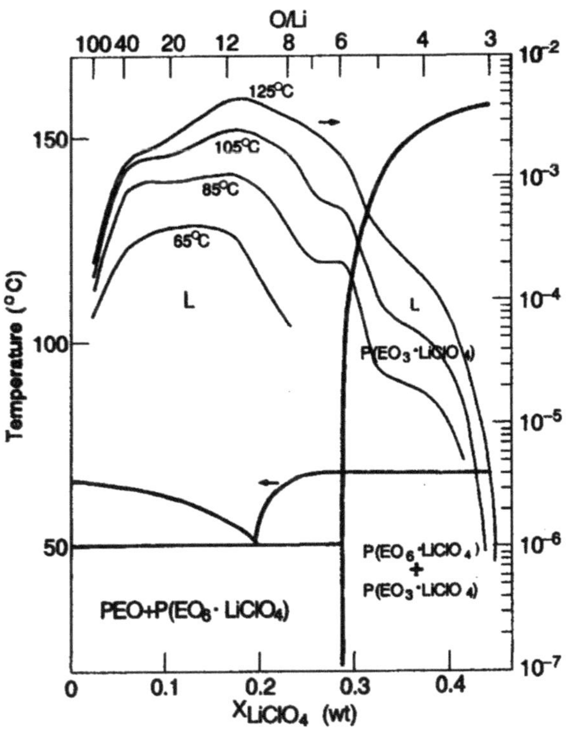

Fig. 3

Phase diagram (left scale) and conductivity isotherms for the PEO-$LiClO_4$ system. From ref. [6] with permission

crystalline polymer, but this material has not been studied. Polymerization of racemic PO with non-specific catalyst yields an atactic elastomer (tg = -65°C) with random orientation of the methyl groups. Crystallization is impossible in such case and the complexes form a continuous solid solution up to a saturation concentration (which has not been studied and should be dependent upon the salt). There is also a possibility that a solubility limit exists on the dilute side, as could be deducted from a study on the mechanical relaxation of the PPO- Li ClO_4 system [64].

PEI illustrates the dependence of properties on chain regularity. The cationic polymerization of EI yields a branched polymer of approximative structure :

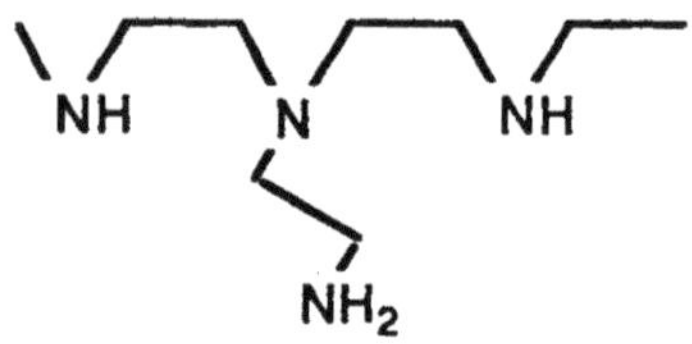

which is completely amorphous and hygroscopic solid with a ratio of ≈ 1 :2 : 1 of primary, secondary and tertiary amines. On the other hand, the hydrolysis of poly(oxazolines) yields linear PEI :

a highly crystalline polymer, with a melting point of 55°C, insoluble in water at room temperature, and a t_g of -21°C [24]. Such polymer is expected to give phase diagrams comparable to those of PEO and to be more solvating than BPEI.

2.3 Proton-Polymer Complexes :

The proton represents a special case, of interest considering the ubiquitous involvement of hydrogen in nature, and its unique chemical behaviour. H^+ differs markedly from other cations in size making it highly polarizing, leading ultimately to covalent bonds. It thus corresponds to low coordination numbers, usually two. While solid electrolytes like ion exchange membranes and compounds like uranyl hydrogen phosphate are well known, they transport the hydrated proton, H_3O^+, xH_2O, (x up to 20).

The interest for anhydrous proton conduction is recent, and new polymeric materials are emerging. Ethers form hydrogen bonds, even if they are less basic than water towards protons (pK_a ≈ -2.5) and they also tend to be cleaved by strong aqueous acids, via an oxonium intermediate according to the scheme :

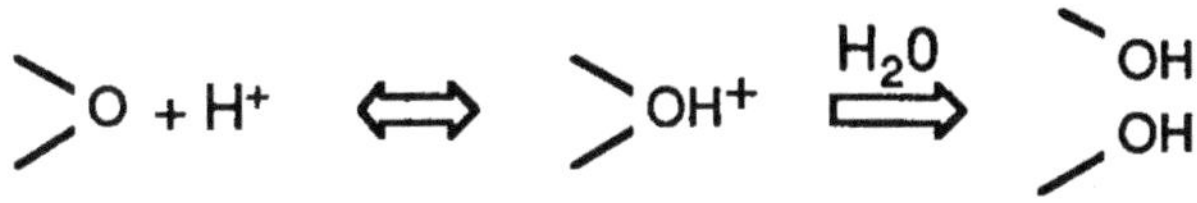

Such chemical degradation is however suppressed in the absence of moisture. Thus, PEO and orthophosphoric acid form well defined complexes $PEO_nH_3PO_4$ [25,26]. Again, a phase diagram controls this system in which a stoichiometric compound ($n = 1.33$, $t_f = 73°C$) exists ; two eutectics melting close to room temperature are formed between this compound, pure PEO on one side, and pure H_3PO_4 ($t_f = 40°C$) on the other (Fig. 3).

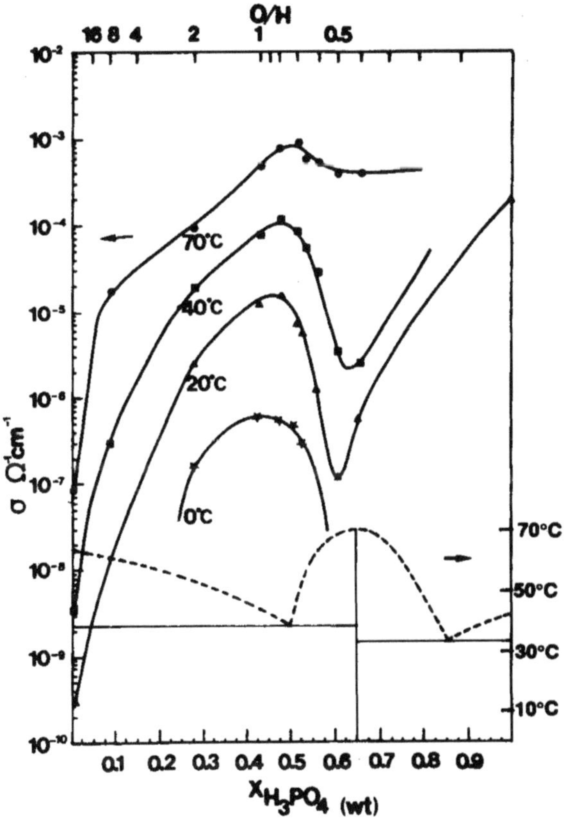

Fig. 4 : Phase diagram and conductivity isotherms for the $PEO-H_3PO_4$ system.

Polyamides, like polyvinylpyrolidone (PVP) and polyacrylamide form similar complexes with strong inorganic acids, i.e. H_3PO_4, H_2SO_4 and $HCF3SO_3$. They possess a basicity comparable to that of water or ethers, and can be protonated at either oxygen or nitrogen :

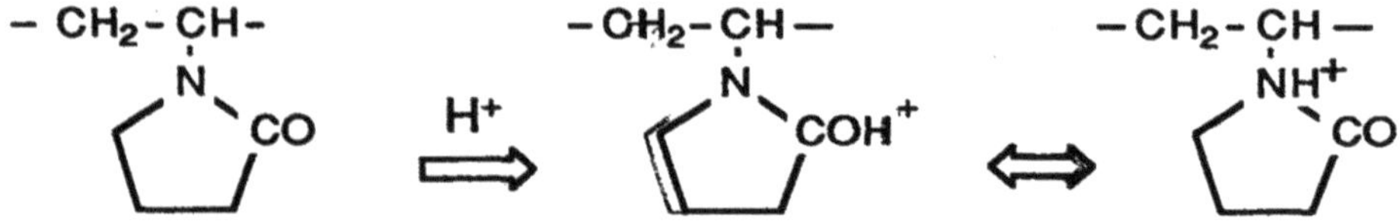

However, polyamides, due to their highly polar nature have strong interchain coupling. PVP is thus a glass with a transition t_g at ~ 105°C . Addition of the acid progressively plasticizes the macromolecule to an elastomer (e. g. PVP,2H_2SO_4).

As expected, the protonation of polyamines takes place easily. Either linear L [27] or branched B [28] PEI have been doped with inorganic acids to yield electrolytes. At low protonation levels (x < 0.15), the compounds are amorphous elastomers, and conductivity is expected to involve proton hoping between neighbouring nitrogen sites. Glasses (BPEI) or crystals (LPEI) are obtained at higher acid contents, and in this case, the motion of the proton may take place on the anion lattice ($HSO_4^-/SO_4^{=}$).

A new class of "Ormosils" (Organically Modified Silicates) has been described by Charbouillot et al. (18). These authors prepared proton conducting glasses from the sol-gel process through the hydrolysis of aminopropyl trialkoxysilanes, according to :

$$(CH_3O)_3SiCH_2CH_2CH_2NH_2 + 1.5H_2O + xHX \Rightarrow 3CH_3OH + SiO_{1.5}CH_2CH_2CH_2NH_2,xHX$$

The structure of these compounds appears to that of a *non-crosslinked* ladder polymer, as :

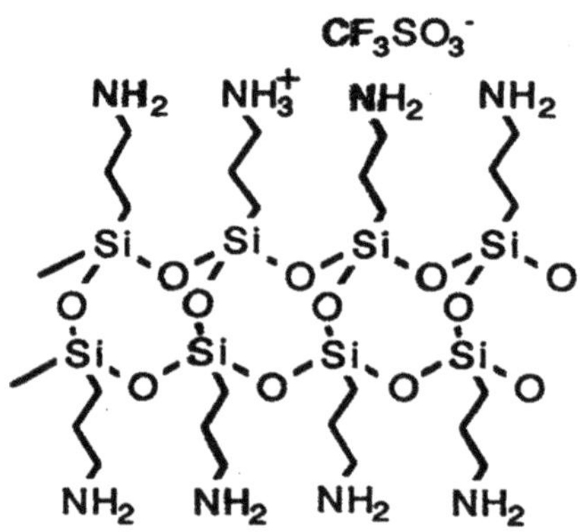

These materials form continuous solid solutions for doping levels x < 0.5 obtained with strong *mono* protonic acids ($HClO_4$, HCF_3SO_3...). They are quite interesting on both theoretical (§ 4.2) and practical basis, especially considering the simplicity and versatility of the silicon sol-gel chemistry.

Polak et al. [29] have disclosed some protonic conductors obtained from poly(vinyl alcohol) and phosphoric acid, but these materials contain water. Anhydrous acids, including H_3PO_4, catalyse the deshydratation of PVA into polyacetylene :

$$(\text{-}CH_2CHOH)_n\text{-} \quad \Rightarrow \quad (\text{-}CH{=}CH)_n + H_2O$$

3 POLYMER ARCHITECTURE :

We have shown in §2 the necessity of an optimal distribution of solvating groups, mainly O and N for cation solvation and the importance of chain regularity on properties exemplified by PEI (§2.2). Yet, many possibilities are offered to assemble the corresponding sequences. As discussed in §4, these variations have been used to increase the amount of amorphous phase in the polymers for a given solvating sequence. Another possibility is to improve the mechanical properties through crosslinking and block copolymerization. We shall distinguish three main groups :

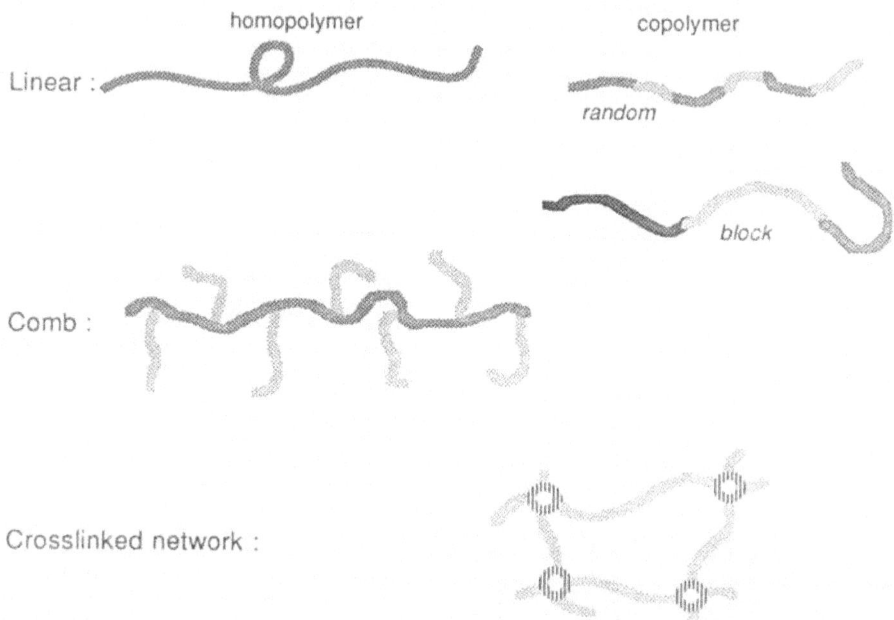

Copolymerization, comb polymers with different subunits and interpenetrated networks extend even further the choices of host matrices.

3.1 Examples of Polymers Studied :

	backbone		*Ref.*
Linear :	PEO $-[CH_2CH_2O]_n-$		1,31)
	PPO $-[CH_2CH(CH_3)O]_n-$		32)
	PEI $-[CH_2CH_2NH]_n-$		33,34)
	Poly(dioxolane)		
	$-[CH_2CH_2OCH_2O]_n-$		107)
	Polyesters		
	$-[(CH_2)mCO_2(CH_2)_p]_n-$		35-37)
Comb polymers:			
	backbone	*pendant chains*	
	$-[P{=}N]_n-$	$CH_3(OC_2H_4)_rO-$	38)
	$-[Si(CH_3)O]_n-$	$CH_3(OC_2H_4)_rO-$	39)
	$-[CH_2C(CH_3)CO_2]_n-$	$CH_3(OC_2H_4)_rO-$	40-42)
	$-[CH_2C(CH_2CO_2)CO_2]_n-$	$CH_3(OC_2H_4)_rO-$	43)
	styrene $-[CH_2CH(C_6H_5)]_n$	$CH_3(OC_2H_4)_rO-CH_2-$	56)
	styrene-block-butadiene	$CH_3(OC_2H_4)_rO-$	44)
Networks :			
	crosslink	*chain*	
	$R(NCO)_p$ (p=2,3)	$H(OC_2H_4)_rOH-$	45,46)
	(+ triol for p =2)	EO-PO copolymer	
		$H[OCH_2CH(CH_3)]_rOH-$	81)
		$-[SiO(CH_3)C_3H_6O(C_2H_4O)_r]H_q$	47,61)
	γ rays	$-(CH_2CH_2O)_r$	48)
	$-[Si(CH_3)O]_n-$	$-(OC_2H_4)_rO-$	49)
	$PO_4\equiv$	$-(OC_2H_4)_rO-$	50)

A more extensive review of these materials has been given recently by Vincent [5]. A new and interesting class of solvating polymers was disclosed by Shi-bi Fang et al. [51]. These materials possess cyanoethyl $N{\equiv}C-CH_2CH_2-$ groups attached to a methylsiloxane backbone, with a trifunctional siloxane as crosslinking agent. The cyano group confers a high polarity, while dipole orientation is still possible from the flexibility of the siloxane backbone.

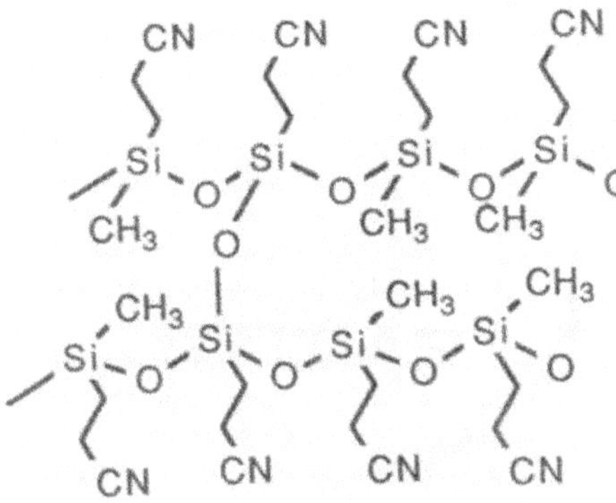

Another type of polymer electrolyte was proposed by Kaplan et al [52] : considering the highly polar propylene carbonate molecule, these authors prepared a macromolecular homologue by radical polymerization :

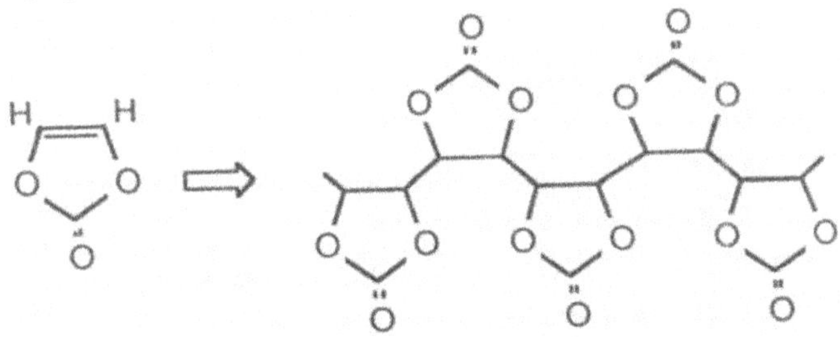

Polyvinylene carbonate is however a high-T_g glassy material and, as expected, no ion motion is observed with the simple addition of salts ; nevertheless, the system becomes conductive when a crown ether is added in stoichiometric ratio with the salt. Since the quantities of salt + crown possibly acting as plasticizer are too small (<10%) to substantially reduce the T_g, these compounds could be the organic equivalent of doped glasses (e.g. P_2O_5-Li_2S-LiI). Caution must also be exercised since in the described procedure [52], there is no proof of a total polymerization, and the possibility of a platicization by unreacted monomer is a definite possibility, especially with the low reactivity of the monomer for homopolymerization. The conduction may also involve the transport of *solvated* cations.

The polymeric cation $-[(CH_2)_pN(CH_3)_2^+(CH_2)_q]_n$, corresponds to delocalized positive charges which tend to reduce the formation of clusters. The spacing between the charges is easily controlled, thus the chain flexibility, and the onset of ion motion at T_g was shown on this interesting family [53].

Mention should be made of metals salts dissolved in the polyelectrolytes complexes formed between the same polymeric cation (e.g. $-[(CH_2)_pN(CH_3)_2^+(CH_2)_q]_n$) and a polymeric anion (e.g. sulfonate on a polyacrylamide backbone) [54].

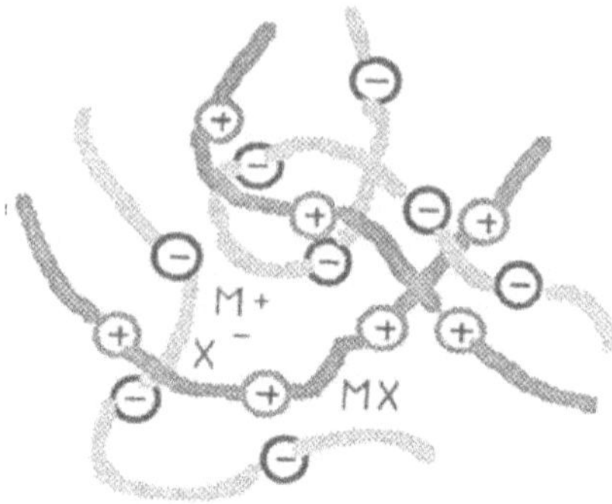

The same possibilities exist with poly(zwitterions) [55]. These two families should be compared with molten salts, even if the conductivitites are low due to high Tg's.

4 TRANSPORT PROPERTIES

4.1 Conductivity :

Total conductivity has routinely been used to characterize the polymer electrolytes , with the value of 10^{-5} $(\Omega cm)^{-1}$being considered as the threshold for applications, and a variety of materials reaching such value between room temperature and 100°C are now known. In Fig. 5 for alkali metals and Fig. 6 for protons, we have shown the Arrhenius plots of conductivity for some of the above mentioned macromolecules. *The important point in the understanding of ionic polymers was the discovery that any conductivity is restricted to the amorphous phase, when above its glass transition temperature* [19,14]. In fact, a conductivity maximum occurs at the eutectic composition, and the transport properties are governed by the phase diagram, as shown in Fig. 2, 3 & 4 where the two parameters are shown superimposed . There is yet no example of crystalline polymer electrolytes, and further characterization of the materials of reference [52] needs to be made prior to consider ion motion in vitreous organic matrices. Thus, the main effort has been directed towards the synthesis of polymers for which crystallization is impossible or limited. Besides intrinsically disordered chains (atactic PPO) [32], either comb [38-44, 56] or network (crosslinked) architectures [45-50] are effective in this sense. The effect is to markedly improve the transport properties at low temperature, when the corresponding simple chain PEO is mostly crystalline. However, these modifications correspond to a finite volume (crosslinks, backbone) which is lost for conduction, and at high temperature >100° where all systems are amorphous, comparison is in favour of PEO, still the best conductor as shown on Fig. 5 Conductivities tend to be higher for the heavier alkali ions, Li < Na or K < Rb or Cs, the relative order depending *for a given stoichiometry* upon the counter anion [57].

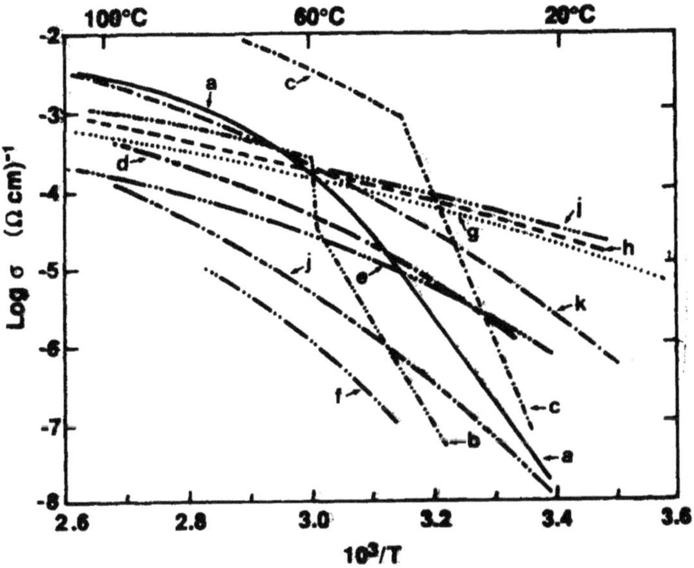

Fig. 5 : Conductivity vs. temperature plots for :

a) P(EO)-$LiClO_4$ 10 : 1 [31, 17]

b) P(EO)-NaI 20 : 1 [1, 80]

c) P(EO)-$NH_4CF_3SO_3$ 12 :1 [23]

d) P(EO)-$LiClO_4$ 8 : 1 γ rays Xlinked[48]

e) P(PO) $LiCF_3SO_3$ 9 :1 [31]

f) Poly(ethylene succinate) $LiBF_4$ 6 :1[36]

g) Poly(w-methyloligooxyethylene methacrylate) $LiCF_3SO_3$ 18 :1 [40-42]

h) Polyphosphazene ME2P, $LiCF_3SO_3$ 4 :1[38]

i) PEO-PPO-PPO block copolymer Xlinked with $R(NC{=}O)_3$ 20 : 1 [45]

j) PEI-$LiClO_4$ 8 : 1 [33]

k) Poly(methyl cyanoethylsiloxane), Xlinked with trifunctional siloxane, 10w.% $LiClO_4$ [51]

The proton conducting polymers show higher conductivities, as appearing on Fig.6 with $\sigma > 10^{-5}$ $(\Omega cm)^{-1}$ already at room temperature for various systems, and a maximum for a given composition, again corresponding to the eutectic of the phase diagram for the PEO systems.

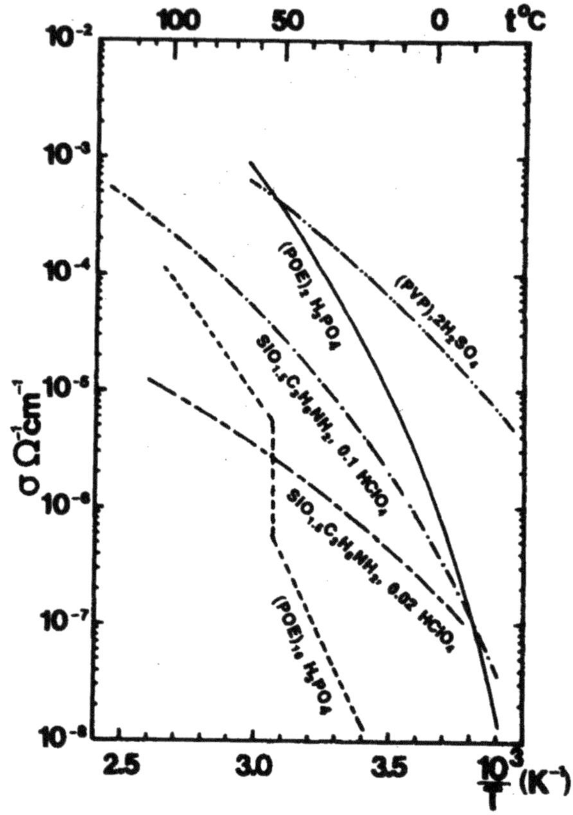

Fig. 6 : Conductivity vs. temperature plots for :
- *PEO-H_3PO_4 system.*
- *Polyvinylpyrolidone (PVP)-H_2SO_4 system.*
- *"Aminosils"$SiO_{1.5}CH_2CH_2CH_2NH_2.xHClO_4$*

4.2 Macroscopic Laws :

It is beyond the scope of this chapter to detail the approach towards the laws governing the transport properties; an excellent review has recently been given by Ratner in ref. [7]. The idea is now well accepted that the variation of conductivity with temperature for all the systems *which are completely amorphous* depart from classical Arrhenius law. The typical curvature of the $\ln\sigma$ vs $1/T$ has been treated in terms of a derivative of the Vogel-Tamman-Fulcher equation [58] :

$$\sigma = \sigma_0 \exp[- Ea/(T-T_0)]$$

More generally, any mechanical relaxation process $R = f(T)$ can be expressed from an universal scaling law [59] valid also for polymer electrolytes:

$$R(T)/R(T_{ref}) = \exp\frac{-C_1(T-T_{ref})}{(C_2+T-T_{ref})}$$

called WLF (from Williams, Landel & Ferry) ; the two laws are identical if : $C_1C_2 = Ea$ and $C_2 = T_{ref} - T_0$. T_{ref} can be chosen as the experimentaly accessible T_g, and then $C_1 = 17.5$, $C_2 = 52$. This allows master curves to be drawn as a function of T- Tg, as shown in Fig. 7. For both models, the implication is that of a liquid-like motion at the microscopic level, coupled with some of the mechanical relaxation times (modulus) [60], yet decoupled in time from the reptation of the chains which control macroscopic flow properties ; the decoupling might take place with the onset of chain entanglement, in the 10^3 - 10^4 MW range.

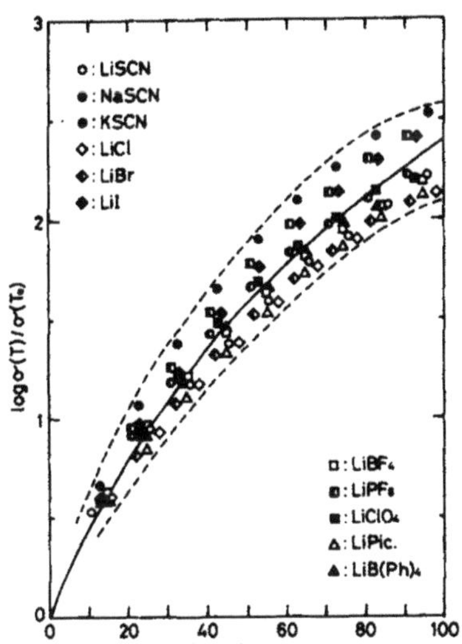

Fig. 7

Master curve of the WLF plots of ionic conductivity for various PEO Alkali metal salts complexes. The reference temperature is :

$$T_{ref} = T_0 = T_g + 50\,^{\circ}C.$$

From Ref. [109] *with permission.*

It is now clear that the ions do not see the polymer as an average mixture of all its chemical components, but are restricted to the vicinity of the polar groups (EO units) which are most favourable to solvation. For example, the comparison between various comb polymers with the same oligo EO segments attached to chemically different polymers shows that the conductivity depends only slightly upon the stiffness (T_g) of the backbone (compare for instance the flexible -P=N- [38] with rigid methacrylate [42]). Microphase segregation may take place so that the ions move in a similar environment, independent of the backbone, resulting in some cases to two distinct T_gs [61]. In other terms, there is probably a limit to the conductivity (for a given salt) close to 10^{-4} $(\Omega cm)^{-1}$ at room temperature of "ideally amorphous" PEO, for which no volume is lost with other moeties to prevent crystallization, and retaining the same bond flexibility. In the case of the proton conductors "aminosils" shown in § 2.3, there is an obvious decoupling between the mechanical properties imposed by the silica-like backbone, those of a hard glass, and the conductivity related to the local motion of the tethered amino groups, which corresponds to T_0 = 190 K.

The Dynamic Bond Percolation theory proposed recently by Druger et al. [62] considers both the segmental motion of the chains and the possible direct hopping of the ions when the conditions are favourable. This model is quite valuable considering the high concentration of charges (up to 5 M/l.) and the short mean distance (< 8 Å) separating the solute particles (coulombic liquid similar to a molten salt).

4.3 Ionic Dissociation :

There is a standing controversy concerning the extent of dissociation in the polymer electrolyte. On one hand, the heritage from solid-state ionics is to consider a total dissociation, the conductivity being limited by the high activation energy for the jumps, short hopping distances giving low mobilities. However, the recent Hall measurements on silver based glasses indicate on the contrary small fraction (<0.1%) of carriers with a high mobility [63]. On the other hand, established theories for liquids predict strong association for low dielectric constant medium, with the existence of ion pairs, clusters and multiplets ions of the type M_2X^+, MX_2^- [65]. Polyethers fall in this category with $\varepsilon \sim 5$. Yet, the effective dielectric constant seen by individual ions is much higher due to the high concentration of ion pairs acting as dipoles (redissociation effect). It has been impossible to measure directly the mobility or the dissociation coefficient α, as measurements like Hall effect cannot be performed yet at these low levels of conductivity. Indirect evaluations from the redissociation effect suggest $\alpha \sim 10\%$ [84]. Curiously, when the conductivity of amorphous networks *measured at the same reduced temperature* $T-T_g$ (constant mobility) was plotted *vs* concentration, LeNest and coworkers observed a slope of unity [66]. Rather than an implication that α is close to 1, such data can be interpreted as a clue for the existence of clusters :

$$n\langle MX\rangle \Leftrightarrow \langle M_nX_n\rangle \Leftrightarrow M^+ + X^- + \langle M_{n-1}X_{n-1}\rangle$$

or :

$$n\langle MX\rangle \Leftrightarrow \langle M_nX_n\rangle \Leftrightarrow M^+ + MX_2^- + \langle M_{n-2}X_{n-2}\rangle$$

If the mobilities for the + and - species are in a fixed ratio (§4.4), thus :

$$\sigma \sim [M^+] \sim [\langle M_nX_n\rangle]^{1/3} \sim [\langle MX\rangle]^{n/3}$$

Clusters with ≈ three salt units are present in the system to explain the linear dependence of conductivity with concentration. The same authors interestingly remarked that the increase in T_g with salt concentration was 1/3 of that due to the crosslinks density :

$$\partial T_g/\partial c_{MX} = 1/3\,\partial T_g/\partial c_{Xlink}$$

This again suggests that the stiffening of the network is the same from a cluster of three MX solvated by the polymer chains and from a covalent crosslink, corresponding to the model shown below (§4.4).

This situation is intermediate between that of polyelectrolytes dissolved in a highly polar solvent (extended chain conformation, high degree of dissociation and high conductivity) and the copolymers of alkenes with for example sodium acrylate for which clusters containing up to $\approx 10^3$ units are present, separated by the hydrocarbon chains[83]. In the latter case, no conductivity is observed as the activation energy for ionic dissociation and motion across the non-polar lipid layer is extremely large.

4.4 Transference Numbers :

In their complexity and conductivity mechanisms, PEs parallel solutions, especially those in low to medium polarity solvents. Liquids lack selectivity in the sense that there is no restriction on the motion of species, whatever charge they carry ; polymer electrolytes correspond to an intermediate situation discussed below, as obviously at least the anions and cations are mobile.

Different techniques have been used to measure the tranference numbers t^+ and t^-. In no case any electronic conductivity is apparent, except when delocalized radical anions are present in the matrix, as in PEO-$TCNQ^{\bullet -}$ salts[67] or PEO-polyiodide I_n^- systems [68]. We shall compare here six techniques wich have been mainly used for polymer electrolyte to get access to t^+ :

Techniques :	Remarks :
Hittorf-Tubandt *HT* [69]	Sensitive to the weight of the ions ; unsensitive to ion pairs and uncharged species.
D.C. Polarization *DCP* [70, 71]	Unsensitive to uncharged species ; response influenced by multiplets if present.
Potentiometric *P* [72]	Unsensitive to uncharged species ; may differentiate multiplets from single ions.

AC Impedance *AC I* [73)	Requires a model of ideal solution ; unsensitive to uncharged species ; sensitive to the quality of the interfaces [82)].
NMR Pulsed Field Gradient *PFG* [74-78)]	Sensitive to the diffusion of nuclei irrespective of their charge ; very precise.
Radioactive Tracer *RT* [79,80)]	Sensitive to the diffusion of nuclei irrespective of their charge.

The two last techniques, *PFG* and *RT*, measure the diffusion coefficients and have to rely on the Nernst-Einstein relation which is not valid when uncharged species are also mobile. All techniques however agree to values of :

$$0.2 < t^+ < 0.4 \quad \text{for PEO-LiClO}_4$$

This result is quite interesting, firstly because this value is quite similar to that of liquids, almost irrespective of the type of solvent (except for proton conduction). Besides, the coincidence of the values found by electrochemical techniques and those from diffusion experiments indicates that *undissociated species are not mobile in the polymer electrolytes,* a complete difference from liquids. Also, there is in this case no indication of multiplet ions, as the *HT* or *P* methods would give different values.

Another salt, $LiCF_3SO_3$, is more puzzling. For systems with PEO, the Nernst-Einstein equation is apparently not obeyed ($D_{NMR} \approx 2D_\sigma$) and the diffusion coefficient for ^{19}F is about twice that of ^7Li. On the other hand, the *P* and *HT* methods yield $t^+_{apparent} \approx 0.7$. Such discrepancies are only explained by the existence of anionic triplets $[CF_3SO_3\text{-}Li\text{-}SO_3CF_3]^-$, the counter ion being Li^+. The schematic representation of the polymer, taking into account the existence of clusters, is suggested schematically as :

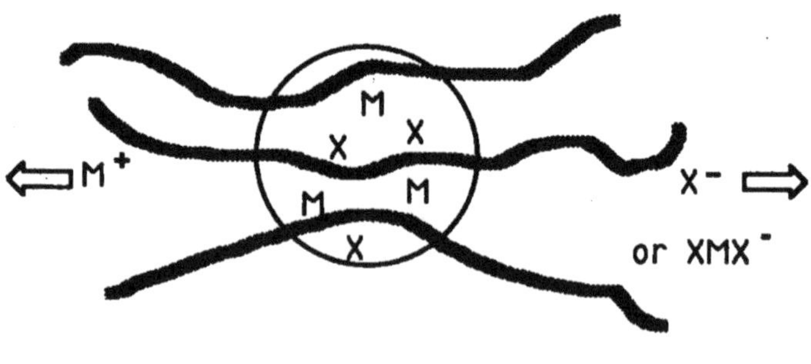

From these clusters, ionic dissociation results in either singlet ions ($LiClO_4$) or triplets ($LiCF_3SO_3$) hopping from site to site.

These examples, taken from the most studied polymer and salts, reflect the complexity of the situation, and the need for different and complementary techniques.

A very interesting study of diffusion coefficients with NMR-PFG was made by Gorecki et al.[78] on homologue perfluorosulfonate salts of various chain length $Li[CF_3(CF_2)_mSO_3]$ with $0 < m < 7$. For these PEO complexes, *the transference number appears independent of chain size*, and comparable to that of the lighter homologue. If the triplet ion model is valid throughout the series, the n = 7 anion is ≈ 20 Å long and ≈ 10 times larger than ClO_4^-. These important results suggest a strong coupling between the motion of ions of opposite charges, and the "free-volume" theory cannot be invoked in this case (coulombic fluid ?). Such experiments, as well as those on high pressure conductivity studies [85, 79], should cast some new lights on the theories of ion transport, including liquids. The case of multicharged cations, alkaline earth and transition metal, needs to be mentioned [86-88]. Cations with a small radius, like Zn and Mg, are very polarizing and the rate of ligand exchange is slow, resulting in $t^+ \approx 0$. Larger or softer cations (Sr, Ba, Cd, Hg, Sn, Pb...) interact less strongly with the chains and show definite mobility ($t^+ > 0.1$) as shown by cyclic voltammetry. As a rule, the cationic mobility should parallel the kinectics of water exchange in the solvation shell [89], considering again the comparable donating properties and orbital hardness of ethers and of H_2O. For A elements, the speed of exchange increases with the radius and decreases with charge ; more dramatic variations are found with transition elements where electronic effect prevails, as shown by the peculiar behaviour of Ni^{++} in PEO [90].

Another interesting case is that of proton conductors ; although no measurements of the transference number has been made yet, there is evidence from neutron diffraction in the "aminosils" depicted in § 2.3 of clusters of ≈ 5 NH_2- groups with equivalent protons, pointing towards a Grotthus-type transport mechanism[77]. Such model is also corroborated by the high conductivities.

5 ELECTROCHEMISTRY AND APPLICATIONS

5.1 Electrochemistry :

While the conduction mechanisms are complex, the electrochemical behaviour of PE is relatively straightforward and governed by simple considerations:

- Chemical inertness of the bonds involved in the macromolecule, *especially polyethers.*
- Absence of convection or even of diffusion of moeties attached covalently to the chains

  for high MW or crosslinked systems.

Thus, for PEO cyclic voltammetry only shows the stability window limited by both anions and cations, as in a molten salt. Anions thermodynamically (I^-, $B(CH_3)_4^-$) or kinetically (ClO_4^-)

stable towards reduction show the plating of the alkali metal, and well defined reversible behaviour is also observed for the cations having a non-zero t^+, Ag, Ba, Cd, Cu ... and shown in Fig. 8 for Pb in $Pb(CF_3SO_3)_2$ [72].

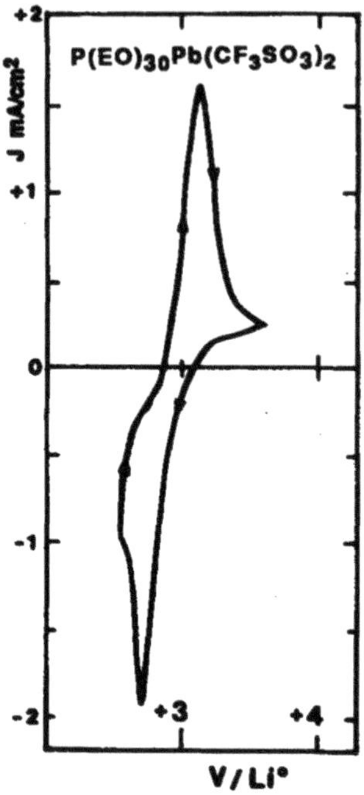

Fig. 8

Cyclic voltammogram of $Pb(CF_3SO_3)_2$ 30 : 1 in PEO at 90°C. Reference Li°. From ref.[72] with permission.

The anodic limit of oxidation-resistant anions (ClO_4^-, $CF_3SO_3^-$) is typically irreversible, probably involves radical species and proton abstraction from the polymer, but the pathways are unknown. It is now quite evident that the interface with lithium metal is covered with a passivating film, as shown from detailed impedance analysis[91]; traces of moisture or the reduction products (LiF, Li_2O...) of the anions are the likely source of this film, especially if we consider the facile reduction of other anions (SCN^-, NO_3^-, AsF_6^-). Scrosati [92] has studied in detail the interfacial properties and shown for example with $PEO_{16}LiCF_3SO_3$ an appreciable exchange current density of ~ 1 mA/cm^2 at 90°C. The main limitation in kinetics seems to arise from the transference number. Derived from the original work of Atlung et al.[93], the limiting current is written as :

$$I_{lim} = \frac{4RT\,\sigma+}{Fd}$$

$\sigma+$ = cationic conductivity
d = cell thickness

The validity is based on an ideal solution model ; this is not the case for the PEs, with the existence of the 3:1 and 6:1 stoichiometric complexes which nucleate during anodic stripping at metal electrode or deintercalation at solid-solution electrodes. The theoretical and practical values are however within the range 2-5 mA/cm^2.

5.2 Applications :

In addition to the above mentioned advantages of electrochemical stability, the PEs have all the desirable properties of plastic materials, and can be easily processed into films. Thus, they have been, almost from their discovery, associated with intercalation electrochemistry for solid state batteries. In fact, this foreseen application has been the driving force for most of the studies [2)]. The simple and elegant electrochemistry of intercalation cathodes transposes well with PEs, including TiS_2, VO_x $(2 < x < 2.5)$, MoO_2, LiV_3O_8, etc... The principle of a solid state battery is shown on Fig. 9.

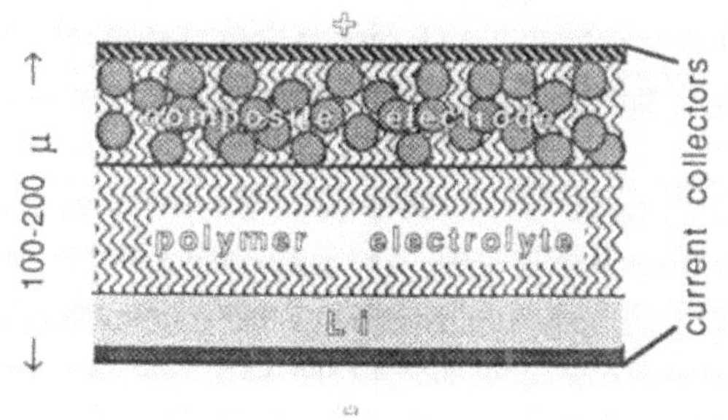

Fig. 9
Schematic diagram of a solid state battery using intercalation electrodes and film technology.

Many groups are now involved worldwide in the R&D of this type of power source [91,92)] considering the advantages in terms of cycle life, safety and energy density. In fact the polymer-based cells appear the most feasible for power sources of any size, with their easy multicell stacking ; competition from glasses where solid state electronics technologies (sputtering, vacuum deposition) can be used is evident only for small or sub-miniature (≈ 1μm) systems. The paucity of the present literature in details concerning the polymer battery technology is probably an indication of their interest. With the simple PEO, the phase diagram has required operation of these systems at 80-120°C. The results meet most of the expectation for a novel energy storage :

Energy density ~ 100- 120 Wh/kg in actual systems.

Power density > 100 W/kg and peaks up to 1kW/kg.

Cycle life > 500 deep cycles at > 50% of original capacity.

Absence of self-discharge < 5% /year.

No scale-up effect from 1 to >1000 cm^2.

These results correspond to the drastic requirements set for electric traction purposes. There has also been a major incentive to lower the operating temperature, with modifications of the polymer to bypass the phase diagram limitations. Recently, Gauthier et al.[93)] have disclosed a room temperature battery using both a modified polyether and a new salt with plasticiser properties resulting in discharge rates up to C/2 (total discharge in two hours), well beyond the requirements of most electronic applications. The solid state cells are also quite resistant to abuse, overdischarge or overcharge ; the latter has been studied in details [102)]. Lithium alloys instead of the pure metal have been successfully used as negative electrode, but also sodium [95)].

There is a recent interest for the use of polymer electrolytes in electrochomic devices, especially windows with controlled visible light absorption and IR reflection. These so-called "smart windows" could save up to 10% energy in either cold (heating)or hot (air conditioning) weathers. Liquid electrolytes or conventional solid electrolytes would be very unpractical for such large surfaces. Examples are :

Glass / ITO/WO3 / H^+ or Li^+ polymer electrolyte/counter electrode / ITO / Glass

Other applications are far less developed. The possibility of introducing redox couples has been applied to photo-electrochemical cells. Either :

Si // PEO I^- / I^{3-} // ITO

CdSe // PEO S_m^{2-}/S_n^{2-} // ITO

systems have been used [103,104)]. The main limitation of these cells is due to the high resistivity of the electrolyte. However, use of extremely thin films obtained by spin-coating techniques has allowed current densities up to 10 mA/cm^2 corresponding to a few % optical to electrical energy conversion efficiency [103)]. In each case the stability of the interface, i.e. the absence of photocorrosion, was noted.

PEs are used increasingly as tools for precise electrochemical measurements. Besides the already mentioned stability, the main advantages are the possibility of working under high vacuum, thus minimizing contamination by water or oxygen, and the absence of co-intercalation in host structures. For instance, graphite-alkali metal intercalates cannot be prepared electrochemically in any solvent, but are easily obtained with PEO[96]. The variety of solutes, the "clean" interfaces and thermal stability up to 200°C, have made the complexes quite ideal for thermodynamic activity measurements in alloys and titration curves for intercalation hosts. The system $Li_xZrSe_{2-\delta}$ for instance shows a semiconductor-to-metal transition for $x_o = f(\delta)$, clearly appearing on extremely slow scan voltammetry ($\approx 10^{-5}$ Vs^{-1}), which is made possible with the absence of side reactions [97]. Fabry et al. [105] have used mixed complexes of NaI and AgI (containing the AgI_2^- anion) in contact with silver as solid reference electrode in ion selective electrodes. Finally, mention should be made of the elegant work of Baudry et al. [106] who have observed *in situ* with a Scanning Electron Microscope the intercalation of lithium into various host structures from a thin-film PEO cell, which enables to follow the morphological changes induced by the progressive reduction.

6 CONCLUSIONS :

The polymer electrolytes form undoubtedly a new class of ionic conductors : while the mechanical properties are those of solids, especially for crosslinked material, the ionic motion at the microscopic level is better described with the concepts of liquid solutions. Yet, the diffusion of charged species, for complex as it appears with the existence of clusters, involves a permanent desolvation- solvation process, assisted by the segmental motion of the polymer. In this respect, the PEs deserve the denomination of "immobile solvent", justified even further by the absence of net macromolecule motion , exemplified by the absence of co-intercalation.

Most of the research has been focused on improvement of the conductivity, especially to avoid the crystallinity problem. The values now reached at room temperature, in the 10^{-5} - 10^{-4} $(\Omega cm)^{-1}$ are probably close to the intrinsic limit, set both by the microscopic viscosity and the uncomplete dissociation. Trying to improve the latter by the introduction of polar groups [51] is likely to raise the T_g, lowering the diffusion, thus resulting in similar conductivities. What is important is the possibility to address the mechanical properties almost independently of the conductivity, as microphase separation allows to preserve a connective conductive path, as exemplified in the "aminosils" with a silica backbone. Also we must insist on the wide possibilities of coordination chemistry ; the list with PEO is already impressive, and we have stressed the properties of PEI which represents a different class of solvating units ("soft" and weakly protic, *vs* "hard" and aprotic for PEO), with completely different selectivities. Many other possibilities can be explored.

Major progress could be accomplished with charge-selective PEs. Present materials with mixed transport behave like liquid solutions, but the problem of transference numbers is more apparent for the solids because of the lower conductivity. The attachment of the negative charge to the chain to reach $t^+ \approx 1$ has resulted in lower conductivities, but no conclusion can be drawn yet concerning the possible limitations as the moeties used as negative charge carriers which were used are not those giving optimum dissociations (when under the form of simple salts). Some structures which have been tested are shown below :

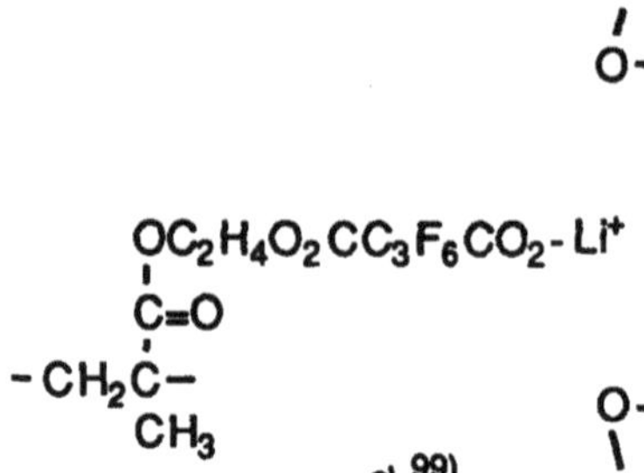

a) 99)

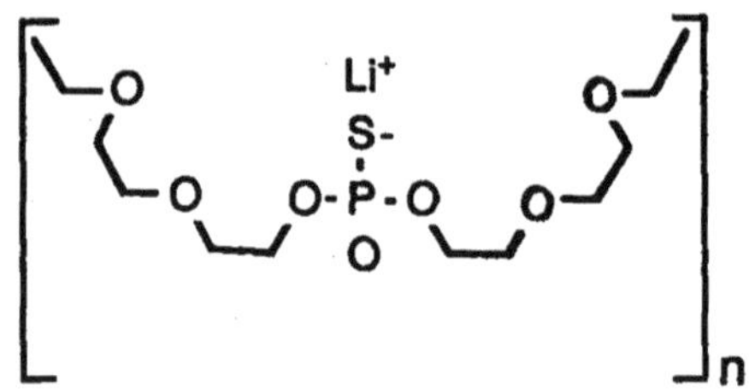

b) 100)

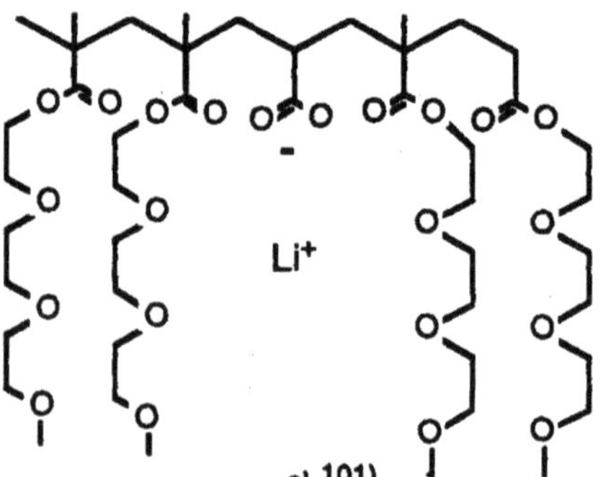

c) 101)

$t^+ = 1$ electrolytes, termed as self-ionizing ion-exchange polymers are yet a challenge and a goal for elaborate polymer chemistry .

Progresses made for the major application, solid state batteries, are quite impressive, probably justifying some commercialization at this stage, such prospect certainly assures the perenniality of this field.

7 REFERENCES :

(1) P.V. Wright *Br. Polym. J.* 7, 319 (1975).

(2) M.B. Armand, J.M. Chabagno & M. Duclot *Second International Meeting on Solid Electrolytes, St Andrews, Scotland* - Extended Abstract 20-22 Sep. (1978).

(3) H.Shirakawa., E. J.Louis, A. G. Mac Diarmid, C.K. Chiang, A.J. Heeger *J. Chem.Soc., Chem Com.* 578 (1977).

(4) M.B. Armand *Ann. Rev. Mater. Sci.* **16** 245 (1986).

(5) C.A. Vincent *Progress in Solid State Chemistry* **17-3** (1987)

(6) M. Gauthier, M. Armand & D. Muller in "*Electroresponsive Molecular and Polymeric systems Vol. 1* " T. Skotheim ed. Marcel Dekker Inc. Publisher New-York, (1988)

(7) "*Polymer Electrolytes Reviews 1* ", J.R. MacCallum, C.A. Vincent ed. Elsevier Applied Science London, (1987).

(8) *Structure and Bonding* N°16 "*Alkali Metal Complexes with Organic Ligands*" Springer Verlag Publisher Berlin, (1973).

(9) D.L. Papke, M.A. Ratner, D.F. Shriver, *J. Electrochem Soc.*, 129, 1694 (1982).

(10) A. Bouridah, F. Dalard & M. Armand Proceedings of SSI 87 Garmish FRG, Sep. 6-11, 1987 to appear in *Solid State Ionics*, (1988).

(11) J. Sandahl, S. Schantz, L.M. Torell & R. Frech Proceedings of SSI 87 Garmish FRG, Sep. 6-11, 1987 to appear in *Solid State Ionics*, (1988).

(12) M.C. Lang, C. Noel, A.P. Legrand J. Polym Sci, Polym. Physics **15** 1318 (1977).

(13) R.C. Allen, L. Mandelkern *J. Polym Sci, Polym. Physics* **20** 1465 (1977).

(14) P.R. Sorensen, T. Jacobsen *Polymer Bull.* **9**, 47 (1982).

(15) M. Minier, C. Berthier, W. Gorecki *J. de Physique* **45**, 739 (1984).

(16) P. Ferloni, G. Chiodelli, A. Magistris & M. Sanesi, *Solid State Ionics* **18-19** 265 (1986).

(17) C.D. Robitaille, D. Fauteux, *J. Electrochem Soc* **133**, 315 (1986).

(18) R. Neat, M. Glass, R. Linford & A. Hooper *Solid State Ionics* **18-19** 1088 (1986).

(19) C. Berthier, W. Gorecki, M. Minier, M.B. Armand, J.M. Chabagno, P. Rigaud *Solid State Ionics* **11**, 91 (1983).

(20) K.J. Adamic, , S.G. Grenbaum, M.C. Wintergill, J.J. Fontanella J. Appl. Phys. 60 1342 (1986).

(21) S. Bhattacharja, S.W. Smoot & D.H. Whitmore *Solid State Ionics* **18-19** 306 (1986).

(22) T. Hibma, *Solid State Ionics* **9-10** 1101 (1983).

(23) M. Stainer, L.C. Hardy, D.H. Whimore & D.F. Shriver J. Electrochem Soc. **131**, 784 (1985).

(24) R. Tanaka, I. Ueoka, Y. Takaki, K. Kataoka, S. Saito *Macromolecules* **16**, 849, (1983).

(25) F. Defendini, M. Armand, W. Gorecki & C. Berthier *Electrochemical Society* San Diego Meeting **86-2** Extended Abstract N°3 (1986).

(26) P. Donoso, W. Gorecki, C. Berthier, F. Defendini & M. Armand Proceedings of SSI 87 Garmish FRG, Sep. 6-11, 1987 to appear in *Solid State Ionics*, (1988).

(27) R. Tanaka, T. Iwase, T. Hori & S.Saito Proceedings of ISPE 1 Saint -Andrews Scotland June 1987 to appear in British Polym. J. (1988).

(28) M.F. Daniel, B. Desbat, F.Cruege, O. Trinquet & J.C. Lassègues Proceedings of SSI 87 Garmish FRG, Sep. 6-11, 1987 to appear in *Solid State Ionics*, (1988).

(29) A. Polak,S. Petty-Weeks,A.J. Beulher *Chem. Engin. News* 28, Nov (1985).
S. Petty-Weeks, A. Polak Sensors & Actuators 11,377, (1987).

(30) Y. Charbouillot, D. Ravaine M. Armand & C. Poinsignon to appear in *J. Non Cryst. Solids.* (1988).

(31) M.B. Armand, J.M. Chabagno, M. Duclot *"Fast Ion transport in Solids"* P. Vashishta, J.N. Mundy, G.K. Shenoy Eds. Elsevier-North Holland New-York 131 (1979).

(32) R.E. Wetton & G. Allen Polymer 7, 331, (1966)

(33) C.K. Chiang, G.T. Davis, C.A. Harding & T. Takahashi *Macromolecules* 18, 825 (1985).

(34) C.S. Harris, D.F. Shriver & M.A. Ratner *Macromolecules* 19, 987 (1986).

(35) R.D. Armstrong, & M.D. Clarke, *Electrochim. Acta* 29, 1443, (1984).

(36) R. Dupon, B.L. Papke, M.A. Ratner & D.F. Shriver, *J. Electrochem Soc.* 131, 586 (1984).

(37) M. Watanabe, M. Togo, K. Sanui, N. Ogata, H. Kato, T. Kobayashi & Z. Othaki *Macromolecules* 17, 2908, (1984).

(38) P.M. Blonsky, D.F. Shriver, P. Austin & H.R. Allcock , *Solid State Ionics* 18-19 258 (1986) ; H.R. Allcock, P.E. Austin, T.X. Neenan, J.T. Sisko, P.M. Blonsky & D.F. Shriver *Macromolecules* 19, 1508, (1985).

(39) P.G. Hall, G.R. Davies, J.E. McIntyre, I.M. Ward, D.J. Bannister & K.M.F. Le Brocq, *Polym. Comm.* 27, 98 (1986).

(40) D.W. Xia, D. Soltz & J. Smid *Solid State Ionics* 14, 221(1984).

(41) D.J. Bannister, G.R. Davies, I.M. Ward & J.E. McIntyre *Polymer* 25, 1600, (1984).

(42) N. Kobayashi, M. Ushiyama, K. Shigehara & E. Tsuchida *J. Phys. Chem.* 89, 987 (1985).

(43) J.M.G. Cowie & A.C.S. Martin Polymer Comm. 26, 298, (1985).

(44) J.R.M. Giles, F.M. Gray, J.R. MacCallum & C.A. Vincent *Polymer* 28 1977 (1987).

(45) A. Killis, J.F. LeNest, A. Gandini, H. Chéradame, *J. Polym Sci. Polym. Phys.* 19, 1073, (1981). A. Killis, J.F. LeNest, A. Gandini, H. Chéradame & J.P. Cohen-Addad *Solid State I Ionics* 14, 231, (1984).

(46) M. Watanabe, K. Sanui, N. Ogata & F. Inoue *Polymer J.* 15, 175, (1984). M. Watanabe, S. Oohashi, K. Sanui, N. Ogata, T. Kobayashi & Z. Ohtaki, *Macromolecules* 18, 1945, (1985).

(47) A.Bouridah, F.Dalard, J.F. LeNest & H. Chéradame,*Solid State Ionics* ,15, 231, (1985).

(48) J.R. MacCallum, M.J. Smith & C.A. Vincent *Solid State Ionics* 11, 307, (1984).

(49) R. Spindler & D. F. Shriver *Macromolecules* to appear (1988).

(50) J.R.M. Giles & M.P. Greenhall *Polymer Comm.* **27** 360 (1988).

(51) Shi-bi Fang, Peng Zhang, Ying-Yan Jiang *Polymer Bulletin* **19**,81, (1988).

(52) M.L. Kaplan, E.A. Rietman, R.J. Cava, L.K. Holt & E.A. Chandross, *Solid State Ionics* **25**, 37, (1987).

(53) W.H. Meyer Proceedings of SSI 87 Garmish FRG, Sep. 6-11, 1987 to appear in *Solid State Ionics*, (1988).

(54) S. Toyota, T. Nogami, H. Mikawa *Solid State Ionics* ,**13**, 243, (1984).

(55) M. Galin, E. Marchal, A. Mathis, B. Meurer, Y.M. Monroy-Soto & J.C. Galin *Polymer*, **28** 1937, (1987).

(56) A. Revillon, T. Hamaide *Polym. Bull.* **6**, 235 (1982).

(57) E.A. Rietman, M.L. Kaplan & R.J. Cava *Solid State Ionics* **25,** 41, (1986).

(58) H. Vogel, *Phys. Z.* **22**, 645 (1921) ; G. Tamman and W. Hesse, *Z. Anorg. Allg. Chem.*, **156**, 245 (1926) ; G.S. Fulcher, *J. Amer. Ceram. Soc.*, **8**, 339 (1925).

(59) M.L. Williams, R.F. Landel & J.D. Ferry, *J. Amer. Chem. Soc.* **77**, 3701 (1955).

(60) C.A. Angell *Solid State Ionics* **9-10** 3, (1983).

(61) M. Watanabe, S. Nagano, K. Sanui & N. Ogata *J. Power Sources* **20**, 327 (1987).

(62) S.D. Druger, M.A. Ratner & A. Nitzan *Solid State Ionics* **9-10** 258 (1986) ;
S.D. Druger,M.A. Ratner & A.Nitzan, *Phys. Rev. B* **31,** 3939 (1985) ; C.S. Harris
S.D. Druger, M.A. Ratner & A. Nitzan, *Solid State Ionics* **18/19**, 151, (1986).

(63) V. Clément, D. Ravaine, C. Déportes & R. Billat Proceedings of SSI 87 Garmish FRG, Sep. 6-11, 1987 to appear in*Solid State Ionics*, (1988).

(64) J. Moacanin & E.F. Cuddihy, *J. Polym. Sci. Poly Symposium* **14**, 313, (1966).

(65) L.D. Pettit & S. Bruckenstein *J. Am. Chem. Soc.* **88**, 4783 (1966).

(66) A. Killis, J.F. LeNest, A. Gandini,& H. Chéradame *Macromolecules* **17**,63 (1984).

(67) J.A. Siddiqui & P.V. Wright *Polymer Comm.* **28**, 90 (1987).

(68) Ph. Rigaud Thesis University of Grenoble (1980) ; D.F. Shriver 2nd Symposium on Electroresponsive Polymers Brookhaven, NY October 4-7 (1987).to appear in *Molecular Crystals & Liquid Crystals* (1988).

(69) M. Levèque, J.F. LeNest, A. Gandini & H. Chéradame *J. Power Sources* **14**, 27 (1985).

(70) H.R. Allcock, P.E. Austin, T.X. Neenan J.T. Sisko, P.M. Blonsky & D.F. Shriver *Macromolecules* **19**, 1508 (1986).

(71) P.G. Bruce and C.A. Vincent J. Electroanal. Chem & Interfacial Electrochem. to appear (1988).

(72) A. Bouridah, F. Dalard, D. Deroo & M. Armand *Solid State Ionics*, **18/19**, 287, (1986). A. Bouridah, F. Dalard, D. Deroo & M. Armand *J. Appl. Electrochem.* **17**, 625 (1987).

(73) P.R. Sorensen & T. Jacobsen *Electrochimica Acta* **27**, 1671 (1982).

(74) M. Mali, J. Roos, D. Brinkman Proc XXII Congress Ampere K.A. Muller, R. Kind & J. Roos Eds. Zurich (1984).

(75) W. Gorecki, R. Andréani, C. Berthier, M.B. Armand, M. Mali, J. Roos & D. Brinkman *Solid State Ionics*, **18/19**, 295, (1986).

(76) S. Bhattacharja, S.W. Smoot & D.H. Whitmore *Solid State Ionics*, **18/19**, 306 (1986).

(77) P. Donoso, W. Gorecki, C. Berthier, F. Defendini & M. Armand Proceedings of SSI 87 Garmish FRG, Sep. 6-11, 1987 to appear in*Solid State Ionics*, (1988).

(78) W. Gorecki, P. Donoso, C. Berthier, M. Mali, J. Roos, D. Brinkman & M.B. Armand, Proceedings of SSI 87 Garmish FRG, Sep. 6-11, 1987 to appear in*SolidState Ionics*, (1988).

(79) A.V. Chawick, J.H. Strange & M.R. Worboys *Solid State Ionics*, **9-10**, 1155 (1983).

(80) D. Fauteux, M.D. Lupien & C.D. Robitaille *J. Electrochem. Soc.* **134** 2761, (1987).

(81) M. Watanabe & N. Ogata in "*Polymer Electrolytes Reviews 1* ", J.R. MacCallum, C.A. Vincent ed. Elsevier Applied Science London, (1987).

(82) P.G. Bruce, J.Evans & C.A. Vincent Proceedings of SSI 87 Garmish FRG, Sep. 6-11, 1987 to appear in*Solid State Ionics*, (1988).

(83) A. Eisenberg & M. King in *"Ion-Containing Polymers"* Academic Press New York (1977).

(84) G.G. Cameron, M.D. Ingram & G.A. Sorrie J. Chem Soc. Faraday Trans. I. **83**, 1115 (1983).

(85) J.J. Fontanella, M.C. Wintergill, J.P. Calame, M.K. Smith & C.G. Andeen *Solid State Ionics*, **18/19**, 253 (1986) ; J.J. Fontanella, M.C. Wintergill, M.K. Smith, J. Semancik & C.G. Andeen *J. Appl. Phys.* **60** 2665 (1986).

(86) L.L. Yang, R. Huq, G.C. Farrington & G. Chiodelli *Solid State Ionics*, **18/19**, 253(1986) ; L.L. Yang, A.R. McGhie & G.C. Farrington *J. Electrochem .Soc.* **133**, 1380 (1988).

(87) A. Patrick, M. Glasse, R. Latham & R. Linford *Solid State Ionics*, **18/19**, 1063 (1986).

(88) T.M.A. Abrantes, L.J. Alcacer & C.A.C. Sequeira *Solid State Ionics*, **18/19**, 315 (1986).

(89) M. Eigen Pure & Appl. Chem. **6** , 97 (1963).

(90) R. Huq, G.C. Farrington *J. Electrochem .Soc.* **135**, 524 (1988).

(91) A. Hooper & J.M. North *Solid State Ionics*, **9/10**, 1161 (1983) ; A. Hooper,& B.C. Tofield, *J. Power Sources* **11**, 33 (1984) ; A. Hooper, J.M. North, M. Oliver & B.C. Tofield 31st Power Sources Symposium (1984).

(92) M Gauthier, D. Fauteux, G. Vassort, A. Bélanger, M. Duval, P. Ricoux, J.M. Chabagno, D. Muller, Ph. Rigaud, M.B. Armand & D. Deroo *J. Electrochem .Soc.* **132**, 1333 (1985).

(93) G. Vassort, M. Gauthier, P.E. Harvey, F. Brochu & M. Armand *Electrochemical Society* Honolulu Meeting **87-2** October 1986. Submitted to *J. Electrochem .Soc.*

(95) K. West, B. Zachau-Christiansen, T. Jacobsen Proceedings of ISPE 1 Saint-Andrews Scotland June 1987 to appear in British Polym. J. (1988).

(96) R. Yazami & Ph. Touzain *J. Power Sources* **9**, 365 (1983).

(97) Y. Chabre, P. Deniard & R. Yazami Proceedings of SSI 87 Garmish FRG, Sep. 6-11, 1987 to appear in*Solid State Ionics*, (1988).

(98) H. Benisty, Ph. Colomban, J.N. Chazalviel *Appl. Phys. Letters* **51**, 1121 (1987).

(99) D.J. Bannister, G.R. Davies,, I.M. Ward & J.E. MacIntyre *Polymer* **25**1291, (1984).

(100) J.F. LeNest Thesis University of Grenoble (1985) ; J.F. LeNest, A. Gandini, H. Chéradame & J.P.Cohen-Adad *Polymer Comm.* **28**,302 (1987).

(101) N. Kobayashi, M. Uchiyama & E. Tsuchida *Solid State Ionics* **17**, 307 (1985).

(102) R. Gopaliengar, M.Z.A. Munshi & B.B. Owens *Electrochemical Society* Honolulu Meeting **87-2** October 1986. Submitted to *J. Electrochem .Soc.*

(103) T.A. Skotheim, S.W. Feldberg & M.B. Armand *J. de Phys. C3* **44** 615 (1983) ; T.A. Skotheim & O. Ingranas *J. Electrochem .Soc.* **132** 2116 (1985).

(104) B. Marsan Ph.D. Dissertation Université du Québec INRS Energie (1988).

(105) P. Fabry, C. Montero-Ocampo & M. Armand to appear *Sensors & Actuators* (1988).

(106) P. Baudry, M. Armand, M. Gauthier & J. Massounave Proceedings of SSI 87 Garmish FRG, Sep. 6-11, 1987 to appear in *Solid State Ionics*, (1988).

(107) J.S. Foos &S.M. Erker *J. Electrochem. Soc.* **134** 1725 (1987).

(108) Y. Chatani, S. Okamura *Polymer* **28**1815, (1987).

(109) M. Watanabe, M. Itoh, K. Sanui & N. Ogata *Macromolecules* **20**, 569 (1987).

Composite Solid Ion Conductors

by

J. Bruce Wagner, Jr.
Center for Solid State Science
Arizona State University
Tempe, Arizona 85287-1704, U.S.A.

Since the discovery of the dispersed second phase effect by Liang[1,2] in 1973, there have been numerous investigations on composite electrolytes. Most of the studies have utilized submicron particles of alumina but silica, fly ash, wax, zirconia, and other insulator materials have been used to increase the ionic conductivity of electrolytes. Requisites for a solid state battery separator or a chemical sensor include a high ionic conductivity and a low electronic conductivity. Methods to achieve these goals include 1) aliovalent or homogeneous doping such as the introduction of CaO into ZrO_2 [3]; 2) the development of materials with open structures or channels for rapid ion migration such as the beta and beta" aluminas[4,5], and 3) the development of thin polymeric[6] and glassy materials[7,8], which are ionic conductors.

The study of composite electrolytes is, thus, one approach to the development of high ionic conductivity materials. Such studies also offer opportunities to investigate surface chemistry and solid state "internal" absorption. Furthermore, as will be described, such heterogeneous doping* can be utilized in conjunction with aliovalent or homogeneous doping to change not only the magnitude of ionic conduction, but also the type and number of lattice defects responsible for transport.

---

* The term, heterogeneous doping, was first introduced by Maier[24].

Figure 1 shows a schematic diagram of the conductivity of the composite, σ, relative to that of the pure matrix, σ_0. The enhancement is usually a factor of fifty to one hundred, but it can be as large as a thousand. Enhanced transport in composite electrolytes is due to interfacial transport. Interfacial transport in solid electrolytes can be divided into three categories. The first is the purposeful introduction of an insoluble second phase, such as Al_2O_3 into silver halides. The second involves the enhancement found in a mixture of two immiscible phases which are at equilibrium with one another, such as a mixture within the miscibility gap of the silver iodide-silver bromide system[9,10]. The third involves the enhancement in a polycrystalline, but in single-phase material such as polycrystalline silver halide. It has long been known that transport in polycrystalline single phase electrolytes may be larger than that in large grained or single crystal electrolytes. This effect has recently been treated by Maier[11], and it must be remembered that a polycrystalline heterogeneously doped or composite electrolyte may also have single phase grain boundaries, which may also contribute to the conductivity.

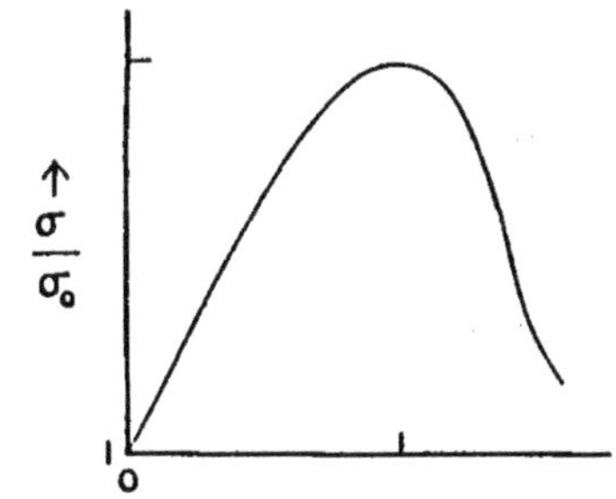

Fig. 1. Schematic diagram of the conductivity of the composite, σ, divided by the conductivity of the pure electrolyte, σ_0, as a function of the concentration of added insulator particles. The maximum enhancement is generally a factor of 50 to 100 and the concentration at maximum enhancement is usually at 0.1 to 0.4 volume fraction. The decrease in conductivity following the maximum is due to the "blocking effect," i.e., the insulating dispersoids occupy such a large volume, that transport is decreased.

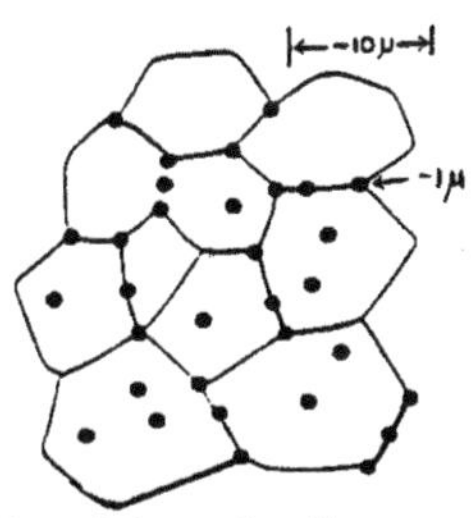

Fig. 2. Schematic diagram of a solid electrolyte matrix consisting of polycrystalline grains, in this example of the order of ten microns in diameter. The dispersoid particles are represented by spheres of the order of one micron in diameter. The distance between particles depends on the volume fraction and the particle distribution. Assuming the particles are distributed uniformly, the distance between particles is of the order of a tenth of a micron.

Figure 2 shows a schematic representation of a polycrystalline matrix containing a dispersed second phase. The grain sizes of the electrolyte are on the order of 10μ. The dispersoid particle sizes are on the order of 1μ or less, and they are represented as spheres although, at least for particles of Al_2O_3, they are more nearly of platelet geometry. The distance between particles is on the order of a tenth of a micron.

The maximum conductivity usually occurs between 10 and 40 volume percent insulator particles. Table 1 below provides information about the surface area of the added particles. The stated particle size is two times the radius, r_1, in microns, and the BET determined surface area is in m^2/g. Calculations assuming spheres yield about one-third of the values for the surface area.

TABLE 1

DISPERSOID SURFACE AREA
(ADOLPH MELLER ALUMINA)
AS DETERMINED BY THE BET METHOD

Stated particle size = $2r_1$ (microns)	Undried Al_2O_3(m^2/gm)
0.06	84
0.3	23
1.0	7
3.0	2.8
8.0	0.4-0.7

Table 2 shows examples of the concentration of alumina at maximum enhancement at room temperature. The first is lithium iodide containing alumina, and the second is silver iodide containing alumina.

TABLE 2

CONCENTRATION OF ALUMINA AT MAXIMUM ENHANCEMENT

Composite	mole %	weight %	volume %
LiI(Al_2O_3, submicron)	35	29	32
AgI(Al_2O_3, 0.06 μ)	30	15.6	23

Most of the mechanisms suggested involve a highly conductive layer surrounding an insulator particle. At any rate, the mechanism must be consistent with the experimental observations. In the case of composite electrolytes, the experimental observations are:

1. The enhancement, $\Delta\sigma$, ($\Delta\sigma=\sigma - \sigma_0$, where σ denotes the conductivity of the composite and σ_0 that of the pure matrix) increases as the surface area of the dispersoid increases. Therefore, for a given concentration, the smaller the dispersoid particles, the larger the enhancement.

2. The enhancement is larger at the lower temperatures than at higher temperatures.

3. The maximum enhancement occurs between about 0.1 and 0.4 volume fraction of added dispersoid.

4. At larger volume fractions of added dispersoid, the conductivity decreases from its maximum value.

5. The addition of a second phase may change the mechanism of conduction.

6. Sorbed materials in the dispersoid can change the magnitude of the enhancement. For example, wet alumina yields a larger enhancement than dried alumina.

7. The addition of a second phase may change the temperature of a first order phase transition of the matrix electrolyte.

At the outset is must be pointed out that the dispersed phase effect or heterogeneous doping may involve more than one mechanism, and these may be operating simultaneously. An example mentioned above was in the transport via single phase grain boundaries, which can exist in the composites. Furthermore, one mechanism may be dominant in one concentration regime while another may be dominant in another concentration regime. The following mechanisms or models have been suggested:

1. Change in the concentration of lattice defects in a space charge layer adjacent to the dispersoid.

2. Deformation induced defects, i.e. defects introduced during the processing of the composite material.

3. Transport via dislocations induced by the thermal expansion mismatch between the dispersoid and the matrix.

4. Formation of a new phase adjacent to the dispersoid particle.

5. The stabilization of fine grains of the matrix electrolyte by the dispersoids, which allow grain boundary transport as the primary transport mode,

and

6. Effective medium model

There is evidence for each of the above in some systems. For example, Dudney[12] has studied the effect of deformation on the conductivity of $AgCl(Al_2O_3)$ composites at room temperature. Dudney used alumina filter paper which was dipped into molten AgCl to impregnate it. The samples were then cooled and conductivity measurements were made at room temperature. It was found that $AgCl(Al_2O_3)$ composite samples which had not been subjected to intentional deformation by pressing exhibited the same conductivity as pure polycrystalline AgCl (with no Al_2O_3). On the other hand, when this composite was deformed by the application of a 900 kg load, the conductivity was enhanced over the composite or the pure AgCl which were not deformed. Likewise, if pure AgCl (without Al_2O_3) was subjected to the same deformation process, its conductivity was also increased. On aging at 250°C overnight, the conductivity measured at room temperature of the pure AgCl decreased and returned to nearly its original (un-deformed) value. The conductivity of the composite (also measured at room temperature) retained its enhanced value. The experiments are evidence that deformation induced defects in composites contribute to the enhanced conductivity.

In addition to these deformation-induced defects, there can be defects introduced by the thermal expansion mismatch between the dispersoid and the matrix. During the 1950's, J.W. Mitchell and co-workers[13] added small glass spheres to silver halides and studied the prismatic dislocations generated because of this thermal contraction mismatch. The research effort was focused on the mechanism of dislocation generation and transport measurements were not reported. Still, it is known that dislocations can offer a rapid transport path[14] and dislocations in electrolytes may have a space charge associated with them. Therefore, this is another source for enhanced transport. Recently Dudney has reported[15] that Al_2O_3 fibers or particles act to stabilize the grain structure in composites or in some cases to stabilize a "metastable phase which probably accounts for the enhanced conductivity." In some composites, in particular composites of lithium halides which are very hygroscopic, it has been suggested that a new phase may form adjacent to the dispersoids which, in turn, results in enhanced conductivity[16]. Dupree et al.[17] have utilized NMR and reported the existence of a "radically different environment" for about 50 percent of the lithium ions in lithium iodide containing alumina particles. This is evidence for another phase in addition the the bulk LiI. While this mechanism may be valid for hygroscopic salts, it is unlikely for electrolytes such as the copper and silver halides. Stoneham, Wade, and Kilner[18] have utilized the effective medium model[19] using the lithium iodide composites as an example. From their analysis, the layer adjacent to the dispersoid exhibits a high conductivity which gradually decreases with distance from the dispersoid particle.

Jow and Wagner[20] presented evidence for a space charge layer adjacent to the dispersoid in composite solid electrolytes. They took the concept of a space charge layer at an electrolyte-vacuum interface, as was derived by Lehovec[21] and by Kliewer et al.[22,23] and applied it to the electrolyte-dispersoid interface. When the radius of the dispersoid, r_1, was much smaller than the radius of the electrolyte, r_2, surrounding the particle, the following equation was derived, for the enhancement:

$$\Delta\sigma = 3\Sigma u_i e \langle \Delta n_i \rangle \lambda \left[\frac{1}{r_1}\right] \left[\frac{V_v}{1-V_v}\right] \tag{1}$$

where u_i = mobility of the ith species

e = electronic charge

$\langle \Delta n_i \rangle$ = change in the number of defects in the space charge layer

λ = Debye length

$= \left[\frac{8\pi e^2}{\varepsilon k T} \cdot n_1(\infty)\right]^{-1/2}$

$n_1(\infty)$ = concentration of defects in the bulk electrolyte away from the dispersoid interface.

ε = dielectric constant

k = Boltzmann's constant

V_v = volume fraction of the dispersoid

This equation shows that for a given temperature, the enhancement conductivity is proportional to $(1/r_1)(V_v/1-V_v)$. Of course, the equation is valid only for small concentrations of dispersoid. Moreover, this model implicitly assumed that the dispersoid-electrolyte interface was the same as a vacuum-electrolyte interface. This model agreed well with existing data and was the first to show a quantitative relation for the enhancement. Systematic studies showed that the enhancement was proportional to the reciprocal of the radius of the dispersion particle and the volume fraction.

Maier[24] developed a more sophisticated model, which is:

$$\Delta\sigma = 3\sqrt{2} \left[\frac{V_v}{r_1}\right] \beta \sqrt{\varepsilon\varepsilon_o RT/V^m} \; u_j \sqrt{N_{jo}} \tag{2}$$

where β = constant to allow for the topology of the dispersed particles and usually $\beta \approx 0.5$

$\varepsilon\varepsilon_o$ = dielectric constant of the electrolyte

V^m = molar volume

u_j = mobility of the defect

N_{jo} = defect mole fraction at the interface

Maier's model well represents the data for the enhancement. See Figure 3. Thus the space charge layer or depletion layer model well represents the ionic conductivity enhancement in many systems. According to Maier[24], the dispersoid acts as a nucleophillic surface, which attracts positively charged species. The consequences of this attraction are striking. For example, Shahi and Wagner[25] studied $AgI(Al_2O_3)$ using the thermoelectric effect. They reported that transport in the composites was via cation vacancies. But AgI exhibits a Frenkel disorder and transport in pure AgI (without AL_2O_3) is predominantly via silver ion interstitials. The mechanism in the case of the composite, $AgI(Al_2O_3)$, changes from dominant transport via interstitials to dominant transport via cation vacancies. This switching from one mechanism to another has been observed in a number of composites. On the other hand, an electrolyte such as LiI exhibits a Schottky disorder, and the pure LiI (without Al_2O_3) conducts primarily via cation vacancies. Introduction of the dispersoid causes the positively charged species to be attracted, leaving a depletion zone with a greater concentration of cation vacancies than in the bulk LiI where the concentration of cation vacancies is given by the square root of the Schottky constant. Thus, in LiI, there is no mechanism change but simply an increase in the number of cation vacancies. Figure 4 shows a schematic diagram of the concentration profiles for a compound MX in which the ionic conduction is via cation vacancies in pure MX and also in $MX(Al_2O_3)$. Table 3 lists some composites which have been studied.

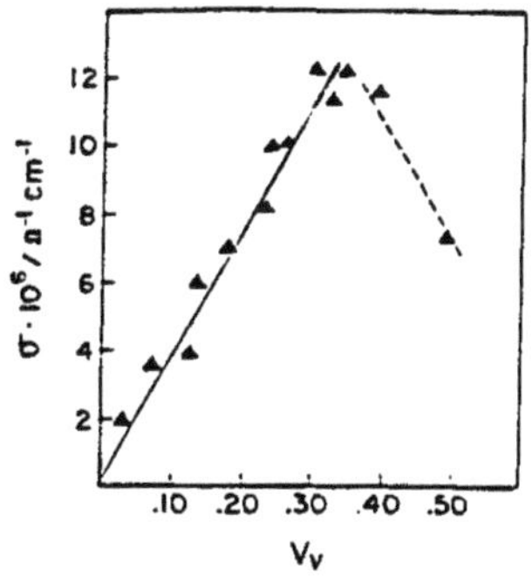

Fig. 3. Plot of conductivity at 25°C versus volume fraction of dispersoid for LiI (Al_2O_3). Data of Liang[1]. Figure adapted from Maier[24] with permission.

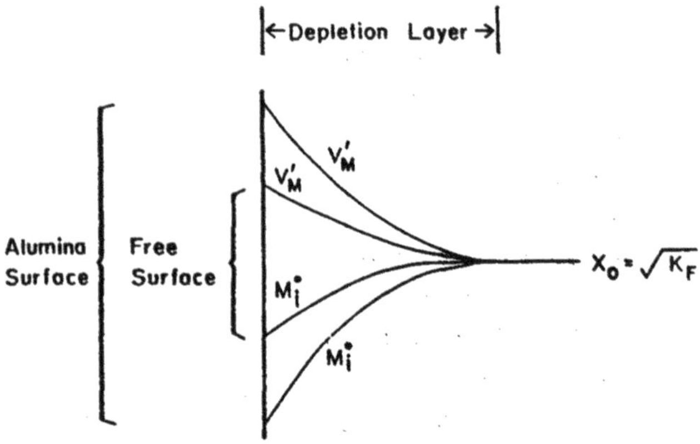

Fig. 4. Schematic diagram of space charge layer or depletion layer for an electrolyte MX and a free surface and for MX containing Al_2O_3. Figure adapted from Maier[24] with permission.

The MX exhibits a Frenkel disorder. At large distances from the surfaces, the concentration of defects is given by the square root of the Frenkel constant.

TABLE 3

Examples of enhanced ionic conductivity in composites containing dispersoids of Al_2O_3. Kroeger-Vink[26] notation is used.

Electrolyte	Disorder Type	Most Mobile Defect in Pure Electrolyte	Species Attracted to Dispersoid	Most Mobile Defect in Composite	Reference*
LiI	Schottky	V_{Li}'	Li^{x}	V_{Li}'	20, 24
LiBr	Schottky	V_{Li}'	Li^{x}	V_{Li}'	32
AgI	Frenkel	$Ag_i^{\bullet}$	Ag_i'	V_{Ag}'	25
AgBr	Frenkel	$Ag_i^{\bullet}$	$Ag_i^{\bullet}$	V_{Ag}'	35, 36
AgCl	Frenkel	$Ag_i^{\bullet}$	$Ag_i^{\bullet}$	V_{Ag}'	33
CuCl	Frenkel	$Cu_i^{\bullet}$	$Cu_i^{\bullet}$	V_{Cu}'	20
CuBr	Frenkel	$Cu_i^{\bullet}$	$Cu_i^{\bullet}$	V_{Cu}'	34
CaF_2	anti-Frenkel	$V_F^{\bullet}$	F^{x}	$V_F^{\bullet}$	27, 28, 29, 30
TlCl	Schottky	$V_{Cl}^{\bullet}$	$V_{Cl}^{\bullet}$	V_{Tl}'	31

*The references are to publications reporting enhancements in ionic conductivity. The mechanisms reported therein are not necessarily those stated in Table 3. $CaF_2(Al_2O_3)$ has been reported to conduct primarily via F_i' in reference 28 and via $V_F^{\bullet}$ in references 29 and 30. Undoped CaF_2 conducts primarily via F_i' above 670°C[37].

In most examples cited, the positive species is attracted to the alumina surface. A negative species is then available for transport. Shulka et al.[38] using PbF_2 as an example suggest the enhancement increases with decreasing oxide dispersoid isoelectric point and increasing normal anion coordination of the oxide cation. Their prediction satisfactorily accounts for the relative enhancement factors measured for PbF_2 containing oxide dispersoids for which the enhancement was in decreasing order, $ZrO_2 > SiO_2 > CeO_2 > Al_2O_3$.

An interesting example is calcium fluoride, CaF_2, which conducts via fluoride ions[27-30]. CaF_2 exhibits the anti-Frenkel disorder with equal concentrations of anion vacancies, $V^{\cdot}_F$, and anion interstitials, F_i'. The activation energy for motion of $V^{\cdot}_F$ is 0.51 eV, while that for motion of F_i' amounts to 0.92 eV[30]. Accordingly, ionic current is attributed to fluorine ion vacancies in undoped CaF_2 at low temperatures[29,30]. Adding Al_2O_3 (or CeO_2 or ZrO_2) increases the ionic conductivity. Khandkar et al.[28] attributed this enhancement to F_i'. More extensive studies of Jacob et al.[29,30] attributed the enhancement to an increase in $V^{\cdot}_F$ near the dispersoid interface. It follows that the fluorine ions must be attracted to the interface. See Table 3.

Recently Brune and Wagner[39] reported results on $PbCl_2$ (Al_2O_3). Pure lead chloride exhibits a Schottky disorder and the ionic current is carried by the chloride ion vacancies, $V^{\cdot}_{Cl}$. When Al_2O_3 is added, positively charged species are attracted to the Al_2O_3. One would expect the dominant transport to change to that via the negatively charged cation vacancies, V_{Pb}', just as in the studies reported on TlCl[31]. However, the mobility ratio, u_{Pb}/u_{Cl}, at 220°C is 4.6×10^{-5}, and Brune and Wagner suggest that this ratio is too extreme to allow the cation vacancies to become the dominant ionic current carrier in the composites. Therefore, additions of Al_2O_3 result in a <u>decreased</u> ionic conductivity in $PbCl_2$. Moreover, for a given volume fraction of Al_2O_3, the smaller the particle (greater surface area), the <u>larger</u> is the <u>decrease</u> in ionic conductivity. These observations are consistent with the space charge layer mechanism.

Recently Petuskey[40] has performed an ingenious experiment on a predominantly electronic conductor, $Ag_{2+x}S$, containing Al_2O_3. Figure 5 shows a schematic diagram of his experimental set up.

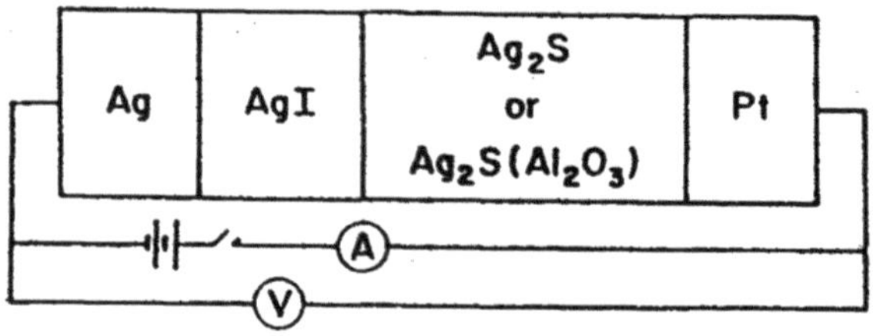

Fig. 5. Schematic diagram of the experimental setup of W. Petuskey[40]. The cell consisted of pellets of pure silver, silver iodide as the electrolyte and a pellet of Ag_2S orAg_2S (Al_2O_3). By means of passing current into or out of the Ag_2S or $Ag_2S(Al_2O_3)$ the amount of silver could be monitored (solid state coulometric titration) and the resulting voltage could provide a measure of the activity of silver. The effective homogeneity range of the composite was larger than that of the Ag_2S. See text.

He used solid state coulometric titration to add (or withdraw) silver into silver sulfide with or without a dispersed second phase of Al_2O_3. In the case of the silver sulfide containing 0.2 μ diameter Al_2O_3 particles, or 3.5 m^2 per cc of composite, it was found that at 158°C, the silver sulfide containing the Al_2O_3 exhibited an effective homogeneity range 120 times greater than silver sulfide without the alumina particles. The analysis was that of silver absorption at solid-solid interfaces. This experiment is consistent with the space charge argument put forth by Jow and Wagner[20] and by Maier [24].

Other evidence for the space charge layer concept is provided by the influence of alumina dispersoids on the temperature of a first order phase transition. In pure AgI, the α (high temperature phase) to β (low temperature phase) occurs at 147.5°C in AgI. Shahi and Wagner[41] reported that the introduction of Al_2O_3 resulted in a change in the α-β temperature, and Chowdhary et al.[42] carried out a systematic study of this effect. It was found that the β to α temperature was unaffected, but the α to β temperature decreased as the concentration of Al_2O_3 increased. See Figure 6. Furthermore, while the introduction of Al_2O_3 particles into the low temperature β phase increased the conductivity, the introduction into the α phase reduced the conductivity. See Figure 6. The hypothesis suggested was that in the β phase, the cation vacancies tend to remain near the interface. Thus, the silver iodide transition is determined by the defects in the bulk AgI and the transition temperature is "normal", occurring at 147.5°C. On the other hand, in the high temperature phase, the cation vacancies are more mobile and react with interstitials. This results in a decreased ionic conductivity. Furthermore, on cooling, there is a smaller concentration of lattice defects than in bulk AgI without Al_2O_3, and according to Rice, Strassler, and Tombs[43], a critical concentration of defects is necessary to cause the transition. Hence, the composite must be cooled below 147.5°C because there is not the concentration necessary to "drive" the transition, α to β. If this suggestion has merit, the change in transition, ΔT, should be proportional to the surface area of the added alumina as is shown in Figure 7. Z.Y. Zhao and C.Y. Wang[44] have carried out similar studies on AgI containing Fe_2O_3 particles with results that are in agreement with the data for AgI (Al_2O_3) described above. These results show clearly that the virtually insoluble dispersoid is not inert, but influences the concentration of defects in the electrolyte.

Most of the arguments described above are consistent with the space charge model.

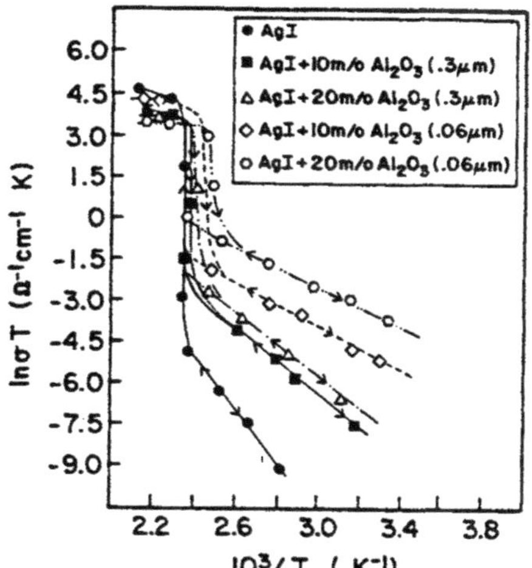

Fig. 6. Plot of lnσT vs $10^3/T$ for AgI(Al_2O_3). Note that there is enhancement for the composite in the low temperature β phase but a decrease in the high temperature α phase. Reprinted by permission of the publisher, The Electrochemical Society, Inc.[42].

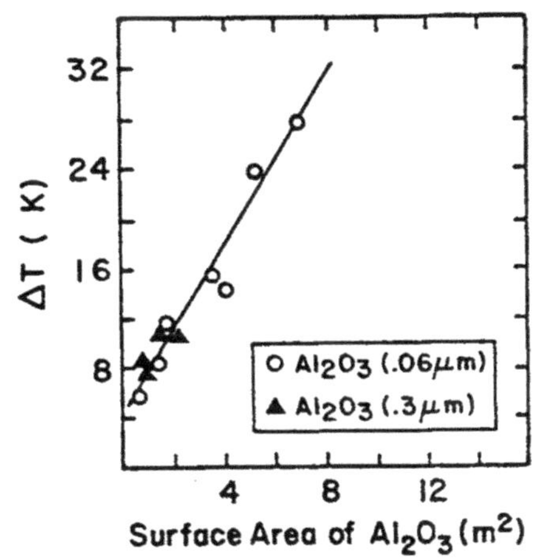

Fig. 7. Depression of the α to β phase transition temperature on cooling AgI(Al_2O_3) as a function of surface area per gram of AgI[42]. Reprinted by permission of the publisher, The Electrochemical Society, Inc.

Accordingly, a combination of both aliovalent or homogenous doping and heterogeneous doping provides additional opportunities to control the number and type of defects in a solid electrolyte. Maier and Reichert[31] used this concept to dope TlCl simultaneously with soluble $PbCl_2$ and insoluble Al_2O_3, both of which create cation vacancies, so the resultant conductivity was a "weighted superposition" of the two effects. Chowdhary[45] doped silver iodide aliovalently with Ag_2S or CdI_2 and simultaneously heterogeneously with alumina. In a more extensive study, Akila and Jacob[30] aliovalently doped CaF_2 with NaF_2 or YF_3 and simultaneously heterogeneously doped the resulting solid solutions with Al_2O_3 or with CeO_{2-x}. Homogeneous doping CaF_2 with NaF yields anion vacancies according to:

$$NaF \xrightarrow[CaF]{in} Na'_{Ca} + F_F + V^{\cdot}_F \qquad (3)$$

while the introduction of YF_3 yields interstitial fluorine ions according to:

$$YF_3 \xrightarrow[CaF]{in} Y^{\cdot}_{Ca} + F'_i + 2F_F. \qquad (4)$$

According to Jacob et al.[29,30], the heterogeneous doping of CaF_2 yields anion vacancies in the space charge layer. Therefore, adding Al_2O_3 to sodium doped CaF_2 results in an increase in ionic conductivity over that of the homogeneously doped sample. Conversely, adding Al_2O_3 to yttria doped CaF_2 results in the annihilation of F'_i owing to the $V^{\cdot}_F$ introduced by the insoluble Al_2O_3 with a consequent decrease in ionic conductivity.

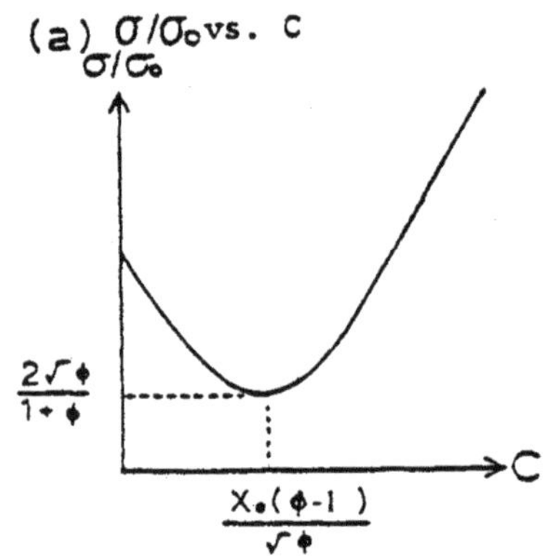

Fig. 8. Schematic diagram of the relative conductivity of homogeneously or aliovalently doped electrolyte, σ, to undoped electrolyte σ_0 for a compound MX exhibiting a Frenkel disorder and in which the mobility of the cation interstitial is larger than that of the cation vacancy. The dopant is a soluble divalent impurity in a monovalent host, for example CdI_2 dissolved in AgI. In this diagram, C denotes the concentration of dopant and ϕ denotes the mobility ratio, $\phi = u_I/u_V$.

K. Kitajima[46], following the studies of Chowdhary[45], combined the homo- and heterogeneous doping AgI containing CdI_2 and CdI_2 plus a dispersion of Al_2O_3. Figure 8 shows a schematic of the conductivity behavior on adding a doubly divalent cation of a Frenkel compound such as AgI, in which the cation interstitial is more mobile than the cation vacancy. According to the mechanism suggested by Shahi and Wagner[25], $AgI(Al_2O_3)$ conducts via cation vacancies. Thus, adding Al_2O_3 to AgI or to AgI doped with CdI_2 increases the concentration of cation vacancies available for transport.

Figure 9 shows the behavior of AgI containing no dispersion and AgI with a dispersion, both as a function of the homogeneous doping concentration of CdI_2. Clearly this is an example of the combined effect of homo- and heterogeneous doping. By means of both effects, the conductivity and conducting species can be controlled.

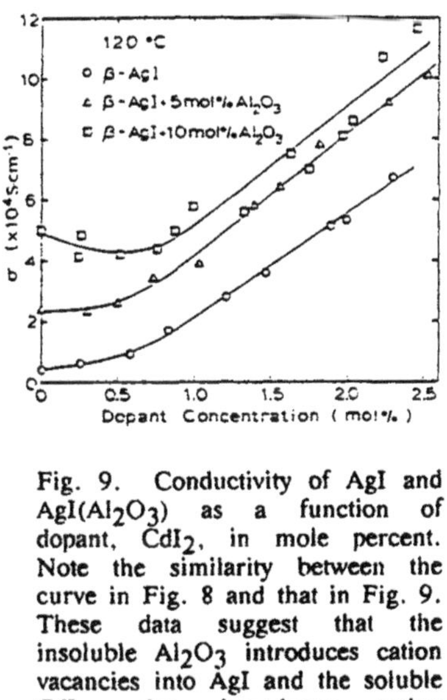

Fig. 9. Conductivity of AgI and $AgI(Al_2O_3)$ as a function of dopant, CdI_2, in mole percent. Note the similarity between the curve in Fig. 8 and that in Fig. 9. These data suggest that the insoluble Al_2O_3 introduces cation vacancies into AgI and the soluble CdI_2 also introduces cation vacancies. This is an example of using both homo- and heterogeneous doping simultaneously[46].

One of the parameters yet to be decided is how the enhanced transport occurs when the space charge mechanism is operative. One suggestion is that the dispersoids line up along grain boundaries.

Thus, in Figure 2, all of the dispersoids would be in the grain boundary regions and none would exist in the interior of the grains. This model was first suggested by Maier [24]. The usual inference has been that the space charge region exists over a length equal to one or two Debye lengths and the concentration of lattice defects in the matrix far away from the dispersoid is that of the bulk matrix given by the square root of the appropriate equilibrium constant. In contrast, it may be that the space charge layers overlap one another. Recently Asai et al.[47] have reported that neither the surface area measured by the BET method nor the particle size showed any correlation with thickness of the interfacial layer in LiI (Al_2O_3) composites. The thickness of this layer and the dependence on acid or basic nature of the dispersoid is an area needing much attention.

In some composites, the addition of dispersoids does not change the transport mechanism, but only the number of lattice defects. This is the case for the lithium halides which exhibit a Schottky disorder. Adding Al_2O_3 particles simply increases the concentration of cation vacancies. Other composites exhibit a change of the predominant transport mechanism such as is the case for the silver salts (Frenkel disorder). In the latter case, the composite consists of the dispersoid, the highly conducting region in the space charge layer wherein transport is via cation vacancies and the bulk matrix in which transport is via the more mobile silver interstitials. One consequence of this is the existence of a V'_{Ag} - $Ag_i^{\bullet}$ junction in analogy to a n-p junction as reported by Maier[48-49].

Finally, we mention the case of two immiscible phases in equilibrium with one another. The system studied was the AgBr-AgI system [9,10,50-52]. At room temperature there is a miscibility gap. Electrical conductivity measurements showed an enhanced conductivity in the two-phase regime. The grain size of the solid solution particles was not determined. The conductivity of the two phase samples at 25°C is shown in Figure 10. Note that the conductivity is enhanced over that of the two terminal solid solutions, the maximum occurring at ~23 m/o AgBr $(I)_{ss}$.

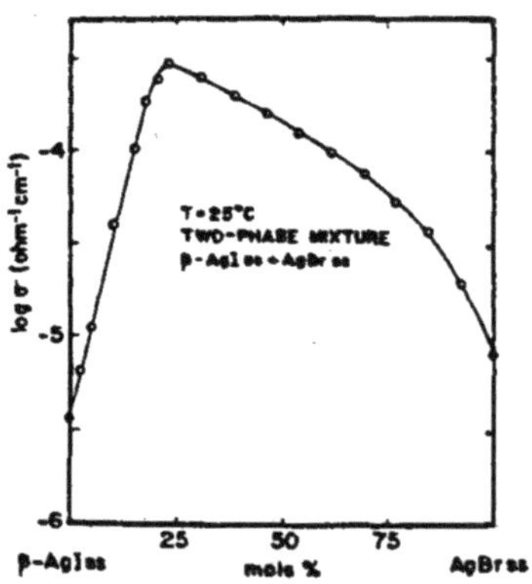

Fig. 10. Logarithm of conductivity as a function of composition (mole %) at 25°C for the two-phase mixtures of β -$AgI_{0.95}Br_{0.05}$ - $AgBr_{0.7}I_{0.3}$. Note the maximum occurs at ~ 23 m/o $AgBr_{0.7}I_{0.3}$. Reprinted by permission of the publisher. Pergamon Press Ltd.[52].

Calorimetric studies were also carried out on these samples. The enthalpy increment, $[H_{518} - H_{298}]$ was determined using drop calorimetry. This enthalpy increment is the sum of:

(i) $[H_{421} - H_{298}]$

(ii) ΔH_{421}

and

(iii) $[H_{518} - H_{421}]$

It was shown from ancillary calorimetric data that the terms (ii) and (iii) were much smaller than (i). The difference between:

$$[H_{518} - H_{298}]_{\substack{\text{melted}\\\text{sample}}} - [H_{518} - H_{298}]_{\substack{\text{mechanical}\\\text{mixture}}} \quad (5)$$

was obtained and is plotted in Figure 11. If only the bulk phases were considered, then the difference in the enthalpy increments given in Equation 5 should be zero. The non-zero difference indicates an influence of the surfaces, an influence that Khandkar et al.[10] attributed to an increase in concentration of cation vacancies. Furthermore, the fact that the enthalpy increment exhibits a maximum at ~ 50 m/o while the conductivity exhibits a maximum at ~ 23 m/o indicates that not only is there an influence of the surfaces (increased concentration of lattice defects), but the mobilities may also be affected.

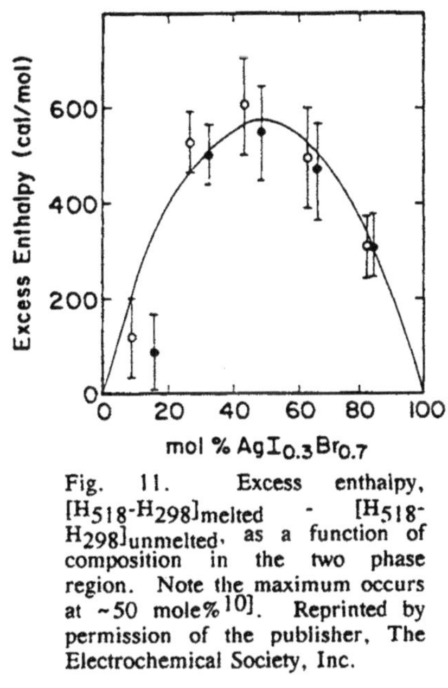

Fig. 11. Excess enthalpy, $[H_{518}\text{-}H_{298}]_{melted} - [H_{518}\text{-}H_{298}]_{unmelted}$, as a function of composition in the two phase region. Note the maximum occurs at ~50 mole%[10]. Reprinted by permission of the publisher, The Electrochemical Society, Inc.

At present, we are testing the generality of this effect by studying the conductivity at 144°C in the CuI-CuCl miscibility gap[53]. So far an enhancement analogous to that for the AgI-AgBr system has been found. More work on the influence of particle size (surface area) in these two-phase systems is needed and an assessment of the temperature range in which negligible grain growth occurs for systems of interest.

The enhancement in ionic conductivity in biphasic composite electrolytes has been clearly demonstrated for a variety of systems. No single mechanism explains all the results. However, the space charge model has been shown to be operative and dominant in many composites. From a practical point of view, the enhanced ionic conductivity means that the electrolytes can be used at lower temperatures or for higher current drains. But if the reaction product of a cell, for example, forms without incorporating the second phase, then the separator involves a composite in series with an electrolyte (the reaction product) without the second phase. Such difficulties might be surmounted by incorporating the second phase into the cathode material (for a reaction product formed by transport of the cation to the cathode).

REFERENCES

1. C.C. Liang, J. Electrochem. Soc., 120 1289 (1973).
2. C.C. Liang, U.S. Patent 3541 124 (1973).
3. K. Kiukkola and C. Wagner, J. Electrochem. Soc., 104 308 (1957; ibid. 104 379 (1957).
4. J.T. Kummer in Prog. Solid State Chemistry edited by H. Reiss and J. O. McCaldin, Pergamon Press, N.Y. (1972) p141.
5. R. Collongues, J. Thery, and J.P. Boilot in Solid Electrolytes General Principles, Characterization, Materials, Applications, edited by P. Hagenmuller and W. van Gool, Academic Press, N.Y. (1978) p. 253.
6. M.B. Armand, J.M. Chabagno, and M.J. Duclot in Fast Ion Transport in Solids edited by P. Vashishta, J. N. Mundy, and G. K. Shenoy, Elsevier North Holland Inc., Amsterdam (1979) p. 131.
7. D. Ravaine and J.L. Souquet in Solid Electrolytes, General Principles, Characterization, Materials, Applications, edited by P. Hagenmuller and W. van Gool, Academic Press, N.Y. (1978) p. 277.
8. C. Angel, this volume.
9. K. Shahi and J.B. Wagner, Jr., J. Phys. Chem. Solids 43 (8) 713 (1982).
10. A. Khandkar, V.B. Tare, A. Navrotsky and J.B. Wagner, Jr., J. Electrochem. Soc. 131 (11) 2684 (1984).
11. J. Maier, Ber. Bunsenges Phys. Chem. 90 26 (1986).
12. N.J. Dudney, J. Am. Ceram. Soc. 70, 65 (1987).
13. See, for example, D.A. Jones and J.W. Mitchell, Phil. Mag. 3 1 (1958).
14. See, for example, J.S. Kirkaldy and D.J. Young, Diffusion in the Condensed State, Institute of Metals, London (1987).
15. N.J. Dudney, Bull. Am. Phys. Soc. 33 no.3 (1986) Abstr. A11 p. 227.
16. S. Pack, B. Owens and J.B. Wagner, Jr., J. Electrochem. Soc. 127 (10) 2177 (1980).
17. R. Dupree, J.R. Howells, A. Hooper and F.W. Poulsen, Solid State Ionics 9&10 131 (1983).

18. A.S. Stoneham, E. Wade and J.A. Kilner, Mat. Res. Bull. 13 661 (1979).

19. R. Landauer, J. Appl. Phys. 23 779 (1952).

20. T. Jow and J.B. Wagner, Jr., J. Electrochem. Soc. 126 (11) 1963 (1979).

21. K. Lehovec, J. Chem. Phys. 21 1123 (1953).

22. K.L. Kliewer and J.S. Koehler, Phys. Rev. 140 1226 (1965).

23. K.L. Kliewer, J. Phys. Chem. 88 1057 (1984).

24. J. Maier, Ber. Bunsenges. Phys. Chem. 88 1057 (1984).

25. K. Shahi and J.B. Wagner, Jr., J. Electrochem Soc. 121 (1) 6 (1981).

26. F.A. Kroeger and H.J. Vink, Solid State Physics, Vol. 3, edited by F. Sietz and D. Turnbull, Academic Press, New York (1956) p. 307.

27. T.L. Wen, R.A. Huggins, A. Rabenau and W. Weppner, Revue de Chim. Minerale, 20 643 (1983).

28. A. Khandkar, V.B. Tare and J.B. Wagner, Jr., Revue de Chim. Minerale. 23 274 (1986).

29. N. Vaidehi, R. Akila, A.K. Stukla and K.T. Jacob, Mat. Res. Bull. 21 909 (1986).

30. R. Akila and K.T. Jacob, Solid State Ionics 25 (2,3), 217 (1987).

31. J. Maier and B. Reichert, Ber. Bunsenges. Physik. Chem. 90 666 (1987).

32. A. Khandkar and J.B. Wagner, Jr., Solid State Ionics, 20 267 (1986).

33. A. Khandkar and J.B. Wagner, Jr., Abstract 833 p 1236, The Electrochemical Society Extended Abstracts, Vol. 83-1, San Francisco, CA May 8-13 (1983).

34. P.M. Dubec and J.B. Wagner, Jr., Mat. Lett. 2 (6A&B), 502 (1984).

35. J. Maier, Solid State Ionics 18&19, 1141 (1986).

36. J. Maier, Mat. Res. Bull. 20, 383 (1985).

37. W. Bollman and R. Reimann, phys. stat. sol. (a)16 187 (1983).

38. A.K. Shulka, R. Manoharan and J.B. Goodenough, Solid State Ionics, 26 5 (1988).

39. A. Brune and J.B. Wagner, Jr., Solid State Ionics, 25 165 (1987).

40. W. Petuskey, Solid State Ionics 21 117 (1986).

41. K. Shahi and J.B. Wagner, Jr., Solid State Ionics 3/4, 295 (1981).

42. P. Chowdhary, V.B. Tare and J.B. Wagner, Jr., J. Electrochem. Soc. 132 (1) 123 (1985).

43. M.J. Rice, S. Strassler and G.A. Tombs, Phys. Rev. Lett. 32 596 (1974).

44. Z.Y. Zhao and C.Y. Wang, Solid State Ionics, 9&10, 1175 (1983).

45. P. Chowdhary, Masters Thesis, Arizona State University, May 1985.

46. K. Kitajima, Ph.D. Thesis, Arizona State University, December 1987.

47. T. Asai, C-H Hu and S. Kawai, Solid State Ionics 26 1 (1988).

48. J. Maier, Ber. Bunsenges. Phys. Chem. 89 355 (1985).

49. J. Maier, J. Electrochem. Soc. 134(67) 1524 (1987).

50. K. Shahi and J.B. Wagner, Jr., Appl. Phys. Lett. 37(8), 757 (1980).

51. K. Shahi and J.B. Wagner, Jr., Solid State Ionics 3/4, 295 (1981).

52. K. Shahi and J.B. Wagner, Jr., J. Phys. Chem. Solids 43(8) 713 (1982).

53. Z. Y. Zhao and J.B. Wagner, Jr., to be published.

PROTON CONDUCTION IN SOLIDS

FINN WILLY POULSEN
Metallurgy Department
Risø National Laboratory
DK-4000 Roskilde, Denmark

1. TRENDS IN PROTON CONDUCTOR RESEARCH

This chapter deals with solid inorganic materials, in which at least part of the ionic conductivity has been ascribed to protons or protonic species such as oxonium-, ammonium- or hydroxyl ions.

A phenomenological approach to solid state protonic conductors is chosen,- and similarities rather than differences are emphasised. The review should primarily be considered a readers guide to the rapidly growing literature on solid proton conductors.[1)]

Proton conducting solid (SPC) materials were initially studied in their own right. The oldest reference to a solid state electrolysis experiment, where hydrogen was detected as reaction product, seems to be the electrolysis experiment of Kakuichi in 1947 on cetylalcohol,[2)] which has a rather low conductivity, i.e. 10^{-8} S/cm at 40 °C. The first conductivity determination on a SPC could well be the measurements on ice by Ayrton and Perry in 1877.[3)]

Typical materials, studied in the period before 1975, were substances such as ice, KH_2PO_4 in the paraelectric phase[4,5)] and organics like oxalic acid and imidazole.[6,7)] Conferences devoted entirely to the physico-chemical aspects of ice are still important, cf. the proceedings that are published as special issues of Journal de Physique.[8)] The conduction process in ice is still under debate. The research into solid protonic conductors (SPC's) between 1975 and ~1985 was motivated by an urge to find protonic conductors for practical applications, notably for an all solid state fuel cell. The search for high and long term stable conductivities, rather than studies of minute details of the conduction process was now emphasized. The goal was to produce a proton conducting material that could be processed into thin, strong membranes,- the material should at the same time be chemically stable in the 200 - 400 °C region, where relatively cheap (non-noble metal based) fuel- and oxygen electrodes supposedly could be found.

The materials studied in this period were mostly inorganic in nature, such as HUP found by M.G.Shilton and A.Howe,[9)] and the hydronium/ammonium β-aluminas.[11,12,115)] The materials with the highest bulk protonic conductivities were found in the group of heteropolyacid hydrates, e.g. $H_3PW_{12}O_{40}\cdot 28\ H_2O$, which has a conductivity at 25 °C of $8\cdot 10^{-2}$ S/cm.[13,14] The close connection between ion exchange properties of solids and solid electrolyte behaviour was stressed by England et al.,[15)] who were also amoung the first to describe proton conduction of the particle hydrates, such as ZrO_2,nH_2O, cf. Figure 1. Apart from finding high conductivity materials in the heteropolyacid hydrate group,[13,14)] the efforts were by and large not crowned with success: high and stable conductivities above 100 °C are seldomly reported; as proved after careful dc-studies, by for instance E.Skou and coworkers,[10)] proton conduction in many materials could often be attributed to residual liquid electrolyte solution in the grain boundaries. The distinction between solid compounds, where water content can be varied continuously, like in clays, zeolites and particle hydrates, and heterogeneous water containing systems is not easy. The most successful " solid " protonic membrane for fuel cell use is the Nafion membranes used in kW-size fuel cells in the US Gemini space program. Nafion is a perfluorinated polymer with sulphonic acid groups.[16)] The membrane must be soaked with 30 w% water in order to be conductive and is as such not at all a solid electrolyte. The upper operation temperature for Nafion is approximately 150 °C .

The years after 1985 have revealed a more relaxed attitude among the researches. Specific conductivities in the 10^{-2} - 10^{-1} S/cm range are unlikely to be achieved for SPC's in the 200-400 °C regime, one therefor has to think of using SPC's in potentiometric and low current drain devices only.

An exciting development in SPC's has taken place in the field of $SrCeO_3$ - type perovskite materials, which behave as SPC when equilibrated with hydrogen or water containing atmospheres.[17-24] The various research groups have arrived at rather conflicting conclusions concerning the nature of the conductive species, be it H^+,[17-22] or O^{--} and OH^-.[24]

A new type of mixed materials have also appeared in the last few years, examplified by mixtures between anhydrous acids (e.g. H_3PO_4) and PEO or PEI (polyethylene oxide/imide).[25-29]

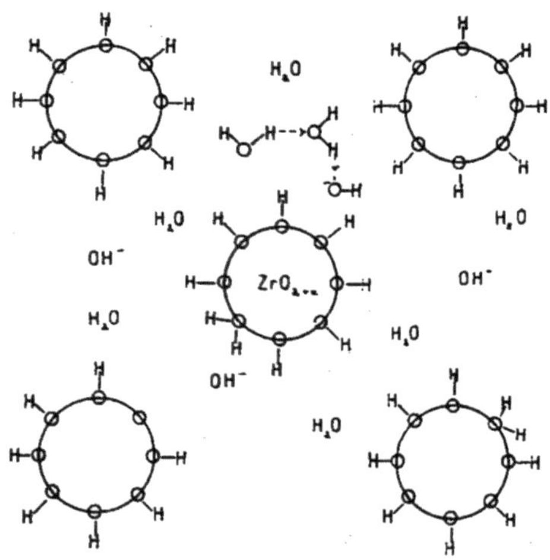

Figure 1. Model of a particle hydrate, $ZrO_2 \cdot nH_2O$. An illustration of the typical schism in proton conductor research: is the proton conduction a solid state process through the zirconia particles or does it only take place in the liquid electrolyte layers between the particles ? The oxide particles may be as small as 25 Å, as deduced from line broadening in x-ray diffraction. England et al.(1980).[15]

2. DEFINITION, ABBREVIATIONS AND SYNONYMS

A number of frequently encountered abbreviations and synonyms,- many of them are special of the solid proton conductor (SPC) literature, is given in Table 1.

A universal definition of a solid state protonic conductor is not at hand. Here, as in the SPC- literature in general, a material is named a SPC as long as protons can be transferred through the (at least from a practical point of view) solid material, and are converted into hydrogen gas at the cathode. Note that metal hydrides, which contain H^- ions, develop hydrogen at the anode! when subjected to electrolysis. It is only when hydrogen evolution can be maintained over long times - with due supply of protons of some form at the anode, that the hydrogen evolution can be taken as a "proof" of protonic conductivity. Crystal water or other protonic species in the vicinity of the cathode can be electrolysed as a parasitic phenomenon, but this side reaction would stop by itself due to depletion of protonic species, or the experiment would stop due to loos of contact at the cathode.

The conducting species may be protons, oxonium ions, ammonium ions, hydrazinium ions or hydroxyl ions, cf. Table 2 for typical examples.

Proton conduction has been found as a bulk property in crystalline, amorphous, inorganic, organic, and polymeric materials; proton conduction is furthermore also a surface property of certain materials, cf. Table 2.

3. BIBLIOGRAPHY

The present literature review is based on literature that has appeared since the comprehensive review (566 references) by L.Glasser in 1975.[64)]

In many cases reference is only given to an authors more recent contributions in a given field. In order to fully appreciate the achievements of the respective authors the reader <u>must</u> consult the quoted literature for further references.

Item	Explanation	Reference
αZrP	α- $Zr(HPO_4)\cdot H_2O$	30-37
β-alumina	ion exhanged Na- or K-β-alumina	11,12,38,42
Bjerrum defects	Two protons (D) or none (L) in O - O bond in ice	39,40
CER	ceramic electrochemical reactor	21
dioxonium	dihydrated proton, $H_5O_2^+$	41
FAIE	Field assisted ion exchange	42
Grotthuus mechanism	proton transfer coupled to reorientation movement of H_2O	43,44
HUAs	hydrogen uranyl arsenate tetrahydrate	9,10,43
HUP	hydrogen uranyl phosphate tetrahydrate	43,45-48
Hydride ion	H^-; ionic radius (Pauling) = 1.54 Å	
Hydrogen bond	proton between two electronegative atoms	
Hydronium ion	pyramidal H_3O^+	
IINS	inelastic incoherent neutron scattering	49,50
KDP	Potassium hydrogen phosphate	4,5
Nafion	polyperfluorosulfonic acid	16
PEC	Photo electro chemical cell	51,52
PEM	Proton exchange membrane (e.g. Nafion)	16
PFG NMR	Pulsed Field Gradient NMR	53,54
PMR	Proton magnetic resonance (NMR)	53,54
proton	elementary particle; radius = 10^{-5} Å	
proton injection	Electrolysis with hydrogen anode	55
SCP	Superconducteur protonique (French)	
SPC	solid proton conductor	
SSPC	solid state protonic conductor	
SPFC	solid proton conductor/polymer fuel cell	
Vehicle mechanism	Migration of H_3O^+ coupled to counter diffusion of water molecules	56

Table 1. Abbreviations, neologisms and synonyms frequently used in the literature on solid state proton conductors.

Characteristic	System	Reference
<u>Bulk conductors</u>		
crystalline	$H_3PW_{12}O_{40}\cdot 28H_2O$	13,14
polymeric	H_3PO_4/PEO	25-29
<u>Surface conductors</u>	α-ZrP,hydroxyapatite	33,60
<u>Liquid like systems</u>	Nafion	16
<u>Suggested mobile species</u>		
H^+	$SrCeO_3$,HUP	17-24,9-10
H_3O^+	β-alumina	11,12,38
OH^- or OH^-/H_2O exchange	$[Zn_2Cr(OH)_6]OH,nH_2O$	61
NH_4^+	β-alumina,zeolites	62,63

Table 2. Diversity of proton conducting solids. A typical example of each type is given.

A rather high number of references to publications, which were presented at the 6'th International Conference of Solid State Ionics, has been included in an attempt to make the review as up to date as possible. These 17 papers will appear in volume 27-28 of the 1988 edition of Solid State Ionics.[1)]

A survey is given in Table 3 of review type articles on solid state protonic conductors that has appeared since 1972. Monographs dealing with SPC's do not exist; SPC have been the subject of numerous thesis studies. The Ph.D. thesis by J.D.deNuzzio [71)] on protonic β-aluminas from 1986 is worth while mentioning in this context.

The small review report[66)] of the author from 1980 has been cited literally[69)] in English and parts have been translated word by word into Italian [72)]. This illustrates the need for up to date (new) reviews.

Proton conductors have had their own series of conferences (I-III), starting with the 1981 meeting in Paris, which was a result of a French-Danish- British initiative. The fourth meeting on SPC's is to be held in Exeter, UK September 1988. The three proceedings published sofar by Odense University Press, Denmark, should be considered as major sources of information in the SPC-area.[57-59]

None of the reviews (including the present) are nearto as thorough as the review of Glasser.[64]

Extensive listings of room temperature conductivities and activation energies for SPC's are to be found in ref.s 57-59, 64, 67, 69.

Reviewer	Year	Number of references	Reference
J.Bruinink	1972	51	65
L.Glasser	1975	566	64
F.W.Poulsen	1980	42	66
W.A.England et al.	1980	24	15
anon	1981	202	57
H.Arribart	1981	92	67
A.Potier	1983	22	41
T.A.Wheat et al.	1985	31	68
S.Chandra et al.	1987	48	69
K.-D.Kreuer	1987	34	70

Table 3. Earlier review type publications in the field of solid state protonic conductors.

4. THEORY

4.1 Bonding situations and transport models

As a result of the small radius of the free proton, ~10-5 Å, "isolated" H^+ ions are not present in solids under equilibrium conditions. Due to its strongly polarising power, H^+ is covalently bonded to one or two electronegative atoms/ ions of the surroundings.

Tunnelling effects may explain some of the low temperature conductivity of SPC's, eg. in $HTaWO_9 \cdot H_2O$,[73] but at temperatures from room temperature and upwards, thermally activated processes will completely dominate the conduction phenomena. An excellent discussion of different transport modes has been given by Kreuer, Rabenau and Messer.[43]

At least three different bonding situations for the proton are relevant:

i) the acceptor site for the proton may be an ion of the immobile lattice, e.g. a hydroxogroup has formed; or

ii) the proton may be attached to a mobile ion, e.g. an oxygen ion, thus forming a hydroxyde ion, OH^-; or

iii) the proton is attached to a mobile molecule, such as the water or ammonia molecule, thus forming a hydronium, respectively an ammonium ion.

Figure 2 shows a calculated conformation of the $H_5O_2^+$ ion.[76] Two positions are indicated for the central proton. The dihydronium ion is found in solid $HClO_4 \cdot 2H_2O$, $HFe(SO_4)_2 \cdot 4H_2O$ and many other acidic substances.[74,75,101]

The crucial problem to be solved (by nature as well as by the scientist) for a typical SPC containing ions such as $H_5O_2^+$, is the following: - can the ion migrate as a whole? If not,- how does an ion deliver a proton to the surroundings in the direction of the field, and how is the ion reconstructed after the two water molecules have accepted a proton from the opposite direction? Even for an isolated hydrogen bonded system as simple as $H_5O_2^+$, the rearrangement steps are quite complicated, involving at least rotation around both oxygen atoms. A.Potier and coworkers have been very active in describing proton conduction using vibrational spectroscopy, structure determinations, and quantum chemical calculations.[41,74,76]

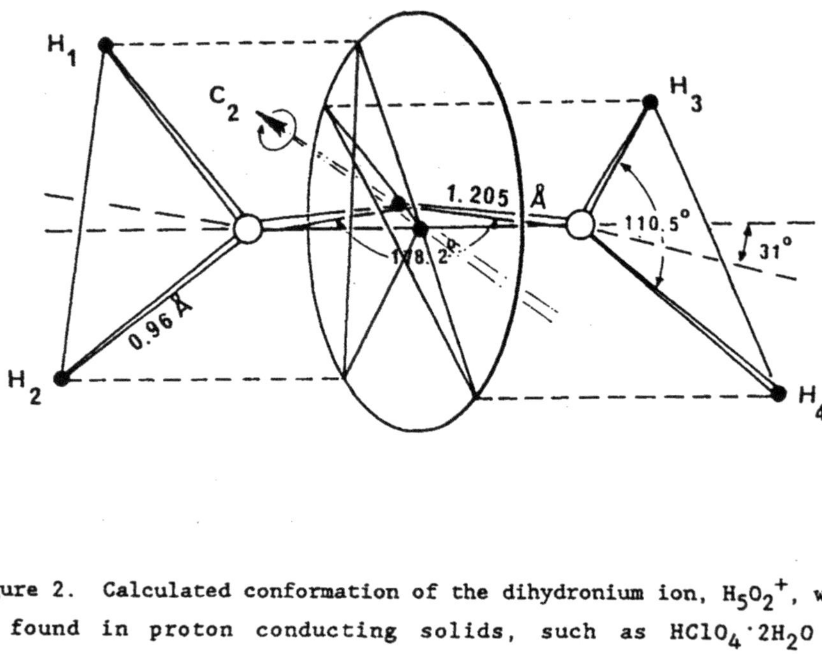

Figure 2. Calculated conformation of the dihydronium ion, $H_5O_2^+$, which is found in proton conducting solids, such as $HClO_4 \cdot 2H_2O$ and $H(Fe/In)(SO_4)_2 \cdot 4H_2O$. If this ion can not diffuse as a unit, how does proton conduction then take place in materials containing such ions? From A.Potier (1983).[76]

4.2 Single proton jumps

In the first bonding situation, i), proton conduction can only take place by a simple hopping mechanism of the proton to an adjacent acceptor site, followed by a rotation of the new H-acceptor bond in the direction of the field. The mechanism may be described as "point defect migration ". In cases with very high conductivities, the SPC's are denoted "superionic protonic conductors", in analogy to systems with a disordered cation sublattice with rapid exchange of cations between filled and empty sites. $CsHSO_4$ in its high temperature phase above 144 °C has a cation conductivity of 10^{-2} S/cm, which is comparable to the conductivity of the molten compound.[77-79] At present it appears, however, that the cesium - ions are responsible for most of this conductivity.

The situation in the cubic high temperature modification of KHF_2 differs in that only potassium and fluorine ions are mobile, whereas the protons are believed to be the majority carriers in the poorly conducting low temperature form.[80]

The proton may be attached to a polyatomic ion in the lattice, e.g. PO_4^{---}, which has an extra degree or rotational freedom. If the jump-reorientation mechanism takes place in a concerted manner, one can imagine a very efficient charge transport over long distances in a single step, a mechanism which is called the Grotthuss mechanism (see Figure 3A). The proton attains its highest degree of "ionicity", eg. being closest to a free H^+-ion state, when it is midway in a jump between the donor site and the acceptor site.

In high temperature (> 500 °C) SPC's such as proton containing $SrCeO_3$,[17-23] and Y_2O_3,[81] simple hopping conduction is the most likely mode of transport.

4.3 Correlated jumps of protons

Grotthuss-type conduction has often been mentioned as a possible explanation for the fact that protons in ice have a higher mobility than in water. Slade et al. have modelled the temperature dependence of their NMR data obtained on 12-tungstophosphoric acid 14-hydrate assuming coupled reorientation of the water molecules.[54]

It is interesting to note, however, that ice is still the only convincing case of Grotthuss-type conduction[44] in solids that has been demonstrated since the idea was advanced in 1805 !

4.4 Migration of polyatomic ions

When the proton is attached to a mobile unit, i.e. case ii) and iii) two modes of transport have been suggested:

Polyatomic ions, such as H_3O^+ and NH_4^+, may migrate by simple jumps from site to site in sufficiently open structures, as has been claimed for the β-aluminas, [62)] and in zeolites.[63,82)] Polyatomic ion transport can only be maintained under dc-conditions if the molecule or molecule ion, which is set free as a side product in the cathodic production of hydrogen, is also supplied at the stoichiometrically required amount at the anode side.

A more involved mechanism based on the simultaneous diffusion of two types of polyatomic units, the socalled " vehicle " mechanism, was suggested by Kreuer et al. in 1982,[56)] see Figure 3B. The proton travels through the lattice on a "vehicle", which may itself be neutral or charged. There must be a counter diffusion of the vehicles, i.e. H_3O^+ in one direction and the vehicle, H_2O, in the opposite direction. Such a mechanism is not likely in cases, where vehicle and vehicle + proton have charges of same sign, as for example in the OH^- /O^{--} case.

Figure 3. Concerted transport of protons in solids (and liquids) as pictured by Kreuer et al. (1982).[56] 3A shows the postulated Grotthuus mechanism; 3B depicts the vehicle transport model, where H_3O^+ ions moved towards the left and the H_2O vehicles moves towards the right. See section 4.4 for further details.

Mechanisms involving polyatomic ions can in many cases be ruled out from structural considerations. The bottlenecks (openings in the structure) in a unit cell and between adjacent cells can be estimated from electron density maps for known structures. Migration of, say protonated phosphate groups in KDP, or a protonated perchlorate group in solid perchloric acid hydrates then proves to be sterically impossible.

Positive indications for transport mechanisms involving polyatomic species are:

1) The absence of an isotope effect, when the conductivity of a protonic and fully deuterated version of a compound is compared. If H^+ and D^+ are the mobile species, then the conductivities should to a first approximation scale as $\sqrt{2}:1$. (If a larger ratio was to be observed, all other things being equal, then tunnelling effects or differences in the defect chemistry must be invoked). If, on the other hand, a very small or no isotope effect is present (ratio of conductivities = 1), then it can be assumed that the proton/deuteron is part of a larger mobile unit, whereby the mass difference " drowns". The absence of an isotope effect in heteropoly tungstic acid $H_3PW_{12}O_{40}\cdot 28\ H_2O$ has been shown by Kreuer et al.[14]

2) The presence of H_3O^+ or NH_4^+ ions in samples made by ion exchange of a metal ion compound. (Hydronium and ammonium ions can also form "in situ", if H_2O or NH_3 are already in the structure.)

4.5 Surface and Liquid Like Transport

Surfaces of crystals and grain boundaries may have disordered regions, where proton transport is facilitated. Adsorbed water or electrolyte solution trapped between the grains will also increase the conductivity and give rise to false conclusions concerning the solid state nature of the conduction. Surface and liquid-like conduction normally has a low activation energy for conduction, in the range from 0.1 -0.4 eV, thus approching the activation energy in acidic aqueous solutions of 0.1-0.15 eV.[83]

4.6 Predictions of Highly Conductive SPC's

Under this heading nothing but handwawing remarks can be offered: There is no quantitative theory available for predicting structures, mobilities of ions, thermal stability etc. of any type of solid electrolyte, including hypothetical SPC's. A few hints to the existence of good proton conductors will be set forward:

Classical hop conductivity should be most effective in structures, where comparable concentrations of bonded protons and empty acceptor sites are present. Correlated Grotthuss -type movement should be able to operate in structures, where all sites are occupied. In general it can be expected that good proton conductors have protons as major constituents of the structure.

Proton conduction in extended hydrogen bonded structures are most likely found in layer structures, leading to 2-D conduction. Effective 3-D conduction is favoured on " the surface" of macro-anions, such as the heteropoly tungstates.

If protons are able to survive in the structures at high temperature (see 6.4), any mechanism (in 1D,2D or 3D) will give rise to a high conductivity as a result of the high temperature. On a speculative basis one could imagine that incommensurate structures would be ideal. Incommensurate structures would originate if the characteristic repeat unit of the hydrogen bonded system does not match with the lattice constants of the immobile part of the structure. Protons could constantly be on the verge of changing positions in such a system.

It is most likely that good SPC's are found among amphoteric and acidic materials. Protons are strongly bonded in basic materials and the hydroxyle ion does not easily move. Only a few basic SPC compounds are known, cf. Table 2. The simple basic materials, NaOH and KOH, are also good proton conductors in their cubic (NaCl-type) high temperature phases, which are stable above 293 and 248 °C, respectively.[84] The conductivities are in the 10^{-2} to 10^{-1} S/cm range, but such materials are extremely difficult to handle.

5. CHARACTERISATION METHODS

Characterisation of a proton conducting solid is a task, where electrochemistry, electron microscopy, spectroscopy, and structure determinations must go hand in hand.[29] A recent example of a multi-instrumental investigation (IR, Raman, DSC, conductivity, as well as inelastic and quasieleastic neutron scattering) is given by Ph.Colomban et al. in their study of the conductivity and dynamics in the various polymorphs of $CsHSO_4$.[77] See also the work on PEO-based electrolytes.[29]

A recent summary of useful methods for the characterisation of solid electrolytes has been given by C.N.R.Rao.[85]

Barboux, Morineau and Livage have demonstrated the usefulness of comparing conductivity isotherms as function of water content with Brunauer adsorption isotherms. HUP, V_2O_5-xerogel and $Ce(HPO_4)_2 \cdot 3H_2O$ are shown to be typical cases of framework hydrates, swelling hydrates and particle hydrates, respectively.[86]

The salient features of SPC's are the electrolytic properties, which however are anything but trivial to measure. There are more studies devoted to characterisations by x-ray and neutron diffraction, NMR and vibrational spectroscopy- methods, where the requirement to a perfect (mechanically) sample can be relaxed somewhat.

Conductivity measurements and two spectroscopical methods, which rely on the intrinsic nature of the proton, are the only methods to be treated here.

5.1 Conductivity Measurements

The measurement of the ionic conductivity of SPC's is plagued by a low mechanical strength of samples, and a high sample porosity (sintering can not be performed). The only advantage of the porous nature of many SPC's is the ease of equilibration with water containing atmospheres (sensor applications). Guard electrode techniques for elimination of surface conductivity are not useful on porous SPC's . The "iso-conductance " method is a fruitful method to measure conductivity of powders and porous materials. The sample is immersed into a liquid electrolyte, the conductivity of which is changed until it matches the conductivity of the sample.[87)]

Heating of a SPC in a water lean atmosphere often leads to an irreversible change of the material.

Table 4 gives an impression of the diversity of electrode materials that has been used for dc-experiments (in ice literature called ohmic electrodes, as opposed to blocking electrodes). Figure 4 shows an experiment designed to determine the amount of water liberated during electrolysis of an acid hydrate. In the case of long term dc-electrolysis on HUP no water was liberated on the cathode side (no weight increase of the P_2O_5- trap,[88)] the proton and not the hydronium ion was therefor appointed the mobile species in HUP. Ac-measurements are usually restricted to frequencies below 1MHz. Conductivities determined by dc and ac -measurements (<1MHz) should therefor be identical.

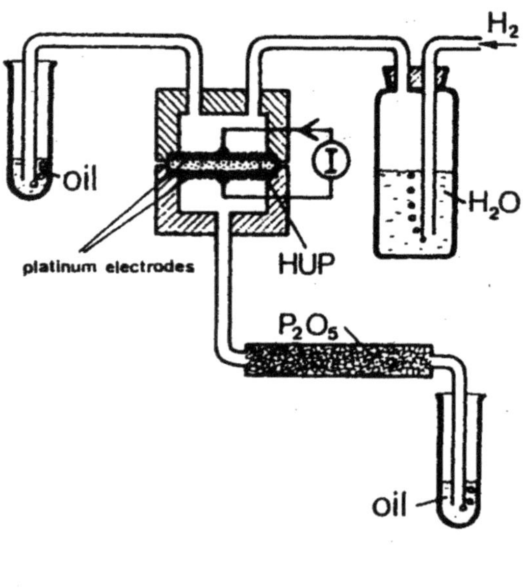

Figure 4. Setup for long term dc-conductivity measurements, according to Lundsgaard et al. (1983).[88)] If the protons are transported in the form of H_3O^+ ions and water is set free at the cathode side, then the P_2O_5-trap will gain in weight.

England et al.[15]) suggested that local movement of protons on a surface of a particle (but no long distance transport) may give rise to a high ac-conductivity but a low dc conductivity. There is no reason for this worry, provided that bulk and grainboundary conductivities are well isolated by equivalent circuit analysis. Reorientation of e.g. hydronium ions as well as oscillation of protons within a hydrogen bond corresponds to localised charge transport. These phenomena are observable in the GHz region and above. Badot et al.[89]) have measured dielectric properties of V_2O_5-1.6 H_2O xerogels from 100 kHz to 12 GHz; the frequency dispersion of the complex dielectric constant was interpreted in terms of proton diffusion, relaxation of strongly and loosely bound water molecules, and finally in the high frequency end, indications for a rapid rotation of hydronium ions were seen with an extrapolated relaxation time equal to $1.8 \cdot 10^{-12}$ s at 23 ^{o}C.

Non-linear behaviour of the Arrhenius-plots is most often due to chemical instability (water loss) during the measurements. Conductivity maxima are often encountered; above a certain temperature one finds that conductivity decreases with increasing temperature. Most materials, when investigated in their stable temperature range, such as H^+NYS and H^+NGS, yield linear Arrhenius plots.[42])

5.2 Neutron and PMR techniques

Neutron scattering techniques are especially suited for investigating the average structure of SPC's by elastic scattering , see for instance the diffraction studies by J.O.Thomas et al. on β-aluminas.[11,12,38,98,114]) Deuterium analogues of SPC's are often used.[102,108]) The quasielastic and inelastic (and incoherent) scattering (IINS) by protons yield information about the dynamics of diffusion, reorientation as well as about the vibrational frequencies (there are no constraints from selection rules, and only energy transfers due to protonic vibrations are observable).[49,50,101] For HUP a separation of three relaxation processes has been demonstrated: individual H^+ jumps, H_2O and H_3O^+ reorientations.[48])

Electrode	System	Reference
DC - experiments		
Ag	$MoO_3 \cdot H_2O$	90
Porous Pt	Al_2O_3	91
Pt-Ni	$SrCeO_3$	22
Pt + $La_{0.4}Sr_{0.6}CoO_3$	$SrCeO_3$	22
H_xMoO_3	protonic Zeolites	82
Zn, MnO_2	HUP	92
$Ti_{1.1}NiH_{1.5}$	HUP	93
AC-experiments		
Stainless Steel	$[Fe(HSO_4)_2] \cdot 4H_2O$	75
Graphite powder	SnO_2-containing	95
In	PWA, zeolites	94,96

Table 4. Electrode materials used for ac and dc measurements on solid state proton conductors.

The vibrational spectrum of the $H_5O_2^+$-ion in $HClO_4 \cdot 2H_2O$ and $H(Fe/In)(SO_4)_2 \cdot 4H_2O$ obtained by IINS has been reported.[101]

Nuclear magnetic resonance (PMR or NMR) is an equally important method.[53,54,67] Pulsed field gradient NMR has been used succesfully to distinguish between local (rotational) and external movements of protonic species, see for example ref.14.

5.3 Use of tagged atoms (D, ^{18}O)

The use of deuterium exchanged material in conductivity and structure investigations has already been mentioned in section 4.4 and 5.2. One spectacular application of hydrogen/deuterium exchange in conjunction with Raman spectroscopy deserves mentioning: Brown and Frech have followed the progression of the deuterium diffusion front into lithium hydrazinium sulphate, when deuterium is injected electrochemically.[55] The experiment and a typical series of Raman spectra are shown in Figure 5. Chandra and coworkers have likewise recently started to apply Raman spectroscopy to studies of the electrolysis of MoO_3 hydrates.[90]

Kreuer and collaborators have used ^{18}O labeled water in their investigation on HUAs.[43] The ^{18}O tracer diffusion coefficient was determined, and the conductivity calculated via the Nernst-Einstein equation matched well with the measured conductivity.

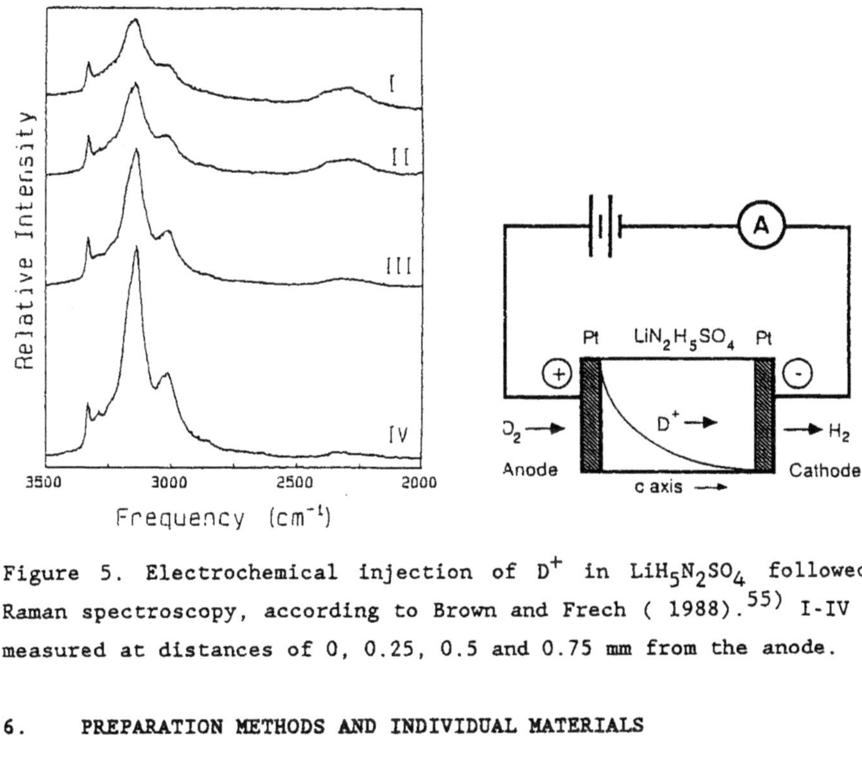

Figure 5. Electrochemical injection of D^+ in $LiH_5N_2SO_4$ followed by Raman spectroscopy, according to Brown and Frech (1988).[55] I-IV were measured at distances of 0, 0.25, 0.5 and 0.75 mm from the anode.

6. PREPARATION METHODS AND INDIVIDUAL MATERIALS

Only materials groups with some scientific importance or some chances of future applicability will be dealt with in the following sections. As can be seen below, the synthetic chemistry aspect is emphasised, at the expense of structural and dynamical aspects.

A pragmatic way of dividing SPC into groups according to preparation method was thus adopted. The preferred preparation method of a SPC has a close relationship to its materials properties: e.g. SPC materials made by ion exchange are all insoluble in water, contains hydronium or ammonium ions, and the precursors are made by high temperature routes, etc. Seven common methods of preparation are listed in Table 5.

Method	Example	Reference
Precipitation from aq. solution	HUP	9,10,45,46
Ion exhange	β-aluminas	11,38,71
Field assisted ion exchange	$(H_3ONa)_5GdSi_4O_{12}$	42
Decomposition of NH_4^+-analogue	H-β''-alumina	12
Complexation between acid and polymer	PEI + H_3PO_4	27-29
Reduction of noble metal-analogue with H_2 or D_2.	(Ag/H)β-alumina	102
Equilibration in water or hydrogen atmosphere	$BaCe_{1-x}Y_xO_3$	17-24
Intercalation-deintercalation	pellicular α-ZrP	32,34,36

Table 5. The common routes of synthesis for solid state protonic conductors.

6.1 Precipitated materials

6.1.1 Heteropoly acid hydrates. A protonic conductivity equal to 0.16 S/cm of polycrystalline $H_3PW_{12}O_{40}\cdot 29H_2O$ was first reported by Nakamura et al. in 1979.[13] Large crystals can be obtained by slow evaporation of solutions of the commercial powders: Merck No.583 (PWA) and 659 (SiWA)),[14] however, the high conductivities originally measured, were obtained from pellets made from crushed large single crystals! [13]

The acidic nature of metal oxides increases as the valency of the metal increases towards its maximum value. Hydrates and "hydroxides" of

WO_3 and MoO_3 are thus solid compounds with acidic character. Silicon and phosphor substituted tungstates and molybdates (host:dopant ion ratio between 6 and 12) are the most studied combinations. The structures contain macroanions (10-12 Å diameter) made form corner and edge sharing WO_6 octahedra; tetrahedrally coordinated P(V) or Si(IV) are at the center of the macro anion. Water and protonic species are on the outside of the macroanion. All heteropoly anions depolymerise completely in basic solution,- i.e. normal tungstate ions, WO_4^{-2}, result. The content of water of hydration varies with precipitation conditions as well as subsequent equilibration.[94] Single crystal conductivities have now been reported and are 0.08 S/cm at room temperature.[14] There are 6-8 charge carriers (H_3O^+?) per formula unit of $H_3PW_{12}O_{40}\cdot 28H_2O$.[14]

6.1.2 <u>HUP and HUAs.</u> These are layer compounds, $H_3OUO_2MO_4\cdot 3H_2O$, where M = P or As. HUP will precipitate, when uranyl nitrate solution is mixed with 2-3 M phosphoric acid.[9] HUP and HUAs are not stable in contact with solutions with higher pH than 2-3. The compounds easily loose water (and conductivity) above 50 °C and HUO_2PO_4 is eventually formed at 157 °C.[103]

HUAs changes from the paraelectric (tetragonal) high temperature phase to an antiferroelectric (orthorhombic) phase at 26 °C. HUP undergoes a similar phase transition around 0°C.[43,104]

The associated changes in conductivity at the transition points are less than one order of magnitude, but the activation energy changes from 0.27-0.32 eV in the high temperature phases to 0.57 and 0.64 eV in the low temperature phases of HUP and HUAs, respectively. [43]

Measurements of the conductivity as a function of pressure for HUAs showed a transition from vehicle to a Grotthuus mechanism at 2.5 GPa.[46]

The most recent ^1H and ^{31}P NMR study of HUP in both its high and low temperature phase confirms that protons (and not, say, H_3O^+) are the mobile species. [104]

6.1.3 Hydrous oxides Interesting materials properties of powders emerge as particle size is reduced down to a few hundred Å or below, cf. Figure 1. Local electroneutrality is not required at this microscopic level,- the net charge of a particle becomes a function of particle size, structure, nature of particle atoms, pH of surrounding solution, etc. The net charge may be negative, neutral (by coincidence or only at the pH-point of zero Zeta potential, pzzp), or positive. The solution in between the particles contains excess charge of the opposite sign. The protonic conduction takes place mostly in this liquid like region. Another consequence of the charged nature of the particle hydrates is their ion exchange properties, which can be used to determine pzzp's.[15] Textbooks on powder technology and surface science must be consulted for further information on the consequences of surface charges etc.

Antimonic acid, $HSbO_3 \cdot nH_2O$ (or $Sb_2O_5 \cdot (2n+1)/2\ H_2O$) is sofar described in three modifications, derived from: the defect pyrochlore structure made by hydrolysis of $SbCl_5$,[105] the ilmenite structure,[106] and ion exchanged $KSbO_3$.[106,107] Antimonic acid with various content of water seems to be obtained as a well-crystallised material as indicated by the sharpness of the diffraction peaks. The room temperature conductivity of pyrochlore type antimonic acid decreases from 10^{-3} S/cm to 10^{-7} S/cm, when the water content per mole of $HSbO_3$ decreases from 1.5 to 0.25.[105] The activation energies for conduction in antimonic acid and α-stannic acid are low, 0.112 - 0.238 eV.[105]

A different cubic version of $Sb_2O_5 \cdot nH_2O$ is obtained by ion exchange of $KSbO_3$ in acid solution. It also has a room temperature conductivity in the 10^{-3} S/cm range for n = 0.8 (n was determined both by thermogravimetry and by structure refinement).[107] Bell and Weller found OH, H_2O and H_3O^+ units in this structure by neutron diffraction.[108]

Hydrated SnO_2 and ZrO_2, on the other hand, are obtained with mean particle sizes from 25 Å to a few hundred Å, as deduced from the broadening of the diffraction peaks using the Scherrer equation. Submicron $ZrO_2 \cdot nH_2O$ is made by hydrolysis of $ZrOCl_2$ and has a cubic (stabilised ZrO_2) structure , and submicron $SnO_2 \cdot nH_2O$ shows diffraction lines characteristic of rutile-type SnO_2.[15]

Hydrous tin oxide has been made by hydrolysis of $SnCl_4$ with ammonia solution,[109] or thermal decomposition and oxidation of $SnCl_2 \cdot 2H_2O$ or $SnSO_4$.[95] Knudsen et al. have deposited hydrous tin oxide inside mordenite material (a zeolite), with the purpose of preventing graingrowth of the tin oxide phase.[95] Such material showed a conductivity of $2.5 \cdot 10^{-2}$ S/cm at 100 °C and 100% RH.

Hydrated forms of $HTaO_3$ and $HNbO_3$ are also proton conductors and propably also belongs to this group of materials.[110]

6.1.4 α-$M(HPO_4)_2 \cdot H_2O$. Alberti, Costantino, Casciola and coworkers in Perugia have thoroughly investigated this family of layer structures since 1978.[30-37] $Zr(HPO_4)_2 \cdot H_2O$ and Ti(IV) and Ce(IV) analogues exist in an α ,γ, and in a colloidal, delaminated form (called pellicular), which is easily processed into thin sheats.[32,34,36] A new modification, polyhydrated α-$ZrP_{(sl)}$ is made by ultrasonic treatment (7 MHz) and lyophilisation; at 20°C the water content will adjust itself in the range 1.2-3.5 moles of water/ Zr as a function of the relative humidity (5-90 % RH).[111] The phosphates are formed by reaction between zirconyl chloride and phosphoric acid,[112] or by decomplexation of hexafluoro zirconate ions with phosphoric acid.[113] The anhydrous form of the materials are obtained by drying at 250°C for 1 day, and they are poor conductors, 10^{-4}- 10^{-6} S/cm at 300 °C for the Ti-analogue.[37] In hydrated form, conductivities around $3 \cdot 10^{-4}$ S/cm at room temperature, are quoted,[30-36] whereas the polyhydrated form has a conductivity approaching 10^{-3} S/cm, since this material contains optimally alligned crystallites as well as a high water content.[111]

For α-ZrP the influence of relative humidity on conductivity,[34] porosity and density,[35] surface and bulk contribution to conductivity,[33] have been systematically investigated.

6.2 Ion Exchanged Material

Ion exchange has frequently been used to produce protonic conductors from sodium or potassium ion precursors. It was realised early that ion exchange in aqueous acids should preferentially start with the potassium analogue, because of the similar in size of the hydronium and potassium ion (diameter of K^+ is 2.66 Å.). The exchange can be speeded up by applying an electric field across the sample,- the FAIE - method.[42)]

It is possible by the ion exchange process to obtain materials that are kinetically stable but unstable in the thermodynamical sense. This may pertain to the protonic β-aluminas, which possibly are not present as equilibrium phases in the Al_2O_3 - MgO - water phase diagram.

The ion exchange conditions vary from rather mild treatments, of for instance $Na_5GdSi_4O_{12}$ in dilute acetic acid at 25-90 °C used by Yamashita and Nicholson,[42)] to treatment in 280 °C hot sulphuric acid for 24 h of Na-β''-alumina.[12)] Mixed hydronium -ammonium β-alumina has been made by treating sodium β-aluminas in molten NH_4NO_3 at 200 °C.[71,98)] Ion exchange into an ammonium-analogue can be made, and this is subsequently deammoniated by heating and a H_3O^+-analogue is obtained. This strategy has been used for H_3O^+-β''-alumina.[12)]

Ion exchange in molten $AgNO_3$ to the Ag^+-analogue (100% yield) followed by reduction in hydrogen has been used by Tofield et al. to produce protonic β-alumina samples for neutron diffraction.[102)] The silver metal atoms diffuse to the grain boundaries and do not disturb the diffraction experiments.

6.2.1 <u>β-aluminas.</u> The structures of the precursors: Na and K-β-, and β''-aluminas are described in detail in other parts of this book. The protonic β''-aluminas (H^+, H_3O^+, NH_4^+ - and mixed forms) have been the subject of numerous studies since G.C.Farrington in 1978 announced that protonic β''-alumina had a conductivity of 10^{-2} S/cm at 25 °C.[115)] However, 10^{-5} S/cm was shown later on to be a more realistic value for this material.[116)]

The large difference in conductivity between the β and the β'' form is noteworthy, the β-form has only a conductivity of 10^{-11} S/cm at 20 °C.[11)] The highest (reproducible) room temperature conductivity reported for a protonic β-alumina is 10^{-3} S/cm for the mixed hydronium-ammonium β''-alumina made by deNuzzio.[12,71,98)] The material has the approximate composition: $(NH_4)_{1.4}(H_3O)_{0.27}Mg_{0.67}Al_{10.33}O_{17}$ and the activation energy for ionic conduction in the range -40 to 200 °C is 0.26 eV. A mechanism for the proton conduction has been suggested and invokes both classical hopping of ammonium and hydronium and Grotthuus-type conduction.[71)]

It is questionable, however, if a mixed hydronium-ammonium SPC material like this can be used in fuel cells in the temperature range 100-400 °C, since both the ammonia and water partial pressure must be sufficiently high in order to avoid decomposition.

Colomban et al. have worked with vibrational spectroscopy on β-aluminas.[117)] Nicholson et al. in Canada put emphasis on ion exchange kinetics, stability and applicability.[116,119-124)] The literature on protonic β-aluminas is large and review articles by Frase et al. in 1984 and 1986 should be consulted.[11,38)]

The synthetic route to protonic aluminas is via ion exchange (see also 6.2). Normal β-alumina gives ammonium- whereas β'' gives a mixed ammonium-hydronium β''-alumina, when sodium- or potassium precursors are ion exchanged in molten NH_4NO_3.[98)] The hydronium to ammonium ratio can be controlled in the range 0.17 to 0.67 by increasing the ion exchange time from 3 to 7 days.[71)] The typical composition for the β-form is: $M_{1+x}Al_{11}O_{17+x/2}$ (x= 0.3), where M is hydronium and/or ammonium ions. The more conductive β''-form has the typical overall composition : $M_{1+x}Mg_xAl_{11-x}O_{17}$ (x= 0.67), and has a structure belonging to the rhombohedral space group $R\bar{3}m$ with a hexagonal unit cell, a = 5.631 Å and c = 34.378 Å.[98)] This unit cell contains approximately 112 atoms, depending on the hydronium/ammonium ratio and content of extra water. The high number of atoms plus the fact that the structure is disordered with respect to protonic species underline the formidable problems that has been overcome in structure analysis. The contributions of J.O.Thomas et al. in protonic as well as sodium β-aluminas are responsible for our understanding of the structural and dynamical features of β-aluminas[11,12,38,98,114)]

It was considered an impossible task in the early 1980-ies to perform protonic ion exchange on ceramic β-alumina products (up to mm-thick tubes and plates) and still maintain mechanical integrity of the ceramic body. The internal stress that develops due to expansion of the crystallites in the c-axis direction was the cause of the disruption of ceramic bodies subjected to ion exchange. This problem has since been solved by Nicholson and coworkers after an impressive effort in ceramic engineering.[119-124] The ideal composition of the precursor ceramic is $Na_{0.97}K_{0.65}Mg_{0.67}Al_{10.33}O_{17}$. The ceramic piece is shaped and sintered with this composition. Part of the Na^+ is then exchanged with more K^+ in a NaCl-KCl melt. When the β''- alumina lattice has expanded sufficiently one carries out the ion exchange in aqueous acid solution.

Complete ion exchange may lead to physical degradation of the ceramic , since the " binding" phase in the partly exchanged material then disappears. The presence of this Na-rich β-alumina fraction however reduces the protonic conductivity of the material.

The ion exchanged material has been tested successfully in steam electrolysis cells running at 100 °C.[119,120] The conductivity at 300°C is still only 10^{-3} S/cm,- the present material can therefor only be used in special applications- and not in fuel cells or normal water electrolysis systems, where other systems have better energy economy.

6.2.2 Nasicons. Protonic nasicon has been made and patented by Hofer, Rabenau and Bell in 1981-1982.[125,126] The compositions being used are approximately: $(H_3O)_{2.5}Zr_2Si_{1.5}P_{1.5}O_{12}$.[125-128] Gulens et al.[127] have demonstrated the use of protonic nasicon in water electrolysis and Canaday et al. have used protonic nasicon as a membrane in a hydrogen sensor.[128] The specific conductivity of the protonic nasicons are in all cases as low as 10^{-6}-10^{-7} S/cm at room temperature.

Ion exchanged $Na_5GdSi_4O_{12}$ (H^+NGS) also belongs to this group.[42]

6.2.3 Zeolites. Contributions to the study of protonic zeolites have been made by I.G. and E. Krogh Andersen and coworkers.[63,96] Ammonium exchanged zeolites have proven to be more conductive than zeolites exchanged in acid. This fact has been rationalised in terms of the acid-base properties of the aluminosilicate host lattice of the zeolites. Water is in general a too weak base to attract the protons from the hydroxyl groups of the host lattice, whereas ammonia succeeds. One of the highest room temperature conductivities, i.e. $2 \cdot 10^{-3}$ S/cm, has been reported for fully hydrated NH_4-zeolite-A,[82] the composition of which can be approximated by: $(NH_4)_4Ca_3Si_{14}Al_{10}O_{48}\cdot 200\text{-}300H_2O$. The proton conductivity of protonic zeolites is extremely dependent on the water content: for ammonium-zeolite-A a variation of the content of water of hydration from 10.5 to 14.5 w% increases the conductivity a little more than a factor 1000.[82]

6.3 Polymeric and Organic Materials

In this category we find systems of the type NH_4HSO_4 mixed with PEO or polyacrylic acid, (PAA), which exhibit conductivities around 10^{-5} S/cm at room temperature.[25] The acid hydrogensulphate material thus has higher conductivity than the more neutral NH_4SCN or $NH_4CF_3SO_3$ mixtures with PEO.[26]

Also mixtures of polyvalent acids like sulphuric or phosphoric acid with polyethyleneimine, PEI,x H_2SO_4, (x = number of moles of acid per PEI repeat unit) have been studied,- here conductivites at 25 °C in the range from $4\cdot 10^{-5}$ down to 10^{-7} S/cm were observed.[27] The best compositions are found for x= 0.35, where partial neutralisation of the amine-groups has taken place. The protons are assumed to jump between these amine groups. At higher acid contents the amine groups are fully protonated, and the only possibility for proton transfer is between sulphate and hydrogensulphate anions, phosphates, respectively. The phosphoric acid - poly ethylene oxide system has also attracted interest.[29]

6.4 High Temperature Systems

It is a relatively new finding, that protons are present in oxides at temperatures well above 1000 °C. Protons are among the majority carriers in Y_2O_3.[81] Protons are also conducting in La_2O_3 and Sm_2O_3 doped ThO_2 in parallel with the mobile oxide ions.[129]

The defect chemistry is complicated even in simple and stable oxides as, for example Al_2O_3 , where El-Aiat and Kroger have suggested the presence of three different forms of hydrogen defects: immobile $H^{\cdot}_{O}$, and mobile positively and negatively charged interstitials.[91]

The most prominent high temperature materials are the perovskites, of doped ABO_3 type, examplified by ytterbium doped $SrCeO_3$. These have been studied intensively by Iwahara and coworkers since 1980.[17-22,130] Sr- and Ba- analogues are made by solid state reaction between the proper oxides or carbonates at 1250 - 1500 °C.

When trivalent ions are replacing Ce^{4+}, the material may be formulated as :

$$SrCe_{1-x}Yb_xO_{3-\delta} \text{ or } SrCe_{1-x}Yb_xH_yO_{3-\delta}$$

where δ is < or eq. to x/2 ; and y < or eq. to x.

The preferred dopant concentration is x= 0.05. The amount of protons as a function of preparation conditions, temperature and concentration of dopant remains to be determined. The specific conductivity is 10^{-2} S/cm at 900°C and 10^{-3} S/cm at 600°C in hydrogen.[20]

The conclusions of transport number determinations are that protons are the majority carriers in all cases. A hole conductivity , a few % of the total conductivity, is present when samples are equilibrated in oxygen.[20] A second, strong proofs for protonic conduction was that water formed only at the cathode side when discs were operated as small fuel cell with oxygen and hydrogen as feed gases.

The third argument for the interpretation of protonic conduction was, that there was no EMF , when a gradient in oxygen partial pressure was set up across the pellet. This was interpreted as sign of lack of oxygen-ion conduction.

The major conclusions have been substantiated by the work of Sheriban and Nowick: they also found that protons are the majority carriers.[23] This is also supported by the observation of a strong isotope effect.[23]

Bonanos on the other hand had no other choise than concluding that mixed proton and oxide ion conduction is present in the same type of ytterbium doped $SrCeO_3$.[24] The main finding of the latter investigators was that water was formed both on the anode and the cathode side of a cell operated as a fuel cell. Bonanos stressed that all partial pressures but one must be known, when the concentration cell test is performed and used as proof of mobility of a certain ionic species.

Norby has described a fully automated gas controle system for performing such tests on oxides.[131]

It will be interesting to follow the development in this particular field of proton conducting materials in the future.

Suggested Device		Proton conductor employed or suggested	Reference
Fuel cells		HUP	100
		$SrCeO_3$	21
Batteries		Zeolites	82
Steam electrolyser		β-alumina	119,120
		Protonic nasicon	127
		$SrCeO_3$	17-22
Sensors	H_2	HUP	93
		Protonic nasicon	128
		$PVA+H_3PO_4$	132
	pH	α-ZrP	99
	NH_3	$HSbO_3 \cdot H_2O$	133
	CO_2	hydroxo apatite	118
Photo electro chemical cell		HUP	51,52
Hydrogen pump (gas purification, tritium removal)		protonic nasicon	128
Electrochromic display		PEI+acid	97
Ceramical electrochemical reactor (CH_4 to C_2H_6)		$SrCeO_3$	21

Table 5. Suggested applications of solid state protonic conductors.

7. APPLICATIONS

Solid state protonic conductors have not yet gained industrial importance. This is due to a combination of shortcommings of the materials , which sofar have been at hand. The various applications, listed in Table 5, must be considered as potential areas of application of SPC, when/if suitable proton conducting candidates are found.

The lack of industrial experience with SPC is the rationale for the brevity of the present section.

Most SPC's can be discarded due to low conductivity, water solubility, lack of thermal stability, low mechanical strength, and high water vapor pressure. Compounds of basic nature would not survive in CO_2 containing ambient.

One of the more sofisticated applications of a solid protonic conductor in an energy convertion device has been suggested by Schoonman, cf. the photo electrochemical cell.[51,52]

8. ACKNOWLEDGEMENTS

The author wishes to thank all who provided reprints and preprints during the preparation of the review. Special thanks in this respect are due to prof.s P.Nicholson, J.O.Thomas , and H.Iwahara and to Dr. E.Skou. U.Smidt and Östen Svensson are acknowledged for stimulating discussions.

9. REFERENCES

1) Proceedings of the 6'th International Conference on Solid State Ionics, September, 1987, Garmisch-Partenkirchen FRG. To be published as volume 27/28 of Solid State Ionics (1988)
2) Y.Kakuichi, Proc. Jap. Acad., 23, 65 (1947)
3) W.E.Ayrton and J.Perry, Proc.Phys.Soc., 2, 171 (1877)
4) L.B.Harris and G.J.Vella, J.Chem.Phys., 58, 4550 (1973)

5) A.L. de Oliveira, O. de O. Damasceno, J. de Oliveira and E.J.L.Schouler, Mat. Res. Bull., 21 , 877 (1986)

6) J.M. Pollock and A.R. Ubbelohde, Trans. Far. Soc.,52, 1112 (1956)

7) A. Kawada, A.R. McGhie and M.M. Labes, J.Chem.Phys.,52 ,3124 (1970)

8) VII'th Symposium on the Physics and Chemistry of Ice, Journal de Physique, Colloque C1, 1-707 (1987)

9) M.G.Shilton and A.T.Howe, Mat. Res. Bull. 12 (1977) 701

10) E.Skou, I.G. Krogh-Andersen, K.E.Simonsen and E.Krogh-Andersen, Solid State Ionics, 9&10 (1983) 1041

11) K.G.Frase, G.C.Farrington and J.O.Thomas, Ann.Rev.Mater.Sci.,14 279 (1984)

12) J.O.Thomas, G.J.McIntyre and J.DeNuzzio, Solid State Ionics, 18&19 642 (1986)

13) O.Nakamura, T. Kodama, I. Ogino and Y. Miyake, Chem. Lett. (Chem.Soc. Japan) 1 , 17 (1979)

14) K.D.Kreuer,M.Hampele, K.Dolde and A. Rabenau, Ref.1 (1988)

15) W.A. England, M.G. Cross, A. Hamnett, P.J. Wiseman and J.B.Goodenough, Solid State Ionics, 1,231 (1980)

16) P.C.Rieke and N.E.Vanderborgh, J.Electrochem.Soc., 134, 1099 (1987)

17) H.Iwahara, T.Esaka, H.Uchida and N.Maeda, Solid State Ionics, 3&4 ,359 (1981)

18) H.Uchida, N.Maeda and H.Iwahara , Solid State Ionics, 11 , 117 (1983)

19) H.Uchida, S.Tanaka and H.Iwahara, J.Appl. Electrochem., 15 , 93 (1985)

20) H.Iwahara, T.Esaka, H.Uchida , T.Yamauchi and K.Ogaki, Solid State Ionics, 18&19 , 1003 (1986)

21) H.Iwahara, H.Uchida and S.Tanaka, J.Appl. Electrochem., 16, 663 (1986)

22) H.Iwahara, H.Uchida and I.Yamasaki, Int.J.Hydrogen Energy,12, 73 (1987)

23) T.Scherban, W.-K. Lee and A.S.Nowick, Ref. 1 (1988)

24) N.Bonanos, B.Ellis and M.H.Mahmood, Ref. 1 (1988)

25) M.F.Daniel,B.Desbat and J.C.Lassegues , Ref. 1 (1988)

26) M.Stainer, L.C.Hardy, D.H.Whitmore and D.F.Shriver ,J.Electrochem.Soc.,131 , 784 (1984)

27) M.F.Daniel, B.Desbat, F.Cruege, O.Trinquet and J.C.Lassegues, Ref.1 (1988)

28) J.J.Zupancic, R.J.Swedo and S.L.Petty-Weeks, US Pat. 4.664.761 (1987)

29) P.Donoso,W.Gorecki, C.Berthier, F.Defendini, C.Poinsignon and M.B.Armand, Ref. 1 (1988)

30) G.Alberti, M.Casciola,U.Costantino, G.Levi and G.Ricciardi, J.Inorg. Nucl. Chem., 40 ,533 (1978)

31) G.Alberti, M.Casciola,U.Costantino and R.Radi, Gazzetta Chimica Italiana, 109 ,421 (1979)

32) G.Alberti, M.Casciola,U.Costantino and M.Leonardi, Solid State Ionics , 14 ,289 (1984)

33) M.Casciola and D.Bianchi, Solid State Ionics, 17 ,287 (1985)

34) M.Casciola and U.Costantino, Solid State Ionics 20, 69 (1986)

35) D.Bianchi and M.Casciola , Solid State Ionics, 17 ,7 (1985)

36) M.Casciola,U.Costantino and S.D'Amico, Solid State Ionics, 22,127 (1986)

37) G.Alberti, M.Bracardi and M.Casciola, Solid State Ionics, 7, 243 (1982)

38) K.G.Frase, J.O.Thomas, A.R. McGhie and G.C.Farrington, J.Solid State Chem.,62, 297 (1986)

39) N.Bjerrum, Kgl. Danske Videnskab. Selsk., Mat.-Fys. Medd. 27, 56 (1951)

40) N.Bjerrum, Science, 115, 385 (1952)

41) A.Potier In: De la Microelectronique a la Microionique (Ed. P. Colomban), Palaiseau , 149 (1983)

42) K.Yamashita and P.S.Nicholson , Solid State Ionics, 20, 147 (1986)

43) K.-D.Kreuer, A.Rabenau and R.Messer, Appl. Phys., A32, 45,155 (1983)

44) C.J.D.von Grotthuss, "Mémoire sur la décomposition de l'eau et des corps, qu'elle tient en solution, a l'aide de l'éléctricité galvanique, Rome, Italy (1805), and Ann.Chim., LVIII,54 (1806)

45) A.T.Howe and M.G.Shilton, J.Solid State Chem., 34 (1980) 149

46) K.D.Kreuer, I.Stoll and A.Rabenau, Solid State Ionics, 9/10, 1061 (1983)

47) Y.T.Tsai, W.R.Halperin and D.H.Whitmore, J.Solid State Chem., 50, 272 (1983)

48) C.Poinsignon, A.Fitch and B.C.F.Fender, Solid State Ionics, 9/10, 1049 (1983)

49) G.L.Squires," Introduction to the theory of thermal neutron scattering", Cambridge Univ. Press, London, (1978)

50) S.M.Shapiro and F.Reidinger, In:" Physics of Superionic Conductors" (M.B.Salamon,Ed.) Springer-Verlag, Berlin, p 45 (1979)

51) J. Schoonman, Chem.Phys.Lett., 86, 515 (1982)

52) J. Schoonman, Ber. Bunsenges. Phys.Chem., 86 , 660 (1982)

53) P.M.Richards, In:" Physics of Superionic Conductors" (M.B.Salamon,Ed.) Springer-Verlag, Berlin, p 141 (1979)

54) R.T.Slade,I.M.Thompson,R.C.Ward and C.Poinsignon, J.Chem.Soc., Chem.Commun., 1987 ,726 (1987)

55) S.H.Brown and R.Frech , Ref. 1 (1988)

56) K.D.Kreuer, A.Rabenau and W.Weppner, Angew.Chem.,Int.Ed.Engl., 21, 208 (1982)

57) J.Jensen and M.Kleitz (Eds.) : "Solid State Protonic Conductors I- for fuel Cells and Sensors", Odense University Press, pp 339 (1982)

58) J.B.Goodenough, J.Jensen and M.Kleitz (Eds.): "Solid State Protonic Conductors II- for fuel Cells and Sensors", Odense University Press, pp 419 (1983)

59) J.B.Goodenough , J.Jensen and A. Potier (Eds.) : "Solid State Protonic Conductors III- for fuel Cells and Sensors", Odense University Press, pp 284 (1985)

60) M. Nagai and T. Nishino , Ref. 1 (1988)

61) M.Lal and A.T.Howe , J.Chem.Soc., Chem. Commun., 1980 , 737 (1980)

62) H.Arribart, H.Carlos and B.Sapoval, Solid State Ionics, 5, 253 (1981)

63) E.K.Andersen, I.G.Krogh Andersen, E.Skou and S.Yde-Andersen, Solid State Ionics, 18/19, 1170 (1986)

64) L.Glasser, Chem. Rev., 75 ,21 (1975)

65) J.Bruinink, J.Appl. Electrochem., 2, 33 (1972)

66) F.W.Poulsen, Risø-M-2244 , 23 pp (1980)

67) H.Arribart, Ref. 57, 111 (1982)

68) T.A.Wheat, A.Ahmad, A.K.Kuriakose and J.D.Canaday, J.Can.Ceram.Soc., 54, 32 (1985)

69) S.Chandra, N.Singh and S.A.Hashmi , Proc. Indian Natn. Sci.Acad., 52A, 338 (1986)

70) K.D.Kreuer, to be published in J.Mol.Struct.," Horizons in Hydrogen Research"(1988). Presented at VIII'th Workshop, Wroslow (1987)

71) J.D.DeNuzzio, Thesis,University of Pennsylvania , 201 pp (1986)

72) C.M.Mari and D.Narducci, La Chimica e l'Industria, 68, 70 (1986)

73) M.A.Butler and R.M.Biefeld, Phys.Rev. B ,19 , 5455 (1979)

74) D.Rousselet, J.Roziere, N.H.Herzog-Cance, M.Allavena, N.Pham Thi, J.Potier and A.Potier, Ref. 57 ,65 (1982)

75) I.Brach, D,J,Jones and J.Roziere , Ref. 1 (1988)

76) A.Potier, Ref.58, 173 (1983)

77) P.Colomban, C.Lassegues, A.Novak, M.Pham Thi and C.Poinsignon I: Dynamics of Molecular Crystals (J.Lassegues Ed,.) Elsevier, Amsterdam, p 269 (1987)

78) B.Baranowski, M.Friesel and A. Lunden, Z. Naturforsch., 419, 733 (1986)

79) B.Baranowski, M.Friesel and A. Lunden, Z. Naturforsch., 419, 981 (1986)

80) J.Bruinink, J.Inorg.Chem., 35, 1698 (1973)

81) T.Norby and P.Kofstad, Solid State Ionics, 20, 169 (1986)

82) K.-D.Kreuer, W.Weppner and A.Rabenau, Mat.Res.Bull., 17, 501 (1982)

83) B.E.Conway, In:" Modern aspects of electrochemistry", (J.Bockris and B.E.Conway,Eds.) , Butterworths, London, p 43 (1964)

84) K.-H. Hass and V.Schindwolf, J.Solid State Chem., 54 , 342 (1984)

85) C.N.R.Rao, In: "Materials for Solid State Batteries", (B.V.R.Chowdari and S.Radhakrishna, Eds.) , World Science Publ. Co.,Singapore , p 3 (1986)

86) P.Barboux,R.Morineau and J.Livage, Ref.1 (1988)

87) M.C.Sauer Jr., P.F.Southwick, K.S.Spiegler and M.R.J.Wyllie, Ind.Eng.Chem.,47, 2187 (1955)

88) J.S.Lundsgaard, J.Malling and S.Yde-Andersen, Ref.58, 400 (1982)

89) J.C.Badot, A.Fourrier-Lamar and N.Baffier, J.Physique, 46, 2107 (1985)

90) S.Chandra,H.Singh,B,Singh,A.L.Verma,S.S.Khartri and T.Chakrabarty, J.Phys. Chem. Solids, 48, 1165 (1987)

91) M.M.El-Aiat and F.A.Kroger, J.Appl.Phys., 53, 3658 (1982)

92) H.Kahil, E.J.L.Schouler, M.Forestier and J.Guitton, Solid State Ionics, 18/19, 892 (1986)

93) P.Lamberterie, M.Forestier, J.Guitton, A.Roualt, R.Fruchart and D.Fruchart, C.R.Acad.Sc.Paris, t.300, Serie II, no 14 (1985)

94) E.Skou and S.Yde-Andersen, Ref.58, 44 (1982)

95) N.Knudsen, E.Krogh Andersen, I.G. Krogh Andersen and E. Skou, Ref. 1 (1988)

96) E.Krogh Andersen, I.G.Krogh Andersen, E.Skou and S.Yde-Andersen, Ref.59, 100 and 247 (1985)

97) M.Akhtar, R.M.Paiste ans H.A.Weakliem, J.Electrochem.Soc., 135, 1597 (1988)

98) J.O.Thomas and G.C.Farrington, Acta Cryst., B39, 227 (1983)

99) R.Palombari and M.Casciola, J.Electroanal. Chem., 216, 283 (1987)

100) P.E.Childs, A.T.Howe and M.G.Shilton, J.Power Sources, 3, 105 (1978)

101) J.Penfold, J.Tomkinson, D.J.Jones and J.Roziere , (submitted to J.Mol. Struct. (1987)

102) J.Newsam,B.C.Tofield, W.A. England and A.J.Jacobsen, in : " Fast Ionic Transport in Solids", (P.Vashishta, J.N.Mundy and G.K. Shenoy, Eds.) North-Holland, Amsterdam p 405 (1979)

103) F.Weigel and G. Hoffmann, J.Less-Common Metals, 44 ,99 (1976)

104) K.Metcalfe and T.K.Halstead, Solid State Ionics, 26,209 (1988)

105) C.Forano, J.P.Besse, J.P.Battut,J.Dupuis and A.Hajjimohamad, Ref.1 (1988)

107) H.Watelet, J.-P. Picard, J.-P. Besse and R.Chevalier, Mat. Res. Bull., 16 , 1131 (1981)

106) U.Chowdhry, J.R.Barkley, A.D. English and A.W.Sleight, Mat.Res.Bull.,17, 917 (1982)

108) R.G.Bell and M.T.Weller, Ref.1 (1988)

109) E.W.Giesekke, H.S.Gutowsky, P.Kirkov, H.A.Caitinen, Inorg.Chem., 6, 1294 (1967)

110) P.G.Dickens and M.T. Weller, Solid State Comm., 59, 569 (1986)
111) G.Alberti, M.Casciola, U.Costantino and F.Di Gregorio, Ref.1 (1988)
112) A.Clearfield and G.D.Smith, Inorg.Chem.,8, 431 (1969)
113) G.Alberti and E.Toracca, J.Inorg.Nucl.Chem., 30, 317 (1968)
114) J.O.Thomas , Ref. 59 , 185 (1985)
115) G.C.Farrington and J.L.Briant, Mat. Res. Bull.,13,763 (1978)
116) A.Teitsma, M.Sayer, S.L.Segel and P.S.Nicholson, Mater.Res.Bull., 15, 1611 (1980)
117) Ph.Colomban and A.Novak, Solid State Ionics , 5, 241 (1981)
118) M.Nagai, T.Nishino and T.Saeki, Ref. 1, (1988)
119) M.Nagai and P.S.Nicholson, Mat.Res.Bull.,17, 1131 (1982)
120) P.S.Nicholson, M.Nagai, K.Yamashita, M.Sayer and M.F.Bell, Solid State Ionics, 15, 317 (1985)
121) P.S.Nicholson, M.Z.A.Munshi, G.Singh, M.Sayer and M.F.Bell, Solid State Ionics, 18/19, 699 (1986)
122) K.Yamashita and P.S. Nicholson, Solid State Ionics, 20 ,147 (1986)
123) M.Z.A.Munshi and P.S.Nicholson, Solid State Ionics, 23, 203 (1987)
124) M.Z.A.Munshi and P.S.Nicholson, Solid State Ionics, 25, 51 (1987)
125) H.H.Hofer, M.F.Bell and A.Rabenau, Ger. Offen. DE. 3013677 (1981)
126) A.Rabenau, H.H.Hofer and M.F.Bell, Ger. Offen. DE. 3110571 (1982)
127) J.Gulens, T.H.Longhurst, A.K.Kuriakose and J.D.Canaday,Ref.1 (1988)
128) J.D.Canaday, A.K.Kuriakose, A.Ahmad and T.A.Wheat, J.Can. Ceram. Soc.,55 ,34 (1986)
129) D.A.Shores and R.A.Rapp, J.Electrochem.Soc., 119, 300 (1972)
130) H.Iwahara ,Ref. 1 (1988)
131) T.Norby, Ref. 1 (1988)
132) J.J.Zupancic, R.J.Swedo and S.L.Petty-Weeks, US-Pat. no. 4.664.757 (1987)
133) N.Miura and W.L.Worrell, Chem. Lett. Jpn., 2, 319 (1987)

CRYSTALLINE LITHIUM ION CONDUCTORS

JTS Irvine and AR West
University of Aberdeen
Department of Chemistry
Meston Walk, Old Aberdeen
Aberdeen AB9 2UE, UK

ABSTRACT

The occurrence of high lithium ion conductivity in crystalline materials is reviewed. Materials discussed include Li_3N and its derivatives, Li_4SiO_4 and its derivatives, γ phases (LISICONS), Li_2SO_4 and its derivatives, lithium halide spinels, lithium beta-alumina, rutiles and ramsdellites, lithium zirconium phosphates and framework silicates. Factors responsible for high lithium ion conductivity are discussed.

1. INTRODUCTION.

Since the feasibility of using solid electrolytes in battery systems was demonstrated by Takahashi and Yamamoto[1], there has been intense interest in the development of solid electrolytes. Lithium-based systems offer the advantages of high energy density and high open circuit potentials. Systems containing lithium ion conducting solid electrolytes have been developed for use in both high temperature, high energy density cells[2] and ambient temperature, low energy density cells[3,4]. Any solid electrolyte to be used in these cells should have high ionic conductivity, low electronic conductivity and stability towards the electrode materials at the temperature of use.

Several types of lithium ion conducting solid electrolyte have been investigated; these can be classified as crystalline materials, glasses, composites (e.g. LiI/Al_2O_3) and polymers. As glasses, polymers and composites are discussed elsewhere in this book, the coverage in this chapter is restricted to crystalline conductors.

2. SURVEY OF CRYSTALLINE Li^+ ION CONDUCTORS

2.1 Lithium Nitride, Li_3N and Related Compounds

Lithium nitride (Li_3N) is a crystalline ionic compound that forms when lithium is heated in nitrogen. As a powder it is unstable in air[5]; however, it is relatively stable as either a single crystal[6] or as a sintered ceramic[7]. The lithium ion conductivity of Li_3N, figure 1 is rivalled only by that of Li beta-alumina at ambient temperatures, being about 1×10^{-3} $\Omega^{-1}cm^{-1}$ at 25°C for a single crystal[8]. Li_3N, which has a very low electronic conductivity[8], would be an excellent solid state battery electrolyte if problems associated with its low thermodynamic decomposition potential (0.44 V) could be avoided.

The crystal structure of Li_3N is hexagonal and can be described as consisting of layers of Li_2N alternating with layers of Li[9]. Each nitrogen is surrounded by six equivalent lithiums, Li(2), in the Li_2N planes. The lithiums in the Li plane, Li(1), lie between two nitrogens in the Li_2N planes, figure 2. The conductivity of Li_3N is highly anisotropic; at 25°C, the conductivity of a single crystal perpendicular to the c-axis (parallel to the layers) is 100 times greater than that parallel to the c-axis[8]. The activation energy for conduction is much higher parallel to the c-axis, 0.45 eV, than perpendicular to it, 0.26 eV. In good polycrystalline samples, the activation energy and the bulk ionic conductivity can both be comparable to that of a single crystal perpendicular to the c-axis.

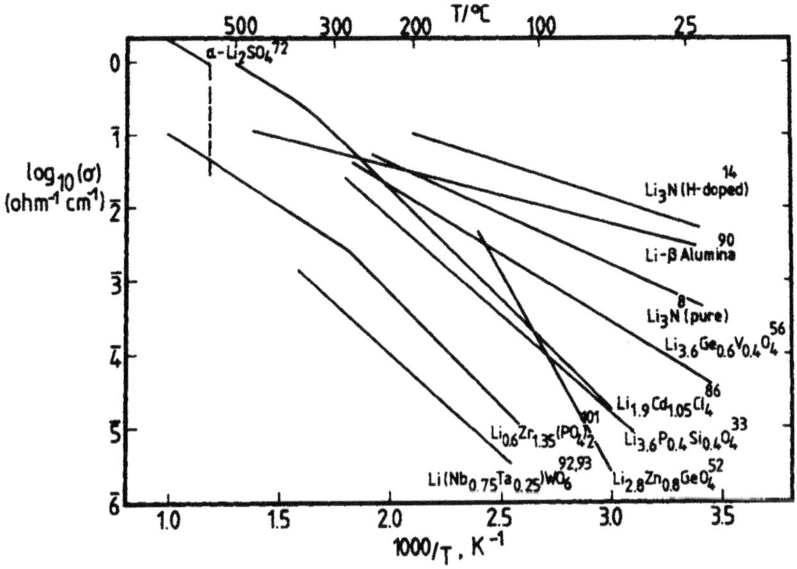

Figure 1 Lithium Ion Conductivity of Various Crystalline Lithium Ion Conductors.

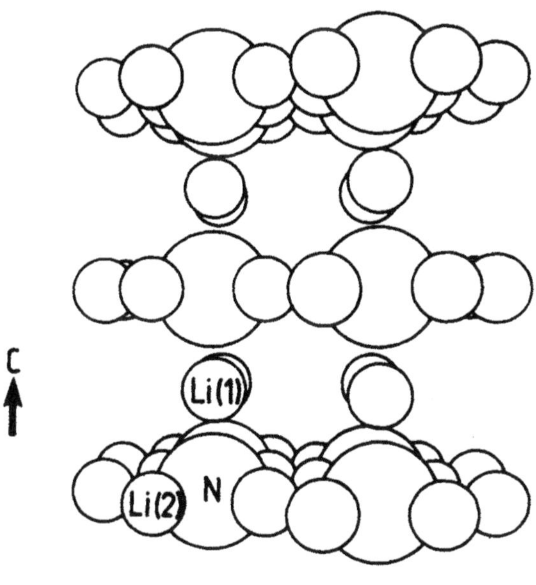

Figure 2 Crystal Structure of Li_3N (9).

NMR experiments[10] indicated that the Li(2) ions in the Li_2N layers, are mobile rather than the Li(1) ions in the Li layers. X-ray data tend to confirm this[11], showing that 1-2% of the Li(2) positions are vacant and that the Li(1) sites are filled.

The vacancies in the Li(2) sites appear to be associated with the presence of hydrogen as an impurity. Wahl[12] has shown that hydrogen is usually present in Li_3N as NH^{2-} ions by infrared spectroscopy and has demonstrated that the conductivity perpendicular to the c-axis increases greatly (x1000) with increasing hydrogen content. It appears that a small amount of hydrogen (<2%) is essential to the high ionic conductivity of Li_3N. Transport measurements[13] have shown that the hydrogen can be mobile, with transport numbers as high as 0.1, and that hydrogen can be discharged at the cathode.

Two mechanisms have been proposed for the enhancement of conductivity by hydrogen doping. One assumes that the hydrogens block the Li sites and so attributes the enhanced conductivity to facilitation of the formation of Frenkel defects due to local distortions in the Li_2N layers[14]. The second assumes that the hydrogens do not block the Li sites and so the enhancement of conductivity is due to extrinsic vacancy formation. The presence of extrinsic defects has been confirmed by positron annihilation experiments, which indicated that vacancies are present below 10K[15]. Doping Li_3N with hydrogen to levels much beyond 2% was found to be impossible[14]. Attempts to improve the conductivity of Li_3N with dopants other than hydrogen have generally proved unsuccessful; for example, partial replacement of Li with aliovalent ions (Mg,Cu,Al) was found to reduce conductivity[14]. Enhanced conductivities have been reported, however, caused by the presence of oxygen, either as O^{2-} or as OH^-, in the Li_3N structure[16].

The problems associated with the low thermodynamic decomposition potential of Li_3N are a major obstacle to its utilisation in ambient temperature batteries. Decomposition was observed when intimate mixtures of Li_3N and cathode active materials (of potential greater than 1.75 V vs Li) were ground together[17]; however, some kinetically stable battery systems have been developed[18,19]. Solid state cells consisting of $Li/Li_3N/LiCF_3SO_3$-PEO/TiS_2[18] and $Li/Li_3N/(Li_3N+TiS_2)$[19] have been constructed and cycled at $\sim$ 150°C successfully.

A number of compounds that are related, compositionally, to Li_3N have been investigated, in order to find good Li^+ conductors that are more stable than Li_3N. Many of these compounds have antifluorite related structures. $Li_{1.8}N_{0.4}Cl_{0.6}$, for example, has a defect antifluorite structure with 10% of the Li sites vacant and a decomposition potential of 2.21 V, but its conductivity is about 1000 times lower than that of Li_3N at 25°C[20]. Phases such as $LiMgN$[21], Li_3AlN_2[21,22], Li_6NBr_3[23] and Li_8SiN_4 [24] have much lower conductivites (0.01% - 1%) than Li_3N. The conductivities obtained in the Li_5NI_2 - LiOH quasi-binary system, with a decomposition potential of $\sim$1.6 V, are reported to be much higher, $\sim 10^{-3}\ \Omega^{-1}cm^{-1}$ at 25°C, being almost comparable with that of Li_3N[25]. Li_5NI_2 has a defect antifluorite structure and in Li_5NI_2 - LiOH, the OH^- ions substitute for the N^{3-} ions[26]. The enhanced conductivity has been attributed to the introduction of Li vacancies and an increase in the Li-N^{3-} (OH^-) distance on substitution[26].

2.2 Li_4SiO_4 - Related Phases and γ Phases (LISICONS)

Two closely-related structure types, based on Li_4SiO_4 and so-called γ phases, have formed the basis of a large number of lithium ion conducting materials, some of which have been used in prototype solid state cells[27,28]. Most of these are ceramic materials, stable to temperatures of 1000°C, or higher and are atmospherically stable. The large number of conducting phases found in these systems arises because a variety of ion substitutions into the parent structure are possible, especially ones which give rise to non-stoichiometric materials.

2.2.1 Li_4SiO_4 and related phases. Stoichiometric Li_4SiO_4 and Li_4GeO_4 have modest lithium ion conductivity, 2×10^{-5} and 4×10^{-6} $ohm^{-1}cm^{-1}$ at 300°C, respectively [29-32]. The conductivity is increased somewhat for 'stoichiometric' solid solutions between Li_4SiO_4 and either Li_4GeO_4 or Li_4TiO_4; these have formulae such as $Li_4(Si_{1-x}Ge_x)O_4$[30-32].

A more dramatic increase in conductivity arises on partial substitution of Si by aliovalent species. This effect was discovered by substitution of pentavalent P for Si[33-36] and has been seen also with substitution of As or V for Si[37-39]. The resulting solid solutions, with formulae such as $Li_{4-x}(P_xSi_{1-x})O_4$, have lithium ion vacancies whose concentration increases with x. At some intermediate x value(s), a change in structure from monoclinic Li_4SiO_4-type to orthorhombic γ-type occurs. These γ solid solutions are also good Li ion conductors and are discussed in the next section. In the $Li_{4-x}(P_xSi_{1-x})O_4$ series[40], the Li_4SiO_4 solid solution extend to x=0.12 under equilibrium conditions, but by melt quenching can be extended to x=0.60. In the corresponding As and V series, the stable solid solutions extend to x=0.15 and 0.40, respectively[37.38].

The increase in conductivity which occurs with increasing x is attributed to a combination of an increase in the concentration of lithium ion vacancies, which are mobile and a decrease in the activation energy for migration. The significance of the vacancy concentration term was shown by consideration of the conductivity preexponential factors; these pass through a broad maximum for intermediate compositions in fair agreement with theoretical predictions[37,38].

The crystal structure of Li_4SiO_4 is, in principle, simple with lithium ions occupying sites between isolated SiO_4 tetrahedra[41], but some crystals were shown to possess a complex weak superstructure associated with ordering of lithium ions into 19 out of 42 possible sites in the supercell[42]. A further complexity is that Li_4SiO_4 shows complex polymorphism at high temperature which may be associated with redistribution of the lithium ions[43]. The high conductivity of the $Li_{4-x}(P_xSi_{1-x})O_4$ solid solutions was attributed to the good 3D connectivity between the Li sites, as well as their partial occupancy[42].

Li_4SiO_4 can be partially substituted with a range of other aliovalent cations, including Al,Ga[36,44,45] and Zn,Mg[43,46]. The Al,Ga-containing materials are particularly interesting because the host Li_4SiO_4 structure can be substituted in two ways, giving the series, $Li_{4+x}(M_xSi_{1-x})O_4$ and $Li_{4-3y}M_ySiO_4$: M=Al,Ga. For Al, the solid solution limits are x=0.40, y=0.06[44]; for Ga, they are x=0.30 and y=0.25[45]. The first series leads to the creation of interstitial lithium ions, whereas the second leads to vacancies; in both, dramatic increases in conductivity occur. Highest conductivities were found in the second series, M=Ga, eg $5x10^{-5}$ $ohm^{-1}cm^{-1}$ at 150°C for y=0.25.

The crystal structures of Li_4GeO_4[47] and Li_4TiO_4[48,49] are different to that of Li_4SiO_4[41,42] and no aliovalent doping of these structures has been reported. Instead, when attempts are made to dope them, γ phases form, as dicussed in the next section.

2.2.2 <u>The γ phases</u>. These phases are sometimes also called LISICONS. The first to be discovered, $Li_{14}Zn(GeO_4)_4$[50], is an interstitial solid solution based on the high temperature γ-Li_2ZnGeO_4 structure[51], with the substitution mechanism, 2Li ⇌ Zn. $Li_{14}Zn(GeO_4)_4$ has very high conductivity at high temperature, eg. 0.13 $ohm^{-1}cm^{-1}$ at 300^oC, but lithium ion trapping at lower temperatures leads to a much reduced conductivity at room temperature[52,53] of 10^{-7} $ohm^{-1}cm^{-1}$. This composition is also thermodynamically unstable below about 630^oC[51,54], reacts with atmospheric CO_2 and was reported to react violently in contact with molten lithium[55]; other γ solid solution compositions within the Li_4GeO_4-Li_2ZnGeO_4 system have been studied, specifically those around $Li_3Zn_{0.5}GeO_4$, which retain the high conductivity at high temperatures but which are thermodynamically stable down to room temperature[52].

A variety of other γ solid solution systems have high Li^+ ion conductivity[33-40,56-60]. These are derived from phases with tetrahedral structures, such as Li_3XO_4:X=P,As,V. The substitution mechanisms are of the type, X ⇌ (Si,Ge,Ti) + Li and lead to the creation of interstitial lithium ions which are responsible for the high conductivity. The highest conductivity at room temperature, $4x10^{-5}ohm^{-1}cm^{-1}$, was found in the series, $Li_{3+x}(Ge_xV_{1-x})O_4$[56] and in related systems containing As and Ti[57]. These phases are generally, thermodynamically stable at all temperatures and do not show the marked decrease in conductivity on ageing at ~25-200^oC that was found with the lithium zinc germanates.

The crystal structures of the stoichiometric γ phases such as γ-Li_3PO_4, contain an array of oxide ions, which is intermediate between hexagonal close packed and tetragonal packed, with the cations distributed over various tetrahedral sites[61-63]; all the LISICONrelated phases have the same parent structure, but contain lithium ions in interstitial sites[50,62,64-67]. Recently, neutron diffraction has been used to locate the interstitial lithium ions, determine their site occupancies and correlate the high conductivity of the lithium ions with the connectivity of the interstitial sites [65,67].

A related γ tetrahedral structure is that of γ-$LiAlO_2$, but it is characterised by a different distribution of cations[63,68-70]. This structure type has also been shown to take up interstitial lithium ions; the parent structure is $Li_5GaSi_2O_8$ and the replacement mechanism is Ga $\rightleftharpoons$ 3Li, giving the series, $Li_{5+3x}Ga_{1-x}Si_2O_8$, with a maximum conductivity at the solid solution limit, $x = 0.5$[45,71]. This solid solution series is interesting as it undergoes a smooth change from a γ structure at low x to a Li_4SiO_4 structure, section 2.1.1, at high x. This is the first high conductivity, interstitial solid solution to be reported that is based on the γ-$LiAlO_2$ structure type, and it is anticipated that others may also be prepared.

2.3 Li_2SO_4 and Related Phases

These materials have been studied for many years by Kvist, Lundén and co-workers in Gothenberg. The high temperature, cubic, α form of Li_2SO_4 has very high conductivity, around 1 $ohm^{-1}cm^{-1}$ between 575°C and its melting point 870°C[72,73], which makes this potentially, a very interesting material. The conductivity of the low temperature monoclinic β form is much less, however, and this structure is of little interest as a solid electrolyte, especially since it has not been possible to dope it and generate either lithium vacancies or interstitials[74,75].

Many attempts have been made, successfully, to introduce dopants into the α-Li_2SO_4 structure but with none of these has it been possible to reduce greatly the temperature of the β-α transition. Consequently, the family of Li_2SO_4-type materials, which also includes Li_2WO_4[76], contains good Li^+ ion conductors only above about 500°C. Some studies on quenched materials have been reported in eg. the system Li_2SO_4-Ag_2SO_4[77], with evidence for enhanced conductivity at room temperature, but the nature of the conducting species was not determined.

The conductivity of Li_2SO_4 is sufficiently high that little change in conductivity occurs on melting; from this, it was deduced that the Li^+ ions in crystalline α-Li_2SO_4 are in essentially liquid-like motion. Comparison of conductivity and diffusion data[72,73] lead to the suggestion that cooperative rotation or ring mechanisms were involved in conduction; this topic has been resurrected more recently, under the guise of paddle wheel mechanisms or rotator phases and indeed, the precise mechanism of conduction is a topic of some current controversy[78-82]. The key issue concerns the SO_4^{2-} groups: are they fixed rigidly in the structure, are they able to undergo some small rotational movements or can they rotate freely? Undoubtedly, there is rotational disorder of the sulphate groups[78,79], but the extent to which this is a dynamic process, responsible for high Li^+ ion conductivity, appears not to be clear. Similar questions have been raised about the mechanism of Na^+ ion conduction in Na_3PO_4.

2.4 Lithium Metal Halide Spinels

A number of lithium metal halides (Li_2MX_4), that have spinel or spinel-related structures, exhibit high lithium ion conductivity[83,84]. The parent structure of these phases is generally an inverse spinel, in which half of the lithiums occupy tetrahedral sites with the remaining lithiums and the metal (M) ions occupying octahedral sites[85]. The highest conductivities have been observed in solid solutions of formula $Li_{2-2x}M_{1+x}Cl_4$ (M=Cd,Mn) with values of 1.5 x $10^{-5}ohm^{-1}cm^{-1}$ at 25°C for M=Mn and 0.35 $ohm^{-1}cm^{-1}$ at 400°C for M=Cd[83,86]. Some bromide spinels (M=Mg,Mn) are also good lithium ion conductors with conductivities up to 7.3 x $10^{-2}ohm^{-1}cm^{-1}$ at 400°C[87].

A number of phase transitions can occur when these halide spinels are heated. The chloride spinels that have the cubic spinel structure at ambient temperatures (M=Mg,Mn,Cd) undergo an order - disorder transition, which is associated with a decrease in the activation energy of conduction from ~ 0.6 to ~ 0.4 eV, at 280°C[83,86,88]. Other spinel chlorides (M = Fe,Cr,Co) have distorted spinel structures, with reduced lithium ion conductivities, at ambient temperatures and transform to the cubic spinels on heating[89]. Several compounds with the cubic spinel structure (M=Mg,Cr,Mn,Cd) transform on heating to a defect rock salt structure[89]. The bromide spinels (M=Mg,Mn) transform on heating from a distorted cubic spinel to a cubic spinel and finally to a defect rock salt structure just before melting[87].

2.5 Other Crystalline Li^+ Ion Conductors

2.5.1 Li beta-alumina. This material apppears to have the highest Li^+ ion conductivity reported to date in crystalline materials, $3x10^{-3}$ $ohm^{-1}cm^{-1}$ at 25°C in a single crystal sample[90]. Li beta-alumina is very sensitive to water pick-up and extreme precautions were required during the synthesis, by ion exchange of Na beta-alumina in LiCl at 650°C, and prior to conductivity measurements, by heating in dried He at 400 - 500°C for 2-4 days, in order to avoid water contamination. Earlier reports[91] had indicated that a mixed Li/Na beta alumina had higher conductivity than that of Li beta alumina, but this was shown not to be the case when pure, dry Li beta alumina was subsequently prepared and measured[90]. Following on from the definitive study on Li beta-alumina[90], which resolved many discrepancies in the literature but which also showed the difficulty in obtaining and using pure, dry material, interest in this material as a practical lithium ion conducting electrolyte appears to have ceased.

2.5.2 Trirutile phases, $Li(Nb,Ta)WO_6$ and ramsdellites. The trirutile materials have cation ordered structures derived from that of rutile and are modest Li^+ ion conductors at high temperature[92,93]. Highest conductivity was found for the solid solution $Li(Nb_{0.75}Ta_{0.25})WO_6$ with values of $1x10^{-5}$ $ohm^{-1}cm^{-1}$ at 150°C rising to $5x10^{-4}$ at 300°C.

The ramsdellite, MnO_2 structure is closely related to that of rutile but is built of double columns of octahedra instead of the single columns present in rutile. $Li_2Ti_3O_7$ has the ramsdellite structure and contains interstitial lithium ions within the channels formed by the framework of linked octahedra[94-96]. The lithium ion conductivity of single crystal $Li_2Ti_3O_7$ is anisotropic by a factor of about 10 to 25, depending on temperature[97]. In the more highly conducting c direction, along the channels, the conductivity varies from $7x10^{-6}$ $ohm^{-1}cm^{-1}$ at 25°C to $1.8x10^{-2}$ $ohm^{-1}cm^{-1}$ at 300°C. The existence of significant conductivity perpendicular to the channels is attributed to partial occupancy of the octahedral framework sites by lithium since these lithiums are able to take part in conduction perpendicular to c.

Related ramsdellite structures occur with solid solutions of $Li_2MgSn_3O_8$ which form by the substitution mechanism $Mg \rightleftharpoons 2Li$[98]. These have formula, $Li_{2+2x}Mg_{1-x}Sn_3O_8$; the conductivity of ceramic samples increases with x, achieving a maximum at x = 0.3 with a value of $3x10^{-4}$ $ohm^{-1}cm^{-1}$ at 300°C.

2.5.3 Lithium zirconium phosphates. The NASICON solid solution phase, $Na_{1+x}Zr_2Si_xP_{3-x}O_{12}$: x=2 is a good sodium ion conductor comparable to beta-alumina, but little is known about possible corresponding lithium materials. The end-member phase $LiZr_2P_3O_{12}$ has been studied[36,99] and fairly high conductivity found in Ta-substituted material, $Li_{0.8}Zr_{1.8}Ta_{0.2}P_3O_{12}$, with a value of $4.7x10^{-6}$ $ohm^{-1}cm^{-1}$ at room temperature rising to $1.5x10^{-3}$ $ohm^{-1}cm^{-1}$ at 200°C. The conductivity of stoichiometric $LiZr_2P_3O_{12}$ is low, rising from $3x10^{-9}$ $ohm^{-1}cm^{-1}$ at 25°C to $3x10^{-5}$ $ohm^{-1}cm^{-1}$ at 300°C[100].

The phase $Li_2ZrP_2O_8$ and its solid solutions, $Li_{2-4x}Zr_{1+x}P_2O_8$[101] have much higher conductivity, 10^{-7} $ohm^{-1}cm^{-1}$ at 25°C rising to 0.001 - 0.01 $ohm^{-1}cm^{-1}$ at 300°C[100]. The conductivity is rather insensitive to composition, with a slight maximum at x=0.35, but is more dependent on sample processing treatment; in addition, the conductivity Arrhenius plots are markedly non-linear. Similar conductivity values were obtained during a study of the system $LiZr_2P_3O_{12}$ - Li_3PO_4, although the samples were not well-characterised from their powder X-ray diffraction data[102].

2.5.4 Framework silicates Eucryptite, $LiAlSiO_4$ has attracted considerable interest as a one dimensional lithium ion conductor with a conductivity along the channel direction of $\sim10^{-9}$ $ohm^{-1}cm^{-1}$ at $\sim$ 25°C rising to 10^{-3} $ohm^{-1}cm^{-1}$ at 300°C[103]. Perpendicular to the channels the conductivity is lower by about three orders of magnitude. The conductivity of eucryptite and high quartz solid solutions on the join $LiAlSiO_4$-SiO_2 has been measured[104]. The conductivity was found to depend more on the structure of the sample, whether it was disordered or had a superstructure, than on the composition. Solid solutions based on the β-spodumene, $LiAlSi_2O_6$ structure, with partial substitution of P,V have been studied, but the conductivities are generally low, 10^{-5} to 10^{-4} $ohm^{-1}cm^{-1}$ at 400°C[105]. The conductivities of corresponding lithium gallium silicates are also low[106]; indeed, that of $LiGaSiO_4$ is so low as to show predominantly residual electronic conductivity.

2.5.5 Other materials Various other crystalline materials have been investigated for possible lithium ion conductivity. Ones which have been reported with reasonably high conductivities include:

- lithium boracites:- $Li_4B_7O_{12}(Cl,Br)$ and $Li_5B_7O_{12}S$, have a 3D cage-like structure built of BO_4 tetrahedra and BO_3 triangles with lithiums in the channel sites with partial occupancies[36,107,108]. Highest conductivities are $\sim 10^{-7}$ $ohm^{-1}cm^{-1}$ at 25°C rising to about 5×10^{-4} at 200°C.
- doped lithium carbonate:- $Li_{2+x}(C_{1-x}B_x)O_3$ has the Li_2CO_3 structure with partial substitution of carbonate by borate and creation of interstitial lithium ions. Conductivity values are similar to those of the boracites[36].
- lithium aluminium chloride, $LiAlCl_4$: room temperature conductivity 1×10^{-6} $ohm^{-1}cm^{-1}$ [109,110]
- antifluorite-related oxides and sulphides. Li_2O and Li_2S, which have the antifluorite structure, are moderate lithium ion conductors at high temperatures[35,111]. The compounds Li_5AlO_4, Li_5GaO_4 and Li_6ZnO_4

have antifluorite-related structures with high vacancy contents and have slightly better conductivity than Li_2O, $6x10^{-4}$ $ohm^{-1}cm^{-1}$, at 400^oC[35]. The conductivity of Li_5GaO_4 has been improved somewhat in the solid solution $Li_{5+x}Ga_{1-x}Zn_xO_4$, to give a conductivity of 4.1×10^{-3} $ohm^{-1}cm^{-1}$ at 300^oC[112].

3. DISCUSSION

The existence of very high lithium ion conductivity, e.g. 0.1 to 1.0 $ohm^{-1}cm^{-1}$ at 25^oC, in crystalline materials remains an elusive dream. The two materials which most closely approach this are H-doped Li_3N and Li beta-alumina, with conductivities around 0.003 $ohm^{-1}cm^{-1}$ at 25^oC. For practical applications, however, both of these materials suffer from severe disadvantages.

Very high Li^+ ion conductivity does occur in some materials, e.g. Li_2SO_4 above 575^oC, but it has not proved possible, as yet, to achieve this state of optimised conductivity at temperatures below about 500^oC. In some non-stoichiometric oxides, such as the γ phases (LISICONS), conductivities as high as 0.01 $\Omega^{-1}cm^{-1}$ at 300^oC are obtained, but the activation energies for conduction are still quite high especially at lower temperatures, indicating a considerable activation barrier to ion migration.

There are no obvious structural features that a material must have in order for it to be a good lithium ion conductor. Thus, the reasonably good lithium ion-conducting materials include 1D structures (eucryptite), layer structures (boracite) and close packed structures (γ phases, Li_2SO_4, halide spinels). There are indications that a more 'soft', covalent bonding leads to higher Li^+ conductivity than hard, ionic bonding, e.g. LiI is the most conducting lithium halide and sulphide glasses have higher Li^+ ion conductivity than oxide glasses.

Most good Li^+ ion conductors are non-stoichiometric materials (α-Li_2SO_4 is the exception, but it has a disordered structure) and undoubtedly, partial occupancy of the 'mobile' sites is a critical factor. This is shown with e.g. Li_4SiO_4, a rather modest Li^+ ion conductor, whose conductivity can be enhanced greatly by the introduction of either lithium vacancies or lithium interstitials.

Site occupancy is not the only factor, however, since often, the activation energy for conduction decreases at the same time as the mobile ion concentration increases. It appears likely that conduction mechanisms involve cooperative interactions between the mobile ions, although as yet, it has not been possible to study these quantitatively. The possibility of cooperative interaction between the mobile ions and the framework, as in the proposed paddle wheel mechanism for Li_2SO_4, is intriguing and requires further study. The occurrence of mixed alkali effects in crystalline as well as glassy systems (in which the presence of a second alkali hinders greatly or poisons the conductivity of the first alkali) is not well-explained but again, must be related to the details of the conduction mechanism and the cooperative nature of ion transport.

ACKNOWLEDGEMENT

ARW thanks the Royal Society of Edinburgh for a Support Fellowship.

4 REFERENCES

1. Takahashi, T. and Yamamoto, O., Electrochim Acta, 11, 779 (1966)
2. Scholtens, B.B. and Van Gool, W., Solid Electrolytes, Eds Hagenmuller, P. and Van Gool, W., Academic, N.Y., 463 (1978)
3. Shahi, K., Wagner, J.B. and Owens, B.B., Lithium Batteries, Ed Gabano, J-P., Academic, London, 407 (1983)
4. Scholtens, B.B. and Van Gool, W., Solid Electrolytes, Eds Hagenmuller, P. and Van Gool, W., Academic, N.Y., 451 (1978)
5. Rabenau, A., Solid State Ionics, 6, 277 (1982)
6. Schonherr, E., Muller, G. and Winkler, E., J. Cryst. Growth, 43, 469 (1978)
7. Boukamp, B.A. and Huggins, R.A., Mat. Res. Bull., 13, 23 (1978)
8. Alpen, U. von, Rabenau, A. and Talat, G.H., Appl. Phys. Lett., 30, 621 (1977)
9. Rabenau, A. and Schulz, H., J. Less Common Metals, 50, 155 (1976)
10. Messer, R., Birli, H. and Differt K., J. Phys., C14, 2731 (1982)
11. Schulz, H. and Thiemann, K.H., Acta Cryst., A35, 309 (1979)
12. Wahl, J., Solid State Commun., 29, 485 (1979)
13. Alpen, U. von, and Bell, M.F., Solid State Ionics, 3/4, 259 (1981)
14. Lapp, T., Skaarup, S. and Hooper, A., Solid State Ionics, 11, 97 (1983)
15. Linderoth, S., Hansen, H.E., Shishkin, A.V., Skaarup, S. and Bentzon, M., Solid State Ionics, 18/19, 897 (1986)
16. Nishida, T., Asai, T. and Kawai, S., Solid State Commun., 48, 701 (1983)
17. Rea, J.R. and Foster, D.L., Mat. Res. Bull., 14, 841 (1979)
18. Knutz, B. and Skaarup, S., Solid State Ionics, 9/10, 371 (1983)
19. Knutz, B. and Skaarup, S., Solid State Ionics, 18/19, 783 (1986)
20. Hartwig, P., Weppner, W. and Wichelhaus, W., Mat. Res. Bull., 14, 493 (1979)
21. Schoch, B., Hartmann, F. and Weppner, W., Solid State Ionics, 18/19, 529 (1986)

22. Yamane, H., Kikkawa, S. and Koizumi, M., Solid State Ionics, 15, 51 (1985)
23. Hartwig, P., Weppner, W., Wichelhaus, W. and Rabenau, A., Solid State Commun., 30, 601 (1979)
24. Yamane, H., Kikkawa, S., Koizumi, M., Solid State Ionics, 25, 183 (1987)
25. Obayashi, H., Nagai, R., Gotoh, A., Mochizuki, S. and Kudo, T., Mat. Res. Bull., 16, 587 (1981)
26. Asai, T., Kawai, S., Nagai,R., and Mochizuki, S., J. Phys. Chem. Solids, 45, 173 (1984)
27. Kanehori, K., Matsumoto, K., Miyauchi, K. and Kudo, T., Solid State Ionics, 9/10, 1445 (1983)
28. Pietro, B.Di and Scrosati, B., J. Power Sources, 2, 387 (1977)
29. Gratzer, W., Bittner, H., Nowotny, H. and Seifert, K., Z. Krist. 133, 260 (1971)
30. West, A.R., J. Appl. Electrochem., 3, 327 (1973)
31. Hodge, I.M., Ingram, M.D. and West, A.R., J. Amer. Ceram. Soc., 59, 360 (1976)
32. Shekhtman, I.G., Burmakin, E.I. and Stepanov, G.K., Inorg. Mater., 20, 613 (1984)
33. Hu, Y-W., Raistrick, I.D. and Huggins, R.A., Mat. Res. Bull., 11, 1227 (1976)
34. Hu, Y-W., Raistrick, I.D. and Huggins, R.A., J. Electrochem. Soc., 124, 1240 (1977)
35. Huggins, R.A., Electrochim. Acta, 22, 773 (1977)
36. Shannon, R.D., Taylor, B.E., English, A.D. and Berzins, T., Electrochim. Acta, 22, 783 (1977)
37. Khorassani, A. and West, A.R., Solid State Ionics, 7, 1 (1982)
38. Khorassani, A. and West, A.R., J. Solid State Chem., 53, 369 (1984)
39. Kuwano, J., Nagamine, M., Higuchi, M. and Kato, M., Denki Kagaku 49, 667 (1981)
40. Khorassani, A., Izquierdo, G. and West, A.R., Mat. Res. Bull., 16, 1561 (1981)
41. Völlenkle, H., Wittmann, A. and Nowotny, H., Monat. Chemie, 99,

1360 (1968)

42. Tranqui, D., Shannon, R.D., Chen, H.-Y., Ijima, S. and Baur, W.H., Acta Crystallogr., B35, 2479 (1979)
43. West, A.R. and Glasser, F.P., J. Mater. Sci., 5, 557 (1970)
44. Jackowska, K. and West, A.R., J. Mater. Sci., 18, 2380 (1983)
45. Quintana, P., Velasco, F. and West, A.R., Solid State Ionics, to be submitted
46. West, A.R. and Glasser, F.P., J. Mater. Sci., 6, 1100 (1971)
47. Völlenkle, H. and Wittman, A., Z. Krist., 128, 66 (1968)
48. Dubey, B.L. and West, A.R., Nature Phys. Sci., 235, 155 (1972)
49. Dubey, B.L. and West, A.R., J. Inorg. Nucl. Chem., 35, 3713 (1973)
50. Hong, H.Y-P., Mat. Res. Bull., 13, 117 (1978)
51. Bruce, P.G. and West, A.R., Mat. Res. Bull., 15, 379 (1980)
52. Bruce, P.G. and West, A.R., J. Solid State Chem., 44, 354 (1982)
53. Bruce, P.G. and West, A.R., J. Solid State Chem., 53, 430 (1984)
54. Kamphorst, J.G. and Hellstrom, E.E., Solid State Ionics, 1, 187 (1980)
55. Alpen, U. von, Bell, M.F., Wichelhaus, W., Cheung, K.Y. and Dudley, G.J., Electrochim. Acta, 23, 1395 (1978)
56. Kuwano, J and West, A.R., Mat. Res. Bull., 15, 1661 (1980)
57. Rodger, A.R., Kuwano, J. and West, A.R., Solid State Ionics, 15, 185 (1985)
58. Li-quan, C., Lian-zhong, W., Guang-can, C., Gang, W. and Zi-rong, L., Solid State Ionics, 9/10, 149 (1983)
59. Ivanov-shits, A.K., and Sigaryov, S.E., Solid State Ionics, 27, 89 (1988)
60. Burmakin, E.I., Cherei, A.A. and Stepanov, G.K., Inorg. Mater., 17, 1363 (1981)
61. Zemann, J., Acta Cryst., B13, 863 (1960)
62. West, A.R. and Bruce, P.G., Acta Cryst., B38, 1891 (1982)
63. West, A.R., Z. Krist., 141, 422 (1975)
64. Plattner, E. and Völlenkle, H., Monatsch. Chem., 110, 693 (1979)
65. Abrahams, I., Bruce, P.G., West, A.R. and David, W.I.F., J. Solid State Chem., in press

66. Baur, W.H., Inorg. Nucl. Chem. Letts., 16, 525 (1980)
67. Fitch, A.N., Fender, B.E.F. and Talbot, J., J. Solid State Chem. 55, 14 (1984)
68. Lejus, A.-M., and Collongues, R., C.R. Acad. Sci. [Paris], 254, 2005 (1962)
69. Bertaut, E.F., Delapalme, A., Bassi, G., Durif-Varambon, A. and Joubert, J.-C., Bull. Soc. fr. Miner. Crist., 88, 103 (1965)
70. Marezio, M., Acta Cryst., 19, 396 (1965)
71. Quintana, P. and West, A.R., J. Solid State Chem., submitted
72. Kvist, A. and Lunden, A., Z. Naturforsch., 20a, 235 (1965)
73. Kvist, A., Z. Naturforsch., 21a, 1221 (1966).
74. Lunden, A., Solid State Ionics, 25, 231 (1987)
75. Deshpande, V.K. and Singh, K., Solid State Ionics, 6, 151 (1982)
76. Kvist, A. and Lunden, A., Z. Naturforsch., 21a, 1509 (1966)
77. Singh, K., Chandrayan, V.R. and Deshpande, V.K., Solid State Ionics, 27, 57 (1988)
78. Nilsson, L., Thomas, J.L. and Tofield, B.C., J. Phys. C., 13, 6441 (1980)
79. Aronsson, R., Knape, G., Lunden, A., Nilsson, L. and Torell, L. Solid State Ionics, 5, 445 (1981)
80. Frech, R. and Cazzanelli, E., Solid State Ionics, 18/19, 491 (1986)
81. Borjesson, L. and Torell, L.M., ibid, 18/19, 582 (1986)
82. Secco, E. vs Lunden, A., Debate at Solid State Ionics conference Garmisch, September, 1987
83. Kanno, R., Takeda, T. and Yamamoto, O., Mat. Res. Bull., 16, 999 (1981)
84. Lutz, H.D., Schmidt, W. and Haeuseler, H., J. Phys. Chem. Solids, 42 , 287 (1981)
85. Van Loon, C.J.J. and Jong, J. de, Acta Cryst., B31, 3549 (1975)
86. Kanno, R., Takeda, T., Tokado, K. and Yamamoto, O., J. Electrochem. Soc., 131, 469 (1984)
87. Kanno, R., Takeda, T., Yamamoto, O., Cros, C., Gang, W. and Hagenmuller, P., J. Electrochem. Soc., 133, 1052 (1986)

88. Lutz, H.D., Schmidt, W. and Haeuseler, H., Z. Anorg. Allg. Chem., 453, 121 (1979)
89 Kanno, R., Takeda, T. and Yamamoto,O., Solid State Ionics, 27/28 (1988) in press
90. Farrington, G.C., Dunn, B.S. and Briant, J.L., Solid State Ionics, 3/4, 405 (1981)
91. Farrington, G.C. and Roth, W.L., Electrochim. Acta, 22, 767 (1977)
92. Ohtsuka, H. Yamaji, A. and Okada, T., Solid State Ionics, 14, 283 (1984)
93. H. Ohtsuka and T. Okada, Solid State Ionics, 20, 141 (1986)
94. Lundberg, M. and Andersson, S., Acta Chem. Scand., 18, 817 (1964)
95. Morosin, B. and Mikkelson, J.C. Jr., Acta Cryst., B35, 798 (1979
96. Abrahams, I., Bruce, P.G., West, A.R. and David, W.I.F., J. Solid State Chem., in press
97. Boyce, J.B. and Mikkelsen, J.C. Jr., Solid State Comm., 31, 741 (1979)
98. Grins, J. and West, A.R., J. Solid State Chem., 65, 265 (1986)
99. Taylor, B.E., English, A.D. and Berzins, T., Mat. Res. Bull., 12, 171 (1977)
100. Chavez, M. de L., Quintana, P. and West, A.R., Mat. Res. Bull., 21, 1411 (1986)
101. Chavez, M. de L., Quintana, P. and West, A.R., Brit. Ceram. Trans. J., 85, 167 (1986)
102. Basu, B., Sundaram, S.K., Maiti, H.S. and Paul, A., Solid State Ionics, 21, 231 (1986)
103. Alpen, U. von, Schultz, H., Talat, G.H. and Bohm, H., Solid State Comm., 23, 911 (1977)
104. Nagel, W. and Bohm, H., Solid State Comm., 42, 625 (1982)
105. Ziadi, A., Mercurio, J.P. and Frit, B., Mat. Res. Bull., 19, 1015 (1984)
106. Quintana, P. and West, A.R., Solid State Ionics, 23, 179 (1987)
107. Levasseur, A., Lloyd, D.J., Fouassier, C. and Hagenmuller, P., J. Solid State Chem., 8, 318 (1973)
108. Reau, J.M., Levasseur, A., Magniez, G., Cales, B., Fouassier, C.

and Hagenmuller, P., Mat. Res. Bull., 11, 1087 (1976)
109. Weppner, W. and Huggins, R.A., Phys. Lett., 58A, 245 (1976)
110. Weppner, W. and Huggins, R.A., J. Electrochem. Soc., 124, 35 (1977)
111. Chadwick, A.V., Flack, K., Harding, J.R. and Strange, J.H., Solid State Ionics, (1988) in press
112. Esaka, T. and Greenblatt, M., Solid State ionics, 21, 255 (1986)

SIGNIFICANCE OF NONSTOICHIOMETRY OF BETA ALUMINA

Lin Zu-Xiang

Shanghai Institute of Ceramics
865 Chang-ning Road, Shanghai 200050 China

1. INTRODUCTION

A whole family of aluminates described by the general formula $M_2O\cdot xAl_2O_3$ is understood under the name beta alumina, where M represents a series of monovalent cations such as Ag^+, Li^+, Na^+, K^+, Rb^+, Cs^+, Tl^+, Ga^+, H_3O^+, $NH_4{}^+$, etc. However, among them sodium beta alumina is the most remarkable one owing to its important application as solid electrolyte in advanced sodium sulfur battery(1).

Numerous works have been contributed to investigations of this family of materials in which scientists are interested not only for their potential applications but also for their special crystal structure for theoretical considerations. Apparently, it is not practical to touch all aspects of these materials in a limited space. In addition, there have been many comprehensive reviews on this topic(2-6).

In this paper the stress will be put on the nonstoichiometry of sodium beta aluminas. As known, the nonstoichiometry of beta alumina is an important feature showing significant influence on its properties. In this text, the term " beta alumina " is used to represent sodium beta alumina including β and β" if no special notes are present.

2. STRUCTURE OF BETA ALUMINA

It is well known that beta alumina has two main modifications, i.

e., β alumina and β'' alumina. The structures of both are well established.

2.1 The Structure Of β Alumina

β alumina was first reported as early as 1916 by Rankin and Merwin[7] but it was considered as a modification of alumina ignoring the presence of sodium oxide. This is also why the name " beta alumina" was assigned. Later, Bragg et al.[8] conducted the first crystal structure analysis and found that it is actually a sodium aluminate.

The main feature of the structure of β alumina was deduced by Bragg et al. in 1931. Later, Beevers and Ross[9] redetermined the structure and arrived at the formula $Na_2O \cdot 11Al_2O_3$. The structure was afterwards refined by Peters et al.[10]

According to Beevers and Ross, the unit cell of β alumina consists of two $Al_{11}O_{16}$ spinel type blocks related to each other by two-fold screw axis parallel to c_o . Each block contains four layers of cubic close-packed oxygen ions. Aluminum atoms are sandwiched between each pair of oxygen layers in both octahedral and tetrahedral intersticial sites as Mg^{2+} and Al^{3+} ions in the structure of $MgAl_2O_4$ spinel. The adjacent spinel-type blocks are held apart by Al-O-Al spacer units. Between the adjacent blocks is the loosely packed conduction plane through the center of the spacer oxygen atom. There are one sodium atom in each plane and two in the unit cell. The resulting symmetry is hexagonal and the space group $P6_3/mmc$. The ideal structure of β alumina is shown in Fig.1. The lattice constants are found to be a=5.594 Å and c=22.53 Å for β alumina with 29% excess soda in comparison with stoichiometric formula[10]. The atomic arrangement in the unit cell can be seen in Tab.1.

Theoretically there are three types of positions for sodium occupation in the conductive plane: (1) The BR site has octahedral coordination to the neighboring oxygen ions in the spinel-type blocks above and below the conductive plane, (2) the aBR site with a coordination number of two lies directly between oxygen ions of the two adjacent spinel type blocks, and (3) the intersticial mO site situated between two neighboring oxygen ions on the conduction plane and has a coordination number of

Tab.1 Atomic arrangement of β and β" aluminas

Atom	Number of equivalent position		Type of site*		Coodination number	
	β	β"	β	β"	β	β"
O(1)	12(k)	18(h)	T	T		
O(2)	12(k)	18(h)	T	T		
O(3)	4(f)	6(c)	T	T		
O(4)	4(e)	6(c)	T	T		
O(5)	2(c)	3(b)	L	L		
Al(1)	12(k)	18(h)	O	O	6	6
Al(2)	4(f)	6(c)	T	T	4	4
Al(3)	4(f)	6(c)	T	T	4	4
Al(4)	2(a)	3(a)	O	O	6	6
Na(1)	2(c)	6(c)	BR	BR/aBR	6	4
Na(2)	6(h)	18(h)	mO	mO	8	8

* T=Tetrahedral, O=Octahedral, L=Linear.

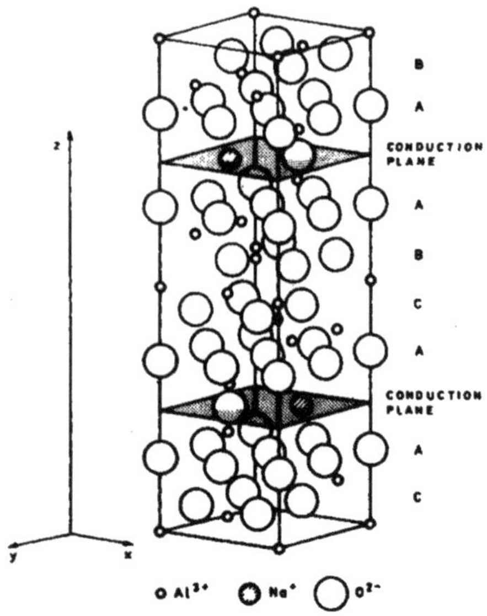

Fig.1 The idealized structure of β alumina(6).

8. Their relations are shown in Fig.2.

Practically, the BR sites accommodate the greater part of sodium ions. The remainder was found by Peters et al.[10] slightly displaced from the mO position. No electron density at aBR position was detected. The distribution of electron density around BR site is triangular and moved somewhat from it. This picture is also true for potassium β alumina. However, it is not the case in silver and thallium β alumina with aBR sites occupied in some degree and mO sites empty.

The difference between Na or K and Ag or Tl β aluminas may be related to the M-O distances on the conduction plane which are 2.87, 2.38 and 2.8 Å for BR, aBR and mO sites respectively. Since the difference 1.47 Å is large enough for all M^+ ions up to Rb^+, it is understandable that most of the mobile ions reside in BR positions. On the other hand, the polarizability of the mobile ions also play an important role. The alkali ions appear to prefer the high coordination number

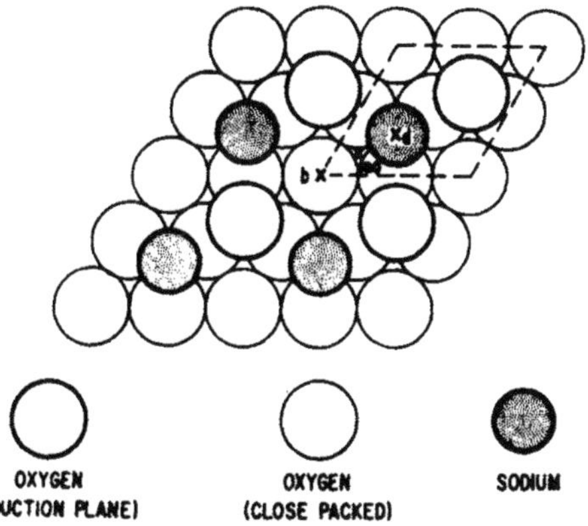

Fig.2 Conduction plane of β alumina[11].

sites as in the case of sodium or potasium β alumina while the more covalent, highly polarizable ions can partially occupy low coordination number sites as in the case of silver or thallium β alumina.

2.2 The Structure Of β" Alumina

β" alumina is a variant of β alumina. The compound $Na_2O{\cdot}5Al_2O_3$ was first reported by Yamaguchi and Suzuki[12] in 1943 and was designated as β" alumina. They also solved the principal features of its structure with powder diffraction patterns[13]. Later, Bettman and Peters determined the structure of a MgO containing ternary β" alumina compound using a single crystal[14]. The idealized composition was $Na_2O{\cdot}MgO{\cdot}5Al_2O_3$. Their work confirmed the β" alumina structure proposed by Yamaguchi et al. and carried out further refinement.

The ideal structure of β" alumina is shown in Fig.3. The difference between the structures of β and β" aluminas consists in that β" alumina has three spinel blocks related by a threefold screw axis parallel to c. in the unit cell comparing two such blocks related by a twofold screw axis in β alumina. As in β alumina, the blocks are seperated by Al-O-Al spacer units, but in β" alumina, alternate sodium sites lie above and below the plane through the center of O(5) atoms. The sodium ions thus reside in an definite volume rather than a plane as in the β

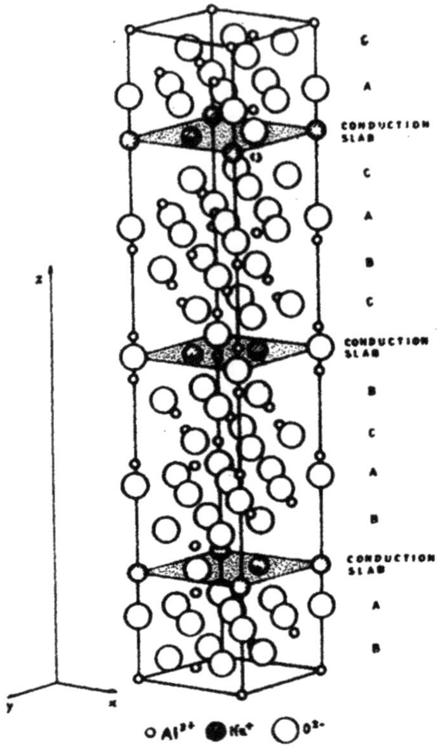

Fig.3. The idealized structure of β" alumina[6].

alumina structure. The theoretical number of sodium ions per conduction layer is two. The resulting symmetry is rhombohedral with R$\bar{3}$m space group and the constants of the triply primitive hexagonal cell of Li_2O stabilized β" alumina with composition $Na_{1.72}Al_{10.66}Li_{0.30}O_{17}$ are: a=5.610 Å the same as in β alumina and c=33.463 Å which is 1.5 times as long as that in β alumina[25]. Its atomic arrangement is shown in Tab.1 in comparison with β alumina.

The important result of the different stacking sequence of oxygen layers in β" alumina is that in the structure of the latter there is no more mirror plane as in the case of β alumina and all sodium sites become consequently crystallographically equivalent as seen in Fig.4. The differences between β and β" aluminas are summarized in Tab.2.

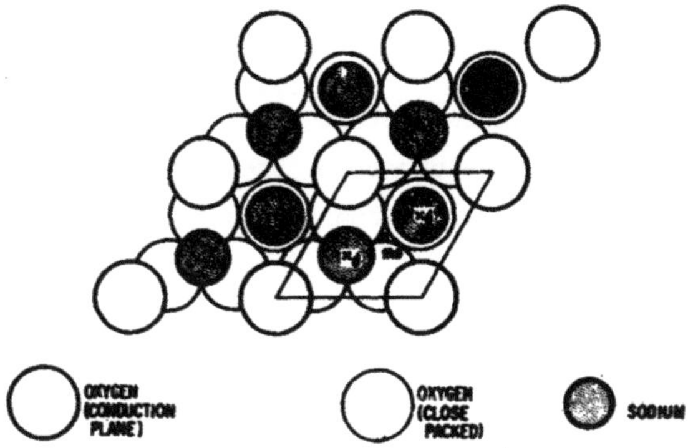

Fig.4 The conduction plane of β" alumina(11)

3. NONSTOICHIOMETRY AND CHARGE COMPENSATION

The structure discussed above are idealized ones. However, beta

Tab.2 Difference between β and β" aluminas

		"
Stoichiometric formula	$Na_2O{\cdot}11Al_2O_3$	$Na_2O{\cdot}5.33Al_2O_3$
Crystalline symmetry	Hexagonal	Rhombohedral
Space group	$P6_3/mmc$	$R\bar{3}m$
Number of spinel blocks in the unit cell	2	3
Cell constants	a	a
	c	1.5c
Mirror symmetry of the conduction layer	Yes	No
Axis of symmetry	Twofold screw axis	Threefold screw axis

alumina are well known by their nonstoichiometry. The problems concerning charge compensation mechanism naturally evolved and were actively investigated in recent years.

3.1 β Alumina

3.1.1 Undoped β alumina

Although β alumina has an idealized formula $Na_2O \cdot 11Al_2O_3$ the more appropriate representation should be $(Na_2O)_{1+x} \cdot 11Al_2O_3$, or $Na_{1+x}Al_{11}O_{17+x/2}$ because it is never stoichiometric as prepared and an excess of sodium with different x is always observed. Thus, the problem that how the remaining positive charges are compensated in the structure is raised.

A straightforward resolution of this problem seems that the charges of excess sodium can be compensated by aluminum vacancies in the spinel type block or extra oxygen occupying the interstitial positions in the conduction layer. The Al vacancies were identified by Peters et al. with X-ray analysis[10]. They found that the occupation of the 12-fold Al(1) atoms is ~3% below the mean percentage of the remaining aluminum atoms and proposed that the excess sodium was stabilized by counterion defects of the aluminum vacancies. In further work completed by Roth et al.[15] by means of neutron diffraction, however, a Flenkel type defect with the formula $V_{Al}Al_i$-O_i-Al_iV_{Al} was found where $V_{Al}Al_i$ is an Al vacancy-interstitial pair. It is formed by the displacement of an Al(1) atom into an adjacent interstitial site. The $V_{Al}Al_i$ is attached through an interstitial oxygen atom O_i to its mirror image in the next spinel block giving the complex linear defect $V_{Al}Al_i$-O_i-Al_iV_{Al}. The presence of each interstitial oxygen allows two extra sodium ions into the conduction plane forming the complex $V_{Al}Al_i\text{-}\underset{Na}{\overset{Na}{O_i}}\text{-}Al_iV_{Al}$.

The Frenkel defects were also suported by the X-ray diffraction study on the structures of K and Tl β aluminas[16]. It was found that the occupation factor of the interstitial Al site is nearly identical with the lack of Al in the Al(1) site. Thus, the $V_{Al}Al_i$-O_i^{2-} -Al_iV_{Al} seems to be the principal means of accommodating excess sodium in undoped β alumina. However, this mechanism can only account for compositions with x up to about 0.35 belonging to usual nonstoichiometric β alumina. For higher sodium level, the substitution of aluminum by cations with lower valence is needed.

3.1.2 Doped β alumina

It is known that the composition of β alumina can be varied over a large range when Al^{3+} in the spinel blocks are substituted by Mg^{2+} ions. Electrical neutrality in this case can be reached with an increased amount of conducting ions and virtually without interstitial oxygen in the conductions layers. Therefore, contrary to the Frenkel type defects, this mechanism allows more departure from the stoichiometry. The chemical composition of this type of compounds can be represented by the general formula $Na_{1+x}Al_{11-x}Mg_xO_{17}$. Up to x=0.66 the β alumina structure could tolerate while only β" alumina structure could be obtaind if x=0.8[17].

In order to study the upper limit of the homogeneity range of β alumina, Collin et al[18] has determined the crystal structure of Mg doped K rich β alumina. Since the distribution of potassium ions in the diffusing planes of K β alumina is very similar to that of its sodium isomorph[16,19], the results can be taken for the consideration of the latter. They have obtained a crystal with a composition of 1.66 sodium ions per average plane unit cell which is the upper limit of the β alumina type structure. For the ion rich composition they observed a $\sqrt{3}\cdot$ a x $\sqrt{3}$ a superlattice where a is the parameter of the parent cell. The model corresponding to the superstructure cell has a high ionic occupation with 3 mO ions surrounding an empty BR site and 2 BR ions on the two remaining ternary points as shown in Fig.5. The ideal composition of this model is 1.66 ions per unit plane cell of β alumina.

According to the two types of charge compensation mechanism, the maximum sodium content which may be accommodated by the β alumina structure was calculated by Newsam.[20] The results showed that for doped alumina x_{max}=0.667 and for undoped β alumina x_{max}=0.571. The atomic distribution deduced for x=0.667 composition represents a scheme with long-range order. The associated two-dimensional supercell has dimensions $\sqrt{3}$ a x $\sqrt{3}$ a. When charge compensation is effected by mO oxygen interstitials in undoped β alumina, there is no requirement for

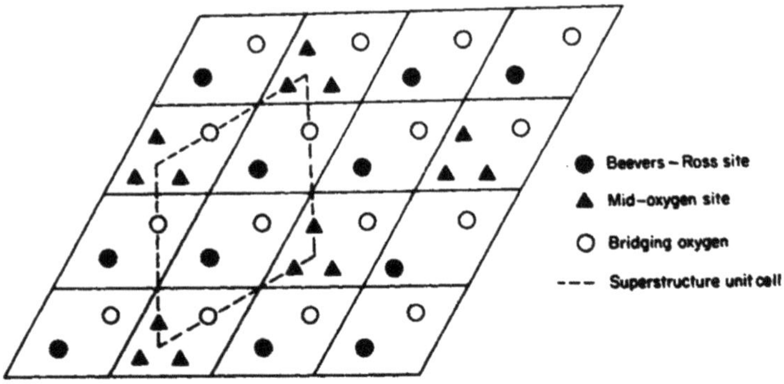

Fig.5 The superlattice of β alumia.[6]

a superlattice ordering. The limiting composition of undoped β alumina deduced above can be exceeded if sodium ions adopt substitutional positions in the spinel blocks. The formula for this type of mechanism will be $Na_{1+x}Al_{11-x/2}Na_{x/2}O_{17}$. Then, according to x_{max}=0.667 for substitution mechanism, the limiting composition will be $Na_{2.0}Al_{10.667}O_{17}$.

It might be seen from above that nonstoichiometry is rather an intrinsic character of β alumina. What are the origins of the driving force for it? West[21] indicated that stoichiometric β alumina, $NaAl_{11}O_{17}$, shows large departure from local electroneutrality. There is an aggregation of positive charges at the center of the spinel blocks and negative charges at its outer sides. Thus, departure from stoichiometry will improve the local electroneutrality by adding extra positive charges to the conduction plane or displacing Al^{3+} ions from the spinel blocks into it and hence stabilize the compound. Walker et al.[22] also showed by an atomistic computer simulation study of β alumina that the formation of all defect species in β alumina is energically favorable and the stoichiometric compound on the contrary would not be expected to be stable.

Nevertheless, stoichiometric β alumina could be prepared though by an indirect route. For example, Boilot et al.[23] approached this composition by the decomposition of NH_4^+ β alumina through a series of chemical reactions to remove the interstitial oxygen and the excess amount of conducting ions. Tofield[24] also reported stoichiometric $DAl_{11}O_{17}$ which was obtained by the action of hydrogen on silver β alumina.

3.2 β" Alumina

It is well known that the β" structure is stable only in the presence of ions with a valence less than three. The excessive positive charges are balanced by the substitution of aluminum with these mono- or divalent ions in the spinel blocks as in doped β alumina. The dopant ions generally used for stabilizing the β" structure are Li^+ and Mg^{2+}. They are found to substitue for the tetrahedrally coordinated Al on Al(2) site[25,26]. Therefore, β" alumina as well as doped ion rich β alumina has the general formula $Na_{1+x}Al_{11-x}M_xO_{17}$, if M is a divalent cation. The compositions with the lowest sodium content are: $Na_{1.67}Al_{10.33}Mg_{0.67}O_{17}$ for Mg^{2+} stabilized β" alumina and $Na_{1.67}Al_{10.67}Li_{0.33}O_{17}$ for Li^+ stabilized one. It is evident from above that at the composition x=0.67 both β and β" structures are possible in spite of the unstability of the β structure. This composition thus represents the maximum Na content for the β structure and the minimum for the β" one. Above this value only the β" structure is stable.

Boilot et al.[27] in X-ray scattering study of β" alumina found that the number of sodium ions per conduction plane is generally less than two indicating the existence of vacant sites. The vacant sites are also ordered leading to a $\sqrt{3}$ a x $\sqrt{3}$ a superlattice in the MgO stabilized β" structure with 1.66 sodium ions per unit plane cell. The local ordering of the cations as a two-dimensional superlattice in β" structure has also been reported by Dudney et al.[28] with polarized Raman measurements for single crystals of Na, K, Ag and Li β" aluminas.

A somewhat different mechanism of Mg stabilized β" alumina was proposed by Harbach[29]. He found two β" alumina phases from the peak splitting on the high resolution X-ray diffraction pattern. They were referred to fully and partially Mg stabilized phases respectively. For the partially stabilized β" alumina, the excess sodium ions are considered as charge balanced by oxygen interstitials in the conduction plane.

Summarizing all mentioned above, it appears that the charge compensation mechanism of beta alumina is rather complex. A general formula $Na_{1+x+y+z}Al_{11-y-z/3}M_yO_{17+x/2}$ can be given if all the mechanisms are taken into account. In this formula the charge of x sodium is balanced by interstitial oxygen, y sodium by divalent cation, and z sodium by aluminum vacancy respectively. Collin et al.[30] have presented an overall summary concerning this topic.

4. INFLUENCE OF NONSTOICHIOMETRY ON PROPERTIES OF BETA ALUMINA

4.1 Electrical Conductivity

4.1.1 Conduction mechanism

The ionic conductivity of beta alumina may be expected from the crystal structure. Since the loosely packed conduction layer are not fully occupied and the M-O distances therein are large enough for most of the monovalent cations, a two-dimensional diffusion of sodium ions is inherently possible, and the conductivity is further enhanced by doping.

Whittingham and Huggins[31] proposed that β alumina probably conducted via an interstitialcy mechanism. It was supported by the potential energy calculation of Wang et al[32,33]. Their results showed that in nonstoichiometric β alumina crystals the extra carrier ions may be paired to form interstitialcy pairs. For the interstitialcy mechanism, a M^+ ion residing in an interstitial site moves into a regular M^+ lattice site, displacing the ion there to another neighboring interstitial site. A mechanism like this gave comparable activation energies to the experimental results. Thus, the conductivity of β alumina is mainly controlled by sodium ion pairs on mO sites. Contrary to Wany's results, the calculation of Walker[34] using computional lattice simulation techniques showed that the stable paired configuration is aBR-aBR. Further calculation completed by Bocquet et al.[35] confirmed that the conductivity mechanism proposed by Wang et al. was quite adapted to β alumina. The role of sodium ion pairs at mO sites in conduction has been tested experimentally by Hattori et al[36]. They substituted sodium ions in β alumina of $1.16Na_2O \cdot 11Al_2O_3$ by ion exchange with Cd^{2+} ions and observed a sharp increase of activation energy after 35% replacement at which nearly all sodium ion pairs at mO sites were exchanged. The interstitialcy mechanism has also been tested by Sutter et al.[37] using similar method as above. Ther also found a sharp increase in activation energy at about 40% sodium replacement. These experimental results are consistent with the conclusions of Wang et al.

Referring to the conduction mechanism of β'' alumina, since BR and aBR sites are identical in this structure, the stoichiometric formula will be $Na_2Al_{10}MgO_{17}$ in Mg stabilized case if all these sites would be occupied. However, the practical number of sodium ions per conduction plane is known less than two. Therefore, a vacancy mechanism is expected for the charge transport.

4.1.2 Influence of doping

From the charge compensation mechanism discussed above, it is not difficult to predict that the conductivity of beta alumina will be increased by doping with cations of lower valence than 3 such as Li^+, Mg^{2+}, Ni^{2+}, Co^{2+}, Cu^{2+}, Zn^{2+}etc. This has been approved by many works.[38-41] The effect is generally explained by the substitution of these cations for Al^{3+} ions in the spinel blocks and consequently the increase of the concentration of sodium ions. If so, the size of the cations must be involved. It has been demonstrated that the cations with ionic radii smaller than 0.97 being able to substitute Al^{3+} in the spinel block will increase the ionic conductivity of beta alumina and the larger ones will have negative effect[42,4]. Harbach[43] has carried out investigations on the influence of the kind of the cations on the conductivity of Li_2O, MgO and NiO stabilized β''alumina. It seems that the variation of the stabilizing agent has no effect on the conductivity if the degree of stoichiometry is kept constant.

4.1.3 Influence of composition

Although beta aluminas are nonstoichiometric compounds, the charge compensation mechanism must be considered in optimization of ceramic compositions. In other words, not any nonstoichiometric composition can give high ionic conductivity. This is especially important in ternary beta aluminas with Li_2O or MgO as dopants. The works[43-45] carried out by Harbach provided favorable evidence referring to this point. The author investigated β" aluminas with chemical compositions corresponding to $Na_{1+x}Mg_xAl_{11-x}O_{17}$ in which x=0.5-1.0 and found that the minimum resistivity occured at x=2/3 which is just the composition of the two-dimensional superstructure model deduced for β" alumina. Therefore, in order to obtain beta alumina ceramics with high conductivity, it is important to rationalize the ratio of the components according to the formula based on theoretical consideration.

4.1.4 Influence of host lattice

It is a well known fact that the β" alumina always has higher ionic conductivity than that of β alumina. Considering that β" alumina has higher sodium content than β, the question raised is whether the different conductivity between β and β" aluminas comes from the difference in sodium concentration only, or host lattice also. The synthesis of ion rich β alumina with similar composition and charge compensation mechanism as β" alumina allows to discuss the influence of the host lattice.

Comparisons have been made between ion rich β and β" aluminas both with compositions $Na_{1+x}Al_{11-x}Mg_xO_{17}$ with x from 0.6 to 0.66 by different authors[46-48]. They have the same charge compensation mechanism of Mg substitution for Al in spinel block. The results obtained demonstrated that β" alumina reveals itself to be a much better host lattice for ionic conduction than β alumina. This is understandable in view of the difference of their structures. As already shown in Fig.4, the high energy aBR site in β alumina becomes energically identical with BR site in β" alumina. In addition, the increase of sodium level in ion rich β alumina is accompanied by a contraction in the c axis[18]. The channel along the conduction path thus becomes narrower than in β" alumina.

4.2 Water Sensitivity

Beta alumina, especially β" alumina, exposed to air tends to pick up moisture and increase in resistivity. Since beta alumina can accommodate water molecules in their conduction layer, the water uptake property seems unavoidable. It is believed that water molecules are located in the conduction plane occupying mO and vacant sodium ion sites[49,50].

It is also known that the kind of mobile ions is an important factor affecting their water uptake property. Li-Na β" alumina is more sensitive to moisture than Na β" alumina. On the other hand, Ag β alumina is unaffected by moisture. It suggests that the water uptake property depends on the size of the monovalent mobile ions and is more evident for the smaller ions. This phenomenon was explained in view of binding force between water molecules and mobile ions[51,52]. The force is the strongest for Li^+, then for Na^+ and the weakest for Ag^+ ion.

For beta alumina ceramics there are two water uptake processes: a fast one controlled by interactions between water molecules and crystal boundaries and a slow one by water diffusion away from the boundary[53]. Since the diffusion of water molecule in β" alumina is considerably higher than in β alumina, the former exhibit higher water sensitivity as known.

It is also essential to emphasize the important influence of chemical composition on the water uptake property of beta alumina. Harbach has investigated β" alumina ceramics of different compositions exposed to 100% relative humidity[54]. It was concluded that partially stabilized β" alumina ceramics are more sensitive to water than fully stabilized ones with compositions $Na_{1+x}Mg_xAl_{11-x}O_{17}$. Here, as in ionic conductivity, appears again the important role of the reasonable nonstoichiometry based on charge compensation mechanism.

4.3 Degradation Property

Sodium sulfur battery is the most important application of beta alumina ceramics and life time of the battery in a high degree depends on the properties of the latter such as chemical stability, ionic conductivity, mechanical strengh etc., which in turn on the first place are determined by the chemical composition of beta alumina.

Investigations on surface of β" alumina revealed that the outer layer of the ceramics is different from its interior. Generally, the surface film is sodium rich and contaminated with impurities. It was formed as sintered[55-57] and is developed during exposure of the ceramics to atmosphere containing mositure and carbon dioxide[58-60]. In the latter case, Na_2CO_3 and $NaHCO_3$ will be formed in addition to water uptake. The surface layer may also contains $NaAlO_2$ if a high level of sodium exists in composition. All these impurities will have a deleterious effect on the ionic conductivity, chemical stability and mechanical strengh of the electrolytes.

The degraded surface is considered as the cause of asymmetric polarization of β" alumina electrolytes in sodium sulfur cells[56] and must be removed by surface treatments[56,61] prior to use in order to obtain reasonable performance of the electrolytes.

5. SUMMARY

In conclusion it is worth to mention that the optimization of the composition of beta alumina for practical application is complicated due to the nonstoichiometry of beta alumina. It is evident from the above discussion that the degree of nonstoichiometry of beta alumina, i.e., whether it is consistent with the charge compensation mechanism, has great influence on most of its properties and also the appearence of second phases. The unsuitable nonstoichiometry may lead to degradation of beta alumina ceramics in practical applications. Therefore, it must be taken into account in the whole fabrication process of beta alumina ceramics from the proportion of the starting components till the final sintering or annealing.

(1) Weder, N. and Kummer, J.T., Proc. Annu. Power Sources Conf. 21, 37 (1967).

(2) Kummer,J.T., Progr. Solid State Chem. 7, 141 (1972).

(3) Kennedy, J.H., in "Solid Electrolytes" (Ed. Geller, S.), Springer-Verlag, p.105 (1977).

(4) Collongues, R., Thery, J. and Boilot, J.P. in "Solid Electrolytes" (Eds. Hagenmuler, P. and Van Gool, W.), Academic Press, p.253 (1978).

(5) Tofield, B.C. in "Intercalation Chemistry" (Eds. Whittingham, M.S. and Jacodson, A.J.), Academic Press, p.181 (1982).

(6) Moseley, P.T. in"Sodium Sulfur Battery" (Eds. Sudworth, J.L. and Tilley, A.R.), Chapman and Hall, p.19 (1985).

(7) Rankin, G.A. and Merwin, H.E., J. Am. Chem.Soc.38, 568 (1916).

(8) Bragg, W.L., Gottfried, C. and West, J.,Z. Krist. 77, 255(1931).

(9) Beevers, C.A. and Ross, M.A.S., Z. Krist 97, 59 (1937).

(10) Peters, C.R., Bettman, M., Moore, J.W. and Glick, M.D., Acta Cryst. B27, 1826 (1971).

(11) Farrington, G.C. and Briant, J.L., Mat. Res. Bull. 13, 763(1978).

(12) Yamaguchi, G., J. Electrochem. Soc., Japan, 18, 260 (1943).

(13) Yamaguchi, G. and Suzuki, K., Bull. Chem. Soc., Japan, 41, 93 (1968).

(14) Bettman, M. and Peters C.R., J. Phys. Chem. 73, 1774 (1969).

(15) Roth, W.L., Reidinger, F. and La Placa, S., in "Superionic Conductors" (Eds. Mahan, G.D.and Roth, W.L.), Plenum Press, p.223 (1976).

(16) Collin, G., Boilot, J.P., Kahn, A., Thery, J. and Comes, R., J. Solid State Chem. 21, 283 (1977).

(17) Collin G., Colomban Ph., Boilot, J.P. and Comes, R., in "Fast Ion Transport in Solids" (Eds. Vashishta, P., Mundy, J.N. and Shenoy, G.K.), Elsevier North Holland, p.309 (1979).

(18) Collin, G., Comes, R., Boilot J.P. and Colomban, Ph., Solid State Ionics 1, 59 (1980).

(19) Dernier, P.D. and Remeika, J.P., J. Solid State Chem. 17, 245 (1976).

(20) Newsam, J.M., Solid State Ionics 6, 129 (1982).
(21) West, A.R., Mat. Res. Bull. 14, 441 (1979).
(22) Walker, J. R. and Catlow C.R.A., Nature 286, 473 (1980).
(23) Boilot, J.P., Colomban Ph., Collin, G. and Comes, R., in "Fast Ion Transport in Solids" (Eds. Vashishta, P., Mundy, J.N. and Shenoy, G.K.), Elsevier North Holland, p.243 (1979).
(24) Tofield, B.C., Jacobson, A.J., England, W.A., Clarke, P.J. and Tomas, M.W., J. Solid State Chem. 30, 1 (1979).
(25) Jorgensen, J.D., Rotella, F.J. and Roth, W.L., Soid State Ionics 5, 143 (1981).
(26) Frase, K.G., ibid. 9/10, 307 (1983).
(27) Boilot, J.P., Collin, G., Colomban, Ph. and Comes, R. Phys. Rev. B22, 5912 (1980).
(28) Dudney, N.J., Bates, J.B. and Brundage, W.E., Solid State Ionics 9/10, 207 (1983).
(29) Harbach, F., J. Mat. Sci. 18, 2437 (1983).
(30) Collin, G., Boilot, J.P., Colomban, Ph. and Comes, R. Extended Abstracts of 6th International Conference on Solid State Ionics, p.71 (1987).
(31) Whittingham, M.S. and Huggins, R.A., J. Electrochem. Soc. 118, 1 (1971).
(32) Wang, J.C., Gaffari, M. and Choi, S., J. Chem. Phys. 63, 772 (1975).
(33) Wang, J.C, Bates, J.B., Dusney, N.J. and Engstrom, H., Solid State Ionics 5, 35 (1981).
(34) Walker, J.R.and Catlow, C.R.A., J. Phys. C. Solid State Phys. 15, 6151 (1982).
(35) Bocquet, J.F. and Lucazeau, G., Solid State Ionics 24, 235 (1987).
(36) Hattori, T., Kurokawa, H. and Mitsuishi, A., ibid. 9/10, 215 (1983).
(37) Sutter, P.H., Cratty, L., Saltzberg, M. and Farrington, G.C., ibid. 9/10, 295 (1983)
(38) Jones, W. and Miles, L.J., Proc. British Ceram. Soc. 19, 161 (1971).

(39) Kennedy, J.H. and Sammells, A.F., J. Electrochem. Soc. 119, 1609 (1972).

(40) Imai, A. and Harata, M., J. Appl. Phys., Japan , 11, 180(1972).

(41) White, D.R., Chen, S., Sankararaman, M. and Sato, H., Solid, State Ionics 18/19, 608 (1986).

(42) Boilot, J.P. and Thery, J. Mat. Res. Bull. 11, 407 (1976).

(43) Harbach, F., Solid State Ionics, 13, 53 (1984).

(44) Harbach, F., ibid. 5, 279 (1981).

(45) Harbach, H., Silicate Industriels 5-6, 81 (1986).

(46) Baffier, N., Badot, J.C. and Colomban, Ph., Mat. Res. Bull. 16,259 (1981).

(47) Boilot, J.P., Collin, G.,Colomban, Ph. and Comes, R. Solid State Ionics 5, 157 (1981).

(48) Collin, G., Comes, R., Boilot, J.P. and Colomban, Ph. ibid. 9/10, 311 (1983).

(49) Kaneda, T., Bates, J.B., Wang, J.C. and Engstrom, H., Mat. Res. Bull. 14, 1053 (1979).

(50) Bates, J.B., Wang, J.C., Dudney, N.J. and Brundage, W.E., Solid State Ionics 9/10, 237 (1983).

(51) Garbarczyk, J., Jakubowski, W. and Wasiucionek, M., ibid. 9/10, 249 (1983).

(52) Garbarczyk, J., Jakubowski, W. and Wasiucionek, M., ibid. 18/19, 653 (1986)

(53) Flor, G., Marini, A., Massarotti, V. and Villa, M., Solid State Ionics 2, 195 (1981).

(54) Harbach, F., ibid. 9/10, 231 (1983).

(55) Singh, R.N., J. Amer. Ceram. Soc. 67, 637 (1984).

(56) Singh, R.N., ibid. 70, 221 (1987).

(57) Yin, Y.B., Chen, K.G. and Lin, Z.X., J. Inorg. Mat., China, 2, 157 (1987)

(58) Heavens, S.N., J. Mat. Sci. 17, 965 (1982).

(59) Bates, J.B. and Anderson, R.L., Solid State Ionics 18/19, 682 (1986).

(60) Roth, W.L., Benenson, R.E. and Chengzhou, J., ibid.27,89(1987).

(61) Singh, R.N., J. Amer. Ceram. Soc. 67, 696 (1984).

SODIUM ION CONDUCTORS WITH NASICON FRAMEWORK STRUCTURE

Klaus-Dieter Kreuer*+, Heinz Kohler* , Joachim Maier+

+ Max-Planck-Institut für Festkörperforschung
Heisenbergstr.1 7000 Stuttgart 80 , F.R.G.

* CONDUCTA GmbH+Co
Dieselstr.24 7016 Gerlingen, F.R.G.

1. INTRODUCTION

The discovery of fast sodium ion transport in ß- and ß''-aluminas more than twenty years ago[1] has stimulated activities in the use of solid electrolytes in electrochemical cells (especially in Na/S secondary batteries which are going to be commercialized as power sources for electric vehicles [2]). The high Na^+ conductance and its low activation enthalpy were unique among all other solid Na^+ conductors known at that time. Accordingly much of the work in the field of solid state ionics was focused on the ß-aluminas from the most different viewpoints. In the meantime the activities associated with Na^+ conductance in ß-aluminas have been moving from research to more engineering development [3], and other interesting features such as mobilty of higher valent cations have attracted attention. For ß-aluminas, sometimes the terms "skeleton" or "framework" structure are used meaning the "rigid" anionic subarray which forms an interconnected space containing the mobile cations on partially occupied lattice sites of identical or similar site binding energies.

Consequently the object in the search for other fast Na^+ conductors were framework structures meeting the above mentioned features such as cristobalite and pyrochlore type compounds, Im3 phases, zeolites, etc. [4]. Some of their Na^+ conductivities are close to those of Na ß-aluminas, but only at elevated temperatures, corresponding to generally higher activation enthalpies (>0.5 eV compared to ~0.2 eV for some ß''-aluminas). During that search the extraordinarily high Na^+ conductance in compounds of the solid solution series $Na_{1+x}Zr_2Si_xP_{3-x}O_{12}$ (in the following termed Nas(x), x=0-3) was discovered [5,6]. The composition with the highest Na^+ conductivity (Nas(2)) has been termed NASICON for NA Super Ionic CONductor. In spite of the controversy about the use of the term "super conductor" for fast ionic conductors we will keep on using the acronym for historical reasons. In the meantime, it is generally being used for any compound with the same principal structure shown in fig.1 .

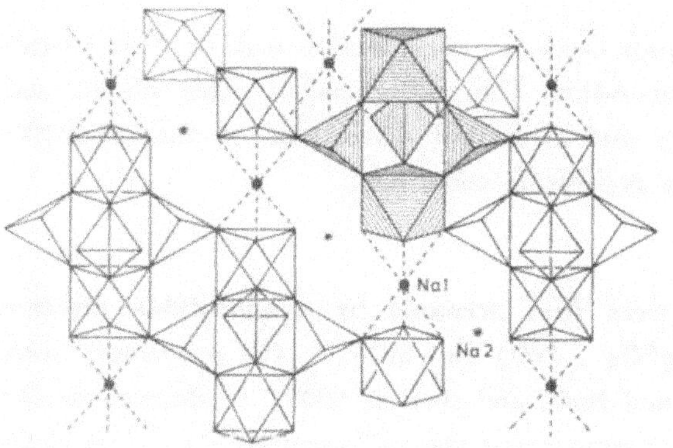

Fig.1
The principle structure of the NASICON framework including the two interstitial sodium sites. One "lantern" is shaded.

Its anionic framework consists of AO_6 octahedra and BO_4 tetrahedra which share corners. In the case of rhombohedral symmetry ($R\bar{3}c$), it provides two cation sites: the octahedrally coordinated Na1 site and the irregularly eight-coordinated Na2 site. Details of the structure are discribed in section 3.

The general approximate formula assuming rhombohedral symmetry can be written as $(Na1)_n(Na2)_m[A_2B_3O_{12}]$ with n=0-1 and m=0-3. As the object of this contribution is Na^+ conductivity we will consider only compounds with sodium ions on the first two sites although many other especially monovalent cations may form compounds with NASICON structure. The octahedral framework site can accept cations (A) of different charge: monovalent (Na^+ partially substituting for higher valent cations), divalent (Mg,Fe,Co), trivalent (In,Sc,Y,Ti,rare earth elements), tetravalent (Zr,Hf,Ti,Sn) and even pentavalent (V,Nb,Ta).

The tetrahedral framework site (B) is generally occuppied by Si and/or P but substitutions by the corresponding homologues Ge and As are also known[7,8]. The formula given above implies a complete occupation of all framework sites, the different charges of the anionic framework are compensated for by varying occupations of Na1 and Na2.

The purpose of this contribution is, rather than to reflect the main interests in the literature, to survey and critically discuss preparation methods, crystal structures, Na^+ conductivity in these structures, deviations from the ideal structures, thermodynamics and finally the resulting consequences for the application of NASICONs in electrochemical cells (see also ref.9.).

The literature about NASICON is abundant. Therefore, we will focus on selected papers in which the interested reader may find further references.

2. PREPARATION

For many NASICON compounds there is a considerable scatter of properties depending on the route of preparation. The appearance of other phases and variations of the stoichiometry and structural distortions of the NASICON phase are a direct result of the peparation conditions.

2.1. Solid State Reaction

Parent NASICONs (fig.3) were first prepared by a subsolidus reaction starting from Na_2CO_3, $NH_4H_2PO_4$, ZrO_2 and SiO_2 [4]. The precursors were kept at 170°C to release NH_3 and H_2O, and then at 900°C to decompose the carbonate. A careful study [10] demonstrated that all components except ZrO_2 are involved in a sequence of reactions leading to different reactive intermediates. At temperatures not far below 1000°C the remaining ZrO_2 starts to dissolve and to react slowly with the other components which have already formed a melt at that temparature. Above 1000°C intermediately formed metaphosphates decompose into orthophosphates and Nas(x) starts to form. Even after several cycles of ballmilling and firing, complete conversion into Nas(x) has never been achieved by this procedure. There was always some monoclinic ZrO_2 left over.

Therefore, more reactive starting materials have been chosen as a source for ZrO_2. The use of $ZrSiO_4$ [11,12], Zirconacetylacetonate [13] or $(NH_4)_3ZrOH(CO_3)_3.2H_2O$ [14] could reduce but not completely avoid the presence of ZrO_2 in the final product.

A common observation of all processes is that along with the conversion of tetragonal ZrO_2 (which usually forms as a reactive intermediate) into its monoclinic form this phase is excluded from further reaction. As a result the final product is contaminated by essentially a glassy sodium silicate grain boundary phase and monoclinic ZrO_2.

The second problem is the volatilization of P_2O_5 and Na_2O in open systems, which slowly sets in at about 900°C and 1300°C respectively (e.g., refs. 10,15). Thus, the total composition is shifted and ZrO_2, $Na_2ZrSi_2O_7$ and/or $Na_5ZrP_3O_{12}$ form as extra phases (fig.2).

The third difficulty is the incongruent melting of all Nas(x). The peritectic reactions involve Nas(x), melt, $ZrSiO_4$ and/or inert monoclinic ZrO_2 (fig.18).

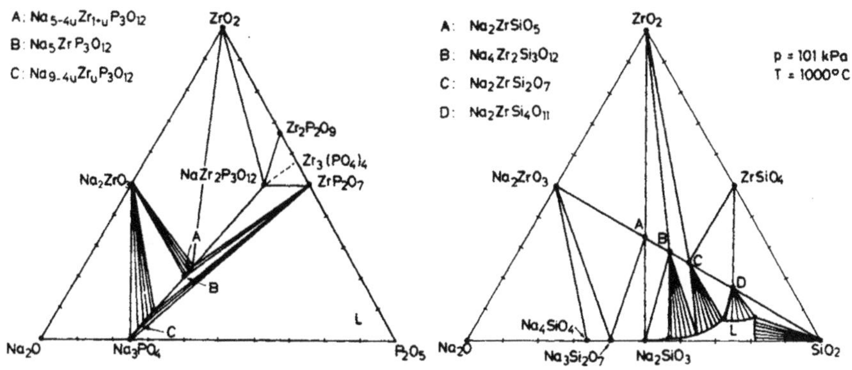

Fig.2 Phase relationships in the ternary systems to which the parent NASICON end members Nas(0) and Nas(3) belong [15].

Therefore, one has to avoid heating above the solidus temperature in any case.

As a consequence of this understanding a simple method has been developed recently for preparing monophase Nas(x) ceramics in the composition range $1.5 \le x \le 3$ starting from $ZrSiO_4$, ZrP_2O_7, Na_2CO_3 and Na_2SiO_3 [15]. Important for a good result are meeting the required stoichiometry, good homogeneity of the precursor and reacting the materials at relatively low temparatures to avoid local precipitation of monoclinic ZrO_2. The dried starting powder is kept at 800°C for one day. The resulting hard product is ground, homogenized again and heated to 1000°C for one hour to volatilize CO_2 completely. The decomposition of Na_2CO_3 goes along with some loss of Na_2O which can be monitored by the corresponding weight loss and which has been compensated by adding an adequate amount of Na_2CO_3. Keeping the powder at 1100°C for one day leads to the desired monophase material (within the resolution of x-ray powder diffractometry and optical microscopy). After isostatic pressing and firing 100°C below the solidus temperature for 1-3 hours 90-95% of the theoretical density are achieved.

Elements substituting for Zr are introduced into the preparation as oxides.

2.2 Sol/Gél Techniques

A near at hand route to attain the desired homogeneities of the NASICON precursors are sol/gel techniques which have become very much en vogue as a low temperature preparation method for ceramics.

The crucial point of this approach is the condensation of all NASICON constituents from clear solutions to form a gel in such a way that no selective precipitation takes place thus achieving homogeneity on a molecular scale. A convenient way is to hydrolyse alcoxides (methoxides, ethoxides, propoxides) from alcoholic solutions (ethanol, propanol). As the rates of polymerization are different for the different alcoxides one has to form soluble copolymers prior to the total hydrolysation. This can be achieved by adding the different alcoxides in a partially prehydrolysed state [16,17,18] or just by stirring [19] or refluxing them [17]. Expensive alcoxides can be partially and even completely substituted by appropriate inorganic compounds [15,18,20]. From the viscous gels the solvents (water and alcohols) are released by careful heat treatment. In the case of gels with low solvent concentrations one may obtain a compact transparent product which is referred to as NASIGLASS [16]; in the case of higher solvent concentrations the product is a fine white amorphous powder. NASICON can be directly formed from these precursors by further heating. Nas(0) and Nas(3) already start to crystallize at 700°C, Nas(x) at about 900°C. Most products are free of ZrO_2 (tetragonal ZrO_2 appears only as an intermediate). At this stage the products are poorly crystallized with a grain size of <1 μm (suitable for silk screen printing [20]), however, sintering under pressure below 800°C allows one to obtain densities higher than 95% of the theoretical value [21]. At ambient pressure 97% density was achieved below 1000°C [17], but 1200°C was required to reach the final conductivity. This demonstrates that all intermediates formed below this temperature are still structurally disordered, which, however, can have certain advantages as will be discussed below. The problems occuring at the temperatures of the final preparation step are similar to those discussed in 2.1..

The use of mixtures of different precursors (hybrid methods) renders possible similar densities at even lower temperature (700°C) [22].

Most elements substituting for Zr in Nas(x) can also be introduced into the preparation as alcoxides [19].

2.3. Stoichiometry of Nas(x)

Because of the initial difficulties to prepare the NASICON parent compounds, Nas(x), free of ZrO_2, there has been a continuing debate about their stoichiometry. Even a new solid solution series limited by the end members $NaZr_2P_3O_{12}$ and $Na_4ZrSi_3O_{10}$ has been suggested [23] but in the meantime it has frequently been observed that these compositions essentially yield NASICON plus a glassy

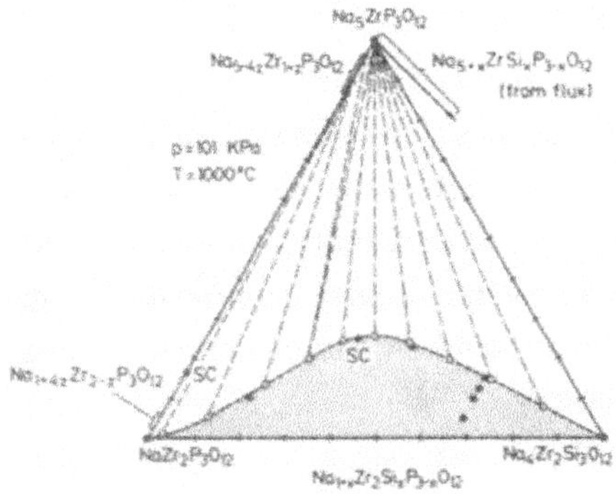

Fig.3
Approximate stoichiometry range of parent NASICON (SC: composition of single crystals [27,54], •: obtained as monophase material [15]) and the compositional extension of other phases in the quasi ternary and quarternary phase diagram respectively (see text).

phase (e.g. ref.24) which is likewise a good ionic conductor. ZrO_2 deficiency has also been claimed for NASICONs prepared by the decomposition of hydrothermally grown precursors [25,26]. Chemical analysis [25] and powder diffraction experiments [26], which will be discussed later, suggest a 0.32 ZrO_2 deficiency with respect to one formula unit (x=1.84). This, however, requires some condensation of framework polyhedra, which is not compatible with the crystal structure (see also 3.3.1.) and could not be verified by ir spectroscopy [15].

In the same contributions the substitution of Zr^{4+} on the octahedral framework position by 4 Na^+ has been suggested, which is in accordance with earlier crystal structure determinations of other authors [27] (see 3.). Equilibrium experiments at 1000°C in sealed platinum tubes and subsequent phase analysis yielded the approximate extension of this substitution represented by the formula $Na_{1+x+4y}Zr_{2-y}Si_xP_{3-x}O_{12}$ (in the following termed Nas(x,y)) (fig.3). The figure also shows compositions obtained from single crystal structure determinations [27,54] (sc) and the stoichiometry ranges of $Na_{1+4x}Zr_{2-x}P_3O_{12}$ (x=0.02-0.06)[28], $Na_{5-4x}Zr_{1+x}P_3O_{12}$ (x=0.04-0.11)[29] and $Na_{5+x}ZrSi_xP_{3-x}O_{12}$ (x=0-0.5)[30]. There is some discrepancy about the stoichiometry of pure phosphate NASICON. It has been suspected that Nas(0,0) actually does not exist, and, instead, the composition range Nas(0,z) with z=0.02-0.06 has been suggested from equilibrium experiments at 1000°C [28].

It should be noted that these results which were obtained in different laboratories, have been doubted [31]. From the results of synthesis experiments in open crucibles the authors claim that NASICON only exists in the initially proposed composition range. However, the preparation temperature of 1150°C was probably higher than the decomposition temperature of Zr deficient NASICON [32].

2.4. Growth of Single Crystals

Most of the uncertainties in the understanding of the properties of NASICONs arose from the lack of single crystals. As pointed out above the ternary end members, i.e. the pure silicate Nas(3) and phosphate Nas(0) already start to crystallize from the corresponding gel derived precursors around 700°C whereas the formation of the solid solutions Nas(x) require significantly higher temperatures. Indeed, crystals of Nas(3)[33)] and Nas(0) [34)] had been grown already before the discovery of their Na^+ conductivity.

A general trend is that phosphate NASICONs can be grown easily (up to a size of several milimeters), silicates crystallize a little harder and solid solutions of both had been obtained only from powders by grain growth at elevated temperatures for a long time [15,35)] (see 3.3.1.). Some attempts have been made to grow Nas(x) from fluxes [15,36)] or hydrothermally [37)] but always pure Nas(3) has been found.

The conditions for the hydrothermal growth of Nas(3) are: a molar ratio of ZrO_2 and SiO_2 of 2/3, NaOH as a source for Na_2O in large excess and a concentration higher than 30% by weight. At pressures greater than $4x10^7$Pa and a temperature higher than 330°C single crystals up to a size of 0.5 mm can be obtained after several days [37)]. Single crstals with a composition close to Nas(3) (see 3.) can also be grown in a flux of NaF and V_2O_5 in which the starting compounds $NaPO_3$, SiO_2, ZrO_2 and Na_2CO_3 are dissolved and react at temperatures around 900°C [36)].

A straightforward way to grow phosphate NASICONs is to keep an adequate powder precursor at a temperature not too far below the solidus temperature for several hours. Thus the preparation of Nas(0) and $Na_3Sc_2P_3O_{12}$ single crystals suitable for x-ray diffraction experiments succeeded [4,38)].

Lower temperatures are required for flux methods. Using an excess of B_2O_3 as a solvent Nas(0) can be crystallized at 1200°C within one week up to a size of 1mm [15,39)]. Flux techniques have also been applied for the growth of the related compound $Na_5ZrP_3O_{12}$ using $Na_4P_2O_7$ as a solvent between 1150°C and 1190°C [40)]. The most favorable technique for preparing $Na_3Sc_2P_3O_{12}$ is also a flux technique [38,41)] around 1100°C. By the classic Kyropolous technique, this composition could be obtained according to the reaction:

$$8ScPO_4 + 3Na_4P_2O_7 \longrightarrow 4Na_3Sc_2P_3O_{12} + P_2O_5$$ [41)]

Bigger crystals of phosphate NASICONs containing rare earth elements can be precipitated from aqueous solutions at even lower temperatures. Because of the negative temperature coefficient of their solubility in H_3PO_4 single crystal growth sets in upon heating adequate solutions slowly (5 °C h^{-1}) from 250-300°C to 500-600°C. In this way crystals of $Na_2(R,Co)ZrP_3O_{12}$, (R= La,Ce, Pr,Nd) have been obtained [42]. In ref. 42 a hydrothermal technique operating between 200°C and 300°C using an autoclave with teflon liners is discribed. As a curiosity the readers attention should be directed to single crystals of $NaTi_2P_3O_{12}$ which were just found in an autoclave with titanium liners initially filled with CdO, $(NH_4)_2HPO_4$, NaCl and H_2O [43].

For all NASICON compositions the precise structure and the corresponding properties of the crystals depend significantly on the preparation temperature which for example in the case of the phosphates may range from 200°C to 1600°C. The anionic framework at ambient temperature usually reflects a quenched state in which the mobile Na^+ is detained in a relative minimum of the free enthalpy at a given temperature. In the following sections we therefore will always refer to the preparation method resulting in specific crystal structures and properties.

3. CRYSTAL STRUCTURES

This section summarizes and discusses the results from x-ray and neutron diffraction experiments. After a brief description of some features common for all NASICONs deviations from the "ideal structure" will be presented for different compositions. Octahedral framework site substitutions, phase transitions, interstitial occupancies and order/disorder phenomena will be discussed.

3.1. Description of the NASICON Structure

In fact, the NASICON framework structure is a three dimensional network of AO_6 octahedra and BO_4 tetrahedra which share corners. It exhibits A-O-B bonds only. For reasons of a plainer discription many authors gather up two octahedra and three tetrahedra to a polyeder group called "lantern". Fig.1 shows such a lantern together with the interstitial sites Na1 and Na2 and the bonds to the neighbouring lanterns. If the symmetry is rhombohedral ($R\bar{3}c$) the two octahedra of a lantern (with A (12c) on the three fold axis parallel to c) have two faces parallel (001). Faces opposite one another are bridged by three tetrahedra with one edge approximately parallel to c and the opposite edge perpendicular to c.

The Na1 site (6b, point symmetry $\bar{3}$) is located in an octahedron formed by two opposite AO_6 octahedron faces. Between neighbouring lanterns the Na2 site is located. Both sites, B and Na2, are special (18e) and they are located on a common twofold axis in (001). Na2 is coordinated by eight oxygens. The coordination polyhedra of both interstitial sites are unusually wide and are connected with one another by a more or less wide channel (see 4.1.).

3.2. The monoclinic/rhombohedral Phase Transition

Most compounds with NASICON structure are rhombohedral at high temperatures only. Upon cooling their framework undergoes a distortion. The resulting lowering of the cell symmetry (fig.6 in ref.6) gives rise to a splitting of diffraction peaks, which are indexed on a monoclinic cell (C2/c) according to the relations: $h_m=h_{hex}-k_{hex}$, $k_m=h_{hex}+k_{hex}$, $l_m=1/3(l_{hex}-h_{hex}+k_{hex}$. In ref.44 the relationships between the rhombohedral (hexagonal) and the monoclinic notations are given.

X-ray powder diffraction data of $Na_3Zr_2Si_2PO_{12}$ compiled at room temperature (C2/c) and at 520K ($R\bar{3}c$) are listed in ref.45. In many cases the peak splitting is too small to be recognized in diffraction experiments with conventional X-ray sources. Using synchrotron radiation the monoclinic symmetry of the pure silicate Nas(3,0.2) at room temperature was demonstated [46] by peak profile recordes below and above the transition temperature. (Silicate rich compositions of Nas(x) (x>2.2) were initially thought to be rhombohedral [6] (see also fig.6)).

The framework distortion gives rise to a splitting of Na2 and B ($R\bar{3}c$) into two sites each (i.e. 3Na2(hex) > Na2(m)+2Na3(m)). For the reason of better comparability the same notation of sites (Na1, Na2) will be used for both phases in the following discussion.

3.3 Structural Variations

So far we have considered the structure of solid solutions confined by the end members $NaZr_2P_3O_{12}$ and $Na_4Zr_2Si_3O_{12}$ (Nas(x)). However, as already pointed out in 2.3., this solid solution range has to be extended according to the substitution Zr^{4+}<> $4Na^+$. Even other compositions within the quarternary system have been found with the same principal structure (e.g. $Na_5ZrP_3O_{12}$). In the following the structural consequences for two compositional variations will be described.

3.3.1. Zr-deficient NASICON. The first structure determination using a single crystal within the composition range of high Na^+conductivities was published in refs. 27,48. The tiny crystal had been selected from a batch of crystals which were all triple-twinned at room temperature. It was actually the only one which was found to be a single crystal above the temperature of the phase transition. The material had been obtained by grain growth of a ceramics ($1.5Na_2O$, $1.33ZrO_2$, $2SiO_2$, $0.5P_2O_5$) in the presence of NaF. The material had been sealed in a platinum tube and kept at 1200°C for three days. The composition of the crystal was determined from the Si/P ratio (calculated from the (Si,P)-O bond length) and the cation site occupancies. It was found to be $Na_{3.3}Zr_{1.8}Si_{1.4}P_{1.6}O_{12}$ (Nas(1.4,0.2)), i.e. 0.2 Zr^{+4} are substituted by 0.8 Na^+ with respect to Nas(1.4). As an interesting structural consequence (rhombohedral high temperature phase) some of the Na^+ was found within the framework octahedron (A) approximately 0.85Å off-centered along the c-axis, the sum of the occupancies of both, Na^+ and Zr^{4+}, amounting to 95%.

Whereas this crystal was shown to be triple-twinned at room temperature even within the high peak profile resolution of synchrotron radiation diffraction[49)], other crystals of Nas(2.1,0) used for structure determination (grown from a sol/gel precursor) were found to be systematically hexa-twinned [50)].

Extended structure investigations on several NASICON powders based on Rietfeld refinements of X-ray and neutron time of flight diffraction data [51)] also revealed considerable Zr-deficiencies. Part of the Na^+ required for charge compensation was found on the octahedral framework site (A)[26)] in agreement with the results presented in 3.4. (table1). Assuming fully occupied A and B sites all refinements demonstrated an occupancy for the framework oxygen site remarkably close to 100% according to 12 oxygen per formula unit. But, because of the cation deficiency with respect to the oxygen found in the same studies, a ZrO_2 deficient composition ($Na_{1+4y+x}Zr_{2-y-z}Si_xP_3O_{12-2z}$) has been proposed. This, however, is not compatible with the structure of the NASICON framework (octahedra and tetrahedra sharing corners only) which had been assumed in the refinements. The observed cation deficiencies may rather be explained by:

1) the use of solely anisotropic, harmonic temperature factors in the refinements of the sodium sites. This might have pretended too low site occupancies because of anharmonic contributions to the temperature factors which have recently been shown to be significant already at room temperature [50)].

2) the presence of protons (in the form of OH-groups) in the hydrothermally and sol/gel derived samples. This has already been suggested for a Nas(3,0.2) single crystal [36)].

3.2.2. Compositions in the vicinity of $Na_5ZrP_3O_{12}$. $Na_5ZrP_3O_{12}$ (Nas(0,1)) does not fit into the composition range of parent NASICON (fig.3). A single crystal structure determination at 200°C, however, revealed the same principal NASICON structure [40]. As for Zr-deficient parent NASICON sodium was found on the octahedral framework site (new Na3 site), however well ordered with Zr according to the sequence Na1-Na3-Na3-Na1- Zr-Zr-Na1 along the c-axis (instead of Na1-Zr-Zr-Na1-Zr-Zr-Na1 for Nas(x)). The space group R32 (instead of $R\bar{3}c$) accounts for a slight splitting of the Na2 and the B-site into two sites each.

Although all cation sites are (almost) completely occupied for this composition, NASICONs with even higher sodium contents have been prepared [30] (fig.3). At room temperature in Nas(0.5,1) the additional sodium was found on a new Na4 site located in an octahedron formed by the two opposite ZrO_6 octahedron faces in the center of the lantern (fig.1). Off-centering of Zr from the special Zr site (0,0,z) and a considerable occupancy of the Na4 site (~20%) were interpreted with different super structures based on the superposition of monoclinic domains [30]. But above the first-order phase transitions at 86°C and 129°C there exists no super structure and a neglegible Na4 occupancy which is not compensated by additional Na on another site. This has been observed for both compositions, Nas(0,1) and Nas(0.5,1), and may raise doubts about the completeness of these structure descriptions.

3.4. Interstitial Site Occupancies

On account of their particular relevance for the Na^+ conductivity the sodium site occupancies as a function of temperature and composition are presented separately. The total sodium content is determined by the substitutions $Si^{+4}+Na^+ <> P^{+5}$, $M^{(4-z)+}+zNa^+ <> Zr^{+4}$ and $4Na^+ <> Zr^{4+}$. Its distribution over the different sodium sites is still controversial as can be seen from some representative data compiled in table 1.

But some of the data is not directely comparable because its refinement has been carried out together with the refinement of a so-called mid-Na occupancy, which must lead to lower occupancies of Na1 and Na2. However, the presence of a mid-Na site [52] in Nas(x,y) located between the "bottlenecks" (see 4.1.2.) within the diffusion path is being doubted. First of all the average mid-Na-O distance is 0.2Å shorter than the average Na1-O distance. This is not in accordance with the unusually high temperature factors of this site (a factor of two or more higher than for Na1 and Na2). Indeed, the bent diffusion path between Na1 and Na2 can be hardly discribed by anharmonic temperature factors of Na1 and Na2 only (see 4.1.). The introduction of the mid-Na site can, therefore, be considered to be a mathematical aid for an improved diffusion path description [53] rather than a new crystallographic site.

With these restrictions in mind, we can extract from table 1 at least some general trends for the sodium interstitial site occupancies in rhombohedral Nas(x,y).

In both end members Nas(0,y) and Nas(3) the Na1 site is fully occupied. At room temperature the Na2 site is empty for Nas(0,y) and successively filled with increasing x. But at the same time the increasing Na^+/Na^+ electrostatic repulsion is believed to destabilize the Na1 site leading to an additional occupancy on the Na2 site. This interpretation is consistent with the results of single crystal (SC) data rather than with those from powder (P) diffraction experiments (table 1).

In the monoclinic phase of Nas(x,y) a fully occupied Na2 site was found in all studies according to partially occupied Na1 and Na3 sites. But systematic trends are beyond the consistency of the data of table 1.

However, the correlation between framework distortion and sodium site occupancies is more evident for some NASICON analogues (e.g. $Na_3M_2P_3O_{12}$, M=Sc,Cr,Fe). Two phase transitions associated with a stepwise transfer of sodium from Na1 to Na2 were observed (M=Sc,Fe) [60] . In the meantime details of the structural changes of $Na_3Sc_2P_3O_{12}$ have been described [61]. In the low temperature phase (<50°C) the sodium ions are strongly displaced from their ideal positions (table 1). Above the first phase transition the rhombohedral symmetry is restored, 8% of the sodium is transferred from Na1 to Na2 and the appearance of a super-structure was observed. The second phase transition at 168°C goes along with a further transfer of sodium from Na1 to Na2 and a decrease of Na^+ order. However, distinct phase transitions are only observed for crystals grown at very high temperature (1600°C). Smoother or even vanishing phase transitions for crystals grown at lower temperature (1470°C, 1100°C) are a result of static tetrahedral orientational disorder, i.e. deviations from thermodynamic equilibrium.

Static framework disorder was also observed for Nas(0,0.17) [54]. Structure studies revealed a splitting of oxygen sites at room temperature which was not observed at T=993K. This behaviour is correlated with a transition of sodium ions from a site located within the framework octahedron (AO_6) to the Na2 site.

Table 1: Sodium Site Occupancies, "Bottleneck" Widths (B1,B2), Free Enthalpies of Na⁺ Diffusion (ΔG_d) and Enthalpies of Na⁺ Conductivity (ΔH_d) for Nas(x,y) and some NASICON Analogues (in both space groups the same Na site notations are used (see 3.2))

		x,y	(M)	Na Cont.	T /K	Space Group	Occupancies [%] Na1	Na2	Na3	B1 /Å	B2 /Å	ΔG_d /eV	ΔH_d /eV	ref.
$Na_{1+x+4y}Zr_{2-y}Si_xP_{3-x}O_{12}$	SC	0,0		1	295	R3̄c	100	0						(5)
	SC	0,.17		1.3	295	R3̄c	100	0		1.85	1.75 }	.37	.38	(54,46)
	SC	0,.17		1.3	993	R3̄c	100	10		2.10	1.85 }			
	SC	0,1g)		5.	295	R32	95	96						(52)
	SC	.5,1g)		5.5	295	R32	100	99						(52)
	P	1,0e)		2.	295	R3̄c	58(91)	23(36)					.37	(55)
	P	1,0e)		2.	573	R3̄c	49(84)	23(36)					.28	(55)
	SC	1.4,.23d)		3.3	641	R3̄c	70	83		2.30	2.00	.19		(27)
	P	1.6,0		2.6	593	R3̄c	100	53		2.30				(56)
	P	1.84,.32		3.2	295	C2/c	26.5	100	80.4					(26)
	P	1.9,.07		3.0	295	C2/c	27	100	95					(51)
	P	2,0		3.	593	C2/c	100	66		2.35				(56)
	P	2,0e)		3.	295	C2/c	44(81)	100	32(60)				.36	(55)
	P	2,0e)		3.	573	R3̄c	40(57)	56(81)					.195	(55)
	SC	2.05,0a)		3.05	443	R3̄c	36	67						(52)
	SC	2.05,0a)		3.05	623	R3̄c	40	68						(52)
	SC	2.25,0a)		3.25	443	R3̄c	36	75						(52)
	SC	2.25,0a)		3.25	623	R3̄c	37	75						(52)
	SC	2.35,0a)		3.35	300	R3̄c	49	77						(52)
	SC	2.1,0b)		3.1	295	C2/c	22	100	56					(50)
	SC	2.1,0		3.1	623	R3̄c	70	80						(50)
	SC	3,.2		4.	295	R3̄c	96	100		1.72	1.75 }	.54	.62	(27,46)
	SC	3,.2		4.	1013	R3̄c	97	100		1.95	1.90 }			
$Na_{1+x+4y}Zr_{2-y}M_ySi_xP_{3-x}O_{12}$	P	1,1	Sc	3.	295	R3̄c	85.4	69.4					.31	(57)
	P	1.5,.5	Sc	3.	295	C2/c	69	86	73				.23	(57)
	P	0,2	Sc	3.	373	R3̄c	99	67					.45	(41)
	P	0,2	Sc	3.	498	R3̄c	80	73					.20	(41)
	SC	0,2α	Sc	3.	295	Cc	100	100	50					(58)
	SC	0,2αc)	Sc	3.	325	Cc	69	67						(38)
	SC	0,2αc)	Sc	3.	600	Cc	34	67						(38)
	P	0,2	Cr	3.	300	R3̄c	79	73						(59)
	SC	0,2	Cr	3.	295	C2/c	83	83	67					(60)
	SC	0,2	Fe	3.	295	C2/c	100	66	67					(60)

mid-Na occup.: a) 9-11%, b) 39%, c) 6-11%

d) Here a Na1 split site was refined. For comparison with the other data the Na1 site occupancy in ref.27 was multiplied by two.

e) Values in parentheses are normalized to satisfy the composition.

f) The effective "bottleneck" diameter was calculated as the static opening of the triangles (see fig. 10) using a oxygen-ion diameter obtained directly from the structure. Details of calculation are given in the references.

g) The Na2 site is splitted to two independent sites. In the table average values are given.

The order/disorder transition is also indicated by a smooth change of the thermal expansion between 550K and 750K (fig.11).

These are only few examples for the mutual dependence of the degree of static framework disorder and the variation of Na site occupancies which may both be the consequence of variations of the stoichiometry (fig.3) and/or the thermal history of the sample (see also 5.).

4. Na^+ CONDUCTIVITY

Let us now turn to the Na^+ conductivity of NASICON which is the major reason for the interest in this class of compounds. Most of the compositions with NASICON framework structure can be derived from the parent compounds (see 1.) by continuous substitutions which go along with continuous changes of the conduction parameters. Therefore, the particular features of parent NASICON (fig.3) will be presented and discussed first and then the respective trends for the numerous kinds of substitution.

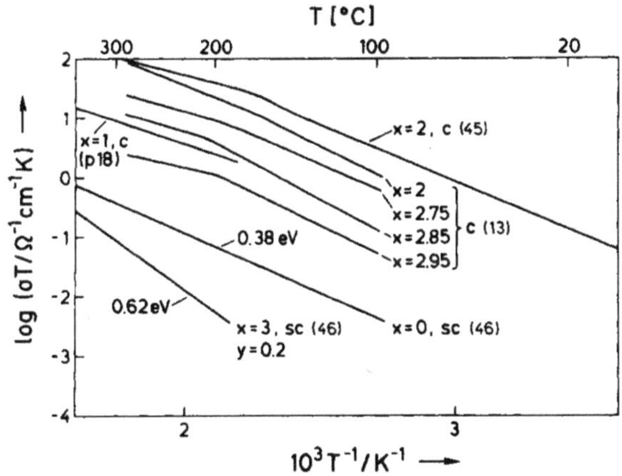

Fig.4
Some Na^+ conductivities of Nas(x) (x=0–3) single crystals (SC) and ceramics (C).

The only single crystal conductivity values available so far are those of Nas(0) and Nas(3,0.2) (the crystal contained some protons [36]) (fig.4). For the solid solutions Nas(x) no such data is available because of the lack of sufficiently large single crystals (see 2.4.). Instead, the conductivities of some individual ceramics are shown. Fig.5 presents the conductivities of Nas(x) ceramics at 300°C including the activation enthalpies in the rombohedral high temperature phase from different sources. Especially for silicate rich compositions there is a considerable scatter of the data which will be treated in 5.. The numbers for intermediate and phosphate rich compositions can be considered to be fairly reliable because of the consistency with the results from

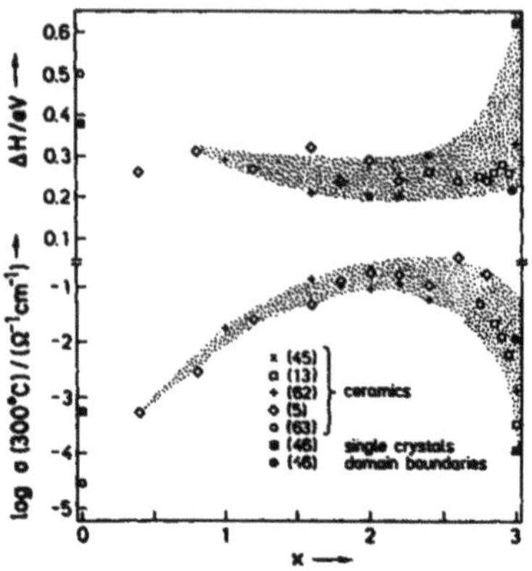

Fig.5
Na^+ conductivities of Nas(x) (x=0-3) at 300°C and the corresponding activation enthalpies (rhombohedral high temperature phase) from different sources.

X-ray diffraction experiments (see 4.1.). The temperatures of the transition to the monoclinic low temperature phase at which the conductivity curves bend to higher activation enthalpies, as first observed in ref.45, is shown in fig.6. At the beginning this transition has been overlooked for silicate rich compositions, but it actually extends to Nas(3), for which, however, the structural changes are very small [46].

In the case of pure phosphates a very popular scheme of substitution is the introduction of di- or trivalent cations (M) for Zr^{+4} on the octahedral framework site along with additional Na^+ $(Na_{1+nx}Zr_{2-x}M_x^{(4-n)+}P_3O_{12})$. In this way the sodium content can be increased up to three Na^+ per formula unit which corresponds to the sodium content of Nas(2), the composition with the highest conductivity. The conductivities develop differently with progressive substitution depending on the cation being introduced. For low values of x, however, the behaviour is rather uniform, the ionic conductivity increases with increasing x without significant change of the activation enthalpy. For M=Mg or In this holds up to x=2 [32,63]. But for other elements the activation enthalpy starts to increase at x<2 (e.g. fig.7) and with the exception of Yb, Mg and Zn [32,63-65] phase transitions appear [41,66-70], which are not observed for Nas(0) above room temperature. Fig.8 shows the conductivities of some examples for substitution levels around x=2. In any case the high conductivity of Nas(2) is not reached. The octahedral framework site also tolerates pentavalent cations such as Nb^{+5}.

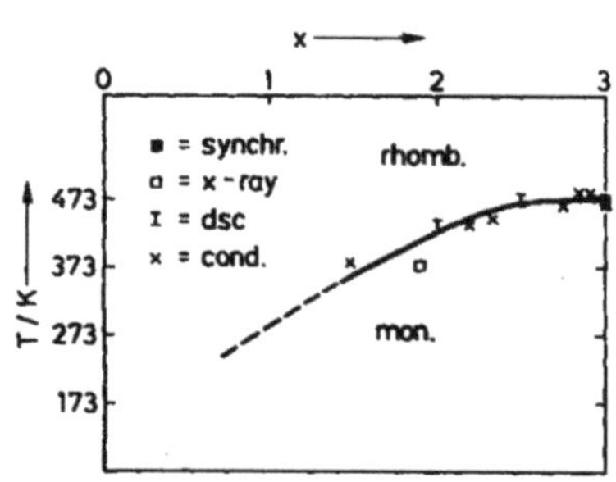

Fig.6
Monoclinic/rhombohedral transition temperature for Nas(x) (x=0-3) [15]

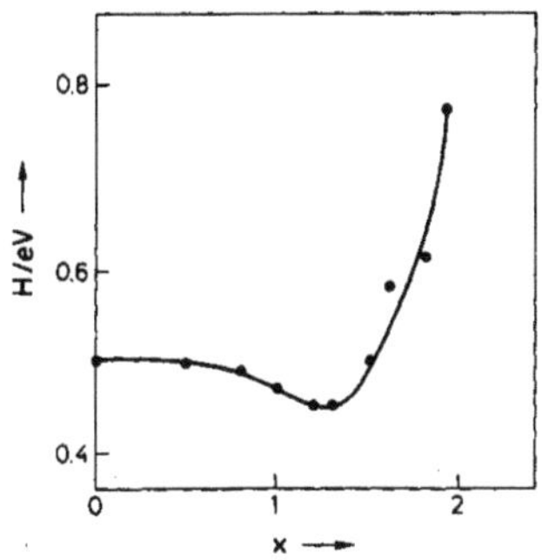

Fig.7
Activation enthalpy of the Na^+conductivity versus the composition of $Na_{1+x}Yb_xZr_{2-x}P_3O_{12}$ [63].

The semi-quantitative conductivity data of $Na_{1-x}Zr_{2-x}Nb_xP_3O_{12}$ at 200°C [71] shows a maximum for x=0.5. Another approach is to start the substitution of Zr^{+4} at Nas(x)(x=1-2) the parent compound compositions with the highest ionic conductivities (figs.4,5). This has been done with leaving the Na content constant ($Zr^{+4} + nSi^{+4} \leftrightarrow M^{(4-n)+} + nP^{+5}$) or keeping the P/Si ratio constant ($Zr^{+4} \leftrightarrow M^{(4-n)+} + nNa^+$). The first sheme of substitution is only known for $Na_3Zr_{2-x}Yb_xSi_{2-x}P_{1+x}O_{12}$ [63]. For x=0.8 a maximum in conductivity, which is about a factor of two higher than the conductivity of Nas(1), has been observed but even more interesting is the absence of any phase transition phenomena for this composition. The latter approach leads to slight increases of the conductivity of Nas(2) (M=Mg, V,Nb,Ta and a low dopant concentration of 2 mol% [72]). However, the effect can be explained by the more effective sintering of the doped ceramic samples rather than by an increase of the bulk conductivity. For higher concentrations of M which have been prepared only for di- and trivalent cations the resulting conductivities are lower than for Nas(2) in any case [72,73].

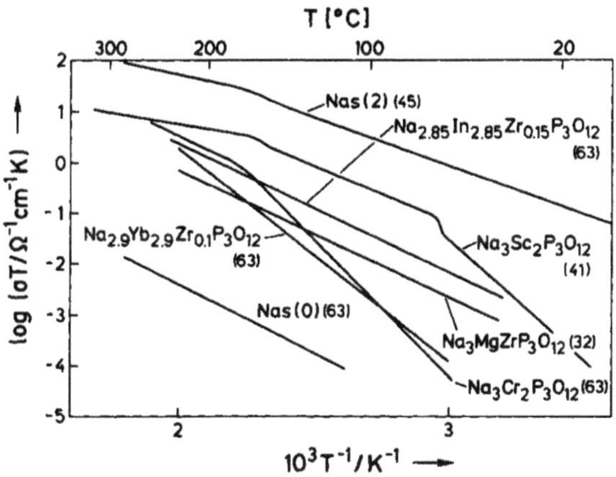

Fig.8
Na^+conductivity of some NASICONs containing approximately three sodium per formula unit and with lower valent cations substituting for Zr^{+4}. The conductivity of Nas(2) is given for comparison.

It should be mentioned that the natural substitution of Zr^{+4} by its homologue Hf^{+4} leads to a slight increase in conductivity for Nas(2) [74].

Substitutions are also possible on the tetrahedral framework site, i.e. the substitution of P^{+5} and Si^{+4} by their homologues As^{+5} and Ge^{+4}. For Nas(0) the complete exchange of P^{+5} by As^{+5} increases the conductivity by about one order of magnitude whereas the activation enthalpy is slightly decreased [8]. For the substitution of Si^{+4} by Ge^{+4} in Nas(3) the opposite behaviour is observed. The pure germanate shows lower conductivities with an activation enthalpy of 0.8 eV [75].

The compositions $Na_5ZrP_3O_{12}$ and $Na_{5-4x}Zr_{1+x}P_3O_{12}$ $(0.04<x<0.15)$ cannot be derived from Nas(0) by continous substitution (see also fig.3), nevertheless their crystal structures are related to that of NASICONs [29] (see 3.2.2.) and their conductivities are similar to those of Nas(0), however, with a change of slope between ~100°C and 150°C indicating a phase transition [29].

The Na^+ conductivities of the different NASICONs are a direct consequence of their crystallographic structures (see 3. and 4.1.). To some extent their variation can be understood by steric considerations (the space provided for the mobile ion to move), the chemical interaction of the mobile cation with the anionic framework along the entire conduction (diffusion) path and the interaction among the mobile cations. The influence of the electronic polarizability of the mobile cation and its environment on the ionic conductivity will not be discussed since it is practically invariant for all compounds under consideration here [76] (Na^+ with oxygen as nearest neighbours).

The object of this section is only bulk conductivity of well crystallized (ordered) NASICON. The conductivity contributions from other sources will be dealed with in 5..

4.1 Structural Aspects of Na^+ Diffusion in Nas(x,y)

In this section we will focus on the microscopic processes taking place. After a description of the diffusion path some steric parameters of the framework, which have a major impact on Na^+ diffusion, are presented.

4.1.1 Discription of the Na^+ Diffusion Path. For a better understanding some basic functions and their physical meaning are introduced first:

The Fourier transform of the temperature factors (including anharmonic expansion) fitted on the intensity data of single crystal X-ray or neutron diffraction experiments yield the average distribution (over space and time) of any atom in the unit cell around its equilibrium positions, i.e. the probability density functions $pdf_i(\vec{r})$. The diffusion path is actually the superposition of the single probability densities $pdf_i(\vec{r})$ of the mobile ion sites, i.e. the joint probability density function [77]:

$$PDF_{Na}(\vec{r}) = \sum_i w_{Na,i}\ pdf_{Na,i}(\vec{r}) \tag{1}$$

where $w_{Na,i}$ is the occupancy of the sodium site Na,i and $\sum_i$ is taken over all Na sites. $PDF_{Na}(\vec{r})$ is the probability of finding a Na^+ in a volume element of the crystal. From $PDF_{Na}(\vec{r})$ the effective one particle potential of the Na^+ (i.e the potential the Na "sees" on its way through the crystal) can be calculated [77,78]:

$$V_{Na}(\vec{r}) = V_o - kT\ln(PDF_{Na}(\vec{r})) \tag{2}$$

$V_{Na}(\vec{r})$ is an average free potential enthalpy of Na^+ originated by the interaction with all other atoms (ions) of the crystal. The scaling value V_O is determined by the absolute maximum of $PDF_{Na}(\vec{r})$.

In fig. 9 the $PDF_{Na}(\vec{r})$ of Nas(0,.17), Nas(1.4,0.23) and Nas(3,0.2) are depicted. They describe a Na^+ diffusion path between Na1 and Na2 for each composition [27]. These results disprove the studies on Nas(1.6), Nas(2.0) [56] and Nas(3) [13]. From geometrical and structure chemical considerations they suggest the Na2-Na2 jump as the predominant transport mechanism. But from the $PDF_{Na}(\vec{r})$, calculated from diffraction experiments at two different temperatures [27,48], the path along Na2/Na1/Na2 was unambiguously extracted as the most favourable. Meanwhile this path has been confirmed for several Nas(x,y) structures [52,55] as well as for some NASICON analogues [59,79].

So far the diffusion path in monoclinic NASICON has not yet been investigated in detail. From Na site occupancies (see 3.4) the Na3/Na1/Na3 path may be assumed as the most favourable [55].

$V_{Na}(\vec{r})$ in fig. 9 represents the "valley floor" of the corresponding diffusion paths. The maximum (A) equals the free enthalpy of diffusion ΔG_d and therefore, according to the relation $\Delta G_d = \Delta H_d - T\Delta S_d$, this potential barrier is a measure for the activation enthalpy ΔH_d ($T\Delta S_d$ is frequently small) as determined for example by conductivity measurements. ΔG_d and ΔH_d are listed in table 1.

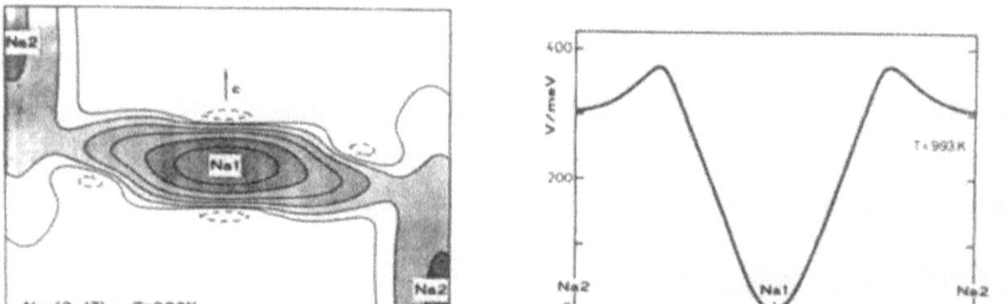

Fig.9

Na^+ diffusion paths presented as contour lines of constant Na^+ probability density ($PDF_{Na}(\vec{r})$) in planes defined by the two sodium sites [27,54]. B1 and B2 are the projections of the "bottlenecks". The corresponding one particle potentials $V_{Na}(\vec{r})$ are shown on the right hand side.

4.1.2. Influence of Steric Framework Parameters on Na^+ Diffusion. The diffusion paths described in fig.9 correspond to isomorphic structures. However, the $PDF_{Na}(\vec{r})$ maps and the corresponding conductivities are distinctly different. Nas(1.4,0.23) exhibits one of the lowest values for ΔG_d ever found for a NASICON type compound (0.19eV) whereas for Nas(0,0.17) and Nas(3,0.2) ΔG_d is relatively high in accordance with their moderate Na^+ conductivity. Therefore, these compounds are quite suitable for detailed studies of the influence of the framework structure on Na^+ conductivity, which are briefly presented now.

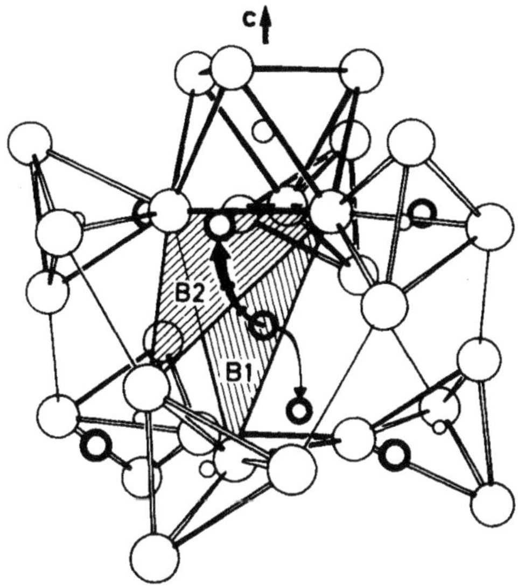

Fig.10
"Bottlenecks", B1 and B2, of the diffusion path depicted by an ORTEP-plot. The edges of framework octahedra and the sodium ions are displayed by bold lines. One diffusion path (Na2/Na1/Na2) is indicated by arrows.

The potential maxima (A) of the diffusion paths in fig.9 are quite distinct, moreover, they are located at the same position. Simple geometrical considerations show that this most unfavourable locus of the diffusion path is related to the "bottleneck" of the diffusion channel[27]. Two triangles formed by oxygen ions (fig.10) determine the effective widths of the "bottlenecks". In fig.9 their locations are given in relation to the diffusion path and the potential maximum.

The effective "bottleneck" widths (B1,B2) may be taken from table 1 and be compared to the corresponding values for ΔG_d. There is a general trend of decreasing values of ΔG_d with increasing diameters of the "bottlenecks". ΔG_d is lowest for the compound, for which B1 and B2 approximately equal to the sodium ion diameter (~2.04Å).

The data in table 1 and figs.9,11 empirically indicate that wide "bottlenecks" (low ΔG_d) are correlated with low Na1 site occupancies and distinctly long c-axes. This behaviour was explained as follows [27]:

A low Na1 site occupancy as a consequence of a high dynamic disorder (high Na^+ diffusivity) induces an increasing O-O repulsion between neighbouring lanterns (fig.1). The resulting anisotropic expansion of the $NaIO_6$ octahedron gives rise to an opening of the bottlenecks (fig.10), i.e. a decreased ΔG_d and, therefore, a higher Na^+ diffusivity. In this way the Na1 site occupancy and ΔG_d are mutually dependent and both are reflected in the length of the c-axis.

Among the compositions in fig.11b it is largest for Nas(1.4,0.23). The related anisotropic stretching of the $NaIO_6$ octahedron goes along with a splitting of the Na1 site $(0,0,\pm\Delta)$[27].

$V_{Na}(\vec{r})$ flattens and ΔG_d decreases with increasing temperature (fig.9 b,c). However, in the Arrhenius slope (ΔH_d) of the Na^+ conductivity $\sigma T \sim \exp(\Delta S_d/k) \exp(-\Delta H_d/kT)$ only the non-linear contributions of ΔG_d are reflected, the linear part is expected to increase the pre-exponential factor only.

Drastically increased pre-exponential factors were observed for the conductivities of $Na_{1+x}A_xZr_{2-x}P_3O_{12}$, (A=In, 1.5 Mg) compared to the one of Nas(0) (fig.8). Indeed, NASICONs with di- or tri-valent cations on the A site have higher thermal expansions than the compounds in fig. 11 [80]. This might explain their particular conductivity behaviour (figs. 7,8) rather than the assumption of a (high-energy) mid-Na (M5) site (see 3.4) involved in the diffusion according to a jump sequence mid-Na/Na2/mid-Na for low values of x [81] (the suggested sequence actually does not form a continuous diffusion path).

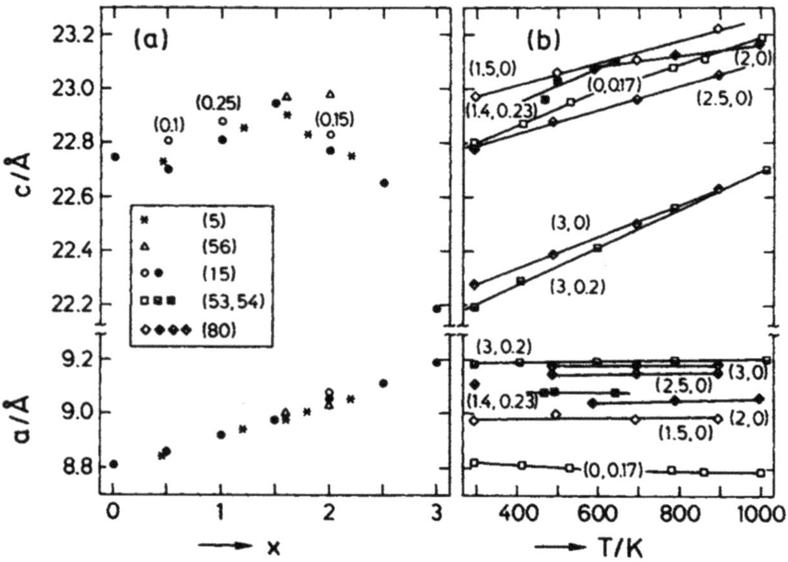

Fig.11
Cell-constants of Nas(x,y) (hexagonal notation) as a function of composition (T=295 K) (a) and temperature (b); in (a) only y-values deviating from zero are indicated.

4.2. Na^+/framework interaction

Changes of the composition, i.e. of the chemistry of the NASICON framework are expected to have an influence on the chemical interaction between the anionic framework and the mobile Na^+. The variation of this interaction, which is predominantly electrostatic in nature [76] along the entire diffusion path is a direct measure of the corresponding free enthalpies of diffusion. These are expected to react especially on the charge distribution at the narrowest locus of the diffusion path, i.e. the bottlenecks (see 4.1.2.). The oxygens forming the bottlenecks between Na1 and Na2 are coordinated with the octahedral framework site which is occupied by Zr^{+4} for the parent compounds. Lower valent ions substituting for Zr^{+4} (see 4.) polarize these oxygen ions less and the attractive interaction between the bottleneck oxygens and the mobile Na^{+} is expected to increase, thus giving rise to higher actvation enthalpies and lower conductivities. Just the opposite is expected for the substitution by higher valent cations.

Indeed, for all substitutions of Zr^{+4} by lower valent cations the conductivities are lower than for the corresponding parent compounds with the same Na^{+} content (fig.8). All these compounds, however, have a shorter c-axis due to a lower electrostatic repulsion between AO_6 octahedra of neighbouring "lanterns" leading to a contraction of the $Na1O_6$ octahedron along the c-axis [32]. This might narrow the bottlenecks (see 4.1.2.) and it cannot be decided whether the change of the bottleneck width or the polarization of the bottleneck oxygens is responsible for the depressed conductivities.

However, the increase of the conductivity for the partial substitution of Zr^{+4} by Nb^{+5} , in spite of the decreased Na^{+} content, support that the degree of polarization of the bottleneck oxygen ions has some effect on the diffusion barrier.

A similar trend is observed for the substitution $Na^{+}+Si^{+4}$ <> P^{+5} on the tetrahedral framework site which shares oxygen ions with the diffusion bottlenecks. The widths of the bottlenecks are comparable for the pure phosphate Nas(0,0.17) and the pure silicate Nas(3,0.2) (see 4.1 and table 1) but the activation enthalpy of the conductivity is significantly lower for the phosphate (0.38 eV) compared to the silicate (0.62 eV) in accordance with a stronger polarization of the bottleneck oxygen, i.e. a loosening of the O/Na interaction.

4.3. Na^{+}/Na^{+} interaction

The transfer of Na^{+} from the Na1 to the Na2 site, i.e. the increasing Na^{+} configurational disorder with progressing sodium content (see 3.) is also a result of the Na^{+}/Na^{+} repulsive interaction. A significant interaction among different alcali ions in NASICON has also been indicated by deviations of their activity coefficients from unity in ion exchange experiments [82]. Highly conducting NASICONs generally have a high symmetry ($R\bar{3}c$) and similar partial occupations on the two sodium sites (table1). But with decreasing temperature the Na^+/framework interaction causes a progessive ordering which finally leads to phase transitions (figs.6,8), i.e. a lowering of the framework symmetry and a splitting of the Na2 sites into two sites (see 3.). These transitions go along with an increase of the activation enthalpy and the pre-exponential factor of the conductivity reflecting an increase of the enthalpy and entropy of Na^+diffusion for the more ordered low temperature phases.

The transition temperatures are actually quite reproducible (fig.6) but their thermodynamics very much depend on the samples. This is not surprising when one keeps in mind that the transition temperatures are only about 1/3 of the decomposition temperature and that quenched non-equilibrium states may appear. These will be discussed in the next section.

5. REAL NASICON

We already have directed the reader's attention to deviations from thermodynamic equilibrium. These can be the result of quenching [38,60,83] or of low temperature preparation methods [16,38,61,84]. They reflect the tendency of NASICON compositions to form structurally disordered and even glassy frameworks as well as a tendency of Si/P ordering [83,85] (see also 6.). Both features have an effect on phase transition and conductivity phenomena.

With increasing tetrahedra disorder the phase transitions are diminished and completely disappear for amorphous NASIGLASS [16,61,84], the conductivities being generally lower than those for well-crystallized samples. Speaking in terms of spectroscopy the static disorder decreases the dynamic disorder (diffusivity)[86]. The difference of the two sodium sites which is, e.g., reflected in their different occupancy (see 3.) is a characteristic feature of well ordered NASICON. Both sites show a trend of becoming degenerate with increasing framework disorder [16,38,60] leading to similar occupancies at an already low temperature.

Si/P ordering, which is only relevant for mixed crystals, has the opposite effect on phase transitions. A quenched sample of Nas(1.9) showed a three times higher transition enthaply than after a two week heat treatment at 500°C [83]. The results suggest that the phase transition is a consequence of a quenched Si/P disorder. It might be suspected that the structural ordering (lowering of the symmetry) during the phase transition substitutes for the compositional ordering (decomposition of Nas(x) into a phosphate-rich and a silicate-rich composition, for which some indication is given by thermodynamic

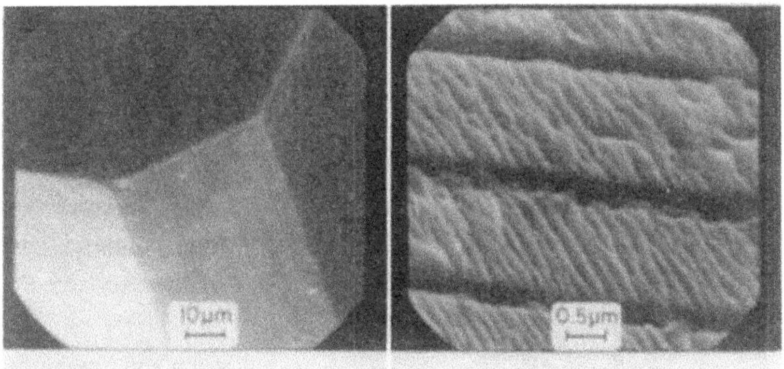

Edging fringes of Nas(3,0.2) "single crystals" visualizing domain (low angle grain boundaries)[46].

data (see 6. and fig.18)). If this is so, the high conductivity of Nas(2) is a property of a quenched state which exists for kinetic reasons. The even more general question arises whether other fast ion conduction phenomena, e.g. in ß-aluminas, are concerned in a similar way [83,87].

Especially for silicate rich NASICONs the appearance of domain and grain boundaries (two dimensional non-equilibrium defects) may affect the macroscopically observed conductivity as already indicated by the discrepancy of the activation barriers for single crystals and ceramics [46] (fig.5). Fig.12 shows edging fringes of a Nas(3,0.2) single crystal which correspond to domain boundaries. They have also been identified by synchrotron radiation diffraction and can be considered to be low angle grain boundaries. The ionic conductivities along these boundaries were found to be higher and less activated than perpendicular to them. Likewise, conductivity enhancement is observed for grain boundaries in Nas(3) ceramics. The mutual dependence of the conduction parameters (pre-exponential factor and activation enthalpy) for domain, respectively grain boundaries and bulk materials, are shown in fig.13. The large scatter of these parameters for the former one (x=3) reflects their dependence from the thermal history of the individual sample. From the ratio $(\sigma_0 T)_{bulk}/(\sigma_0 T)_{domain\ bound.}$ for zero activation enthalpy the extension of the fast conduction paths is estimated to be 2-3 unit cells. The slope of the graphs may be interpreted as the temperature at which some disordered state of the Na^+ sublattice is completed. It is ~1800K for the bulk and ~828K for

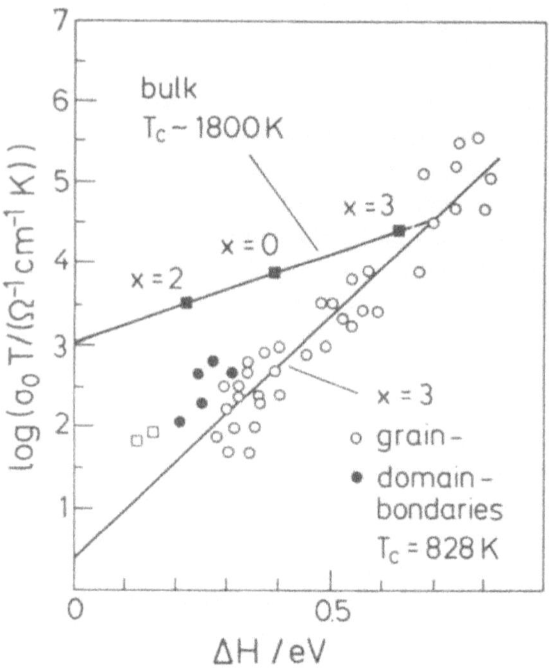

Fig.13
Mutual dependence of the activation enthalpy and the pre-exponential factor for bulk (x=0-3) and domain respectively grain boundary (x=3) Na^+ conductivity [46].

grain and domain boundaries. For the latter this temperature is close to the one at which irreversible changes of the conductivity occur. The observed dependence can be explained by a widening of the NASICON structure in the boundary region rather than by segregation or space charge effects. This probably leads to an opening of the conduction bottlenecks and thus to an increase of the local conductance (see also 4.1.2.).

6.THERMODYNAMICS OF $Na_{1+x}Zr_2Si_xP_{3-x}O_{12}$ (Nas(x))

The knowledge of the chemical thermodynamics of NASICON is of particular interest since:

a) NASICON is considered as a possible candidate for an electrolyte material in Na/S-cells. Thermodynamic stability against the active masses is desirable since these cells are to work at elevated temperatures and are planned for long-term performance. The calculation of reaction equilibria and of phase compatibilities requires the knowledge of thermodynamical data.

b) In addition, such data allows one to make general statements on intrinsic stability, mixing behaviour, miscibility gaps and order/disorder phenomena.

In spite of the need for such values only two publications [88,89] are available in which this problem has been systematically tackled (in contrast to β- or β"- alumina for which different approaches can be found in the literature [90-96]). On these sources the following section is chiefly based.

In ref.88 the experimental techniques and results are given. The preliminary evaluation in ref.88 has been thoroughly refined in ref.89 by introducing novel data and by constructing a mixture model.

5.1 Thermodynamic data for Nas(x)

All these results refer to experiments on Nas(x) (x=0,0.5,1,1.5,2,2.5,3). For the preparation (that may have a slight influence on the thermodynamic data, see below), the reader is referred to refs. 89,15. A complete data set for the specific heats (C_p^o), the entropies (S^o) and the formation enthalpies ($\Delta_f H^o$) has been obtained for T=0K up to 1100K [89].

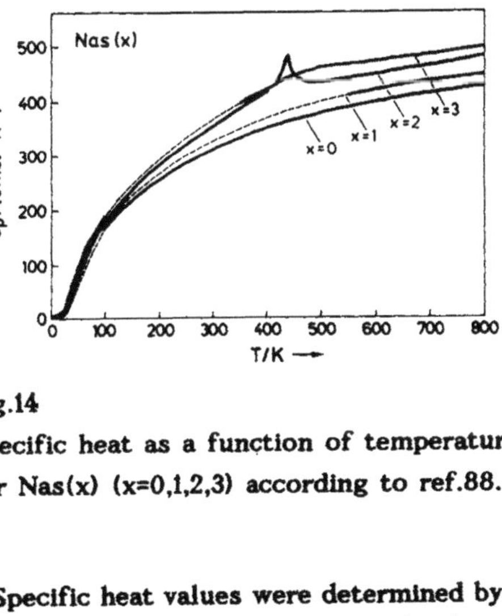

Fig.14
Specific heat as a function of temperature for Nas(x) (x=0,1,2,3) according to ref.88.

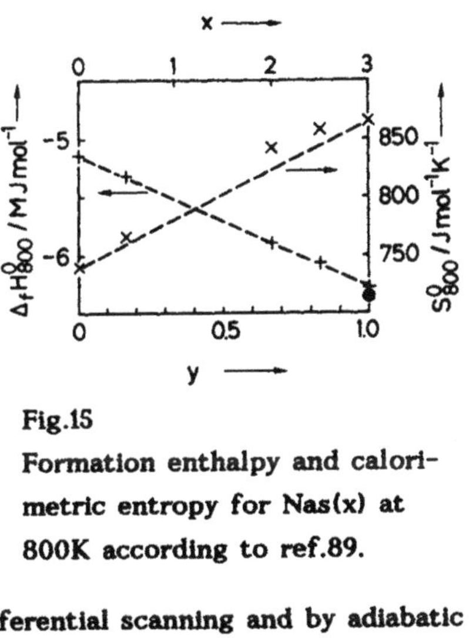

Fig.15
Formation enthalpy and calorimetric entropy for Nas(x) at 800K according to ref.89.

Specific heat values were determined by differential scanning and by adiabatic calorimetry from 1.5K to 800K [88]. The C_p vs. T graphs are plotted in fig. 14. The curve for Nas(2) clearly reveals the monoclinic/rhombohedral phase transition at 420K. (ΔH=1.1 kJmol^{-1}; ΔS=2.5 Jmol^{-1}K^{-1}). The form and the temperature of the maximum of the C_p-curve are quite similar to that presented in ref.45, but the transition heat is about half (2.1 kJmol^{-1} in ref.45). The nature of the phase transition and the reason for the scatter of transition data (dependence on the pretreatment) has been discussed above (see 5.). The transition for Nas(3) is weakly indicated in fig. 14. At low temperatures all graphs exhibit T^3-behaviour allowing the extrapolation to the zero point. At high temperatures the classical limit of $C_p(x)=3N(x)R$, where N is the number of atoms per formula unit, is asymptotically reached (e.g. N(2)=20, $C_p(2)$=500 J $mol^{-1}K^{-1}$). The increase of the high T-values with increasing x impressively reflects the graduation of N: The distances in fig. 14 between all neighbouring curves at high temperatures amount to ~3R ($\Delta C_p/\Delta x = 3R\Delta N/\Delta x = 3R$). The C_p-data above room temperature (besides the anomaly in the transition range) can be analytically comprised as:

$$C_p^o(x,T) = A^o(x) + B^o(x)T + C^o(x)T^{-2} + D^o(x)T^{-1/2} + E^o(x)T^2 \quad (3)$$

The T-independent coefficients are listed in table 2.

Absolute entropies and formation enthalpies have been determined by measuring the emf of electrochemical formation cells in the T-range of 700-1100K [88,97]. In this way S^o and $\Delta_f H^o$ have been obtained for the temperature of the measurement and by using the specific heat values for the whole T-range under consideration. Above room-temperature S^o and $\Delta_f H^o$ can be represented as polynominals:

$$S^o(x,T) = K_S^o(x)+A^o(x)\ln T+B^o(x)T- 1/2C^o(x)T^{-2}-2D^o(x)T^{-1/2}+1/2E^o(x)T^{-2} \quad (4)$$

and

$$\Delta_f H^o(x,T) = \Delta_f K^o(x)+\Delta_f A^o(x)T + 1/2\Delta_f B^o(x)T^2-\Delta_f C^o(x)T^{-1} +2\Delta_f D^o(x)T^{1/2}+ 1/3\Delta_f E^o(x)T^3 \quad (5)$$

The coefficients are also included in table 2. Hereby, attention must be paid to the fact that all coefficients refer to the low T-phase and the formation parameters to the reference states of Na(l), P(red), Si(s),O_2 (g) (valid between 371K (Na(s)/Na(l) and 704K (P(red)/P(g)). For the calculation of $\Delta_f H^o$ values for 298K or 800K, which are listed in table 3, the change in the reference states must be thoroughly considered (Na(s)/Na(l): 2.6 kJ mol^{-1}; P(red)/P(g): 86.4 kJ mol^{-1}). Thus, the distinct difference between $\Delta_f H^o_{800}$ and $\Delta_f H^o_{298}$ mainly arises from

Table 2: Coefficents of the Thermodynamic Functions (see eqs. 3,4,5).

10^{-3} A / J/molK	B / J/molK^2	10^{-7} C / JK/mol	10^{-4} D / J/molK$^{0.5}$	10^3 E / J/molK^3	10^{-4} K_S / J/molK
0.3882	0.1013	-0.4665	-0.0974	-0.0000	-0.2016
-4.5203	4.8522	-5.5903	7.2820	-2.2061	3.2909
0.8059	-0.6685	-1.5617	-0.2120	0.4866	-0.4328
-0.6819	1.1106	-1.9364	1.6671	-0.4300	0.5821
3.8754	-2.8147	3.5091	-5.4749	1.1809	-2.7013

$10^{-3}\Delta_f$A / J/molK	Δ_fB / J/molK^2	$10^{-7}\Delta_f$C / JK/mol	$10^{-4}\Delta_f$D / J/molK$^{0.5}$	$10^3\Delta_f$E / J/molK^3	$10^{-6}\Delta_f K_H$ / J/mol
-0.0467	0.1275	-0.5604	0.1091	-0.0583	-4.9370
-4.9720	4.8797	-5.6871	7.4895	-2.2606	-6.6070
0.3042	-0.6371	-1.6672	-0.0015	0.4436	-5.9290
-1.2004	1.1433	-2.0449	1.8786	-0.4692	-6.4441
3.3402	-2.7807	3.3977	-5.2624	1.1455	-5.2369

Table 3: Thermodynamic Standard Data (298K, 1 bar).

	$\Delta_f H^o_{298}$ / kJmol^{-1}	$\Delta_f H^o_{800}$ / kJmol^{-1}	S^{o*}_{298} / Jmol^{-1}K^{-1}	S^{o*}_{800} / Jmol^{-1}K^{-1}	$S^o_{298,cal}$ / Jmol^{-1}K^{-1}	$S^o_{800,cal}$ / Jmol^{-1}K^{-1}
0	-4889	-5133	370	738	365	733
0.5	-5115	-5320	376	755	379	770
2.0	-5813	-5891	404	826	414	842
2.5	-6044	-6086	422	847	421	857
3.0	-6285	-6286	427	865	425	862

* electrochemically determined

the included energy of sublimation of P. The enthalpy value of the phase itself changes only slightly (see e.g. $\Delta_f H^o(x=3)$). The standard values at 298K are within the range of values expected from estimation procedures (table 3). Also included in table 3 are the entropy values obtained by integrating the specific heats from 0K up to 298K or 800K. The difference to the electrochemically determined values consists, problems of accuracy apart, in not having included zero point entropies which should be mainly due to frozen-in configurational disorder ($S_{cfg}(T=0)$). According to table 3, the difference between S^o_{cal} and S^o is small and does not significantly exceed the range of experimental error (cf. Nas(0) or Nas(3) where S_{cfg} should be zero). Fig.15 shows the formation enthalpy and the (calorimetric) entropy at 800K for the whole composition range according to ref.89. The entropy shows a distinct deviation from the course expected from a linear interpolation of the end-member values. The deviation is similar to that of the molar volume (ref.89, see fig.11). In the resolution chosen in fig.15, deviations of the $\Delta_f H^o$ curve from the ideal behaviour cannot and should not be observed. The only literature value besides these electrochemically determined ones, viz. $\Delta_f H^o$ (3) (determined by solution calorimetry [98]) and extrapolated to 800K with the help of eq.3 lies definitely below the line (by 72 kJ/mol more negative) and is suggested to be incorrect. The accuracy of this entropy data can be considered to be smaller than 1 R. The accuracy of the formation enthalpy, however, can hardly be judged, in particular since the uncertainties of the assisting data necessary for the computation (see ref.15) is not known. The precision should be sufficient for the calculation of reaction equilibria with practically relevant neighbouring phases (see 7.).

5.2 Mixture Model for Nas(x)

As far as the implications are concerned that follow from the thermodynamic values of the elements of the solid solution range relative to the end-member, however, experimental errors have a distinct impact. Thus, the following analysis of the mixing behaviour [89] will have a half-quantitative character. For a discussion of a mixture model Nas(x) is considered as being formed by mixing the end-member in appropriate amounts:

$$y\ \mathrm{Nas}(3) + (1-y)\ \mathrm{Nas}(0) \longrightarrow \mathrm{Nas}(x) \qquad (6)$$

where $y=x/3$ denotes the mole-fraction of Nas(3). Since with increasing x (or y)

P is replaced by Si and simultaneously a vacancy on a Na2-site is occupied by Na, eq.6 reads more clearly

$$y\left[Si(Na2)\right]_3\left[...\right] + (1-y)\left[P(\square 2)\right]_3\left[...\right] \rightleftharpoons \left[Si(Na2)\right]_{3y}\left[P(\square 2)\right]_{3(1-y)}\left[...\right] \quad (7)$$

where $\left[...\right]$ is short for the invariant (Na1) Zr_2O_{12}-unit. Let Z^o be S^o, H^o, C_p^o, G^o or V^o, then $\Delta_M Z^o$ denotes the reaction quantity of the mixing process (eqs.6,7). As suggested from a chemical point of view and also by fig. 11, the mixtures are not expected to be ideal. Thus, it is convenient to consider the deviations (Z_{ex}) from such an ideal behaviour ($\Delta_M Z^o{}_{id}$) according to:

$$Z^o{}_{ex} = \Delta_M Z^o - \Delta_M Z^o{}_{id} \quad (8)$$

In this way $S^o{}_{ex}$, $C^o{}_{pex}$, $V^o{}_{ex}$, $H^o{}_{ex}$ and $G^o{}_{ex}$ are defined. Since the influence of the end members cancels in the difference, $Z^o{}_{ex}(x)$ refers directly to Nas(x). Whereas $\Delta_M H^o{}_{id}$ (and thus also $\Delta_M C_p^o{}_{id}$) is zero, the value of $\Delta_M S^o{}_{id}$ is determined by a random configurational disorder. Since eq. 7 demands a mixing of $\left[Si(Na2)\right]$ and $\left[P(\square 2)\right]$ and since three units are considered in the end--members, it follows:

$$\Delta_M S^o_{id} = k \ln \frac{[\alpha N]!}{[y\alpha N]![(1-y)\alpha N]!} = -R\alpha[y\ln y+(1-y)\ln(1-y)] \quad (9)$$

where $\alpha = 3$ and N = Avogadros number. Eq.9 assumes that Si and (Na2) as well as P and ($\square$ 2) are strictly coupled; i.e., additional mixing processes (e.g. Na2 and $\square$ 2) are not considered.

The excess values calculated according to eqs.8,9 for $\alpha = 3$ are plotted in fig.16 for T=800K. Although G_{ex}^o is positive, the total free enthalpy of mixing (see Fig.16, bottom) is negative (i.e. intrinsic stability with respect to a decay into the end members) and does not exhibit inflection points at 800K (i.e. intrinsic stability with respect to a decay into any other elements of the solid solution range). The excess enthalpy is distinctly positive which is reasonable considering the differences between Si and P with respect to bonding parameters, valences and size [99]. Thus, the Nas(x) compositions are completely entropy stabilized (fig.16 top). Even the excess entropy is positive, which should - if the configurational part is correctly described by eq.9 reflect chiefly vibronic contributions. The destabilizing influence of the energetic interaction should lead to a phase separation at lower temperatures (double minimum shape for $\Delta_M G^o$). The critical point of mixing is predicted as ($T \sim 600K$; $x \sim 1$). This miscibility gap could not be found experimentally.

As obvious in fig. 17, the excess values can be satisfactorily expressed in the form (i.e. subregular model):

$$Z_{ex}(y) = y\,(1-y)\,[W_Z'(T)\,y + W_Z''(T)\,(1-y)] \tag{10}$$

This also holds for the volume data derived from fig.11. Since $C_p{}^o{}_{ex}$ is close to zero (W) for T ≥ 300K and is determining the T-dependence of the parameters W_H', W_H'', W_S', W_S'', we can simply take the values determined at 800K for the whole T-range above room-temperature. Thus, to a first approximation, the whole thermodynamic information at P = 1 bar can be summarized by

$$\begin{aligned} G^o_{ex}(x,T) &= (11\,x - 0.78\,x^2 - 0.96\,x^3)\ \text{kJ mol}^{-1} \\ &\quad - T(11x - 2.9x^2 - 0.26x^3)\ \text{J mol}^{-1}\,\text{K}^{-1} \end{aligned} \tag{11}$$

In this way the isotherms in fig. 16 (bottom) have been obtained. Eq.11 also allows a simple representation of activity coefficients and thus of activity functions [89]. So far all considerations and experiments refer to a normal pressure of 1 bar. By considering volume data, also the pressure dependence may be analyzed in an analogous manner by using the relation $\delta G/\delta P = V$ and considering W_V-parameters (fig.17).

Another important point is related to the low zero-point entropy, that is clearly below the value that must be assumed for a completely frozen-in configurational disorder (eq.8). This is even more so if independent mixing of (Na2) and (□ 2) is considered. Beyond that, by assuming a zero-point entropy which is considerably lower than taken in fig.16 for computing S_{ex} (i.e. α in eq.9 formally lower than 3), the maximum of $S^o{}_{ex}$ would be shifted towards x=2. Thus, the course of $S^o{}_{ex}$ would become more similar to that of V_{ex}, as would be reasonable for a pure vibronic entropy. Assuming an α-value of 1

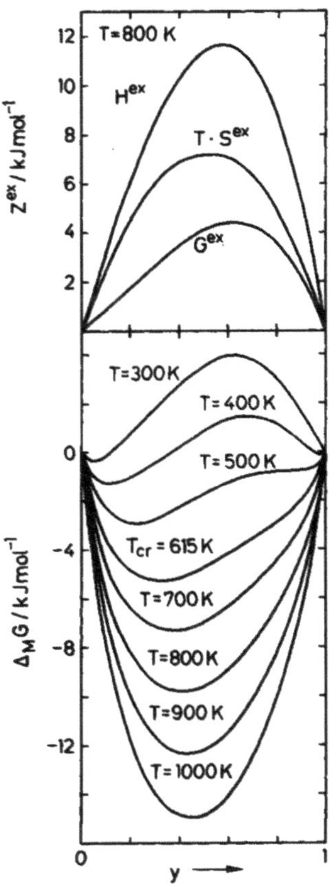

Fig.16
Top: Exess functions for Nas(x) at 800K according to ref.88.
Bottom: Isotherms of the free enthalpy of mixing for Nas(x) according to ref.89.

(see also ref.100) at room-temperature, interestingly a remaining T-dependence of the W_S -values is removed. At 800K however, only "full" α-values give reasonable values for the critical point of mixing.

All these findings and indications of an ordering in the system are totally consistent with recent investigations reported in ref.83. According to this work an ordering at quite moderate temperatures (annealing at 773K) is indeed suggested by C_p-measurements, so that a formal increase of $\alpha \sim 0 ... 1$ at to... 300K to $\alpha \sim 3$ at 800K appears plausible. This ordering has a depressing influence on the sodium conductivity [83], certainly also affects chemical kinetics (e.g. with Na) and has, in retrospect, a slight relativating influence on the thermodynamic values described here. It should be mentioned again that this uncertainty of 3 RT in the thermodynamic values is negligible for most applications.

It is worth noting that the process of ordering might be a first step towards an unmixing into a Si- and into a P-rich composition of the solution range. Obviously the framework does not tolerate too tight periodicities of the minority elements. If the rate limiting steps for both processes are the same, the observed behaviour means that the kinetics have become too slow at about 600K. It should , however, not be forgotten that the critical temperature of unmixing is strongly affected by uncertainties of the basic data set.

Fig.18 shows a tentative phase diagram constructed on the basis of literature results (see ref.15 and legend). The similarity between the predicted miscibility gap and the experimental line of phase transformation is obvious. The question arises whether the system evades an unmixing by changing symmetry (see 5.). Since the G-gain, e.g. for the decomposition of Nas(2) at 300K is about one order of magnitude higher than the G-gain due to the phase transformation

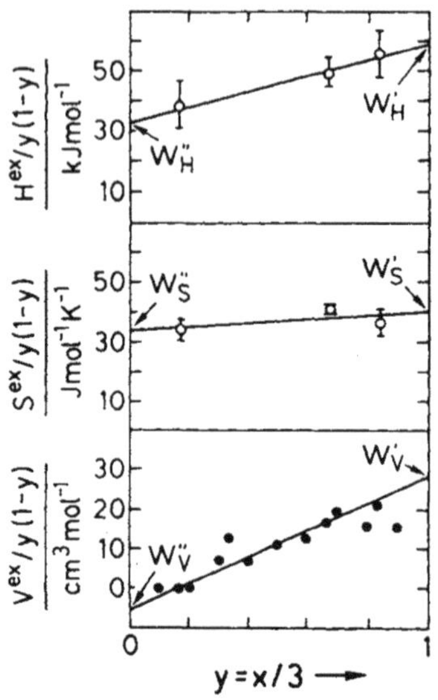

Fig.17

Fulfillment of a simple subregular mixing model (enthalpy and entropy at 800 K; volume at 300 K) according to ref.89.

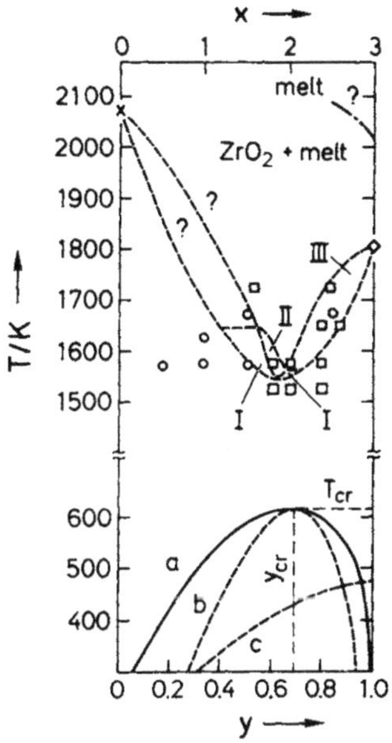

Fig.18
Tentative phase diagram derived from different literature results [15,89].
Top: modification and extension of a diagram proposed in ref. 17.
x,o (ref.102), □ (ref.17), ◇ (refs.103,104).
I: Nas(x)+$ZrSiO_4$+ZrO_2+melt;
II: $ZrSiO_4$+ZrO_2+melt;
III: Nas(x)+ZrO_2+melt.
Bottom: a) calculated miscibility gap [89], b) calculated spinode [89], c) monoclinic/ rhombohedral transition temperature (see also fig.6).

(~10 kJmol^{-1} compared to 1 kJmol^{-1}), this cannot hold completely in a thermodynamic sense but might reflect a tendency. The apparent insensitivity of the phase transition temperature (maximum in the C_p-T-curve) from the prehistory can be ascribed either to a compensation effect of transition enthalpy and entropy or to a relative independence of the transition modes from ordering processes.

For a more detailed discussion further information on the kinetics on unmixing and phase transformation [105,106] (e.g. stress effects) is necessary.

From a crystal chemical point of view, there are considerable similarities of NASICON to mineralogical systems (especially garnets [15]). Therefore, similarities in the thermodynamics (excess thermodynamics, ordering, phase diagram) are not surprising. Significant differences (e.g. unusual high W_H-parameters [15,101]) arise from the distinct differences in chemical bonding properties of phosphorus and silicon.

Owing to immense complexity and to the preliminary status of knowledge, the problem of the phase relations will not be further extended here. The reader is referred to ref.15. Unfortunately, there are no thermodynamic data on NASICON compositions outside the Nas(x)-range, so that no reliable conclusions can be drawn on the mixing properties and stabilities involving those compounds. Particularly, no conclusive statements can be made on intrinsic stabilities of Nas(x) with respect to a decay into compositions outside this composition range. The results of the preparation kinetics, however, (see 2.) as well as long-time stability tests [15] seem to exclude a more complex thermodynamic situation at least for higher temperatures.

Since recent preparations [15] suggest that Nas(x,y) is restricted to small y-values (fig.3), the mixing and ordering behaviour may be qualitatively similar as far as entropic and energetic behaviour is concerned. Similar conclusions hold for reaction equilibria with Na or S (see 7.).

7. NASICON AS A MATERIAL IN ELECTROCHEMICAL CELLS

Most of the initial interest in NASICON was motivated by the potential substitution of ß-aluminas by NASICON as a separator material in secondary Na/S batteries. An essential requirement for this application is the durability of the electrolyte against the active masses, in particular against the liquid sodium anode. Preliminary Na/Na-cell testings [11,107] revealed chemical reactions of Nas(x) ($x \sim 2$) with molten sodium and Na^+ current induced cracking.

The thermodynamic data presented in section 6. actually allowed the calculation of the free enthalpies (ΔG_r) for different reactions of Nas(x) ($x=0-3$) with molten sodium [108]. ΔG_r and the corresponding products for the reactions with the highest affinity are given in fig.19 indicated by 1 and 2. ΔG_r is negative for the entire composition range ($x=0-3$), i.e. in thermodynamic equilibrium any Nas(x) does not coexist with molten sodium. However, the formation of silicon demanded by reaction 2 does not occur in adequate tests. A possible explanation for the deviation of the experiments from thermodynamics may be that breaking of the rather covalent Si-O bond of the SiO_4-tetrahedra involved in the reduction of silicon does not take place at the given temperature for kinetic reasons and, therefore, complete thermodynamic equilibrium may not be attained. Instead, for the reaction of Nas(x) ($2.7<x<3$) with sodium, Nas(3) has been identified as a reaction product. With this boundary condition (inert SiO_4) the reactions 1,3,4,5 were found to be the ones with the highest affinity (kinetic reaction path). They are in accordance with the instability of any phospate containing NASICON ($x<3$) and the apparent durability of the pure silicate ($x=3$) with molten sodium. (The obvious kinetic stability of those ß-aluminas that are thermodynamically unstable [91,92,109] is possibly due to the inertia of the AlO_4-tetrahedra.)

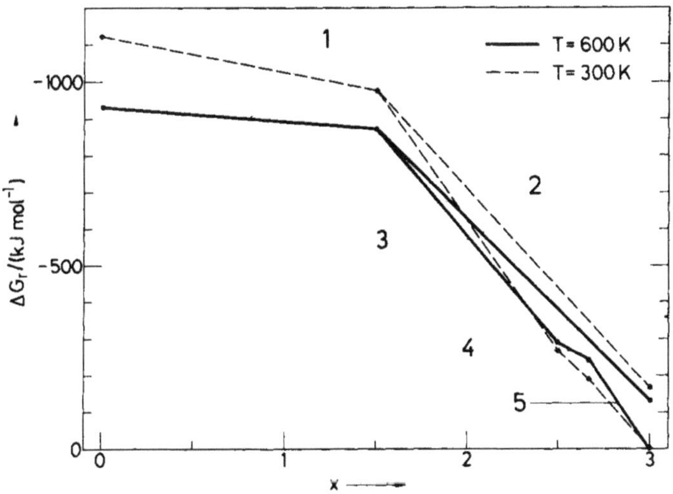

1: $(3-x)Na_3P + 2Na_2ZrO_3 + xNa_4SiO_4 + (6-4x)Na_2O$

2: $(3-x)Na_3P + 2Na_2ZrO_3 + xNa_4SiO_4 + (x-1.5)Si$

3: $(3-x)Na_3P + (5-2x)Na_2ZrO_3 + (3-x)Na_4SiO_4 + (2x-3)Na_2ZrSiO_5$

4: $(3-x)Na_3P + (4x-10)Na_2SiO_3 + (8-3x)Na_4SiO_4 + 2Na_2ZrSiO_5$

5: $(3-x)Na_3P + (6-2x)Na_2SiO_3 + (18-6x)Na_2ZrSiO_5 + (3x-8)Na_4Zr_2Si_3O_{12}$

Fig.19

Calculated free enthalpy (ΔG_r) for the reaction of one mole Nas(x) with sodium at 300K and 600K. The reaction products are specified in the legend [108].

Unfortunately Nas(3) does not have a sufficiently high conductivity for this particular application (see 4.). The use of ceramics with a very low grain size might increase the conductivity to the required level by an increase of the conductivity contribution from fast conduction paths along grain boundaries (see 5.). Another advantage of this approach is the attenuation of the effects of the anisotropic thermal expansion of Nas(3) (fig.11) which can cause cracking of the electrolyte. But one has to keep in mind that the grain boundary properties which have been shown to depend strongly on segregation and the formation of glassy grain boundary phases [110,111] are not yet well understood and might cause problems beyond the bulk properties of NASICON. Therefore, the vitrification of NASICON related phospate free compositions is an interesting alternative [112], but also for these materials the conductivities are not high enough. In conclusion, it seems that NASICON is unlikely to be able to rival β-aluminas as separators in Na/S cells.

The ease of preparing NASICON in the form of thin layers (e.g. by silk screen printing [113]) makes it interesting as an electrolyte material in electrochemical cells, which do not require high current densities or thermodynamic stability of the separator against materials with high sodium activity. This is, e.g., the case for many chemical sensors (potentiometric gas sensors [114], ion selective electrodes [115]), which might benefit from the Na^+ conductivity or in addition from the selectivity of ion exchange reactions at the interface of NASICONs with aqueous solutions. In such applications NASICON might also be used as electrode material (e.g. $Na_{1+x}Ti^{3+}{}_xTi^{4+}{}_{2-x}P_3O_{12}$ [116]) and monolithic devices are conceivable. Thermodynamic stability against sodium might also be dispensable for applications at room temperature, e.g. for low drain solid state batteries [117]. There is much space left for further activities and it is up to the fantasy of the reader to think of other applications.

In any case NASICONs are interesting model compounds for the study of fast ionic conduction phenomena in framework structures because of their compositional, structural and microstructural variety.

ACKNOWLEDGEMENT

The authors thank Udo Warhus who recently died.
This work is dedicated to him.

REFERENCES

1) Y.F.Yao, J.T.Kummer; J.Inorg.Nucl.Chem. 29, 2453 (1967)
2) Internationale Automobilausstellung (IAA), Frankfurt 1987
3) R.Collongues, D.Gourier, A.Kahn, J.P.Boilot, Ph.Colomban, A.Wicker; J.Phys.Chem.Solids 45, 981 (1984)
4) M.Nygen; Proceeding Risoe Int.Symp.Metal Material Sci. (Ed.: F.W.Poulsen) 6, 51 (1985)
5) H.Y.-P.Hong; Mater.Res.Bull. 11, 173 (1976)
6) J.B.Goodenough, H.Y.-P.Hong, J.A.Kafalas; Mater.Res.Bull. 11, 203 (1976)
7) R.G.Sizova, V.A.Blinov, V.A.Kuznetsov, A.A.Voronkov, V.V.Llyukhin, N.V.Belov; Sov.Phys.Dokl. 23, 103 (1978)
8) M.Zahir, R.Olazcuaga, P.Hagenmuller; Materials Letters 2, 234 (1984)
9) J.P.Boilot, E.Collin in: Superionic Solids and Solid Electrolytes. Recent Trends (Eds.: A.Laskar, S.Chandra) Academic Press, in press
10) V.A.Nicholas, A.M.Heyns, A.I.Kingon, J.B.Clark; J.Mater.Sc. 21, 1967 (1986)
11) R.S.Gordon, G.R.Miller, B.J.McEntire, E.D.Beck, J.R.Rasmussen; Solid State Ionics 3/4, 243 (1981)

12) I.K.Lloyed, T.K.Gupta, B.O.Hall; Solid State Ionics 11, 39 (1983)
13) D.Tran Qui, J.J.Capponi, J.C.Joubert, R.D.Shannon; J.Solid State Chem. 39, 219 (1981)
14) B.J.McEntire, R.A.Barlett, G.R.Miller, R.S.Gordon; J.American Ceram. Soc. 66, 738 (1983)
15) U.Warhus; Ph.-D.Thesis, Universität Stuttgart (1986) (may be obtained from the authors)
16) Ph.Colomban; Solid State Ionics 21, 97 (1986)
17) J.Engell, S.Mortensen, L.Møller; Solid State Ionics 9/10, 877 (1983)
18) B.E.Yoldas, I.K.Lloyed; Mater.Res.Bull. 18, 1171 (1983)
19) J.P.Boilot, Ph.Colomban, N.Blanchard; Solid State Ionics 9/10, 639 (1983)
20) H.Perthuis, G.Velasco, Ph.Colomban; Jap.J.Appl.Phys. 23, 534 (1984)
21) Ph.Colomban in: Reactivity of Solids (Eds.: P.Barret and L.C.Dufour) Elsevier, Amsterdam; 487 (1985)
22) O.Bouquin, H.Perthuis, Ph.Colomban; J.Mater.Sci.Letters 4, 956 (1985)
23) U.von Alpen, M.F.Bell, H.H.Höfer; Solid State Ionics 3/4, 215 (1981)
24) A.K.Kuriakose, T.A.Wheat, A.Ahmad, J.Dirocco; J.American Ceram.Soc. 67, 179 (1984)
25) A.Clearfield, M.A.Subramanian, W.Wang, P.Jerus; Solid State Ionics 9/10, 895 (1983)
26) P.R.Rudolf, M.A.Subramanian, A.Clearfield; Mater.Res.Bull. 20, 643 (1985)
27) H.Kohler, H.Schulz; Mater.Res.Bull. 20, 1461 (1985)
28) A.Clearfield, R.Guerra, A.Oskarsson, M.A.Subramanian, W.Wang; Mater.Res.Bull. 18, 1561 (1983)
29) S.J.Milne, J.A.Gard, A.R.West; Mater.Res.Bull. 20, 557 (1985)
30) J.P.Boilot, G.Collin, R.Comes; Solid State Ionics 9/10, 829 (1983)
31) C.L.Tsai, H.Y.-P.Hong; Mater.Res.Bull. 18, 1399 (1983)
32) F.Cherkaoui, J.C.Viala, C.Delmas, P.Hagenmuller; Solid State Ionics 21, 333 (1986)
33) R.G.Sizova, A.A.Voronkov, N.G.Shumyatskaya, V.V.Pyakhim, N.V.Belov; Dokl.Akad.Nank SSSR, Ser.205, 1, 90 (1972)
34) L.Hagman, P.Kierkegaard; Acta Chem.Scan. 22, 1822 (1968)
35) G.Collin, J.P.Boilot, Ph.Colomban, R.Comes; Solid State Ionics (1988), in press
36) H.Kohler, H.Schulz, O.Melnikov; Mater.Res.Bull. 18, 589 (1983)
37) F.Genet,M.Barj; Solid State Ionics 9/10, 891 (1983)
38) G.Collin, J.P.Boilot, Ph.Colomban, R.Comes; Solid State Ionics (1988), in press
39) M.Slujukic, B.Matkovics, B.Prodic, S.Scavnicar; Croat.Chem.Acta 39, 145 (1967)
40) J.P.Boilot, G.Collin, R.Comes; J.Solid State Chem. 50, 91 (1983)
41) S.Susman, C.J.Delbecq, T.O.Brun; Solid State Ionics 9/10, 839 (1983)
42) K.Byrappa, G.S.Gopalakrishna, V.Venkatachalapathy, A.B. Kulkarni; J.Less Common Metals 110, 441 (1985)
43) Yu.A.Ivanov, E.L.Belokoneva, Yu.K.Egorov-Tismenko, M.A.Simonov, N.V.Belov; Sov.Phys.Dokl. 25, 420 (1980)
44) H.Schmid, L.C.DeJonghe, C.Cameron; Solid State Ionics 6, 57 (1982)
45) U.von Alpen, M.F.Bell, W.Wichelhaus; Mater.Res.Bull. 14, 1317 (1979)
46) K.-D.Kreuer, H.Kohler, U.Warhus, H.Schulz; Mater.Res.Bull. 21, 149 (1986)
47) R.G.Sizova, A.A.Voronkov, N.G.Shumyatskaya, V.V.Llyukhin, N.V.Belov; Sov.Phys.Dokl. 17, 618 (1973)

48) H.Kohler, H.Schulz, O.Melnikov; Mater.Res.Bull. 18, 1143 (1983)
49) H.Kohler; Ph.-D.Thesis, Universität Karlsruhe (1984)
(may be obtained from the authors)
50) J.B.Boilot, G.Collin, Ph.Colomban; Mater.Res.Bull. 22, 669 (1987)
51) P.R.Rudolf, A.Clearfield, J.D.Jorgensen; Solid State Ionics 21, 213 (1986)
52) J.P.Boilot, G.Collin, Ph.Colomban; J.Solid State Chem. in press
53) H.Kohler, H.Schulz; Solid State Ionics 9/10, 795 (1983)
54) H.Kohler, H.Schulz; Mater.Res.Bull. 21, 23 (1986)
55) W.Baur, J.R.Dygas, D.H.Whitemore, J.Faber;
Solid State Ionics 18/19, 935 (1986)
56) J.J.Didisheim, E.Price, B.J.Wuensch; Solid State Ionics 18/19, 944 (1986)
57) M.A.Subramanian, P.R.Rudolf, A.Clearfield; J.Solid State Chem. 60, 172 (1985)
58) G.Collin, R.Comes, J.P.Boilot, Ph.Colomban;
J.Phys.Chem.Solids 47, 843 (1986)
59) G.Lucazeau, M.Barij, J.L.Soubeyroux, A.J.Dianoux, C.Delmas;
Solid State Ionics 18/19, 959 (1986)
60) M.de la Rochère, F.d'Yvoire, G.Collin, R.Comes, J.P.Boilot;
Solid State Ionics 9/10, 825 (1983)
61) J.P.Boilot, Ph.Colomban, G.Collin; Solid State Ionics 18/19, 974 (1986)
62) J.P.Boilot, J.P.Salanié, G.Desplanches, D.Le Potier;
Mater.Res.Bull. 14, 1469 (1979)
63) C.Delmas,J.-C.Viala, R.Olazcuaga, G.LeFlem, F.Cherkaoui, R.Brochu;
Solid State Ionics 3/4, 209 (1981)
64) A.Feltz, S.Barth; Solid State Ionics 9/10, 817 (1983)
65) Lin Zu-Xiang, Tian Shun-Bao; Solid State Ionics 9/10, 809 (1983)
66) F.d'Yvoire, M.Pintard-Screpel, E.Bretey, M.de la Rochère;
Solid State Ionics 9/10, 851 (1983)
67) K.Byrappa, A.B.Kulkarni, G.S.Gopalkrishna;
J.Less-Common Metals 111, 359 (1985)
68) J.P.Boilot, G.Collin, R.Comès; Solid State Ionics 5, 307 (1981)
69) L.Boehm, C.J.Delbecq, E.Hutchinson, S.Susman;
Solid State Ionics 5, 311 (1981)
70) D.TranQui, J.J.Capponi, M.Gondrand, J.C.Joubert;
Solid State Ionics 5, 305 (1981)
71) M.Nagai, S.Fujitsu, T.Kanazawa; Solid State Ionics 3/4, 227 (1981)
72) T.Takahashi, K.Kuwabara, M.Shibata; Solid State Ionics 1, 163 (1980)
73) Wenji Wang, Shengban Wang, Lin Rao, Zonglan Lu, Xinzheng Yi;
Solid State Ionics (1988), in press
74) R.J.Cava, E.M.Vogel, D.W.Johnson; Comm.Amer.Ceram.Soc. C-157 (1982)
75) K.-D.Kreuer; unpublished result
76) B.Okray Hall; Solid State Ionics 5, 317 (1981)
77) R.Bachmann, H.Schulz; Acta Cryst. A40, 668 (1984)
78) B.T.M.Willis; Acta Cryst. A25, 277 (1969)
79) V.B.Kalinin, B.I.Lazoriak, S.Yu.Stefanovich; Sov.Phys.Crystallogr. 28, 154 (1983)
80) T.Oota, I.Yamai; J.American Ceram.Soc. 69, 1 (1986)
81) F.Cherkaoui, G.Villeneuve, C.Delmas, P.Hagenmuller;
J.Solid State Chem. 65, 293 (1986)
82) M.Nagai, T.Nishino, T.Kanazawa; Solid State Ionics 18/19, 964 (1986)
83) P.K.Davis, F.Garzon, T.Feist, C.M.Katzan;
Solid State Ionics 18/19, 1120 (1986)
84) M.Barj, H.Perthuis, Ph.Colomban; Solid State Ionics 9/10, 845 (1983)

85) A.Baudry, P.Boyer; Solid State Ionics 9/10, 799 (1983)
86) Ph.Colomban; Stud.Phys.Theor.Chem. 46, 277 (1987)
87) G.S.Rohrer, P.K.Davis, G.C.Farrington; Solid State Ionics (1988), in press
89) J.Maier, U.Warhus, E,Gmelin; Solid State Ionics 18/19, 969 (1986)
89) U.Warhus, J.Maier, A.Rabenau; J.Solid State Chem. 72 (1), (1988), in press
90) M.Barsoum; personal communication
91) N.S.Choudhary; J.Electrochem.Soc. 133, 425 (1986)
92) N.S.Choudhary; J.Electrochem.Soc. 133, 429 (1986)
93) N.S.Choudhary; J.Electrochem.Soc. 120, 1663 (1973)
94) F.A.Elrefaie, W.W.Smeltzer; J.Electrochem.Soc. 128, 1443 (1981)
95) P.K.Davis, G.J.Pfieffer, S.Canfield, Solid State Ionics (1988), in press
96) G.Róg, Kozlowska-Róg; Solid State Ionics 7, 291 (1982);
Electrochim.Acta L8, 43 (1983)
97) J.Maier, U.Warhus; J.Chem.Thermodynamics 18, 309 (1986)
98) E.V.Shibanov, V.G.Chuklantsev; Russ.J.Phys.Chem. 48, 133 (1974)
99) P.K.Davis, A.Navrotsky; J.Solid State Chem. 46, 1 (1983)
100) L.J.Schoiler, B.J.Wuensch, E.Price;
4th Int.Conf.Solid State Ionics, Grenoble (1983)
101) A.Nawrotsky; Reviews in Mineralogy Vol.14 (Eds.: S.W.Kieffer,A.Navrotsky)
Book Crafters Inc., Chelsea (1985), Chapter 7
102) D.K.Agraware, R.Roy; J.Mater.Sci.Letters 20, 4617 (1985)
103) J.D'Ans, J.Löffler; Z.allg.anorg.Chem. 191, 1 (1930)
104) A.Sircar, N.H.Brett; Trans.Brit.Ceram.Soc. 69, 131 (1970)
105) B.Kusatz, H.Kroll, A.Kaiping, H.Pentinghaus; Fortschr.Miner. 65, 203 (1987)
106) J.W.Cahn; Acta Met. 10, 907 (1982);
Met.Soc. AIME 242, 166 (1968)
107) S.Yde-Andersen, J.S.Lundsgaard, L.Møller, J.Engell;
Solid State Ionics 14, 73 (1984)
108) K.-D.Kreuer, U.Warhus; Mater.Res.Bull. 21, 357 (1986)
109) J.Jung, A.Reck, R.Ziegler; J.Nucl.Mater. 119, 339 (1983)
110) V.A.Nicholas, P.J.Johnson, A.I.Kingon; Solid State Ionics 17, 351 (1985)
111) W.Bogusz, F.Krok, W.Jakubowski; Solid State Ionics 9/10, 803 (1983)
112) S.Susman, C.J.Delbecq, J.A.McMillan, M.F.Roche;
Solid State Ionics 9/10, 667 (1983)
113) K.Zucholl; Battelle Institut e.V., Frankfurt, private communication
114) T.Maruyama, Y.Saito, Y.Matsumoto, Y.Yano;
Solid State Ionics 17, 281 (1985)
115) M.Kleitz, J.F.Millon-Brodaz, P.Fabry; Solid State Ionics 22, 295 (1987)
116) C.Delmas, A.Nadiri, J.L.Soubeyroux; Solid State Ionics (1988), in press
120) Denstedt; Varta Kelkheim, private communication

COPPER ION CONDUCTORS

A KASI VISWANATH & S RADHAKRISHNA
Department of Physics
Indian Institute of Technology
Madras 600 036, INDIA

1. INTRODUCTION

Recently there has been a lot of interest in the study of solids with high ionic conductivities. These materials are called Superionic Conductors or Solid Electrolytes or Fast Ion Conductors. In the last two decades many such high ionic conductivity solids have been identified and many applications have been explored. Several practical devices like solid state batteries, fuel cells, display panels, memory devices etc. have been prepared and used. Many excellent conference proceedings[1-3], books[4-8] and reviews[9-19] are available in the literature.

The normal ionic crystals like NaCl, KCl have very low ionic conductivity, of the order of 10^{-12} ohm^{-1} cm^{-1}. The ionic conductivity in these crystals is due to the vacancies inherently present in crystals. Since there is an upper limit for the number of defects that can be created in a crystal by impurity doping, the conductivity could not be pushed beyond 10^{-6} ohm^{-1} cm^{-1}.

Tubandt and Lorentz[20] have observed unusual electrical conductivity behaviour in AgI compared to other silver halides like AgBr and AgCl, over a temperature range of more than 400°C. The conductivity of AgI abruptly increases at a temperature of 147°C. The highly conducting high temperature phase of AgI is called as α AgI. Tubandt etal[21,22] have identified a number of crystals which are highly conducting in their high temperature phases like αAgI. These are αCuBr, β CuBr, αCuI, α Ag_2S, α Ag_2Se and α Ag_2Te. In these crystals the cations form a 'molten sub-lattice' at high temperture and give rise to high ionic conductivity just like in a liquid. Strock[23] and Hoshino[24] did the crystal structure analyses for these crystals and confirmed the above result. Later intensive research has been done by many workers to synthesize high conductivity solids. In 1967, Owens and Argue[25] and Bradley and Green[26] independently discovered $RbAg_4I_5$ which has a very high room temperture ionic conductivity of 0.27 ohm^{-1} cm^{-1}.

Among the solid electrolytes developed so far, Ag^+ ion based conductors were mostly studied[4,9-11]. Since silver is very expensive it is worth while to investigate other ionic conductors which are much cheaper. Pure copper halides which give high ionic conductivity at high temperatures were studied by many workers. Takahashi etal have identified the room temperature copper ion conductors in 1973. In this article we will review the entire literture on copper ion conductors, from the early work on pure copper halides to the more recent work on the highest conducting superionic solid of a copper ternary system. A wide variety of experimental techniques have been discussed. These include the following techniques :

1 Crystal Structure Analysis
2 D.C. Conductivity
3 Heat Capacity
4 Thermo Electric Power
5 Magnetic Susceptibility

6 Nuclear Magnetic Resonance (NMR)
7 Electron Spin Resonance (ESR)
8 Laser Raman and Infrared
9 Microwave Studies & Dielectric Relaxation
10 Miscellaneous experiments, like Photoelectron Spectroscopy.

For each one of the systems listed in the contents all the experimental results known, will be discussed before we pass on to the next compound. Most of the data on copper ionic conductors has been on the d.c. conductivity. Very little work has been done using other techniques like magnetic resonance. We have given an indepth discussion on Laser Raman, infrared and microwave studies, including our latest work to understand the microscopic origin of superionic motion and relaxation.

2. COPPER HALIDES

The importance of crystal structures in the study of solid electrolytes has been stressed by many authors[4,9,27]. In these crystals the number of cations is much less than the number of available sites and these sites are occupied at random. The anions form a rigid lattice in such a way that the local potentials are rather flat along certain lines which interconnect neighbouring sites. Along these lines the difference in potential energy is of the order of the thermal energy. Many α AgI type solids satisfy these requirements.

The cuprous halides have zincblende structure at low temperatures. This low temperture phase is called γ phase. At higher tempertures they transform to β phase having wurtzite structure. The phase transition tempertaures for CuI[28,29], CuBr[30] and CuCl[31] are different. At much higher temperatures CuI and CuBr transform to α phase possessing the body centered cubic structure. The crystal structures of various copper ion conductors are given in Table 1. Very recently Graneli etal[32] did the neutron scattering measurements on CuCl powder between 8 and 658 K. They have confirmed that the low temperature phase with cubic structure is stable below 680K and above 680K, CuCl has the β phase with the hexagonal structure to the melting point at 696K. Above room temperature the copper ion motion appears to be increasingly anharmonic with thermal vibrations along <111> directions in the cubic phase. With increasing temperature Graneli etal have observed that a characteristic diffuse background develops in the neutron diffraction pattern, reflecting the increased disorder in the lattice. They have also related the results with the molecular dynamics simulations. The various phase transitions of the copper halides are given below :

$$\text{CuI} \quad \gamma \xrightarrow{369^\circ C} \beta \xrightarrow{407^\circ C} \alpha$$

$$\text{CuBr} \quad \gamma \xrightarrow{385^\circ C} \beta \xrightarrow{470^\circ C} \alpha$$

TABLE 1

Crystal Structures of some Copper ion Conductors

Crystal	Structure	Temperature in °C	References
CuI	Cubic	> 407	
CuI	Hexagonal	369–407	28,29
CuI	Cubic	< 369	
CuBr	Cubic	> 470	
CuBr	Hexagonal	385–470	30
CuBr	Cubic	< 385	
CuCl	Hexagonal	> 407	31, 32
CuCl	Cubic	< 407	
Cu_2HgI_4	Cubic	> 67	75 - 77
Cu_2HgI_4	Tetragonal	> 67	
Cu_2CdI_4	Cubic	> 308	67
Cu_4CdCl_6	Cubic	> 330	67
KCu_4I_5	Cubic	> 262	67
$RbCu_3Cl_4$	Complicated structure with	> 170	67, 68
$Rb_2Cu_3Cl_5$	low symmetry	> 150	67, 68
$Rb_4Cu_{16}I_7Cl_{13}$	Cubic	< 234	74
Cu_2S	Hexagonal	> 104	100
Cu_2S	Monoclinic	< 104	103
	Orthorhombic	< 104	101, 102

Cu_2Se	Cubic	> 110	112-114
Cu_2Se	Orthorhombic	< 110	
$CuGa_5S_8$	Same as Cu_2HgI_4		117
$CuGa_5Se_8$			
CuTeCl	Tetragonal Symmetry with 'average structure' for copper ions	< 400	
CuTeBr		< 411	127
CuTeI		< 406	
$CuTa_2O_6$	Cubic Perovskite	--	129

$$\text{CuCl} \quad \gamma \xrightarrow{407^\circ C} \beta$$

The defects in cuprous halides are Frenkel type i.e. copper interstitials and copper vacancies and the copper interstitials are the more mobile species[33-37]. Tubandt[21] has measured the ionic transference numbers for all the copper halides by the electrochemical method. Later the ionic transference numbers in these crystals have been determined by various workers[38-41] using both the a.c. electrical conductivity measurements and d.c. polarization methods. It was found that the conductivity in these crystals is mostly due to ionic motion. Figure 1 shows the log conductivity as a function of $10^3/T$ for various inorganic copper ion conductors.

As can be seen from Table 1 CuI exists in three phases. The high temperature phase is the highest conducting state. This can be easily understood, since in this phase it has the α-AgI type crystal structure[28,29] with average structure for cations which are conducting. Wagner and Wagner[37] and Matsui and Wagner[36] have studied the conductivity and activation energy of conductivity for pure and impurity doped CuI. They have explained the high conductivity and low values of activation energy considering the crystal structure. They have also observed abrupt changes in the electrical conductivity at various phase transistions in pure CuI crystals[39,36]. Matsui and Wagner[36] have discussed the conductivity jumps in relation to entropy changes at the phase transition point. In their systematic study Matsui and Wagner[36] have calculated various transport parameters for pure CuI, like the concentration of Frenkel defects(x_o), mobility of copper interstitials (μ_i), mobility of copper vacancies (μ_v), formation energy of Frenkel defects (E_F), migration energy of copper interstitials (E_{mi}) and the migration energy of copper vacancies (E_{mv}).

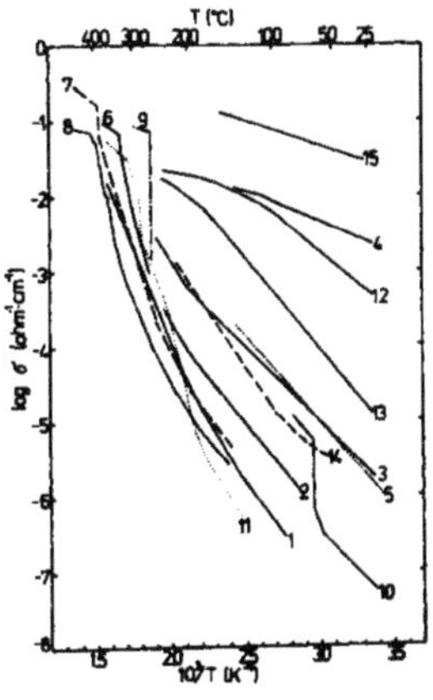

Fig. 1 Conductivity for various inorganic copper ion conductors.

1. CuCl	9. KCu_4I_5
2. CuCl + 0.70 mol % Cu_2S	10. Cu_2HgI_4
3. CuCl + 10 mol % Al_2O_3	11. Cu_2CdI_4
4. $RbCu_3Cl_4$	12. CuTeCl
5. $Rb_2Cu_3Cl_5$	13. CuTeBr
6. Cu_4CdCl_6	14. CuTeI
7. CuBr	15. $Cu_{1.75}Se$
8. CuI	

The references for various systems are given in Table 2. The figure is taken from reference 16.

The electronic transport (electron hole or electron conductivity) properties of cuprous halides have been studied by Wagner and Wagner[39], Joshi[40], Joshi and Wagner[41] and Jow[42]. Cuprous halides are p-type semiconductors[39]. Wagner and co-workers have done the steady state and transient measurements to get the electronic transport properties. They have also used three electro-chemical techniques :

(1) Wagner's method of polarization
(2) Joshi and Wagner's technique of charge transfer[40,41,43]
and (3) The relaxation method of Weiss[44].

A very good treatment of the theory, the experimental procedure and the analysis of the results was given by Wagner[45,46].

Jow[42] has measured various electronic transport properties like electron hole conductivity, concentration of electronic holes and the mobility of electron holes, for pure CuI. At temperatures below 300°C the contribution of electron hole conductivity to the total conductivity was found to be appreciable. Vine and Maurer[47] have measured the Hall coefficient and electron hole conductivity for the non-stoichiometric CuI_{1+x}.

TABLE 2

Electrical conductivity parameters for various inorganic copper ion conductors.

Compound	Conductivity ($Ohm^{-1}cm^{-1}$)	Temperature °C	Activation Energy ($KCal\ mol^{-1}$)	Reference
CuI	1×10^{-1}	450	3.70	36,39, 42
CuCl	4.2×10^{-2}	390	12.5	39,41, 42
CuBr	3.0×10^{-1}	470	--	39,41
Cu_2HgI_4	5.4×10^{-6}	67	13.6	75,76
Cu_2CdI_4	3.2×10^{-2}	308	9.22	67
$CuPb_3Br_7$	3×10^{-2}	200	22	69,70
Cu_4CdCl_6	1×10^{-1}	370	5.99	67
KCu_4I_5	8.1×10^{-2}	262	3.69	67
$RbCu_3Cl_4$	2.3×10^{-3}	25	4.38	67,68
	1.3×10^{-2}	140	3.23	
$Rb_2Cu_3Cl_5$	1.5×10^{-6}	25	11.3	67,68
	2.2×10^{-4}	140		
$Rb_3Cu_7Cl_{10}$	0.3×10^{-2}	25	--	71
$CsCu_9Br_{10}$	0.3	290	--	72
$CsCu_9I_{10}$	0.3	340	--	72
$RbCu_4I_4Cl$	0.28	235	--	74
$Rb_4Cu_{16}I_7Cl_{13}$	0.34	25	--	73
$Cu_{1.75}Se$	3.0×10^{-2}	30	1.3	70
$CuGaS_2$	10^{-6}	20	--	
$CuGaS_5$	10^{-4}	20	--	117
$CuGaSe_6$	10^{-5}	20	--	

CuTeCl	4×10^{-4}	25	8.4	126
	2×10^{-2}	200		
CuTeBr	1×10^{-5}	25	11.7	126
	1×10^{-2}	200		
CuTeI	9×10^{-4}	25	14.5	126
	7×10^{-4}	200		
$CuTa_2O_6$	10^{-3}	300	--	128

In contrast to pure CuI, the CuCl has been found to be mostly an ionic conductor above room temperature and the ionic transport parameters have been determined[36, 39-41]. Joshi and Wagner[41] have also determined the electron hole conductivity, drift and concentration of electron holes in pure CuCl.

CuBr has the highest ionic conductivity among the pure cuprous halides and sudden changes in electrical conductivity at the phase transition temperatures have been observed by Wagner and Wagner[39].

Measurement of the heat capacity $C_p(T)$ is one of the methods of detecting the phase transitions in solids. The heat capacity of pure cuprous halides has been studied by Nolting and Co-workers[48-52]. The heat capacity of CuBr and CuI as a function of temperture is shown in figure 2. The disordering of the cations in the lattice at the phase transition point gives rise to excess heat effects which is reflected in the heat capacity, $C_p(T)$ values. From figure 2 it can be seen that the excess heat capacities due to an increasing degree of cation disorder are found before the transition to the cation-disordered state. Within the highly conducting phases C_p values slowly decrease with almost the same slope. Nolting etal[48] have explained the observed variations in heat capacities by considering the disordering processes in the cation sublattice at higher temperatures. In the high temperature phases the potential barriers are of the order of thermal energy and hence the cation motion will have a transition from the oscillating state to translation state, thus lowering the number of degrees of freedom. The situation is very similar to a hindered rotator.

From figure 2 one can see striking differences in the variation of $C_p(T)$ values for CuBr and CuI in their β phases. For CuI the C_p values keep on increasing in the β phase indicating that the disordering process is still taking place. On the other hand for CuBr C_p values keep on decreasing in the β phase which shows that the complete disorder has already taken place.

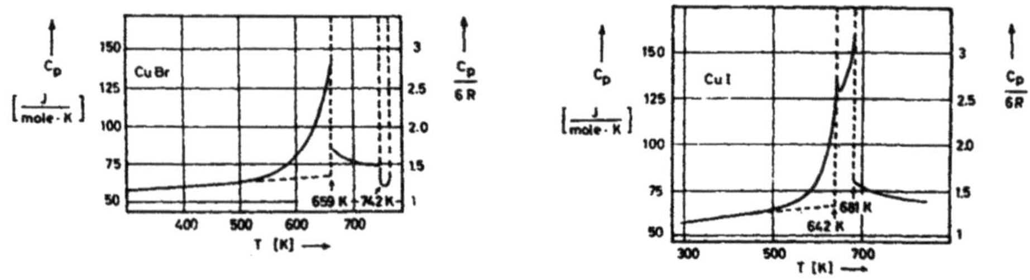

Fig. 2 Variation of heat capacity as a function of temperature for CuBr and CuI. The figure is taken from reference 48.

Hoshino[30] has studied the variations in specific heat for CuBr which were related to the phase transitions. Miyake etal[29a] have made similar measurements on CuI. They have estimated the thermodynamic parameters like transition enthalpy and entropy at the phase transition temperatures.

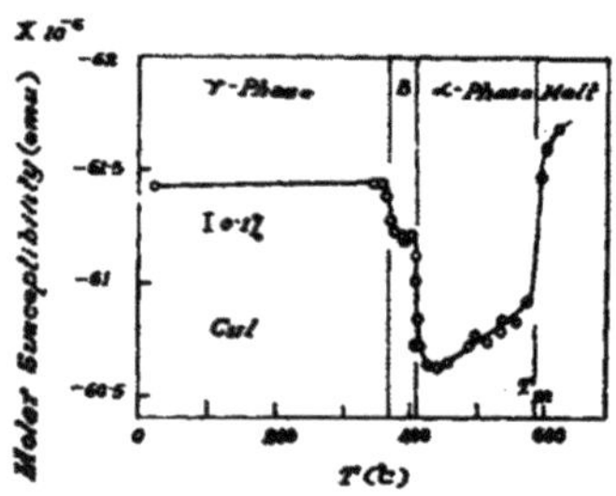

Fig. 3 Diamagnetic susceptibility of CuI as a function of temperature. The figure from reference 53.

Stutius[53] has studied the diamagnetic susceptibility χ of pure CuI from room temperature to melting point. The temperature dependence of χ for pure CuI is shown in figure 3. In semiconductors the diamagnetic susceptibility χ is dependent upon the structural disorder[54]. The changes in the values of χ can be correlated to the changes in the nature of bonding in the material. In superionic conductors the bonding configuration changes with temperature due to defect formation, increasing thermal motion as reflected in the large Debye-Waller factor and ion disorder. These lattice dynamical processes give rise to temperature dependent variation of χ which is generally not observed in other semiconductors. Using the bond orbital model the total diamagnetic susceptibility can be decomposed into three terms[55] :

$$\chi = \chi_c + \chi_L + \chi_{vv}$$

Here χ_c is the diamagnetic core contribution which does not depend upon the chemical bonding and is temperature independent. χ_L is the diagmagnetic term arising from the contribution of valence electrons and χ_L scales approximately with lattice constant. χ_{vv} is the Van Vleck paramagnetic term and depends explicitly upon temperature reflecting the disorder in the material. It can be written as[55,56] :

$$\chi_{vv} = (1 - \alpha^2) \frac{M}{E_g}$$

where α is the ionicity, M is proportional to average interband magnetic dipole matrix element and Eg is the energy difference between the valence band and the conduction band. It was shown by Chadi etal[55] that M depends strongly on the symmetry of bonding configuration. M decreses when structural disorder is introduced. As can be seen in figure 3 in the γ phase χ is constant. In the $\gamma \rightarrow \beta$ and $\beta \rightarrow \alpha$ transition χ decreases only by a small value. This is due to the decrease in the value of M, which arises as a consequence of mobility of Cu atoms in the superionic conducting state. At the melting point χ value increases suddenly because the copper ion loses its tetrahedral coordination only at the melting point. At this temperature M will further decrease and the material becomes more diamagnetic. Stutius[53] argues that the enhanced diamagnetism in other amorphous semiconductors like Si and Ge is also due to the decrease in the interband matrix element M in the Van Vleck paramagnetic contribution to the total diamagnetic susceptibility .

Nuclear magnetic resonance is an excellent experimental technique to study the dynamical processes which are very much temperature dependent[57]. Becker etal[58] have done the NMR line narrowing experiments on pure CuI. In the superionic conducting phase, since the Cu atoms are in the disordered state, the spins are in constant motion like in a liquid state. This gives rise to 'motional narrowing' of the NMR resonance lines. Becker etal[58] have indeed observed such narrow NMR lines in the α phase of CuI giving evidence to Cu diffusion.

In a classic paper Boyce and Huberman[59] have reported the NMR relaxation measurements and disorder processes when sub-lattices melting takes place in CuI. This was also the first indepth study on the defect contribution to magnetic resonance relaxation mechanism. So, we will have an elaborate discussion on these results. Figure 4 shows the various relaxation times as a function of temperature for pure CuI. Boyce and Huberman[59] have measured T_1, the spin-lattice relaxation time, T_2, the spin-spin relaxation time for Cu^{63}, Cu^{65} and I^{127} for the pure poly-crystalline CuI samples, in the temperature range 0 to 800°C, well above the melting point. They have also measured $T_{1\rho}$ the spin-lattice relaxation time in the rotating frame for the low temperature γ-phase for Cu^{63}. T_1 was dominated by qudrupolar interactions throughout the temperature range. This quadrupolar relaxation can be caused by fluctuations in the electric field gradient due to phonons[60] or defects[61,62]. Thus T_1 is written as

$$\frac{1}{T_1} = \frac{1}{T_1 \text{ phonons}} + \frac{1}{T_1 \text{ diffusion}}$$

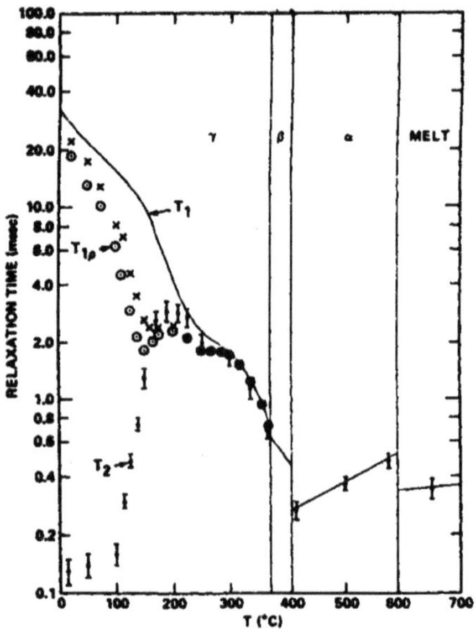

Fig. 4 NMR relaxation times for Cu^{63} in pure CuI
- T_1 spin-lattice relaxation time at 16 MHz
I T_2 spin-spin relaxation time at 16 MHz
O $T_{1\rho}$ spin-lattice relaxation time in the rotating frame at 16 MHz, H_1 = 7 G.
X $T_{1\rho}$ spin-lattice relaxation time in the rotating frame at 16 MHz, H_1 = 14 G.
The figure is taken from reference 59.

The Cu defects, both interstitials and vacancies[39], will affect the relaxation rate of Cu ions on the cubic lattice sites, since they create an electric field gradient which fluctuates, due to the defect motion, at a frequency $1/\tau_d$, where τ_d is the time between hops of the defect. The spin-lattice relaxation rate is very much dependent upon n_d, the atomic fraction of defects. The spin-spin relaxation rate $1/T_2$ and the spin-lattice relaxation rate in the rotating frame $1/T_{1\rho}$ are very much influenced by the fluctuations in both the dipolar interactions and quadrupolar interactions. The dipole-dipole interaction between the nuclei fluctuates at a frequency $1/\tau_d$ where τ_d is the time between hops of a copper atom. In the γ phase, near room temperature, the spin lattice relaxation rate is dominated by phonons. Above 100°C the diffusion becomes important,as can be seen in figure 4, the frequency dependence of T_1 and $T_{1\rho}$ and the increase in the spin-spin relaxation time. The expected minimum in the spin-lattice relaxation time as predicted[61,62] when $\omega_o \tau_d = 1$ is suppressed. At further high temperatures the relaxation rate keeps on increasing with temperature. Above 300°C the relaxation rate is not at all affected by frequency and we can see in the figure 4 that $T_1 = T_2 = T_{1\rho}$. The spin-spin relaxation is very strongly influenced by the copper diffusion in the 100 - 200°C region. Boyce and Huberman[57] have calculated an atomic hopping time which is in agreement with the line narrowing experiments of Becker etal[58]. As the $\gamma \rightarrow \beta$ phase transition is approached the relaxation rate keeps on increasing. This indicates $n_d(T)$ is increasing more rapidly. This shows that at the insulator-to-superionic phase transition the disordering of the sub-lattice takes place[63-65]. In the β phase (369 - 470°C) T_1 rapidly decreases implying that $n_d(T) \longrightarrow 1$ i.e. copper sublattice is melting. In the α phase (T > 407°C), the relaxation rate is similar to what is generally observed in a liquid, as can be seen in the figure 4, the steady decrease of relaxation rate with increasing temperature. Here phonon contribution is negligible and $1/T_1$ is dominated by diffusion. Boyce and Huberman[57] have obtained an activation energy of 0.25 eV for Cu diffusion which is in excellent agreement with the activation energy of 0.2 eV obtained from the conductivity data by Wagner and Wagner[37]. In the α phase $n_d(T)$ is a constant indicating that copper sub-lattice is molten. In the molten phase there is a change in the $1/T_1$ at the phase transition from α to liquid state. This was explained as due to melting of Iodine crystalline cage.

Funke[7] has studied the microwave conductivity and permittivity for β CuBr and α CuI. These results along with the microwave results obtained in our laboratory for Cu_2HgI_4 and Cu_2CdI_4 will be discussed in a latter section.

Dickenson etal[66] have reported the photoelectron spectra for various copper ion conductors. They have observed a single peak for all the copper compounds they have investigated. The binding energy of Cu $2p_{3/2}$ electron in pure CuI was estimated to be 933.6 eV. The binding energies of Cu $2p_{3/2}$ electron for various copper ion conductors are given in Table 3.

TABLE 3

Binding Energy of ESCA Peaks

Compound	Cation electron	Binding energy in eV
CuI	Cu $2p_{3/2}$	933.6
Cu_2HgI_4	Cu $2p_{3/2}$	933.6
Cu_2S	Cu $2p_{3/2}$	934.2

3. BINARY SYSTEMS WITH COPPER HALIDES

After the success of Owens and Argue[25] and Bradley and Green[24] in making high conductivity $RbAg_4I_5$ silver compounds, many attempts were made by different workers to synthesize high conductivity copper compounds. Several approaches were used chiefly by Wagner[34,37,67,68], Takhashi[69-74] and their co-workers for this purpose :

(A) Copper halides were mixed with halides containing divalent cations like CdI_2, HgI_2.
(B) Copper halides were mixed with copper sulphide.
(C) New compounds were prepared by mixing with layered materials like $CdCl_2$.
(D) A novel approach is to disperse a second phase like Al_2O_3.
(E) To stabilize the high temperature average structure at low temperature by mixing with alkali halides like KI,RbI,RbCl etc.

In this section we will review various studies on mixed copper halides.

A. **Copper Halides Doped with Divalent Cations**

(i) <u>CuI + CdI_2 and CuI + HgI_2</u>

The Cu_2HgI_4 compound was synthesized by Ketelaar[75] who observed a high ionic conductivity in these materials. Later on it was studied by Suchow and Pond[76] and Hoshino[77]. Cu_2HgI_4 has a tetragonal structure at room temperature which is the β phase. It differs from its isomorph Ag_2HgI_4 in the Wigner-Seitz cell which has a higher symmetry (D_{2d}) than that of Ag_2HgI_4. Cu_2HgI_4 transforms to α phase above 67°C and has the zincblende structure with average copper ion arrangement. Cu_2CdI_4 was synthesized by Matsui and Wagner[67], who have reported three phases in the crystal. They have observed α→β phase transition at 308°C and β→γ transition at 276°C. The α Cu_2CdI_4 has the αCuI cubic structure. Matsui and Wagner have also studied the CuI,

CdI_2 mixtures for various compositions of CdI_2. They have found that both the $\alpha \rightarrow \beta$ and $\beta \rightarrow \gamma$ transition temperatures decreased with increasing concentrations of CdI_2. The structures of Cu_2HgI_4 and Cu_2CdI_4 are included in Table 1.

Ketelaar[75] and Suchow and Pond[76] have reported the conductivity in α Cu_2HgI_4. They have observed an ionic conductivity of 5.4×10^{-4} at 67°C with an activation energy of 13.6 KCal mol^{-1} in the temperature range 67-80°C. Matsui and Wagner[67] have studied the conductivity of CuI and CdI_2 mixtures for various concentrations of CdI_2. They have noted that the αCuI phase was stabilized in Cu_2CdI_4 only at 308°C in contrast to 67°C in Cu_2HgI_4. They have also pointed out that the activation energies in the α phase of Cu_2CdI_4 increased suggesting that the mobility of copper ions was decreased by cadmium ions at normal copper lattice sites. The electrical conductivities of Cu_2CdI_4 and Cu_2CdI_6 were higher by one order of magnitude than pure CuI in the temperature range 200-300°C which may be due to the contribution of copper vacancies produced by cadmium ions in the divalent state. At low temperatures, the conductivities of Cu_2CdI_4 and Cu_2CdI_6 were lower than pure CuI because of the clustering between copper vacancies and Cd^{2+} ions. Lal and Srivastava[78] also have studied the electrical conductivity and thermoelectric power of $(1-x)CuI.xCdI_2$ mixtures where x = 0, 0.25, 0.33, 0.50 and 0.75. It has been observed that all compounds of this system undergo phase transition from low temperature low conducting β phase to higher temperture high conductivity α phase. Lal and Srivastava[78] have found that the phase transition temperature T_p decreases systematically with x, the concentration of CdI_2. Above T_p all compounds are typical superionic solids with Cu^+ ion as the charge carrier. They have found that Cu_3CdI_5 is the best superionic solid from the stability, temperature span of superionic phase, T_p and σ values point of view. The extended lattice gas model has been shown to explain the conductivity properties. Below the 'Knee temperature' the conductivity was found to be extrinsic. The activation enthalpy, heat of transport and enthalpy for the formation of defects have been evaluated from the conductivity and thermoelectric power data. The conductivity parameters for Cu_2HgI_4 and Cu_2CdI_4 are given in Table 2.

Now we will review the Laser Raman, infrared and microwave results reported on the pure cuprous halides and mixed halides. Funke[9] was the first to study the microscopic ion dynamics in solid electrolytes. The atoms in simple monoatomic liquids have a continuous type of motion. They obey the laws of diffusion represented by Langevin equation and Fick's second law. For this simple diffusion case, the residence time τ_0 is zero. But hydrogen in certain hydrogen - metal systems such as H-Pd and H-Nb definitely performs a jump diffusion from one well defined site to another. In this case the mean time of flight τ_1 is negligible compared to the resident time τ_0. Funke[9] has concluded from his correlated experiments like microwave, far-infrared, Raman and quasi elastic neutron scattering, that the diffusion in superionic solids is an intermediate case between the simple diffusion in liquids and jump diffusion in hydrogen metal systems. The diffusion in superionic conductors can be thought of, according to Funke, as a superposition of jump diffusion and a local kind of motion which is characterized by a large amplitude with strong damping. Here the mean time of flight τ_1 is almost the same order of magnitude as the mean residence time τ_0.

Funke and Coworkers[79] have measured the microwave conductivity σ and permittivity ε' for β CuBr and α CuI. In both the cases they have observed a decrease in σ (ν) with increasing frequency and a maximum in σ in the 20-30 GHz range and negative values of ε' in the low frequency limit. These observations have been explained by considering the 'acceleration effect' and 'start effect'.

TABLE 4

Vibrational Band Assignments of Cu_2HgI_4 Crystals

Band Positions RT (cm^{-1})	LNT (cm^{-1})	Symmetry	Assignment	TO modes cm^{-1}	LO modes cm^{-1}
127	128	A_1	$(Hg\ I_4)^{2-}$ Cu-I Symmetric stretch	144	238
	123				
115	107	A_1	M-I Symmetric stretch	123	111
83	80				
47	43	B_1	$(Hg\ I_4)^{2-}$ deformation	82	56
35	30	E	Cu^+ attempt frequency	32	39
29	24	E		16	
17	13	E	Cu-I deformation		

Radhakrishna and Coworkers[80,81] have studied the far-IR reflection, far-IR transmission and Laser Raman Spectroscopy of Cu_2HgI_4 and Cu_2CdI_4 single crystals. The far-IR reflection spectrum of Cu_2HgI_4 in the wave number range 10-150 cm^{-1} is shown in figure 5. There are six absorption bands at 17, 25,32,52,60 and 120 cm^{-1}. Kramers - Kronig analysis has been done and the bands at 16,32,82,123 and 144 cm^{-1} were assigned to TO modes and 39,56,111,138 cm^{-1} to LO modes. The TO modes were further confirmed by far-IR transmission spectroscopy (see figure 6). In order to have the complete analysis of vibrational modes,

TABLE 5

Raman and far - IR Peaks and their Vibrational assignments for Cu_2CdI_4 Crystals

Raman Cm^{-1}	Far IR trans Cm^{-1}	Assignments	TO modes Cm^{-1}	LO modes Cm^{-1}
14	15	Cu - I deformation	34	40
33	28	$(CdI_4)^{2-}$ deformation	80	106
43	—		128	148
110	85	Cu - I Symmetric		
	93	stretch		
123	128	$(CdI_4)^{2-}$ Symmetric stretch		

Radhakrishna etal[90,91] have also performed the Laser Raman experiments at room temperature and liquid nitrogen temperature. Figure 7 shows the Laser Raman spectra for Cu_2HgI_4 single crystals. There are eight peaks at 16,27,33,49,82,111, and 124 cm^{-1}. The vibrational band assignments for Cu_2HgI_4 and Cu_2CdI_4 crystals are given in Table 4 and Table 5. The observations of Radhakrishna etal on Cu_2HgI_4 were in good agreement with those of Shriver and coworkers[94].

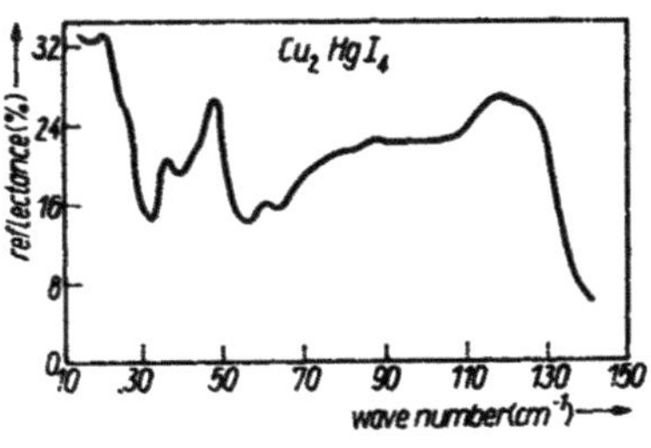

Fig. 5 Far-IR reflection spectrum of Cu_2HgI_4 (from reference 80).

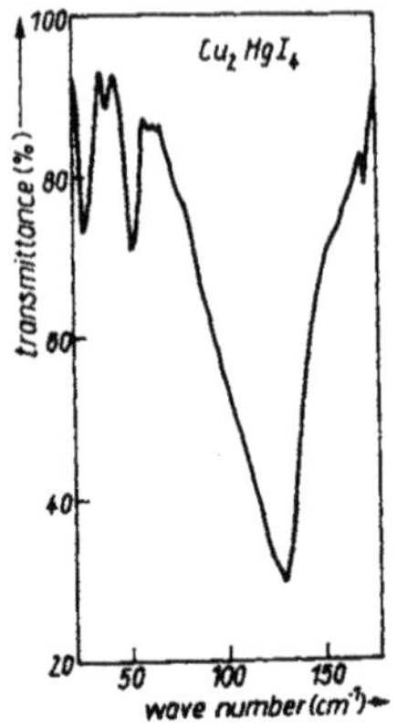

Fig. 6 Far-IR transmission spectrum of Cu_2HgI_4 (from reference 80).

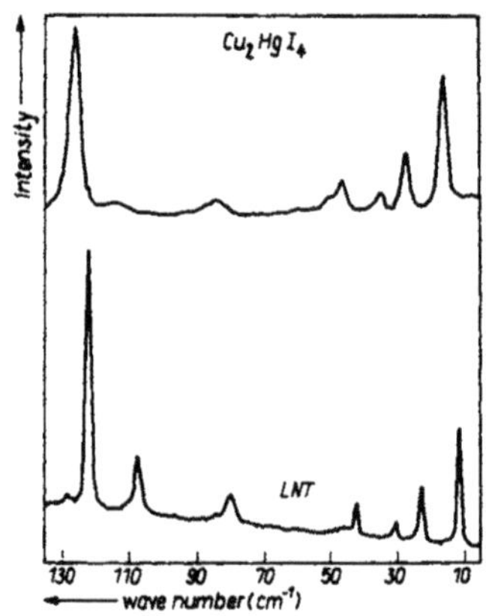

Fig. 7 Laser Raman Spectra of Cu_2HgI_4 at room temperature and liquid nitrogen temperature. (from reference 80)

Radhakrishna etal[80,81] have analysed the optical conductivity behaviour of Cu_2HgI_4 and Cu_2CdI_4 crystals in the frequency range 30–150 cm^{-1}. In order to calculate the conductivity, the reflection spectra were analysed using Kramers-Kronig relations and the values n,k, ϵ' and ϵ'' were calculated. The conductivity was calculated using the relations $\sigma = \epsilon_o \epsilon'' \omega$

where ϵ_o is the absolute permittivity, ϵ'' is the imaginary part of the dielectric constant and ω the radian frequency. Figure 8 shows the far-IR conductivity of Cu_2CdI_4 in the frequency range 30-150 cm^{-1}. There is a peak at around 125 cm^{-1}. In the optical region the moving ions interact with their neighbours which in turn would cause their local potential to fluctuate on a time scale of few picoseconds. Hence, on a short-time scale, the ions can be assumed to be behaving like highly damped forced oscillators. So the interaction of the mobile ions with their neighbours is characterized by a frequency which can be seen in the far-infrared region. In the case of Cu_2CdI_4 the peak at 125 cm^{-1} corresponds to the resonance frequency which is due to the matching of the oscillation frequency of ions inside the potential well to the externally applied frequency. It is well known that the conductivity in Cu_2CdI_4 is due to the mobile Cu^+ ions. Hence the peak at 125 cm^{-1} observed[81] in the far-IR conductivity plot corresponds to the oscillation frequency of Cu^+ ions inside the potential well. Radhakrishna etal[80] have observed an oscillation frequency of 120 cm^{-1} for Cu^+ ions in the case of Cu_2HgI_4. The $\sigma(\omega)$ behaviour was explained by the jump diffusion model. According to this model, the conductivity is written as

$$\sigma(\omega) = \frac{N(ze)^2}{m} \qquad \frac{1}{-i\omega + \Gamma + \dfrac{\omega_o^2}{-i\omega+\gamma}}$$

where N is the number of charge carriers per unit cell, ze the charge, m the mass of the mobile particle, ω_o the resonance frequency, Γ the damping factor, γ the transition time and ω the frequency. Here, $\sigma(\omega)$ has a maximum value when $\omega = \omega_o$, the resonance frequency.

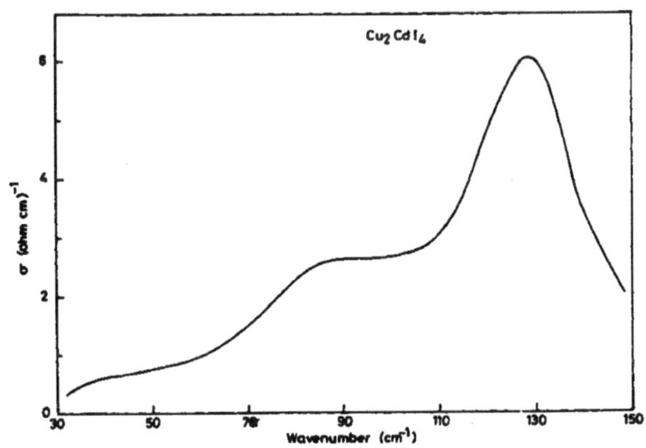

Fig. 8 Far IR conductivity of Cu_2CdI_4 in the region 30 - 150 cm^{-1} (from reference 81).

Radhakrishna and Coworkers[82] have investigated the dielectric dispersion in Cu_2HgI_4 crystals. They have measured the capacitance and dissipation factor in the frequency range 10^2 to 10^4 Hz and at various tempertures. The dielectric constant was calculated from the measured values of capacitance. They have observed a fairly high value of dielectric constant for Cu_2HgI_4, which was attributed to high ionic conductivity. At low frequencies and higher tempertures ionic migration takes place and increases the dielectric constant, ϵ', by the creation of space charge layers at the electrodes. It was also observed that ϵ' suddenly increases after 65°C, which was identified as due to phase transition.

Radhakrishna etal[82] have reported the dielectric loss studies at various temperatures for Cu_2HgI_4 single crystals. Figure 9 shows the dielectric loss for Cu_2HgI_4. The loss was found to increase with frequency and after reaching a peak at certain characteristic frequency, it then decreases. The position of the peak was found to shift to higher frequencies at higher temperatures. Debye type relaxation was considered to explain the loss mechanism. The activation energy was estimated to be 0.42 eV and suggested a hopping type conduction.

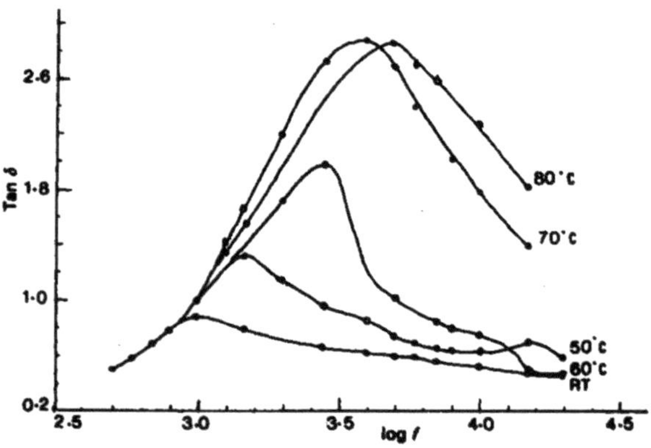

Fig. 9 Tan δ plotted against log f for Cu_2HgI_4 (from reference 82).

Radhakrishna and Sudarsanan[83] have made the microwave measurements on Cu_2HgI_4 in order to check the different models for ion dynamics. They have studied the electrical properties at three different frequencies 36 GHz, 11 GHz and 8 GHz. Figure 10 shows the reflected microwave power versus thickness of Cu_2HgI_4 at 36 GHz. The dielectric constants ϵ' and ϵ'' were calculated at various microwave frequencies. They have considered the Funke's model of complex plane rotating at the angular frequency ω of the applied field and calculated the jump frequency of 6.2 x 10^{-9} sec. for Cu_2HgI_4. The jump frequency was found to be of the same order of magnitude as the oscillation frequency of Cu^+ ions, thus supporting the jump diffusion model.

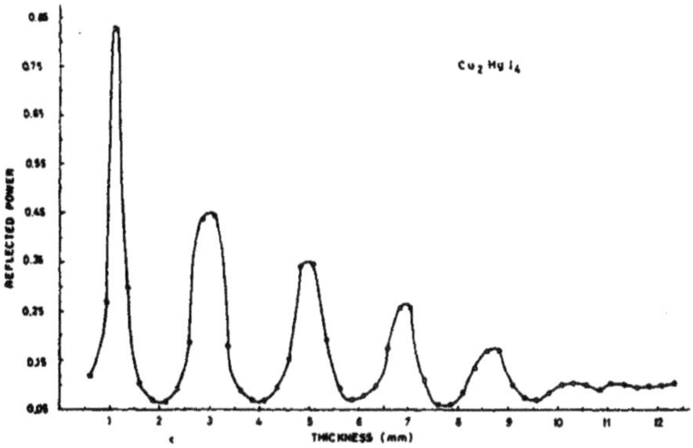

Fig. 10 Reflected microwave power versus thickness for Cu_2HgI_4 at 36 GHz (from reference 83).

(ii) <u>$CuI + PbI_2$ and $CuBr + PbBr_2$</u>

Takahashi and Co-workers[69,70] have studied the $CuI + PbI_2$ and $CuBr + PbBr_2$ systems. In $CuI + PbI_2$ no intermediate compounds were found. In the system $CuBr + PbBr_2$ a compound of the formula $CuPb_3Br_7$ was formed. This was found to be stable between its melting point at 300°C and 160°C. At 200°C it had a conductivity of 3×10^{-2} ohm^{-1} cm^{-1} with an activation energy of 22 kJ mol^{-1}. These values are of comparable magnitude to those observed in average structure compounds. Transport number measurements were also made which confirmed that the conductivity was due to Cu^+ ions. The results reported for $CuBr + PbBr_2$ are included in Table 2.

B Copper Halides Mixed with Chalcogenides

<u>$CuCl + Cu_2S$</u>

Matsui and Wagner[37] have studied the cuprous chloride doped with Cu_2S which increases the number of copper interstitials. The Cu_2S concentration was varied as 0.2, 0.3, 0.4, 0.7, 1.0 mol %. It was found that the ionic conductivity of CuCl increased by increasing the concentration of Cu_2S upto the solubility limit. It should be noted that the ionic conductivity of CuI doped with CdI_2 by small concentrations has not been improved so much by the production of copper vacancies[36].

C Copper Halides Mixed with Layered Compounds

$CuCl + CdCl_2$

Matsui and Wagner[47] have used a novel approach to prepare new copper ion conductors with high ionic conductivity by synthesizing different compounds where other materials with open structures were utilized. Hermann[98] has reported the phase diagram of CuCl + $CdCl_2$ system. He found that only Cu_4CdCl_6 can exist as a single compound and Cu_2CdCl_4 consists of two co-existing phases of Cu_4CdCl_6 and $CdCl_2$. Matsui and Wagner[47] observed that both Cu_4CdCl_6 and Cu_2CdCl_4 exhibited higher electrical conductivity than pure $CuCl_2$ which was due to the contribution of copper vacancies produced by introduction of cadmium ions. At lower temperatures, however, they did not observe any significant change which may be due to the precipitation of $CdCl_2$. The transition temperature of β to γ phase transition for Cu_4CdCl_6 was found to be lowered from 407°C (for pure CuCl) to 330°C. Hermann[98] has reported the transition temperature to be in the range 310–350°C from his thermal analysis. Matsui and Wagner have reported the β to γ transition temperture of 330°C for Cu_2CdCl_4 also and the activation energies obtained for both Cu_2CdCl_4 and Cu_4CdCl_6 were found to be identical. These observations were explained by considering the fact that Cu_2CdCl_4 essentially consists of Cu_4CdCl_6 and $CdCl_2$. The conductivity parameters for Cu_4CdCl_6 are entered in Table 2.

Hsueh and Christy[39] have studied the thermoelectric power of CuCl doped with different divalent impurities like Cd^{2+}, Ca^{2+}, Sr^{2+}, Zn^{2+} and Pb^{2+}. Since these impurities were soluable only to a limited extent, the quantitative evaluation of concentration, mobility and heat of transport of defects was not possible. However, Hsueh and Christy[39], by comparing their results with silver halides, have concluded that cation Frenkel defects are predominant and the copper interstitials are responsible for ionic conduction in these crystals.

D Disperesed Systems

$CuCl + Al_2O_3$

A general approach to increase the ionic conductivity is to disperse a second phase in cuprous halides following the Liang's method[84] who observed enhanced conductivity in LiI mixed with Al_2O_3. Jow[42] has observed an increase in conductivity by two orders of magnitude in CuCl mixed with 10 mol % of Al_2O_3. The improvement in conductivity has been explained as due to the surface effects inside CuCl and the grain boundary effects produced by Al_2O_3.

E Copper Halides Mixed with Alkali Halides

(i) CuI + KI

Bradley and Greene[26] have studied the phase diagram of CuI + KI system. They have reported only one intermediate compound KCu_4I_5 which is stable between 257° and 332°C. Matsui and Wagner[67] have studied the electrical conductivity of KCu_4I_5 and it was about three orders of magnitude higher than pure CuI. From the estimated low values of activation energy it was concluded that the KCu_4I_5 is very similar to MAg_4I_5 (M=K, Rb, NH_4) structure[25,26] and consists of highly disordered copper ions. The conductivity parameters of KCu_4I_5 are given in Table 2. Bonino and Lazzari also studied the conductivity of KCu_4I_5. But their conductivity values and activation energy were much different from those reported by Matsui and Wagner[67].

(ii) CuI + RbI

The CuI + RbI system was originally studied by Bradley and Greene[26] who have reported two compounds $RbCu_2I_3$ and Rb_2CuI_3. Matsui and Wagner[67] have measured the conductivity of the CuI+RbI system in the temperture range 25-250°C. The conductivity of $RbCu_2I_3$ was found to be lower than pure CuI which was explained as due to differences in crystal structures. The conductivity of $RbCu_2I_3$ was little lower than pure CuI. The lower conductivity is due to the coexisting phases of CuI and the low conducting $RbCu_2I_3$.

(iii) CuCl + RbCl

Sandonnini[69] has identified two intermediate compounds Rb_2CuCl_3 and $Rb_2Cu_3Cl_5$ in the CuCl + RbCl system from his thermal analysis. Matsui and Wagner[67,68] have studied the CuCl + RbCl system and reported the conductivity for the two compounds $Rb_2Cu_3Cl_5$ and $RbCu_3Cl_4$. The ionic conductivity of $RbCu_3Cl_4$ is 4 to 5 orders of magnitude higher than that of pure CuCl at room temperature. The ionic conductivities of $RbCu_3Cl_4$ are 2.25×10^{-3} ohm^{-1} cm^{-1} at 25°C and 1.39×10^{-2} ohm^{-1} cm^{-1} at 140°C. X-ray diffraction patterns and thermal analyses indicated that $RbCu_3Cl_4$ was a new compound and not a mixture of the constituents. Very low electron hole conductivity has been observed for $RbCu_3Cl_4$. The conductivity values reported for the RbCl + CuCl system[67,68] are also given in Table 2.

Recently, Kanno etal[71] have studied the phase diagram of CuCl + RbCl system and is shown in the figure 11. This system contains six intermediate compounds, Rb_2CuCl_3, $RbCuCl_2$, $Rb_{11}Cu_{14}Cl_{25}$, $Rb_9Cu_{16}Cl_{25}$, $RbCu_2Cl_3$ and $Rb_3Cu_7Cl_{10}$. Of these intermediate compounds, $Rb_3Cu_7Cl_{10}$ has a high copper conductivity of 0.3×10^{-2} Ohm^{-1} cm^{-1} at room temperature. The other compounds were found to be poor conductors. The $Rb_3Cu_7Cl_{10}$ compound was

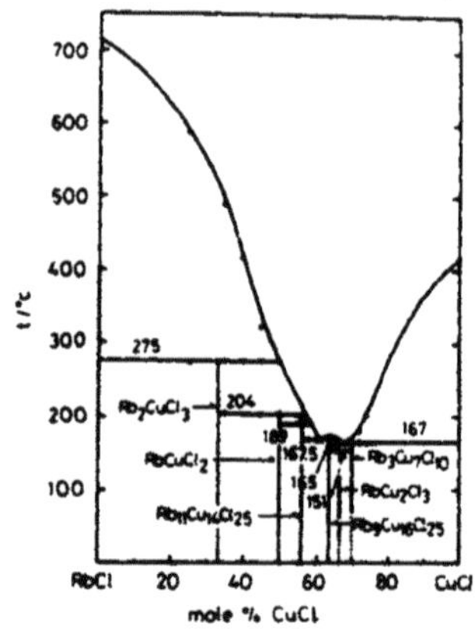

Fig. 11 Phase diagram of the CuCl + RbCl system (from reference 71).

found to be unstable at room temperature as evidenced by the X-ray diffraction measurements. In samples stored in an evacuated pyrex tube at room temperature for several days, a mixture of CuCl, $Rb_4Cu_{16}Cl_{20}$ and $RbCu_2Cl_3$ were detected in slight amounts. The ionic transport number of $Rb_3Cu_7Cl_{10}$ was estimated to be unity by the Tubandt's method.

(iv) <u>CuBr + CsBr and CuI + CsI</u>

Kanno etal[72] have found $CsCu_9Br_{10}$ and $CsCu_9I_{10}$ compounds as the high temperature copper ion electrolytes. $CsCu_9Br_{10}$ is stable between 290 and 300°C and has a conductivity of 0.3 ohm^{-1} cm^{-1} at 290°C. $CsCu_9I_{10}$ is stable between 333 and 347°C and has a conductivity of 0.3 ohm^{-1} cm^{-1} at 340°C. The electronic contribution to the conductivity for these eompounds was found to be very small.

(v) <u>CuI + RbCl</u>

Takahashi etal[74] have reported an intermediate compound $RbCu_4I_4Cl$ in the CuI + RbCl system. It is stable between 231 and 244°C and has an electrical conductivity of 0.28 ohm^{-1} cm^{-1} at 235°C.

4 TERNARY SYSTEMS WITH COPPER HALIDES

$CuI + CuCl + RbCl$

Recently Takahashi etal[73] have discovered a solid in the copper ternary system CuI + CuCl + RbCl with the highest room temperture ionic conductivity known so far. In the compound $Rb_4Cu_{16}I_7Cl_{13}$ they have achieved ionic conductivity of 0.34 ohm^{-1} cm^{-1} at RT. Geller etal[89] have proposed the general formula $RbCu_4Cl_3$ $(I_{2-x}\ Cl_x)$ with $0.0 < x < 0.4$ based on a single crystal structure analysis of the high conductivity solid in the CuCl - CuI - RbCl system. Takhashi etal[74] have studied the compositional dependence of the lattice parameter for the system $Rb_4Cu_{16}I_{7+x}Cl_{13-x}$. The physical properties of $Rb_4Cu_{16}I_7Cl_{13}$ in comparison with $RbAg_4I_5$ are given in Table 6.

TABLE 6

Physical Properties of $Rb_4Cu_{16}I_7Cl_{13}$ and Rb Ag_4I_5
(from reference 17)

	$Rb_4Cu_{16}I_7Cl_{13}$	$RbAg_4I_5$
Crystal Structure	Cubic a = 10.02A°	Cubic a = 11.2A°
X - ray density (g/cm^3)	4.47	5.38
Electrical Conductivity at 25° (ohm^{-1} cm^{-1})	0.34 Hz 0.26 10 KHz 0.31 d.c.	0.28
Activation energy (kJ/mole)	7.0	7.1
Electronic Conductivity (ohm^{-1} cm^{-1})	10^{-12} at 60°C	10^{-11}at 25°C
Transport number for Copper or Silver	1.0	1.0
Decomposition Potential (V)	0.69	0.67
Melting Point (°c)	234	228

5 COPPER HALIDES SUBSTITUTED WITH ORGANIC IONS

Owens[90] had found room temperture silver ion conductors which are double salts of AgI and organic substituted ammonium iodide such as alkyl ammonium iodide. Organic substituted copper ion conductors have been synthesized and studied by Takahashi[12-15, 91-96], Owens[97] and their co-workers. The first room temperature copper ion conductor was reported by Takahashi's group[91]. They have observed a high room temperture conductivity in the N-alkyl - hexamethylenetetramine halide - copper(I) halide system. In fact this must be ragarded as the starting point for search for room temperature highly conducting copper materials. Takahashi's group made the extensive investigations on various organic substituted ammonium halide double salts with cuprous halides[91-96]. This includes tricyclic, bicyclic ammonium halides, substituted and unsubstituted piperidinium, morpholinium and sulphonium halides. The conductivities reported by Takahashi etal for various organic ion substituted cuprous halides are given in Table 7. The results obtained by Owen's group[97] from their conductivity studies on the double salts of substituted organic ammonium halides and cuprous halides are summarized in Table 8. The temperature dependence of the electrical conductivity of some organic substituted cuprous halides is shown in figure 12.

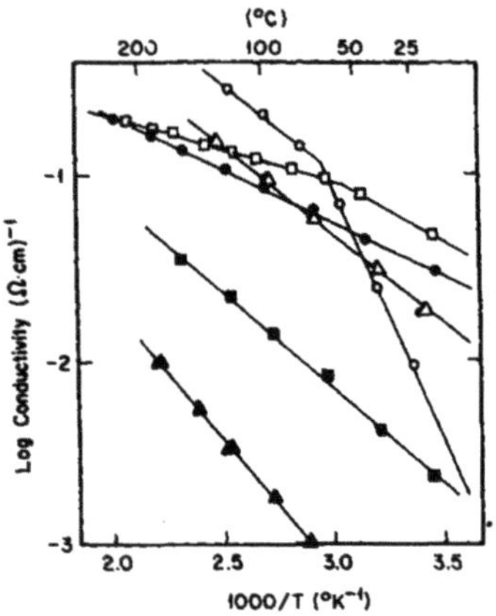

Fig. 12 Temperature dependence of the conductivity of double salts of organic substituted cuprous halides.
O 17 $CuCl \cdot 3C_6H_{12}N_2 2HCl$
7 $CuBr \cdot C_6H_{12}N_2 2HBr$
17 $CuBr \cdot 3C_6H_{12}N_2 2C_2H_5Br$
4 $CuCl \cdot C_6H_{12}N_2 2CH_3Br$
47 $CuBr \cdot 3C_6H_{12}N_2 2CH_3Br$
17 $CuI \cdot 3C_6H_{12}N_2 2CH_3I$
(from reference 92).

The requirement to obtain the high conductivity in the copper organic salts seems to be different from that in silver organic halides for which Owens etal[98] have suggested that the bigger size of the organic ammonium ion as a necessity. In the case of copper conductors, 1-methyl pyridinium iodide did not give high conductivity when mixed with CuI while 1-methyl piperidinium iodide which is having almost the same size as the former gave a high ionic conductivity when a double salt was synthesized with CuI. Takhashi etal[93] have explained these observations by considering the flexibility of the organic ion. The N-methyl pyridinium ions and N-methyl pyrrolidinium ions have a planar structure and tetramethyl ammonium ions have a rigid tetrahedral structure and hence these compounds did not give high conductivity. Of the sulphonium halide systems studied[98], 1-methyl-1-thionacyclohexane iodide / CuI (83.4 mole %) and 4-methyl-1-oxa-4-thionacyclohexane iodide / CuI (83.4 mole %) gave the highest ionic conductivities of 0.14×10^{-2} ohm^{-1} cm^{-1} and 0.072×10^{-2} ohm^{-1} cm^{-1} respectively. From the low activation energies observed for the copper ion conduction, it was predicted that the copper organic double salts should possess the average structure', though no report has been made so far on the crystal structure analysis on these single crystals.

Takhashi etal[18] have studied the thermoelectric power as a function of temperature for the copper organic double salts and calculated the heats of transport. Their results are given in Table 9. They have noted that the heats of transport are smaller than their activation energies and suggested that the conduction mechanism in copper conductors is same as that of silver ions in alkylammonium ion substituted silver iodides.

6 COPPER CHALCOGENIDES

The cuprous chalcogenides are mixed ionic and electronic conductors. The electronic contribution to the total conductivity is always very high even when the ionic conductivity is predominant. The magnitude of conductivity depends very strongly on the deviation from stochiometry.

(i) Cu_2S

Reinhold and Mohring[99] have reported three phases for Cu_2S : α Cubic phase above 470°C, β hexagonal phase in the range 470–110°C and γ orthorhombic phase below 110°C. Later the crystal structure of Cu_2S was modified by Buerger and Wuensch[100], Buerger and Buerger[101], Cook etal[102] and Evans[103]. According to these findings Cu_2S has only two phases, α phase (high chalcocite) above 104°C with hexagonal structure and a β phase (low chalcocite) below 104°C. Buerger and Wuensch[100] have reported the electron density curves of Cu_2S which are smeared and extended between the different copper sites and confirmed the average structure for the cation sublattice. The reports on the β phase crystal structure are controversial. Buerger and Buerger[101] and Cook etal[102] have concluded that the β Cu_2S has an orthorhombic structure while Evans[103] reported a monoclinic structure. The data on crystal structures of Cu_2S in different phases is entered in Table 1.

Hirahara[104] and Tubandt etal[105] have observed a high value of ionic conductivity only for β phase, 0.2 ohm^{-1} cm^{-1} at 400°C and self diffusion coefficient of copper ions, 9×10^{-6} cm^2 sec^{-1} at 420°C which confirmed the cation disorder in the lattice. Okamoto and Kawai[106] have reported that the ionic conductivity of β Cu_2S increases from 0.1 ohm^{-1} cm^{-1} at 110°C to 0.5 ohm^{-1} cm^{-1}

TABLE 7

Ionic Conductivities of the Cuprous halide – organic ammonium halide double salts at 20°C

System: Organic ammonium halide	System: cuprous halide	mole % cuprous halide	Conductivity ohm^{-1} cm^{-1}	Ref.
Azacyclic ammonium halide				
1. N – methyl – piperidinium iodide $C_5H_{11}N \cdot CH_3I$	CuI	84	0.00086	
2. N – methyl – piperidinium bromide $C_5H_{11}N \cdot CH_3Br$	CuBr	84	0.0084	93
3. N – methyl – morpholinium iodide $C_4H_9ON \cdot CH_3I$	CuI	83	0.00058	
Bicyclic ammonium halide				
4. N,N' – dimethyl – triethylene diamine dichloride $C_6H_{12}N_2 \cdot 2CH_3Cl$	Cucl	80	0.018	
5. N,N' – dimethyl – triethylene diamine dibromide $C_6H_{12}N_2 \cdot 2CH_3Br$	CuBr	94	0.035	92
6. N,N' – dimethyl – triethylene diamine diiodide $C_6H_{12}N_2 \cdot 2CH_3I$	CuI	85	0.0024	
Tricyclic ammonium halide				
7. N – hydro-hexamethylene tetramine chloride $C_6H_{12}N_4 \cdot HCl$	CuCl	87.5	0.0043	
8. N – methyl-hexamethylene tetramine bromide $C_6H_{12}N_4 \cdot CH_3Br$	CuBr	87.5	0.017	91
9. N – methyl-hexamethylene tetramine iodide $C_6H_{12}N_4 \cdot CH_3I$	CuI	85	0.0011	

TABLE 8

Conductivity results of Cuprous halide - organic halide double salts at 25°C and at 1 kHZ frequencies (from reference 97)

Electrolyte: Organic halide	Electrolyte: Cuprous halide	mole % cuprous halide	conductivity ohm^{-1} cm^{-1}
Quinuclidine - HBr	CuBr	87.5	4.2×10^{-3}
Quinuclidine - MeBr	CuBr	87.5	6.0×10^{-3}
3 - Quinuclidine - HCl	CuCl	80.0	2.6×10^{-3}
Pyridinium Bromide	CuBr	87.5	7.4×10^{-4}
N,N'- dimethyle tri-ethylene diamine dibromide	CuBr	94.0	2.7×10^{-3}
N - hydrohexamethylene tetramine chloride	CuCl	87.5	5.4×10^{-4}

TABLE 9

Heat of transport, activation energy for Cu^{+} ion conduction and electrical conductivity at 20°C of some copper ion conductors. (from reference 15)

Electrolyte	Heat of transport eV	Activation energy for conduction eV	Conductivity at 20°C $ohm^{-1}cm^{-1}$
$31\ CuBr\cdot 4C_6H_{12}N_4CH_3Br$	0.092	0.147	0.024
$7\ CuBr\cdot 4C_6H_{12}N_4CH_3Br$	0.094	0.178	0.016
$37\ CuBr\cdot 4C_6H_{12}N_22CH_3Br$	0.122	0.140	0.034
$9\ CuBr\cdot 4C_6H_{12}N_22CH_3Br$	0.122	0.155	0.020

at 220°C. The ionic conductivity of $Cu_{1.8}S$ has been measured by Wehefritz[107] who gave a value of 1.6 ohm^{-1} cm^{-1} at 400°C. The copper diffusion in the high chalcocite phase (T > 104°C) of Cw_2S has been studied by Jensen[108], who noticed abrupt changes in diffusion at phase transition tempertures. He arrived at a value of 0.15 eV for activation energy from copper diffusion measurements and confirmed the high cation sub-lattice disorder similar to antifluorite LaF_3. Wagner and Wagner[109] have pointed out that, even though the ionic conductivity is remarkable in βCu_2S, the ratio of the ionic to electron hole conductivity is only a few percent. A large increase in entropy due to cation disorder when $\beta \rightarrow \alpha$ phase transition takes place has been observed in the specific heat measurements on Cu_2S made by Kubaschewski[110]. The thermodynamic properties like transition enthalpy and entropy at phase transition temperatures have been estimated by Hirahara[104] from his calorimetric experiments. The conductivity jumps observed by Hirahara[104] at the phase transition temperatures in Cu_2S have been explained by Matsui and Wagner[36] by considering the entropy changes at the transition points. Jost and Kubaschewski[111] have investigated the heat capacity $C_p(T)$ of β Cu_2S as a function of temperature and found the decreasing values of C_p, with relatively constant slope, which was interpreted by Nolting, Rein and Troe[48] as due to the gradual transition of cationic motion from the oscillatory to the translation state. Dickinson etal[66] have observed only a single peak in the copper photoelectron spectra of Cu_2S. Perhaps high resolution ESCA studies may be necessary to identify different Cu sites in the copper ion conductors.

(ii) <u>Cu_2Se</u>

The Cu_2Se also exists in two phases. At temperatures above 110°C it has the α phase with cubic structure and orthorhombic symmetry in the β phase when T is less than 110°C. The distribution of the cations in the chalcogenide phases of α Cu_2Se and $Cu_{1.8}S$ which has the same structure as α Cu_2Se was studied by Rahlfs[112] and later confirmed by Boettcher etal[113] and Borchert[114a]. Very recently Wuench and co-workers[114b] have made the single crystal neutron diffraction analysis of the cation distribution in the high temperature phase of copper selenide, α-Cu_2Se. They found that high - temperature α Cu_2Se transforms to a monoclinic phase upon cooling to around 123°C. The X-ray powder - diffraction analyses done earlier suggested α= Cu_2Se as having a sphalerite - like structure in space group F43m, with copper distributed among tetrahedral, trigonal and octahedral sites. But the neutron diffraction studies of Wuench etal have revealed that α - Cu_2Se has a space group Fm3m with modified antifluorite - type structure. The thermal motion of Cu^+ ions was found to be highly anharmonic.

Reinhold and Mohring[97] and Reinhold and Seidel[115] have studied the conductivity of Cu_2Se. Though they have observed ionic conductivity in α phase, its value was very small compared to electronic conductivity. Celustka and Ogorolec[116] have measured the conductivity of α Cu_2Se and found that its value increases from 3 ohm^{-1} cm^{-1} to 4 ohm^{-1} cm^{-1} in the temperature range from 580°C to 750°C. Takahashi etal[70] have shown for the first time that the high temperature disordered structure can be achieved at low temperature by controlling the non-stochiometry in α $Cu_{2-\delta}Se$. They have noted that the average structure is stable at room temperature in the composition range $\delta > 0.15$. The ionic conductivity of $Cu_{1.75}Se$ is 3×10^{-2} ohm^{-1} cm^{-1} at room temperture. However, even in this case the electronic conductivity was found to be two orders of magnitude higher than the ionic conductivity.

(iii) $CuGaS_2$, $CuGa_5S_8$ and $CuGa_5Se_8$

Tell's group[117a] have studied a novel series of copper chalcogenides. The $CuGaS_2$, which is isomorphous with $AgGaS_2$, is a chalcopyrite semiconductor and gives some copper ion conductivity at higher temperatures. The $CuGa_5S_8$ and $CuGa_5Se_8$ are isomorphous with $AgGa_5S_8$ and have cation deficient' Cu_2HgI_4 crystal structure with a simple zinc blende lattice. For these crystals only three of the four cation sites are filled in the unit cell. The ionic conductivities in these crystals were also shown to be very low. The conductivity values reported by Tell's group[117a] are given in Table 2.

(iv) $CuInS_2$ and $CuInSe_2$

Recently Kleinfeld and Wiemhofer[117b] have reported the chemical diffusion coefficients of copper in poly - and monocrystalline samples of $CuInS_2$ and $CuInSe_2$. They made the transient measurements with point electrodes in the temperature range 25-150°C. The values were found to be very much dependent on the composition. The crystals which have shown small values of conductivity were found to have highest values for chemical diffusion coefficients. The steady state current - voltage measurements on $CuInS_2$ and $CuInSe_2$ have revealed that the transition from p to n-type semiconduction takes place with increase in copper concentration. It was observed for both the materials, that high polarisation voltages have given copper deposits at the cathode indicating the mobility of copper ions.

(v) Cu_2Se - Ag_2Se

Very recently Yakshibayev etal[118] have investigated the $(Cu_{1-x}Ag_x)_2Se$ quasi-binary system in the region of the solid solution existence as the function of the non-stochiometricity degree and reported their results of ionic conductivity, thermoelectric power and chemical diffusion coefficient. Their ionic conductivity results have shown that in $(Cu_{1-x}Ag_x \pm \delta)_2Se$ solid solutions, at $x > 0.12$ silver ions are the conducting species. An increase in ionic conductivity with increase in silver composition and a decrease in ionic conductivity at a deviation from the stochiometric composition were observed. From the agreement between heats of transport and activation energy for ionic conduction it was concluded that structural disorder exists in the superionic state. For all the solid solutions chemical diffusion coefficient, D, was found to decrease monotonously with an increase in the non-stochiometric degree at $x \leq 0.4$, while for $x > 0.4$ the diffusion coefficient has shown a peak near stochiometric composition.

7 TWO DIMENSIONAL COPPER CONDUCTORS

(i) $CuCrS_2$ and $CuCrSe_2$

The compounds $CuCrS_2$ and $CuCrSe_2$ have sandwich-type structures containing two dimensional layers of Cu and Cr, where the copper ions reside on tetrahedral sites[119]. Yakshibayev etal[120] have recently studied the structural features and ionic transport in $Cu_{1+x}CrSe_2$ superionic conductors in the temperature range 223 to 723°K. From the high temperture X-ray structural studies it has been concluded that, apart from the known antiferromagnetic to paramagnetic transition at 363°K, the $CuCrSe_2$ compound has a second order phase transition at 463°K corresponding to complete disorder of the copper sublattice. It has been noticed that the thermal expansion coefficient and the activation energy for ionic conduction were discontinuously jumping at the phase transition point. It was observed from conductivity experiments that $CuCrSe_2$ becomes a superionic conductor above 463°K with an activation energy for copper conduction of 0.24 eV. The heat of transport for copper ion migrtion was estimated to be 0.22 eV from the thermoelectric power measurements. It was found that the homogeneity range of the Cu_{1+x} $CrSe_2$ system extends in the range of x = 0.0d at 673°K. It is very interesting to compare these results with those obtained by Tell etal for the isomorphous two dimensional $AgCrS_2$ and $AgCrSe_2$. In the case of $AgCrS_2$ a phase transition has been reported at 150°C while in $AgCrSe_2$, it was at 170°C and in both the crystals the phase transition was follwed by abrupt increase in ionic conductivity.

Engelsman etal[119] have shown the existence of order - disorder transitions in $CuCrS_2$. Whittingham[121] has suggested that, since Cu ions reside at tetrahedral sites in $CuCrS_2$ which exhibits phase transitions, the compound would show high ionic conductivity like Ag_2 HgI_4.

(ii) Cu_xTiS_2 and Cu_xTaS_2

Whittingham [122,123] had made the nonstochiometric copper compounds of TiS_2 and TaS_2 by the electrointercalation which show high diffusion. Lazzari et al[124]have found that TiS_2 is an efficient electrode material in rechargeable copper cells using organic based electrolyte. Whittingham[125]has proposed that a cell consisting of a sandwitch-type electrolyte and cathode such as $Cu/CuCrS_2/Cu_xTiS_2$ might minimize interfacial contact problems.

8. TERNARY COPPER COMPOUNDS

CuTeCl, CuTeBr and CuTeI

The copper ternary compounds CuTeCl, CuTeBr and CuTeI have been studied by Von Alpen et al [126]. Conductivity, polarization and transference number results have been reported. In these crystals four types of Cu^{+}ions are populated statistically[127] with tetragonal crystal structure. In CuTeBr and CuTeI the statistical distribution of copper ion occupancy among the four sites has equal probability while in CuTeCl it is not. The conductivity values observed in these crystals are given in table 2. CuTeCl and CuTeBr have a transference number of unity. The electronic conductivity in all these crystals is as low as 1×10^{-8} $ohm^{-1}cm^{-1}$ at room temperature.

9. PYROCHLORES

$CuTa_2O_6$

Goodenough et al [128] have investigated the electrical conductivity of $CuTa_2O_6$ which is a defect perovskite. The crystal structure analysis of $CuTa_2O_6$ was done by Hong[129]. It has a $(TaO_3)^-$ skeleton of corner shared distorted octahedra. The distribution of $(TaO_3)^-$ skeleton creates square-coplanar configurations at 3/4 or the cubic perovskite sites. The Cu^{2+} ions are randomly distributed over the square coplanar sites which are two-thirds filled in $CuTa_2O_6$. It should be noted that the ionic conductivity in this pyrochlore is due to Cu^{2+} ions. The orientations of the square complexes are such that a Cu^{2+} ion must pass through an edge of the square on jumping from site to site. The conductivity observed in this crystal was only of the order of 10^{-3} $ohm^{-1}cm^{-1}$ with an activation energy of 0.36ev.

10. COPPER EXCHANGED β – ALUMINA

The β aluminas were the thoroughly studied solid electrolytes because of their importance in the high energy density batteries. Many excellent reviews[130,131] on β– aluminas are available in the literature. Farrington and coworkers[131] have demonstrated that the Na^{+} in β– alumina can be exchanged either partly or wholly by a number of monovalent, divalent and trivalent ions by ion exchange methods. They have also proposed that the β-aluminas have applications not only in batteries for which conducting properties are important but in very novel optoelectronic devices.

Barrie et al[132] have studied the spectroscopy of Cu^{+} exchanged β– alumina in which the valency of copper could be controlled to a monovalent state. A broad green luminescence has been observed with a peak at 535 nm and the material was found to be suitable for tunable solid state laser application.

Vivien etal[133] have investigated the ESR of Cu^{2+} exchanged β- alumina. They have observed the resonance from the 2d (or Beevers - Ross) and MO (midoxygen) sites. It is remarkable that for one of the sites they could observe the resolution is hyperfine splittings from 63Cu and 65Cu. The ESR spectrum recorded by Vivien et al is shown in figure 13. It can be seen that each set of lines has four resonance transitions which correspond to the hyperfine interaction of $3d^9$ Cu^{2+} electron with copper nucleus with a nuclear spin of I = 3/2.

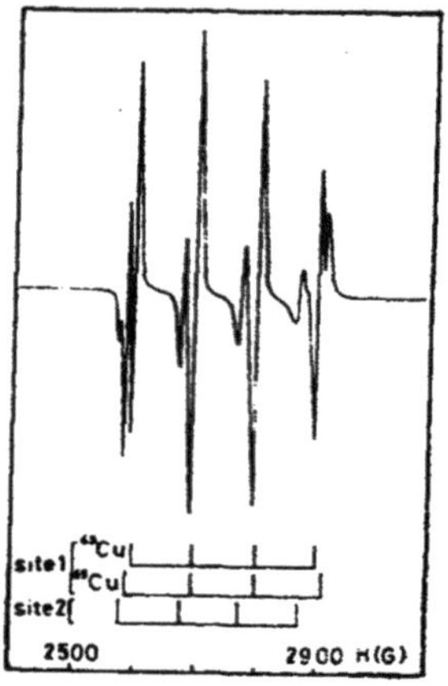

Fig. 13 Room temperature ESR spectrum of Cu^{2+} exchanged βalumina, H∥C Axis (from reference 133).

11. CO-IONIC CONDUCTIVITY

(i) <u>CuI-AgI</u>

Bazan etal[134,135] have studied the co-ionic conductivity in $Ag_{1-x}Cu_xI$ mixtures and the influence of other ionic species in stabilizing the highly conducting structure at lower temperatures. They have found that samples with more than 50 mol% of copper behave as good ionic conductors at temperatures 300 °C and above. The ionic conductivity values are higher than pure CuI which exhibits high ionic conductivity only from 400°C. In a later report Bazan et al[136] have discussed the influence of Cs^+ ion on the co-ionic conductivity in the AgI-CuI solid system. They have observed a decrease in the temperature range in which high conductivity appears together with an increase in conductivity. Both types of change were shown to exhibit a linear relationship with square root of CsI concentration.

Brightwell et al[137,138] have analysed the transitions to the fast ion conducting state in $Ag_{1-x}Cu_xI$, for $x < 0.25$ by studying the electrical conductivity and differential scanning calorimetry as a function of temperature. They have shown that there are two steps at 160°C and 180°C before the fast ion conducting cubic state is reached.

(ii) $CuI-RbAg_4I_5$

The evidence for Cu^+ ion conduction in $CuI-RbAg_4I_5$ was shown by Scrosati et al[139,140] from their electrochemical studies, though no detailed conductivity measurements were made.

12. Cu BASED GLASSES

Li Fake et al[141] have investigated the fast Cu^+ ion conduction in the molybdo phosphate iodide glasses. In the recent years many Ag^+ based glasses have been studied[142] in which high ionic conductivities have been observed. Vitreons Systems have many advantages like isotropic properties, ease of shaping into various forms, ease of thin film formation,wide selection of glass forming systems and wide range of control of properties. The reports on Cu^+ based glasses are very scarce though the amorphous systems have many plus points. A glass system[142] consists of a network former like B_2O_3, SiO_2, MoO_3, V_2O_5, P_2O_5 etc, a network modifier like Li_2O, Na_2O, BaO, MgO, etc and a doping salt like CuI. Li Fake[141] et al have made the studies on the effects of the replacement of one glass former by another upon the conductivities, stability and structure of the glasses. In the system 0.2CuI. $0.8(Cu_2O{\cdot}MoO_3{\cdot}P_2O_5)$ it was shown that the conductivities increased and activation energies decreased when phosphorous was progressively replaced by molybdenum. A room temperature conductivity of the order of 2.8×10^{-3} $ohm^{-1}cm^{-1}$ with an activation energy of 0.16 ev has been observed.

Yuan et al[143] have reported the electrical conductivity of $CuI{\cdot}Cu_2O{\cdot}P_2O_5{\cdot}B_2O_3$ glass. The effect of change in CuI composition on the glass forming regions was studied. It was noticed that when B_2O_3 was added to the ternary $CuI{\cdot}Cu_2O{\cdot}P_2O_5$ glass system, the conductivity improved to a great extent and the thermal stability was also better. Auger and X-ray photoelectron spectroscopic studies have shown that the copper ions were in the monovalent state and mobile. The BO_3, BO_4 and PO_4 groups have been identified from the infrared spectroscopy. The improved conductivity in the borate mixed quarternary glass system was thought to be due to weak bonding between Cu^+ ions and BO_4 tetrahedra. Yuan has also studied the ion transport in surface layers by EPMA,AES and XPS. It was found in the EPMA and AES experiments that under the action of electron beam the mobile cation in the ion conductor moved towards the beam bombardment point and deposited gradually until a saturation limit was attained.

Angell and co-workers[144,145] have reported high conductivities of the order of 10^{-2} $ohm^{-1}cm^{-1}$ at room temperature for glassy samples of $CuI-Cu_2O-P_2O_5$ and all halide CuI-CuCl-RbCl. Minami et al[146,147] also have developed copper ion conducting glasses of $CuI-Cu_2O-MoO_3$ with room temperature conductivity of $10^{-2}ohm^{-1}$ cm^{-1} with a small glass forming region. The Cu^+ ion glasses show higher conductivities than the Ag^+ ion glasses. It should be noted that the direct comparison is difficult because of their different glass forming regions. The difference in conductivity between Cu^{+} and Ag^{+} glasses is found to result from the difference in pre-exponential factors but not in activation energies for conduction. A thorough study of copper ion conductivity glasses seems to be essential to establish the reasons for high conductivity in them.

The components on the surface of the amorphous fast ion conductors $CuCl-Cu_2O-MoO_3-P_2O_5$ have been studied by Xiaoren etal[148] using Auger electron spectroscopy(AES). They found that copper is deposited on the sample during electron bombardment.

Duran et al[149,150] have investigated in detail the d.c. and a.c. electrical properties of the $CuO-BaO-P_2O_5$ glass system. They have measured the thermally stimulated polarization and depolarization currents(TSPC/TSDC) and d.c. conductivity as a function of time and showed that the conductivity in this glass is mostly electronic in character. They have explained the conduction process by the polaron hopping model. The conductivity values were found to be dependent on glass microstructure. The bistable switching phenomena observed in these systems were explained by considering the thermal origion and is associated with the creation of two different structural states. A Debye type dielectric relaxation has been noted in the frequency and temperature dependence of loss tangent and dielectric constant. The activation energies calculated from the dielectric measurements were found to be the same as those estimated from dc conductivity studies and suggested a unique electronic hopping conduction mechanism. They have given a hypothetical model for conduction. The cuprous ions, dispersed and linked to the vitreons network and forming $Cu^{+}-O-Cu^{2+}$ bonds would be responsible for the polaronic and band edge conduction which were activated by phonons. The Cu^{+} ions situated in the other environment, similar to that occuring in Cu_2O crystals, could give rise to band conduction due to delocalized electronic states.

13. APPLICATIONS

Among the various possible applications of solid electrolytes, mostly battery devices have been explored[17] using copper ion conductors. In order that an electrolyte be suitable for use in solid state cell, it should have very high ionic conductivity and very low electronic conductivity. Solid state batteries have many advantages like rugged structure, operation over a wide temperature range, long storage life and miniaturization which makes them suitable for some special applications like heart pace makers etc. Copper ion conductors seem to be the best candidates because of the lowest possible cost. As dicsussed earlier Takahashi et al[73,75] have discovered the $Rb_4Cu_{16}I_7Cl_{13}$ and substituted ammonium halide-copper halide system which show high ionic conductivities of the order of 0.3 $ohm^{-1}cm^{-1}$ at room temperature and very low electronic conductivity which is less than $10^{-10}ohm^{-1}cm^{-1}$ at 100°C. An electrolyte must also have a small polarization at the boundary between the electrolyte and copper electrode for the best performance of the cell. The complex impedance of $Rb_4Cu_{16}I_7Cl_{13}$ and $Rb_4Cu_{16}I_8Cl_{12}$ using copper electrodes has been investigated by Kleitz et al[151]. Takahashi et al[74] have shown that the system with the highest chlorine content, $Rb_4Cu_{16}I_{6.8}Cl_{13.2}$ is the best electrolyte for a cell with copper anode.

The studies by Lazzari et al[124] on a number of cells using the electrolyte $47CuBr\cdot 3(CH_3)_2C_6H_{12}N_2Br_2$ have shown that the cell configuration Cu/electrolyte/TiS_2 or $Cu_{1.8}S$ has the best reversibility. But all these cells have not shown good performance after repeated charge - discharge cycles because of the poor reversibility of copper metal electrode which was explained by Armstrong et al[152] as due to slow charge transfer reaction at the electrolyte/electrode interface.

Shemilt et al[153] have investigated the interfacial polarization for the $47CuBr\cdot 3(CH_3)_2C_6H_{12}N_2Br_2$/copper electrode. The experiments were conducted for two different copper electrodes, a chevral phase copper solid, $Cu_2Mo_6S_{7.8}$ and a copper chalcogenide $Cu_{1.8}S$. The results of scanning electron microscope studies on the electrolyte/copper interface have shown the growth of copper dendrites at the interface. Kleitz et al[151] from their systematic studies by complex impedance spectroscopy, have cocluded that $Rb_4Cu_{16}I_7Cl_{13}$ was the best electrolyte when a chevrel phase copper anode like $Cu_xMo_6S_{8-y}$ was used.

The anodic polarization studies by Kanno et al[154] on different cell configurations with $Rb_4Cu_{16}Cl_{13.2}$ electrolyte have shown that the chevral phase electrodes like $Cu_4Mo_6S_8$ have very high exchange current density and smaller hysterisis and ,the almost linear behaviour of the polarization curve indicated excellent reversibility for the cell.

Takahashi and Yamamoto[74] were, infact, the first to study the performance characteristics of different cells with copper ion conductors. The current-voltage curves of different cells are shown in figure 14, for the electrolyte $7CuBr\cdot C_6H_{12}N_4CH_3Br$. The open-circuit voltages of the different cells shown in figure were found to be 0.448, 0.373 and 0.258V at room temperature and found to be of constant value for a long period of time.

Lazzari et al[124] and Scrosati et al[155] have initiated the studies on rechargeable cells with copper ion conductors, where copper intercalation compounds were used as electrodes. The cell $Cu_{0.8}TiS_2/4HMCH_3Br - CuBr/Cu_{0.4}TiS_2$ has shown the best performance after being subjected to several charge - discharge cycles.

Battery cells of the configuration $Cu_4Mo_6S_8/Rb_4Cu_{16}I_{6.8}Cl_{13.2}$ intercalation compound were throughly studied very recently by Kanno et al[154]. The metal dichalcogenides like TiS_2, NbS_2 were chosen as intercalation compounds. The constant current discharge curves of this cell for different dichalcogenides are

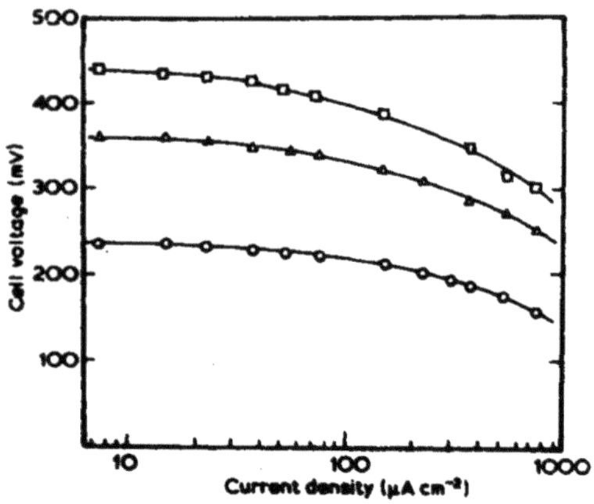

Fig. 14 Current - Voltage curves of different battery cells at room temperature
$Cu, X/X/S, X, Cu_2S$, Graphite
$Cu, X/X/Se, X, Cu_2Se$, Graphite
$Cu, X/X/Te, X, Cu_2Te$, Graphite
Where X is 7 $CuBr \cdot C_6H_{12}N_4CH_3Br$.
(from reference 94).

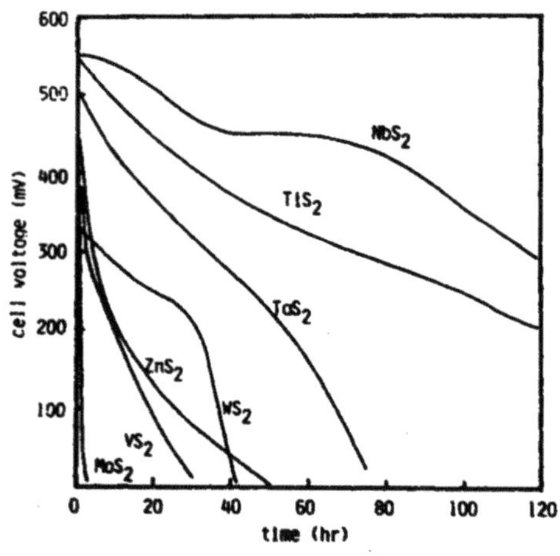

Fig. 15 Constant current discharge curves (100 A) for differerent dichalcogenides in the cell $Cu_4Mo_6S_8/Rb_4Cu_4Cu_{16}I_{6.7}Cl_{13.2}$/Intercalation compound. (from reference 154).

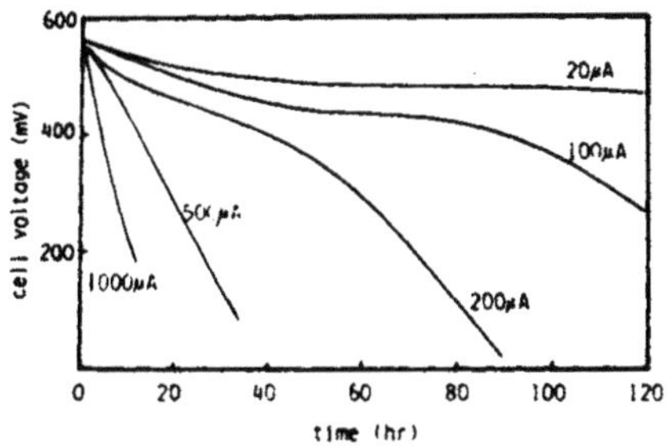

Fig. 16 Discharge curves at various currents for the cell $Cu_4Mo_6S_8/Rb_4Cu_{16}I_{6.8}Cl_{13.2}/NbS_2$ at 25° C (from reference 154)

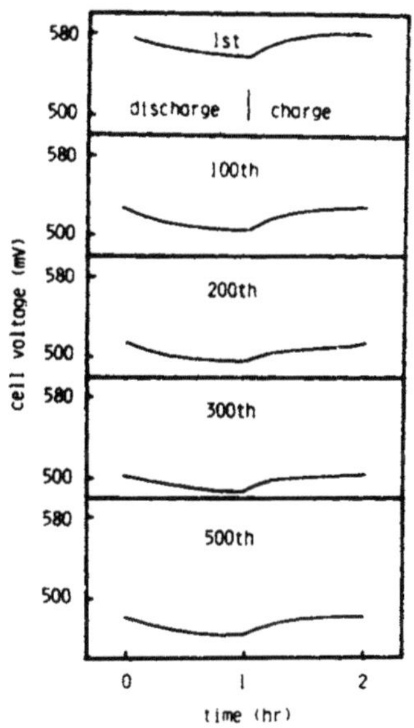

Fig. 17 Charge-discharge cycles at 100 A of the cell $Cu_4Mo_6S_8/Rb_4Cu_{16}I_{6.8}Cl_{13.2}/NbS_2$ at 25° C (from reference 154).

shown in the figure 15. The discharge curves at various currents for the cell with NbS_2 cathode are shown in figure 16. The figure 17 shows the charge - discharge cycles at 100 A of the cell $Cu_4Mo_6S_8/Rb_4Cu_{16}I_{6.8}Cl_{13.2}/NbS_2$ at 25°C. The cells with the chevrel anode and NbS_2 cathode have shown the high rate capability which was thought to be due to the high diffusion rate of intercalated copper metals among Mo_6S_8 blocks of copper cheverel phase and NbS_2 layers. The rechargeable batteries have many potential applications in memory backup for computers, electronic wrist watches and electronic calculators operated by solar cells.

14. CONCLUSIONS

Copper ion conductors have been reviewed. The use of various experimental techniques to understand the different properties of the copper electrolytes has been shown. It is concluded that the copper ion conductors are the cheap and best solid electrolytes which have a potential for a variety of solid state devices.

15. REFERENCES

1. *Solid State Ionic Devices*, Eds. S. Radhakrishna and B. V. R. Chowdari, (World Scientific, 1988).

2. *Materials for Solid State Batteries*, Eds. S. Radhakrishna and B. V. R. Chowdari, (World Scientific, 1986).

3. Extended Abstracts, *6th International Conference on Solid State Ionics*, (Garmisch, PartenKirchen, 1987).

4. S. Chandra, *Superionic Solids*, (North-Holland, 1981).

5. *Fast Ion Transport in Solids*, Ed. W. Van Gool, (North-Holland, 1973).

6. *Solid Electrolytes*, Eds. P. Hagenmuller and W. Van Gool, (Academic Press, 1978).

7. *Superionic Conductors*, Eds. G. D. Mahaw and W. L. Roth, (Plenum, 1976).

8. *Fast Ion Transport in Solids*, Eds. P. Vashista, J. N. Mundy and G. K. Shenoy, (Elsevier-North-Holland, 1979).

9. K. Funke, *Prog. Solid State Chem.* 11, 345 (1976).

10. B. B. Owens, in *Advances in Electrochemistry and Electrochemical Engineering*, Vol. 8, Ed. C. W. Tobias, (John-Wiley, 1971).

11. K. Shahi, *Phy. Stat. Sol.* (a) 41, 11 (1977).

12. T. Takahashi, in *Fast Ion Transport in Solids*, Ed. W. Van Gool, (North-Holland, 1973).

13. T. Takahashi, in *Superionic Conductors*, Eds. G. D. Mahan and W. L. Roth, (Plenum, 1976).

14. T. Takahashi, in *Fast Ion Transport in Solids*, Eds. P. Vashista, J. N. Mundy and G. K. Shenoy, (Elsevier-North-Holland, 1979).

15. T. Takahashi, in *Solid Electrolytes*, Eds. P. Hagenmuller and W. Van Gool, (Academic Press, 1978).

16. T. Matsui and J. B. Wagner, Jr., in *Solid Electrolytes*, Eds. P. Hagenmuller and W. Van Gool, (Academic Press, 1978).

17. O. Yamamoto in *Materials for Solid State Batteries*, Eds. S. Radhakrishna and B. V. R. Chowdari, (World Scientific, 1986).

18. M. D. Ingram and C. A. Vincent, *Chem. Soc. Ann. Rep.* 23, (1977).

19. D. Brinkman, in *Solid State Ionic Devices*, Eds. B. V. R. Chowdari and S. Radhakrishna, (World Scientific, 1988).

20. C. Tubandt and E. Lorenz, *Z. Physik. Chem.* 87, 513, 543 (1914).

21. C. Tubandt, "Leitfahigkeit und Uberfuhrungszahlen in Festen Elektrolyten" in *Handbuch der Experimental Physik XII*, part 1, Eds. W. Wien and F. Harms, (Akadem. Verlagsges, 1932).

22. C. Tubandt and H. Reinhold, *Z. Physik. Chem.* B24, 22 (1934).

23. L. W. Stock, *Z. Physik. Chem.* B25, 411 (1934); B31, 132 (1936).

24. S. Hoshino, *J. Phy. Soc. Japan* 12, 315 (1957).

25. B. B. Owens and G. R. Argue, *Science* 157, 308 (1967).

26. J. N. Bradley and P. D. Greene, *Trans. Faraday Soc.* 63, 424 (1967).

27. a) R. A. Huggins in *Solid Electrolytes*, Eds. P. Hagenmuller and W. Van Gool, (Academic Press, 1978).

    b) *Topics in Applied Physics* Vol. 21, Ed. S. Geller, (Springer-Verlag, 1977).

28. a) S. Miyake, S. Hoshino and T. Takenaka, *J. Phy. Soc. Japan* 7, 19 (1952).

    b) B. Krug and S. Sieg, *Z. Natur.* 76, 369 (1952).

29. W. Buhrer, in abstracts for *Int. Symp. Solid Ionic and Ionic-Electron Conductors*, Rome, Italy, 1976.

30. S. Hoshino, *J. Phy. Soc. Japan* 7, 560 (1952).

31. M. R. Lorenz and J. S. Prener, *Acta Crystallogr.* 9, 538 (1956).

32. B. Graneli, V. Dahlborg and P. Fischer, in Extended Abstracts, *6th Int. Conf. Solid State Ionics*, (Garmisch, PartenKirchen, 1987).

33. J. E. Mayer and B. B. Levy, *J. Chem. Phys.* 1, 647 (1933).

34. J. Teltow, *Z. Phy. Chem. Leipzig* 195, 197 (1950).

35. Y. W. Hsueh and R. W. Christy, *J. Chem. Phy.* 39, 3517 (1963).

36. T. Matsui and J. B. Wagner, *J. Electrochem. Soc.* 124, 300 (1977).

37. T. Matsui and J. B. Wagner, *J. Electrochem. Soc.* 124, 610 (1977).

38. C. Wagner, in *Proc. Int. Committee Electrochem. Thermodynam. Kinet. Meeting*, 7th Lindau, (Butterworths, 1957).

39. J. B. Wagner and C. Wagner, *J. Chem. Phy.* 26, 1597 (1957).

40. A. V. Joshi, in *Fast Ion Transport in Solids*, Ed. W. Van Gool, (North-Holland, 1973).

41. A. V. Joshi and J. B. Wagner, *J. Electrochem. Soc.* 122, 1071 (1975).

42. T. Jow, Ph.D. thesis, Northwestern Univ. (1976).

43. A. V. Joshi and J. B. Wagner, *J. Phy. Chem. Solid* 33, 205 (1972).

44. K. Weiss, *Z. Phy. Chem.* (N. F.) 59, 242 (1968).

45. J. B. Wagner, in *Fast Ion Transport in Solids*, Ed. W. Van. Gool, (North-Holland, 1973).

46. J. B. Wagner, in *Electrode Processes in Solid State Ionics*, Eds. M. Kleitz and J. Dupuy, (Reidel, 1976).

47. B. H. Vine and R. J. Maurer, *Z. Phy. Chem.* 198, 147 (1951).

48. J. Nolting, D. Rein and J. Troe, *Nachr. Akad. Wiss. Gottingen*, Nr. 4, 31 (1969).

49. J. Nolting, *Ber. Bunsenges. Physik. Chem.* 68, 932 (1964).

50. J. Nolting, *Angew. Chem.* 82, 498 (1970).

51. J. Nolting, *Ber. Bunsenges. Physik. Chem.* 68, 939 (1964).

52. J. Nolting, *Ber. Bunsenges. Physik. Chem.* 68, 944 (1964).

53. W. Stutius, *Solid State Commun.* 18, 1275 (1976).

54. S. Hudgens, *Phy. Rev.* 137, 2481 (1973).

55. J. J. Chodi, R. M. White and W. A. Harrison, *Phy. Rev. Lett.* 35, 1372 (1975).

56. S. Hudgens, M. Kastner and H. Fritzsche, *Phy. Rev. Lett.* 33, 1552 (1974).

57. N. Bloembergen, E. M. Purcell and R. V. Pound, *Phy. Rev.* 73, 679 (1948);
A. Abragam, *The Principles of Nuclear Magnetism*, (Oxford University Press, 1961);
H. S. Gutowsky and G. E. Pake, *J. Chem. Phys.* 18, 162 (1950).

58. K. D. Becker, G. W. Herzog, D. Kanne, H. Richtering, and F. Stradler, *Ber. Bunsenges. Physik. Chem.* 74, 527 (1970).

59. J. B. Boyce and B. A. Huberman, *Solid State Commun.* 21, 31 (1977).

60. R. L. Miehr, *Phy. Rev.* 125, 1537 (1962).

61. R. Reif, *Phy. Rev.* 100, 1597 (1155).

62. M. Eisenstadt, *Phy. Rev.* 133, A19 (1964).

63. M. O'Keeffe, in *Fast Ion Transport in Solids*, Ed. W. Van Gool, (North-Holland, 1973).

64. B. A. Huberman, *Phy. Rev. Lett.* 32, 1000 (1974).

65. M. J. Rice, S. Strassler and G. A. Tombs, *Phy. Rev. Lett.* 32, 596 (1974).

66. T. Dickinson, A. F. Povey and P. M. A. Sherwood, *J. Solid State Chem.* 13, 237 (1975).

67. T. Matsui and J. E. Wagner, *J. Electrochem. Soc.* 124, 937 (1977).

68. T. Matsui and J. B. Wagner, *J. Electrochem. Soc.* 124, 941 (1977).

69. T. Takahashi, O. Yamamoto and H. Takahashi, *J. Solid State Chem.* 21, 37 (1977).

70. T. Takahashi, O. Yamamoto, F. Matsuyama, and Y. Noda, *J. Solid State Chem.* 16, 35 (1976).

71. R. Kanno, Y. Takeda, F. Matsuyama, O. Yamamoto, and T. Takahashi, *Solid State Ionics*, 11, 221 (1983).

72. R. Kanno, Y. Takeda, O. Yamamoto, and T. Takahashi (to be published).

73. T. Takahashi, O. Yamamoto, S. Yamada, and S. Hayashi, *J. Electrochem. Sol.* 126, 1654 (1979).

74. T. Takashashi, R. Kanno, Y. Takeda, and O. Yamamoto, *Solid State Ionics*, 3/4, 283 (1981).

75. J. A. A. Ketelaar, *Z. Phys. Chem.* B26, 327 (1934).

76. L. Suchow and G. R. Pond, *J. Am. Chem. Soc.* 75, 5242 (1953).

77. S. Hoshino, *J. Phy. Soc. Japan* 10, 197 (1955).

78. H. B. Lal and O. P. Srivastava, *J. Mat. Science* 19, 303 (1984).

79. K. Funke and A. Jost, *Nachr. Akad. Wiss Gottingen*, Nr. 15, 137 (1969);
W. Jost, K. Funke and A. Jost, *Z. Naturforsch* 2500, 983 (1973);
K. Funke and A. Jost, *Ber. Bunsenges. Physik. Chem.* 75, 436 (1971);

K. Funke and R. Hackenberg, *Ber. Bunsenges. Physik. Chem.* 76, 885 (1722).

80. R. Sudarsan, T. K. K. Srinivasan and S. Radhakrishna, *Phy. Stat. Sol.* (b) 124, 481 (1984).

81. R. Sudarsan, T. K. K. Srinivasan and S. Radhakrishna, *Solid State Ionics* 13, 277 (1984).

82. R. Sudarsan, B. Babuji and S. Radhakrishna, *J. Mat. Sci. Lett.* 4, 830 (1985).

83. R. Sudarsan, Ph.D. Thesis, Indian Institute of Technology, Madras, India (1983).

84. D. R. Greig, G. C. Joy II and D. F. Shriver, *J. Chem. Phy.* 67, 3189 (1977).

85. G. Herrmann, *Z. Anorg. Chem.* 71, 296 (1911).

86. C. C. Liang, *J. Electrochem. Soc.* 120, 1289 (1973).

87. F. Bonino and M. Lazzari, *J. Power Sources* 1, 103 (1976).

88. C. Sandonnini, *Gazz. Chem. Ital.* 44, 311 (1914).

89. S. Geller, J. R. Akridge and S. A. Wilber, *Phy. Rev.* B19, 5396 (1979).

90. B. B. Owens, *J. Electrochem. Soc.* 120, 1431 (1973).

91. T. Takahashi, O. Yamamoto and S. Ikeda, *J. Electrochem. Soc.* 120, 1431 (1973).

92. T. Takahashi and O. Yamamoto, *J. Electrochem. Soc.* 122, 83 (1975).

93. T. Takahashi, N. Wakabayashi and O. Yamamoto, *J. Electrochem. Soc.* 123, 129 (1976).

94. T. Takahashi and O. Yamamoto, *J. Appl. Electrochem.* 7, 37 (1977).

95. T. Takahashi, N. Wakabayashi and O. Yamamoto, *J. Appl. Electrochem.* 7, 253 (1977).

96. T. Takahashi, N. Wakabayashi and O. Yamamoto, *J. Solid State Chem.* 21, 73 (1977).

97. A. F. Sammels, J. Z. Cougoutos and B. B. Owens, *J. Electrochem. Soc.* 122, 1291 (1975).

98. B. B. Owens, J. H. Christie and G. T. Tiedeman, *J. Electrochem. Soc.* 118, 1144 (1971).

99. H. Reinhold and H. Mobring, *Z. Phy. Chem.* B38, 221 (1938).

100. M. J. Buerger and B. J. Wuensch, *Science* 141, 276 (1963).

101. M. J. Buerger and N. W. Buerger, *Amer. Miner* 27, 95 (1944).

102. Cook et al., *J. Appl. Phys.* 41, 3058 (1970).

103. H. T. Evans, *Nature* 232, 69 (1971).

104. E. Hirahara, *J. Phys. Soc. Jpn.* 6, 422 (1951);
E. Hirahara, *J. Phys. Soc. Jpn.* 6, 428 (1951).

105. C. Tubandt, H. Reinhold, C. Jost. *Z. Anorg. Allg. Chem.* 177, 253 (1929).

106. K. Okamoto and S. Kawai, *Jpn. J. Appl. Phy.* 12, 1130 (1973).

107. V. Wehefritz, *Z. Physik. Chem. NF* 26, 339 (1960).

108. M. L. Jensen, Ph.D. thesis, M.I.T., (1951).

109. J. B. Wagner and C. Wagner, *J. Chem. Phy.* 26, 1602 (1957).

110. P. Kubaschewski, *Ber. Bunsenges. Phy. Chem.* 77, 74 (1973).

111. W. Jost and P. Kubaschewski, *Z. Physik. Chem. NF* 60, 69 (1968).

112. P. Ralhfs, *Z. Physik. Chem.* B31, 157 (1936).

113. A. Boettcher, G. Hasse and H. Treupel, *Z. Angen. Phys.* 7, 478 (1955).

114. W. Borchert, *Z. Kristallogr.* 106, 5 (1945);
M. Oliveria, R. K. McMullan and B. J. Wuench, in Extended Abstracts, *6th Int. Conf. Solid State Ionics*, (Garmisch, Partenkircheno, 1987).

115. H. Reinhold and H. Seidel, *Z. Phy. Chem.* B38, 245 (1938).

116. Z. Ogorelec and B. Cleustka, *J. Phy. Chem. Solid* 30, 149 (1969).

117. B. Tell, S. Wagner and H. M. Kasper, *J. Electrochem. Soc.* 124, 536 (1977);
D. W. Murphy, H. S. Chen and B. Tell, *J. Electrochem. Soc.* 124, 1268 (1977);
M. Kleinfold and H. D. Wiemhofer, in Extended Abstracts, *6th Int. Conf. Solid State Ionics*, (Garmisch, Partenkirchen, 1987).

118. R. A. Yakshibayev, R. F. Almukhametov and M. Kh. Balapanov, in Extended Abstracts, *6th Int. Conf. Solid State Ionics*, (Garmisch, Partenkirchen, 1987).

119. F. M. R. Engelsman, G. A. Wiegers, F. Jellinek, and B. Van Laar, *J. Solid State Chem.* 6, 574 (1973).

120. R. A. Yakshibayev, V. N. Zabolotsky and R. F. Almukhametov, in Extended Abstracts, *6th Int. Conf. Solid State Ionics*, (Garmisch, Partenkirschen, 1987).

121. M. S. Whittingham, *Electrochem. Acta* 20, 575 (1975).

122. M. S. Whittingham, Belgium Patent, 819, 673 (issued 1975); USA Patent 4, 040, 917.

123. M. S. Whittingham, *J. Chem. Soc. Commun.* 328 (1974).

124. M. Lazzari, G. Razzini and B. Scrosati, *J. Power Sources* 1, 57 (1976).

125. M. S. Whittingham, in *Solid Electrolytes*, Eds. P. Hagenmuller and W. Van Gool, (Academic Press, 1978).

126. U. Von Alpen, J. Fenner, J. D. Marcoll, and A. Rabenau, *Electrochem. Acta* 22, 801 (1977).

127. A. Rabenau, H. Raw and G. Rosenstein, *Z. Anorg. Allg. Chem.* 374, 43 (1970).

128. J. B. Goodenough, H. Y. P. Hong and J. A. Kafalas, *Mater. Res. Bull.* 11, 203 (1976).

129. H. Y. P. Hong, Lincoln Laboratory, M.I.T., Solid State Res. Rep. (1972).

130. J. T. Kummer, *Prog. Solid State Chem.* 7, 141 (1972).

131. G. C. Farrington, B. Dunn and J. O. Thomas, in *Solid State Ionic Devices*, Eds. B. V. R. Chowdari and S. Radhakrishna, (World Scientific, 1988).

132. J. D. Barrie, B. Dunn, O. M. Stafsudd, and P. Nelson, *J. Lum.* 37, 303 (1987).

133. D. Vivien, J. Antoine, D. Gourier, J. Thery, J. Livage, and R. Collongues, *Coll. Int. du CNRS sur la Spectroscopic des Elements de Transition et des Elements Lourds dans les Solides*, Lyon (1976).

134. J. C. Bazan and J. A. Schmidt, *J. Appl. Electrochem.* 6, 411 (1976).

135. J. A. Schmidt, R. E. Fornari and J. C. Bazan, *Electrochem. Acta* 24, 1131 (1979).

136. J. C. Bazan, J. A. Schmidt and R. S. P. DeMiraglia, *Solid State Ionics* 6, 171 (1982).

137. J. W. Brightwell, C. N. Buckley and B. Ray, *Phy. Stat. Sol.* (a) 69, K11 (1982).

138. J. W. Brightwell, C. N. Buckley and B. Ray, *Solid State Ionics* 15, 331 (1985).

139. B. Scrosati, *J. Electrochem. Soc.* 120, 78 (1973); 120, 519 (1973).

140. B. Scrosati, G. Pistoia, M. Lazzari, and L. P. Bicelli, *J. Appl. Electrochem.* 4, 201 (1974).

141. Li Fake, Chem Dokang and Liao Ninglan, in Extended Abstracts, *6th Int. Conf. on Solid State Ionics,* (Garmisch, Partenkirchen, 1987).

142. T. Minamj, in *Materials for Solid State Batteries,* Eds. B. V. R. Chowdari and S. Radhakrishna, (World Scientific, 1986).

143. Li·Guang Yuan, in Extended Abstracts, *6th Int. Conf. on Solid State Ionics,* (Garmisch, Partenkirchen, 1987).

144. C. Liu and C. A. Angell, *Solid State Ionics* 13, 105 (1984).

145. C. Liu, H. G. K. Sundar and C. A. Angell, *Solid State Ionics* 18/19, 442 (1986).

146. N. Machida, M. Chuscho and T. Minami, *54th Annual Meeting of Chem. Soc. Jpn.; The 87th Annual Meeting of Ceram. Soc. Jpn.*

147. T. Minami, *J. Non. Cryst. Sol.* 95 & 96, 107 (1987).

148. P. Xiaoren, Z. Haibo and B. Yushi, in Extended Abstracts, *6th Int. Conf. on Sol. Stat. Ionics,* (Garmisch, Partenkirchen, 1987).

149. A. Duran, J. R. Jurado and J. M. F. Navarro, *J. Non. Cryst. Sol.* 79, 333 (1986).

150. A. Duran, J. R. Jurado and J. M. F. Navarro, *J. Non. Cryst. Sol.* 79, 353 (1986).

151. M. Kleitz, J. R. Akridge and J. H. Kennedy, *Solid State Ionics* 2, 67 (1981).

152. R. D. Armstrong, T. Dickenson and K. Taylor, *J. Eletroanal. Chem.* 57, 157 (1974).

153. J. M. Schmidt, B. C. H. Steele and J. E. Watson, *Solid State Ionics* 2, 1 (1981).

154. R. Kanno, Y. Takeda, Y. Oda, H. Ikeda and O. Yamamoto, *Solid State Ionics* 18/19, 1068 (1986).

155. B. Scrosati and M. A. Voso, *J. Electrochem. Soc.* 126, 699 (1979).

The Multivalent Beta" Aluminas

Gregory C. Farrington, *Bruce Dunn, and **John O. Thomas

Department of Materials Science and Engineering
University of Pennsylvania
3231 Walnut St.
Philadelphia, PA 19104

*Department of Materials Science and Engineering
University of California at Los Angeles
6531 Boelter Hall
Los Angeles, CA 90024

**Institute of Chemistry
University of Uppsala
Box 531
S-751 21 Uppsala, Sweden

Abstract:
The exceptional ability of Na(I) ß"-alumina to undergo ion exchange reactions in which the Na(I) content is replaced by other monovalent, divalent and trivalent cations, as well as various protonic species, makes the ß"-aluminas arguably the most diverse family of high conductivity solid electrolytes known. This chapter briefly reviews the preparation, structure, ion transport, spectroscopy, and selected applications of the divalent and trivalent ß"-aluminas. Perhaps most interesting are the ß"-aluminas that glow: those containing lanthanide cations and various transition metal cations that luminesce. These materials may result in important new applications of solid electrolytes in lasers and opto-electronic devices, technologies distant from the more traditional applications of solid electrolytes in electrochemistry.

[revised 2/16/88]

I. Introduction

Beta and ß"-alumina are remarkable solid electrolytes whose study has been closely linked to the development of the field of solid state ionics. Yao and Kummer's observation [1], published over twenty years ago, that sodium beta alumina has a sodium ion conductivity at room temperature comparable to that of aqueous NaCl was critical in focusing interest on high conductivity solid electrolytes. It led to extensive efforts to develop the sodium/beta alumina/sulfur battery, and for many years it appeared that this battery would be the first major new technology made possible by the ß-aluminas.

Since Yao and Kummer's seminal work, many solid electrolytes have been explored, but interest in the ß and ß"-aluminas has endured, not because of their historically important role or their possible application in high energy density batteries, but because of one characteristic virtually unique to the beta aluminas among all other solid electrolytes: their ability to undergo ion exchange. The mobile sodium ions present in both structures can be easily replaced with a wide variety of other cations. The ß and ß"-aluminas are thus a family of solid electrolytes, arguably the most diverse family of such materials known.

Kummer [2] first pointed out the remarkable ion exchange properties of sodium ß-alumina by showing that the sodium ions in the structure can be replaced by many different monovalent cations, such as Li^+, K^+, as well as $H(H_2O)_x^+$ and NH_4^+. He also reported that attempts to replace the sodium ions in ß-alumina with divalent cations were not particularly successful. Subsequent investigations by the current authors demonstrated that the ion exchange chemistry of sodium ß"-alumina, whose structure is so similar to that of sodium ß-alumina that the two compounds had been thought to have virtually identical properties, is far richer than that of sodium ß-alumina. In fact, the sodium ion content of ß"-alumina can be exchanged for virtually any +1, +2, or +3 cation in the periodic chart [3, 4, 5]. This work greatly expanded the variety of ß"-alumina compositions and, moreover, demonstrated that fast cation transport in solids is not limited to a few monovalent cations, but, in the right structure, can be observed for many different cations.

The divalent and trivalent ß"-aluminas have proven to be rich and diverse materials. They include the Pb(II) form, which has an unexpectedly high ionic conductivity comparable to that of Na(I) ß"-alumina and develops electronic conductivity upon heating in vacuum; the mixed Na(I)-Eu(II) compositions, which exhibit elegant order/disorder reactions among the mobile cations; Nd(III) ß"-alumina, of great interest as a new solid state laser material; and finally ß"-alumina containing Cu(I), which may prove to be a tunable laser host.

Surely the most fascinating multivalent ß"-aluminas are those of potential use in solid state opto-electronic devices. Exploration of the application of ß"-alumina single crystals in solid state lasers and related techologies is just beginning. Perhaps they will turn out to be a major technological success for solid state ionics, even though optical devices seem distant from the batteries and sensors with which the field has been associated traditionally.

This chapter presents a brief review of the state of research and technology in the divalent and trivalent ß"-aluminas, including their preparation, structure, ion transport, spectroscopy, and potential new applications. We hope that it will convey some of the fascination we have found with these materials and convince the reader that the ß"-aluminas present some of the brightest (literally!) prospects for solid state ionics technology.

II. Beta" Alumina: History and Preparation

In 1916, Rankin and Merwin [6] described a new compound which they thought was an allotropic form of α-alumina. Though it resembled α-alumina, it clearly had a different structure, and they named it ß-alumina. It was not until much later that investigators recognized that ß-alumina is not an alumina at all but a complex sodium aluminate. Subsequent studies demonstrated that beta alumina [ca. $Na_2O \cdot 11Al_2O_3$] is a highly-defective, non-stoichiometric compound that contains a large excess of sodium (about 25%) compared to its ideal formula.

Regardless, the name 'ß-alumina' persisted, and when Yamaguchi and Suzuki [7] reported a closely-related compound [ca. $Na_2O \cdot 5Al_2O_3$] in 1943, they named it ß"-alumina. Yamaguchi and Suzuki correctly deduced the structure of Na(I) ß"-alumina by matching its X-ray powder pattern to that of the isotype, $K_2O{\bullet}5.1Al_2O_3$. Théry and Briançon [8] first described ß"-alumina in the Western literature in 1962. In 1967, workers at Ford Scientific Laboratories discovered that small quantities of MgO stabilize ß"-alumina relative to other alkali polyaluminates [9]. The first complete X-ray structure analysis of ß"-alumina was carried out by Bettman and Peters [10]. Early references to ß-alumina have been reviewed more completely by DeVries and Roth [11] and subsequently by Kummer [2] and Kennedy [12]. Bettman and Peters [10] have discussed the early history of ß"-alumina. It is now clear that ß- and ß"-alumina are particular examples of a larger family of complex oxides having the general formula, $A_2O{\bullet}nM_2O_3$, in which A is most commonly a monovalent alkali cation and M is Al, Fe, or Ga

Beta" alumina is most frequently synthesized in its sodium form. Certainly the greatest synthetic effort has been focused on developing high conductivity polycrystalline ceramics of the ß-aluminas, because of their application in the sodium/sulfur battery. However, fundamental studies of the di- and trivalent ß"-aluminas have been carried out primarily with single crystals, a form which also is of greatest interest for the development of solid state opto-electronic devices. This article will be concerned almost entirely with single crystal materials prepared either in the Mg-stabilized or Li-stabilized forms.

Crystals of Na(I) ß"-alumina are generally grown by the flux evaporation method [10, 13, 14], in which a eutectic mixture of Na_2O and Al_2O_3 containing a small amount of MgO is held at just above its melting point for several days. The composition has the proper MgO/Al_2O_3 ratio for ß"-alumina but an excess of Na_2O. As the excess Na_2O slowly evaporates, the melting point of the mixture increases, and crystals form. Typical reaction conditions call for heating the mixed oxides in a Pt crucible in air at 1700-1710°C for 14-21 days. This method works well for Na(I) ß"-alumina, because Na_2O is much more volatile than either alumina or magnesia at the growth temperature.

The crystals that result are generally in the form of well-defined platelets of Na(I) ß"-alumina, and range in size from about 0.1 mm^3 to 10 mm x 5 mm x 1 mm.

Analyses of the Na content of the samples has been reported by Aldén et al. [15] and Saltzberg [16] and, expressed in terms of the the formula, $Na_{1+x}Mg_xAl_{11-x}O_{17}$, the value of x is typically 0.66±0.03.

Recently, a technique for growing Li-stabilized single crystals of Na(I) ß"-alumina was reported [17]. The addition of $NaAlO_2$ and $LiVO_3$ fluxes to Na(I) ß"-alumina produces a homogeneous melt at temperatures below 1600°C. Upon slow cooling of the melt, well-defined crystals of Na(I) ß"-alumina form. The flux serves the dual role of decreasing the melting temperature and supplying a source of Li(I) for stabilizing the ß"-alumina phase. X-ray diffraction studies have established that Li(I) substitutes primarily in the Al(2) site and that the crystals have the approximate composition of $Na_{1+2x}Li_xAl_{11-x}O_{17}$ with $x = 0.33 \pm 0.03$. The ability to grow crystals by slow cooling from a melt also suggests that it may be possible to pull crystals from the melt, a development that would be of great significance for the technological application of single crystals of ß"-alumina.

III. Preparation of Mutivalent ß"-Aluminas

The first indication that divalent ions could be easily exchanged into ß"-alumina can be found in the work of Toropov and Stukalova [18] in 1939. They reported the existence of $BaO{\cdot}6Al_2O_3$, which is presumably related to ß"-alumina. They claimed that the Ba^{++} ions in the compound could be replaced with K^+ by fusion with K_2O, but the details of their experiments are sketchy. The first clear evidence that divalent ions could diffuse rapidly in ß"-alumina came in 1980 when Dunn and Farrington [3, 4] reported the synthesis and ionic conductivities of Ba(II), Cd(II), and Sr(II) ß"-alumina. Perhaps even more surprising was the revelation in 1983 that trivalent ions can also replace the Na(I) ions in ß"-alumina [5]. As expected, trivalent ions are considerably less mobile than divalent ions in the structure [5].

It is now clear that many different cations can replace the Na(I) of ß"-alumina. They include: the monovalent cations of K, Ag, Tl, and Cu [2]; the divalent cations of Ca, Sr, Ba, Zn, Cd, Pb, Mn, Eu, Ni, Co, Sn, Hg, and Sm [19]; as well as the trivalent cations of Gd, Nd, Eu, Yb, Sm, Tb, Bi, Cr, and Er [19]. Typically, ion exchange is carried out at 200-600°C in a pure molten salt of the exchanging ion, a mixed molten salt of the exchanging cation and sodium ions, or by a gas phase reaction in which the crystal to be exchanged is buried in or suspended over a salt of the exchanging ion and heated to a temperature below the melting point of the salt. The last technique is especially useful for attaining high levels of exchange with mimimal crystal damage. A more complete discussion of the ion exchange process has been presented by Sattar et al. [19].

The extent of exchange can be monitored in a number of ways. They include: measuring the weight change that occurs during the reaction, using a radiochemical method in which the exiting or entering ion is tagged with a radiotracer (typically ^{22}Na), chemical analysis of the crystal before and after exchange, and X-ray absorption mapping, in which the X-ray transmission of the crystal is measured before and after exchange [20]. Each of these methods has its advantages and disadvantages. They are discussed in greater detail in Sattar et al. [19].

IV. Stability of the Multivalent Beta" Aluminas

The multivalent ß"-aluminas are metastable materials. Synthesized at relatively low temperatures, the compounds are susceptible to structural degradation and phase transformations at higher temperatures. The stability of the multivalent ß"-aluminas has been explored in two types of complementary studies. Investigations of high-temperature stability (>800°C) have focused on phase transformations and decomposition reactions [21], while studies at lower temperatures (<700°C) have emphasized the formation of color centers and the characterization of the defect chemistry in the conduction plane [22, 23].

The thermal stability and decomposition reactions of Sr(II), Pb(II), and Nd(III) ß"-aluminas up to 1600°C have been investigated by Yang et al. [21]. Each compound has a distinct behavior. Sr(II) ß"-alumina is clearly the most stable of the three. Upon heating, the Sr(II) ß"-alumina structure remains intact until 1300°C and then co-exists with the decomposition product, Sr(II) ß-alumina, until about 1500°C. At higher temperatures, the ß-alumina structure gradually transfoms to a magnetoplumbite structure. Pb(II) ß"-alumina decomposes in a more complex fashion. Around 1000°C, PbO volatilizes from the material and a second phase forms. As PbO is lost from the structure, the spinel blocks appear to collapse upon each other. Above 1400°C, virtually all of the PbO is volatilized, and the presence of α-alumina can be observed. Nd(III) ß"-alumina initially degrades at 1000°C, and substantial structural collapse occurs by 1050°C. Between 1100 and 1250°C, the ß"-alumina phase is completely transformed into the perovskite-like, $NdAlO_3$, and a δ-type alumina phase. The latter disappears at 1450°C and leads to the formation of α-alumina and a magnetoplumbite phase.

A more subtle form of instability in the multivalent ß"-aluminas involves the *in situ* reduction of the mobile cations in the structure when heated in an atmosphere of low oxygen partial pressure. It appears that samples exposed to these conditions lose oxygen in a quasi-reversible reaction. One example of this reaction can be seen with Eu(III) ß"-alumina which, when heated in vacuum, undergoes spontaneous reduction in which some of the Eu(III) content is reduced to Eu(II) [22]. More recently, Rohrer et al. [23] reported that Ag(I), Pb(II), or Bi(III) ß"-alumina crystals, when heated in an atmosphere with a low partial pressure of oxygen, undergo a color change which is accompanied by an insulator to semiconductor transition. The change in optical and electrical properties can be reversed by heating in air.

The coloration of Pb(II) ß"-alumina has been examined in greatest detail. It proceeds from the edges of the crystal along the high-conductivity conduction planes but not perpendicular to them. At 600°C, the colored region can spread across several millimeters of crystal in a few minutes. This implies that oxygen ions may be able to diffuse rapidly within the conduction planes of the structure. Colored samples of Pb(II) ß"-alumina generally exhibit electronic conductivity, which can vary over six orders of magnitude, the upper limit being approximately 5×10^{-3} (ohm-cm)$^{-1}$. Direct current measurements show that the conductivity is ohmic and impedance spectra show frequency independent conduction at low frequencies.

All of these results are consistent with a proposed model for the coloring reaction that assumes oxygen is lost from the ß"-alumina framework upon heating.

Though not experimentally demonstrated, it is quite reasonable to speculate that the column oxygens (O(5)) are the species principally lost from the structure. In Kroger-Vink notation:

$$O_O \rightarrow V_O^{\bullet\bullet} + 1/2O_2 + 2e^-$$

The two electrons left behind for each oxygen removed presumably are trapped in the conduction plane rather than the spinel block, because electrons can be more easily accommodated in the conduction plane, where high ion mobilities allow greater lattice polarization.

V. Structural Studies

The remarkable ion transport properties of ß"-alumina are the result of its unusual structure [24]. The framework of the material (see Figure 1) consists of close-packed blocks of Al, Mg and O, referred to as 'spinel blocks' because of their resemblance to the mineral spinel ($MgAl_2O_4$), separated by more open regions containing the mobile cations. These regions are about 4Å thick and are generally, albeit imprecisely, referred to as 'conduction planes'. The spinel block layers are connected by Al-O-Al bonds. The linking oxygens, termed O(5), are known as bridging or column oxygens. Ion motion occurs in two-dimensional honeycomb-like pathways around the column oxygens, perpendicular to the *c*-axis.

If the mobile ions are assumed to be distributed randomly, then ß"-alumina has a rhombohedral unit cell of $R\bar{3}m$. Each unit cell spans three conduction planes and three spinel blocks. Structural data for the di- and trivalent ß"-aluminas studied thus far are given in Table I. The various isomorphs show a large range of *c* values, while *a* remains essentially constant. The reason for the wide variation in *c* but not *a* is apparent: the size, charge and polarizability of the mobile cations in the conduction planes have little effect on the nature of the spinel blocks (which determine *a*), but can greatly affect the separation between them. When a small, highly charged cation is introduced into the structure, the column oxygens move off of their normal position towards the cation, bending the Al-O-Al bonds and contracting the structure.

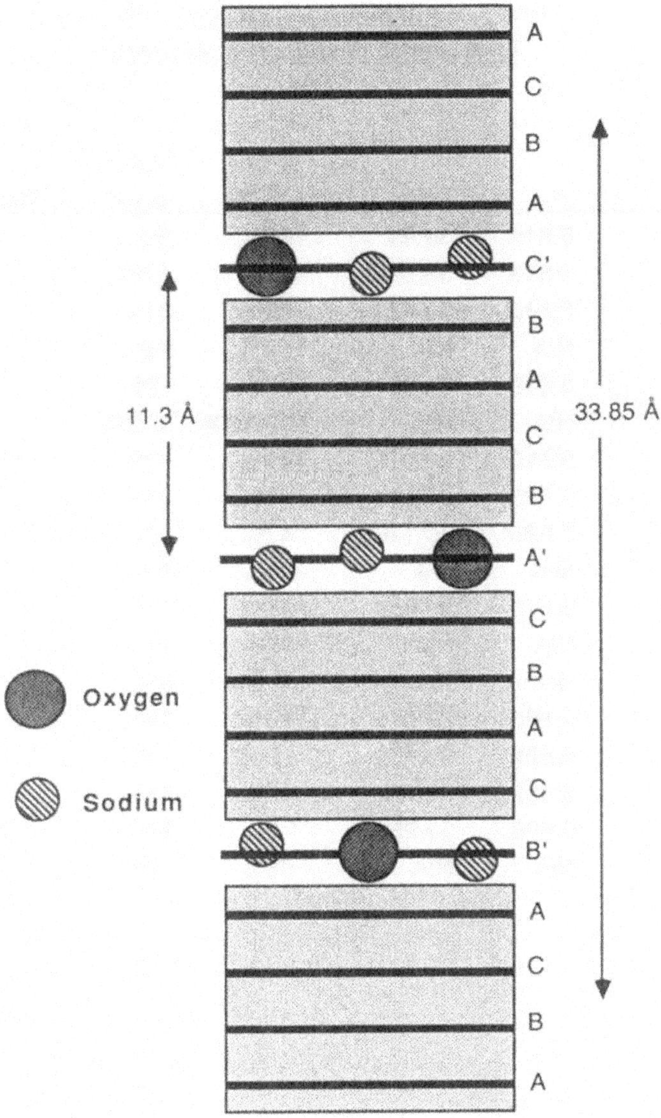

Figure 1. Unit Cell of Na(I) ß"-Alumina. Solid lines indicated by A,B, and C are layers of close-packed oxygens comprising spinel blocks. Solid lines indicated by A', B', and C' are the conduction planes.

Table 1. Summary of Structure Studies of Di- and Trivalent β"-Aluminas

Mobile Cation	*a*	*c*	% ions in BR	%ions in mO	Reference*
Na(I)	5.614	33.85	97%	3%	10
Ca(II)	5.613	33.270	53%	47%	62
Sr(II)	5.610	33.720	39%	61%	63
	N/A	N/A	12%	88%	29
Ba(II)	5.619	34.084	98%	2%	27
	N/A	N/A	90-100%	0-10%	32
Mn(II)	5.618	33.141	40%	60%	64
Zn(II)	5.623	33.517	73%	27%	64
Cd(II)	5.620	33.146	47%	53%	25
	N/A	N/A	37%	63%	29
Pb(II)	5.610	33.967	91%	9%	25
	N/A	N/A	59%	41%	29
	N/A	N/A	100%	0%	32
Eu(II)	5.626	33.575	16%	84%	35
Gd(III)	5.625	33.134	2%	98%	33
Eu(III)	5.627	33.190	4%	96%	35
Nd(III)	5.628	33.259	5%	95%	36
	N/A	N/A	55%	45%	32

The structure of the conduction planes of ß"-alumina is shown in Figure 2. The mobile cations can occupy two non-equivalent sites: the seven-coordinate Beevers-Ross-type site (BR) and the eight-coordinate mid-oxygen site (mO). There are two BR and three mO sites, but only 1.67 monovalent, 0.83 divalent, or 0.56 trivalent ions per conduction plane in each unit cell. Thus the cation vacancy population in the conduction planes is comparable to the number of mobile ions, and ionic conductivity occurs by a vacancy mechanism.

X-ray diffraction studies show that different mobile cations have different site occupations (see Table I) in the ß"-alumina structure. It is obvious that the chemical nature, size and charge of the mobile cation greatly influence which sites it occupies. The mobile ions are also described with large thermal parameters, indicating that the electron density associated with the ions is spread out along the conduction pathways. This is illustrated for two cases (Ba(II) and Ca(II) ß"-alumina) in Figure 3, from which it is clear that the idea that mobile ions simply occupy either BR or mO sites is a serious over-simplification. Such sites are merely a space-time average, and in fact, at any instant, a given ion will have a high probability of occupying an off-site (displaced) position. This tends to make the accurate assignment of site occupations highly tenuous, not the least because site occupations and thermal models are numerically coupled in the refinement procedure. This situation is characteristic of mobile-ion systems.

From this discussion, it is clear that the non-stoichiometric nature of ß"-alumina makes it impossible to assign an integral number of mobile cations to the $R\bar{3}m$ unit cell. This implies that, strictly speaking, the $R\bar{3}m$ description will generally break down if mobile ions are included in the description of the structure. Local correlation in the mobile-ion arrangement will be observed experimentally either as extended regions of diffuse scattering or, if sufficiently large domains of superlattice ordering occur, as 'extra' Bragg peaks in a scattering experiment. The average distance over which local ordering persists in a given direction is known as the *correlation length*. This quantity can be inferred from the form of the diffuse scattering peaks.

Boilot et al. [25] reported X-ray diffuse scattering results from Na(I) ß"-alumina in 1980 . They found that the Na(I) ions have a statistical tendency to order into a

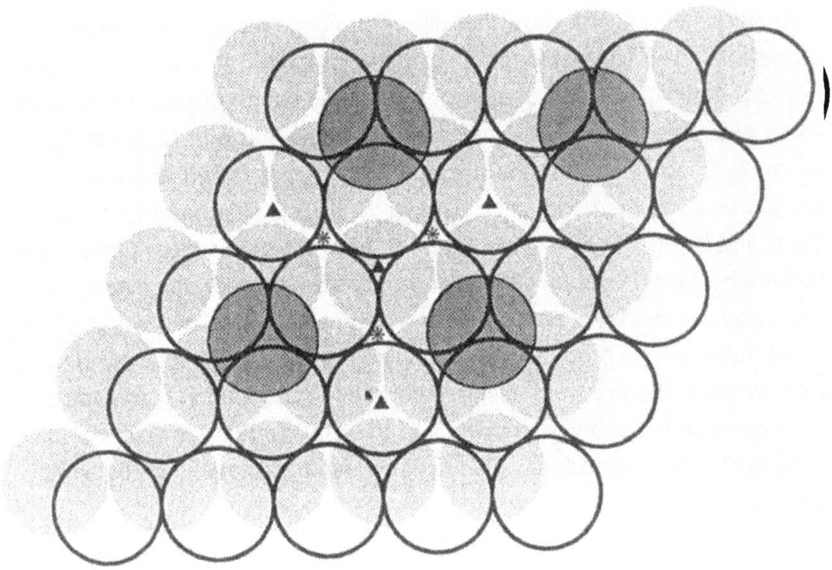

Figure 2. Conduction plane of Na(I) ß"-alumina. View is parallel to *c*-axis, looking down on conduction plane. Open circles are oxygens above plane; dark fill indicates column oxygens within plane; and light fill indicates oxygens below plane. Small black triangles are BR-type sites. Small stars are mO sites.

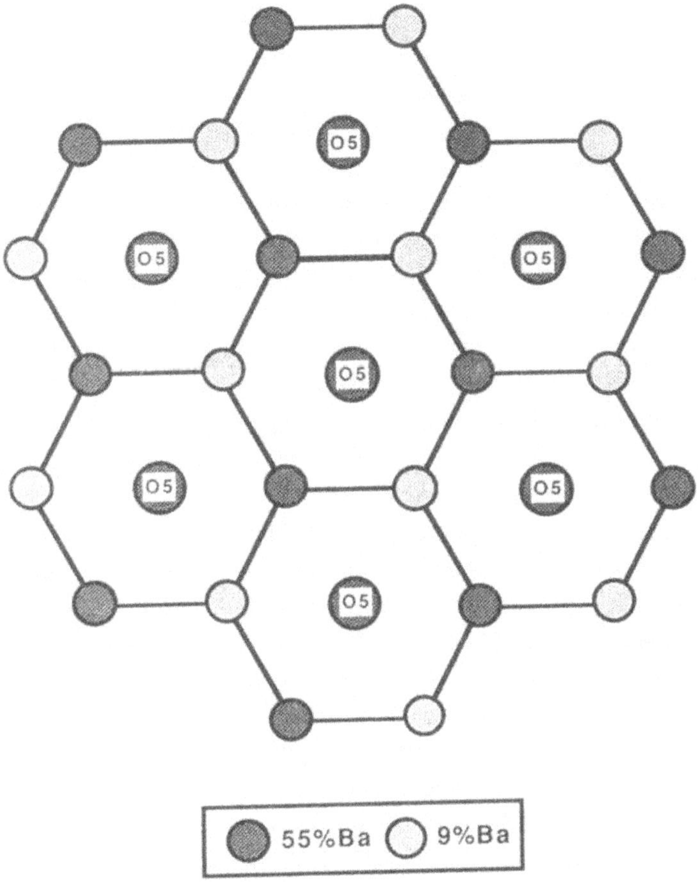

3a

Figure 3. Conduction plane structure (electron density map) of (a) Ba(II) and (b)Ca(II) ß"-alumina

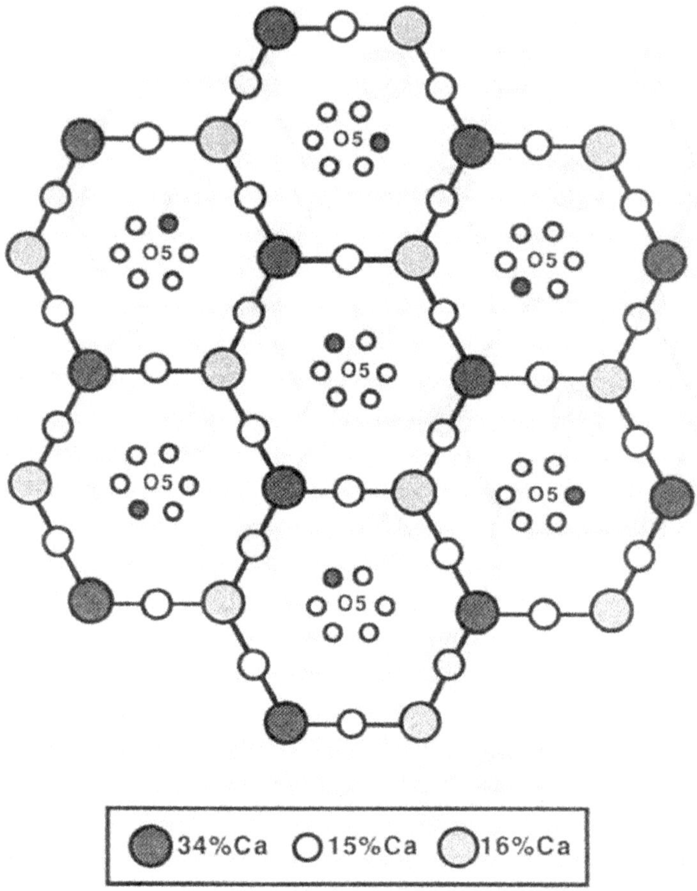

3b

$\sqrt{3}a \times \sqrt{3}a$ lattice with a correlation length in the conduction plane of 70Å at room temperature. Seevers et al. [26] speculated in 1983 that the curvature observed in Arrhenius plots of the conductivities of the divalent ß"-aluminas was the result of an order-disorder process among cations and vacancies in the conduction planes. They suggested that the higher activation energies for conduction seen at low temperatures are the result of local ion ordering involving attractive ion-vacancy interactions, and that, as the temperature is raised, a broad transition occurs to a more disordered state with a lower activation energy for conduction.

Thomas and co-workers [27, 28] found that, within the coherence length of their experiment, many divalent ion arrangements involve some degree of short-range order. The extent of this effect - the *correlation length* - is found to be different from isomorph to isomorph. This is quite logical; in the case of Ba(II), for example, if one ion occupies a BR site, then it is highly unlikely that a second will simultaneously occupy one of the three adjacent BR sites, simply on the basis of coulombic considerations. The direct consequence is that the crystallographic center of symmetry (at mO) is broken locally for the Ba(II) arrangement. A correlation length of the order of the diffraction coherence length was inferred, i.e. several hundred Ångstroms. Thomas et al. [28] also speculated that the short-range ordering of mobile cations in the conduction plane may be related to ordering of Mg(II) ions in the spinel blocks.

Boilot et al. [29] first examined short-range ordering in several of the divalent ß"-aluminas in 1983. They found three-dimensional short-range order in Cd(II) and Pb(II) ß"-alumina, with a correlation length of about 100Å. The authors claimed to observe a tendency towards the formation of a 2a x 2a x 2c cell. Similarly, the Sr(II) and Hg(II) isotopes were found to exhibit a two-dimensional 2a x 2a cell with correlation lengths of 80Å and 70Å, respectively. These types of superstructure are not possible within the quoted stoichiometry, because they require a non-integral number (6.66 for the three-dimensional structure, 3.33 for the two-dimensional structure) of mobile cations in the repeat unit. The same investigators re-examined several of these compounds in 1986 with conventional and diffuse X-ray scattering techniques [30]. Their conventional X-ray diffraction results are included in Table I. Their diffuse scattering results indicate two-dimensional short-range order for Sr(II) and Cd(II), with correlation lengths of 80Å and 40Å, respectively. For the Pb(II) isomorph, three-dimensional ordering was observed, with an in-plane correlation length of greater that 200Å, and an inter-plane correlation length of about 50Å. No mention was made of a long-range superstructure.

Hayes and co-workers have also studied ordering in Ba(II) and Pb(II) ß"-alumina [31, 32]. Their diffuse X-ray scattering results indicate an in-plane correlation length of about 15Å for Ba(II) and about 5Å for Pb(II), which are clearly shorter than suggested by Boilot and (at least for Ba(II)) by Thomas. It is important to remember, however, that the degree of ion ordering may vary with thermal history in these materials and, moreover, that the interpretation of diffraction results for this type of system is by no means straightforward.

Frydrych has studied Ba(II), Pb(II), and Nd(III) ß"-alumina with energy-dispersive X-ray diffraction (EDXD) [33]. Her results for the Ba(II) isomorph confirm those of Thomas, in that the Ba(II) ions occupy predominantly BR sites and have a very low probability of occupying the two adjacent sites. Pb(II) ions were also found to occupy predominantly BR sites, but no evidence was found for short-range ordering.

Very little work has been reported so far on the structures of the ß"-aluminas. Carrillo-Cabrera et al. have examined the structure of Gd(III), Eu(III), and Nd(III) forms by conventional X-ray diffraction [34, 35, 36], and their results are summarized in Table 1. The space-time average distributions are also plotted in Figure 4, where we see that they are virtually identical in the three cases studied, with 98%, 96%, and 95% of the Gd(III), Eu(III), and Nd(III) ions, respectively, at mO sites. It must be stressed, however, that this need not imply the *same* type of local ordering or extended superlattice formation in each of these materials. In this connection, Davies et al. [37] in 1986 published TEM results indicating that long-range ordering of trivalent ions occurs in Nd(III)-, Gd(III), and Eu(III) ß"-alumina with a superstructure of $3a \times 3a \times 3c$. This superstructure, indicated by the presence of extra Bragg reflections in the electron diffraction patterns, *is* stoichiometrically possible, with five trivalent ions per in-plane unit, and forty-five per unit cell. They reported that the ordered structure co-exists with disordered regions and suggested that thermal history is an important factor in determining the microstructure of these materials.

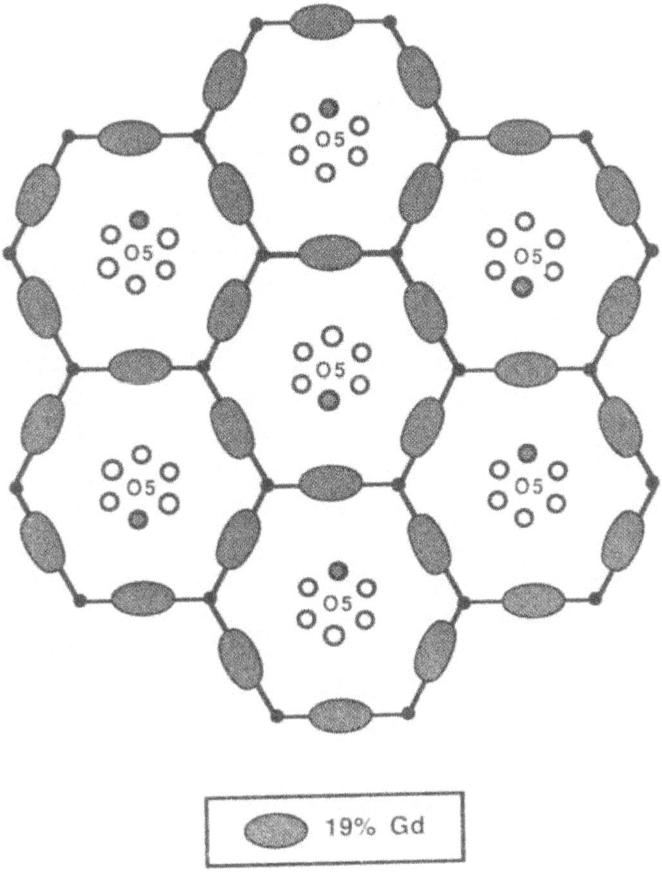

4a

Figure 4. Conduction plane structure (electron density map) of (a) Gd(III), (b) Eu(III), and (c) Nd(III) ß"-aluminas.

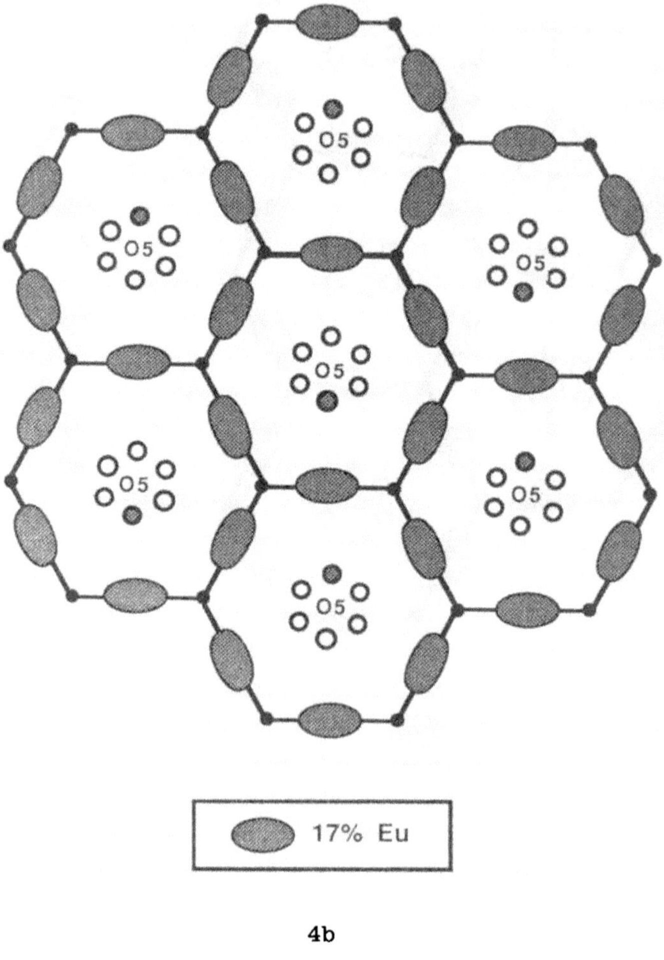

4b

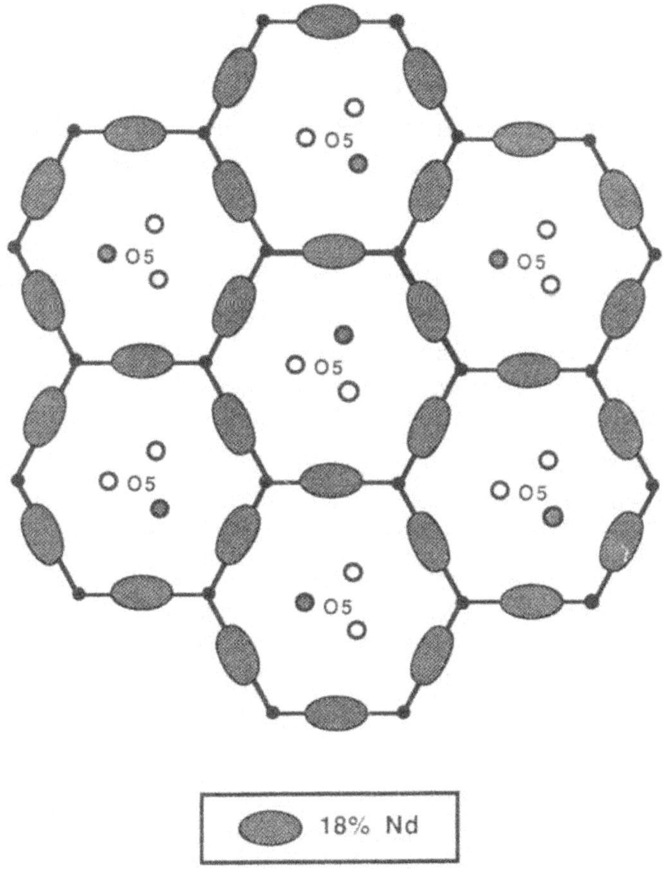

4c

VI. Ion Transport and the Effect of Thermal History

The speed with which divalent cations exchange for Na(I) in β"-alumina at moderate temperatures was the first indication that the divalent beta" aluminas have unusually high ionic conductivities, at least above about 200°C. The first conductivity measurements on single crystals of Ca(II), Sr(II), Ba(II), and Pb(II) β"-alumina, reported by Seevers et al. [26], confirmed this expectation and revealed that Pb(II) β"-alumina has a conductivity that is higher than all other divalent β"-aluminas known. The conductivity of Pb(II) β"-alumina is comparable to that of Na(I) β"-alumina from -100 to 400°C. Figure 5 summarizes conductivity data for various divalent β"-aluminas as a function of temperature.

The curvature in the Arrhenius-type plots for the divalent β"-aluminas shown in Figure 5 led Seevers et al. to suggest that order-disorder transformations among the mobile ions and vacancies might well play a significant role in determining the magnitude of ionic conductivity and the details of the ionic arrangement in these materials [26]. Davies et al. [37] examined mobile ion ordering in β"-alumina in greater detail and suggested that, even though cations can diffuse rapidly, the approach to quasi-equilibrium conduction plane structures may be quite slow, and that high-temperature, disordered structures can be readily preserved upon cooling to lower temperatures. The same investigators reported that long periods of annealing, typically many days at 100-150°C, are generally needed to establish a quasi-equilibrium ion arrangement in the conduction planes [38]. However, more recent investigations [39] have found that the thermal effects once considered evidence of order/disorder processes in the divalent β"-aluminas are, at least in some cases and perhaps in all, the result of hydration/dehydration reactions. It is interesting to note that the conditions prescribed for attaining ordered structures, annealing at 100-150°C for long periods, are also ideal conditions for hydrating the β"-aluminas. Surely, ion-vacancy ordering plays a role in determining the structure and properties of these materials and has been shown to occur in several of the trivalent compositions, but the phenomenology of ion-vacancy ordering in the divalents is not well-explored at this time.

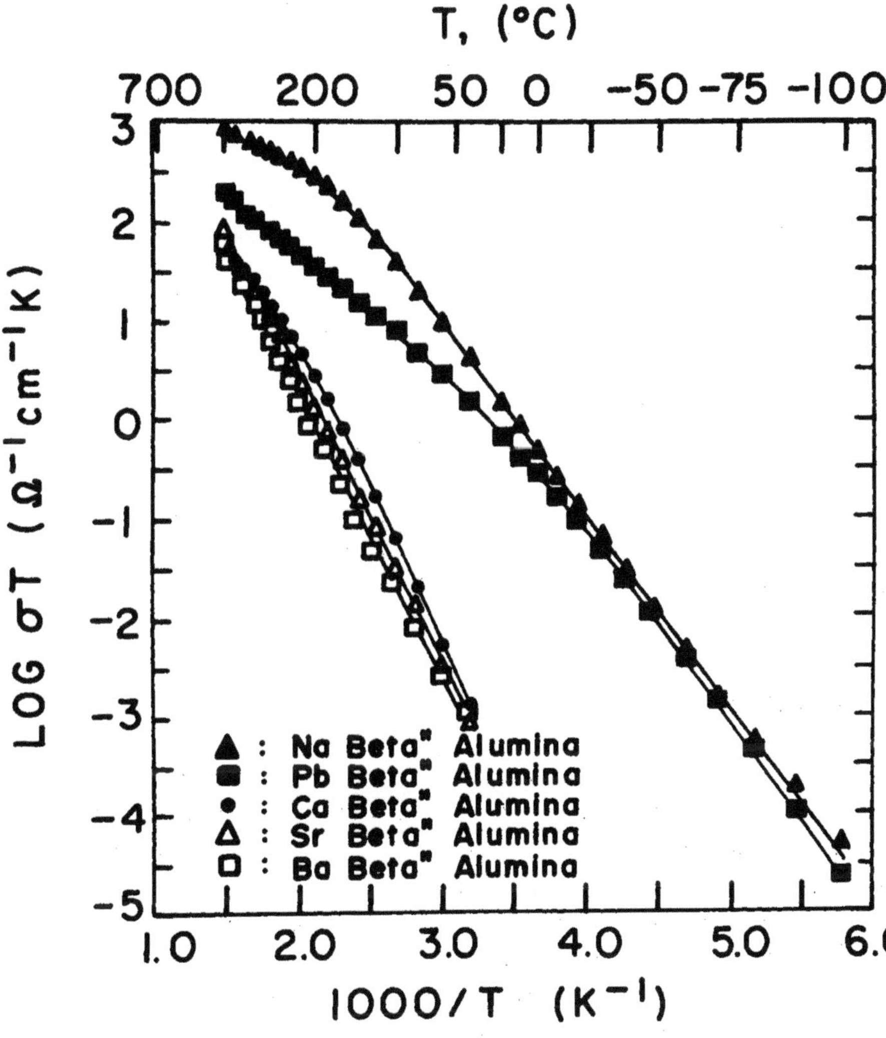

Figure 5. Ionic conductivities of various divalent β"-aluminas.

Saltzberg et al. [40] carried out a particularly thorough investigation of ion ordering in mixed Na(I)-Eu(II) (with X-ray diffraction and spectroscopy) and Na(I)-Eu(III) ß"-aluminas (with spectroscopy only). Their single crystal X-ray diffraction studies and related investigations of Na(I)-Eu(II) ß"-alumina revealed a fascinating relationship between the composition of the conduction planes and the degree of short- and long-range order in the structures. The authors combined information from X-ray diffraction, TEM, and optical spectroscopy studies of this complex, partially disordered system into a single, consistent model in which mobile ion-framework interactions, mobile ion-mobile ion interactions, and kinetic effects all contribute to the evolution of the structure from highly disordered Na(I) ß"-alumina to partially ordered Eu(II) ß"-alumina. They found that at low Eu(II) concentrations, the structure is dominated by Na(I)-Eu(II) repulsions, there is short-range ordering among the mobile cations, and the Eu(II) ions occupy poorly-defined sites between the mO and BR sites. As the Eu(II) concentration increases, mobile ion-framework and Eu(II)-Eu(II) interactions become more important, the Eu(II) ions occupy well-defined mO sites, and a long-range ordered structure among the mobile cations is favored at low temperatures.

VII. Spectroscopy

The ability of ß"-alumina to incorporate lanthanide and transition metal ions has led to the investigation of the spectroscopy and optical properties of divalent and trivalent ß"-aluminas. Spectroscopy studies are of considerable interest because they provide insight into local ion environments. Moreover, the ß"-alumina conduction plane represents a unique structure in which to study optically active ions. The ions are not in close-packed surroundings, and they are able to relax from their equilibrium positions. In addition, there is the opportunity to observe ions in symmetries that are rarely obtained with traditional optical hosts. The optical behavior of the exchanged ß"-aluminas are also of interest for their potential applications. Lanthanide and transition-metal substituted ß"-aluminas have already been found to possess some novel optical characteristics, and both lasing and four-wave mixing have been demonstrated. Ion exchange is particularly advantageous for optical studies because the doping process is separated from crystal growth. As a result, it is possible to control independently the valence of the ion as well as its concentration and distribution within the host material. Following is a summary of some of the more significant results obtained for the lanthanide and transition-metal exchanged ß"-aluminas.

7.1 Lanthanide-Exchanged ß"-aluminas

7.1.1 Nd^{3+}: ß"-alumina

The optical properties of Nd(III) ß"-alumina have been explored in some detail. Optical absorption spectroscopy and ESR have been used to study the site occupations for Nd(III) as a function of Nd(III) concentration [41]. Absorption spectra taken at 4K for the $^4I_{9/2} \rightarrow {}^2P_{1/2}$ transition indicate that Nd(III) is present in two sites. The ESR measurements clearly identify the mid-oxygen site as the predominant one and show the Nd(III) concentration at the second site is rather small. Complementary studies by Queenan and Davies [42] found that the distribution of Nd(III) in mO and BR sites can be influenced by the conditions of thermal annealing. At high Nd(III) concentrations (above 1.5×10^{21} Nd(III)/cm^3), it appears that Nd(III) pairs are present [41]. The concentration above which these pairs form has not been established.

One important feature of the absorption spectrum of significance for potential applications is the presence of an anomalously large absorption coefficient at 575 nm [43]. This absorption arises from the hypersensitive $^4I_{9/2} \rightarrow {}^4G_{5/2}, {}^2G_{7/2}$ transition. The oscillator strength of the transition (6×10^{-5}) is nearly ten times that observed in Nd:YAG (7×10^{-6}), while the other transitions have comparable oscillator strengths (Table 2). One curious aspect of this behavior is that the Nd(III) ions occupy mid-oxygen sites (C_{2h}), which have inversion symmetry, and strong optical absorptions are not expected for such sites. However, using a Judd-Ofelt analysis of the absorption spectrum intensities, Alfrey et al. [44] were able to account for the high oscillator strength of the hypersensitive transition by suggesting that the Nd(III) ions and column oxygens [O(5)] are slightly displaced from their ideal positions. These displacements were calculated to be rather small (0.1 to 0.2Å), but nonetheless serve to break the inversion symmetry at the 9d site.

Table 2: Oscillator Strengths for Various Transitions in Nd:YAG and Nd:β"-Alumina

Wavelength region (nm)	Nd^{3+}: β"-alumina ($\times 10^{-6}$)	Nd^{3+}:YAG ($\times 10^{-6}$)
420		0.032
430	0.7	0.34
460		0.84
475	0.48	0.93
530	8.3	5.6
580	61	7.1
750	5.8	8.0
790	5.5	8.4
880	1.6	1.5

Recent X-ray diffraction results of Carrillo-Cabrera et al. [36] also indicate that the column oxygens are displaced towards BR sites and the Nd(III) ions relax locally along the conduction pathways towards adjacent BR sites. As Nd(III) ions move away from mO sites, their symmetry is lowered. This is consistent with the g-parameters obtained from ESR measurements [41]. The proposed model in which the Nd(III) ions are displaced to various degrees from the mO sites also explains the emission spectra for these materials. The emission spectrum at 1.06 μm is nearly as inhomogeneously broadened as Nd(III) in glass, with an effective linewidth of about 250 cm^{-1} [45]. Although it is possible that the breadth of this line arises from Nd(III) ions in distinctly different lattice sites, it is more likely the result of a distribution of Nd(III) ions around the mO site.

Another novel optical property of Nd(III) ß"-alumina is its fluorescence lifetime [43]. It has two distinctive features. First, the lifetimes, which are in the range of 400 μsec, are nearly twice as large as is commonly observed for Nd:YAG ($Y_3Al_5O_{12}$). Secondly, there is no evidence of fluorescence quenching even at concentrations as large as 1×10^{21} Nd(III)/cm^3. In contrast, Nd: YAG exhibits concentration quenching at Nd(III) concentrations of 1 to 2×10^{20} ions/cm^3. It is interesting to note that the Nd(III) lifetime is much longer than the Na^+ hopping frequency, which means that, in mixed Na(I)-Nd(III) samples, the Na^+ ions can undergo many hops (>1000) to near-neighbor sites while the Nd(III) remain in the excited state. Measurements of lifetime as a function of temperature indicate that the non-radiative recombination rate fits a multiphonon model, where the number and energy of the phonons are associated with the spinel block and not with mobile Na^+ ions [46].

The spectroscopic properties of Nd(III) ß"-alumina suggest that this material may be of considerable interest for optical devices. The combination of high oscillator strength at 575 nm and long lifetime at high Nd(III) concentrations has already been exploited in experimental optical devices. Jansen et al. [43] were able to achieve lasing in both pulsed and cw modes using small single crystals of ß"-alumina containing from 4 to 7×10^{20} Nd(III)/cm^3. Despite a path length of only 5 mm and a high-loss resonator structure (the only feedback was provided by Fresnel reflections of ≈8% from the cleaved ends of the crystal), sufficiently high inversion levels and high gain were obtained by pumping the crystals at the strong absorption band. Both pumping and laser emission occurred within the *ab*-plane. In a complementary study, Alfrey et al. [45] reported both calculated and measured values for the stimulated emission cross section for ß"-alumina containing Nd(III). The values are comparable to those of glassy materials (≈2 x 10^{-20} cm^2), although the stored energy/cm^3 is greater than for either Nd:glass or Nd:YAG because of the relatively long lifetime. Other experimental parameters are listed in Table 3.

Measurements of single pass gain along the *c*-direction have shown the gain/cm to be ≈4/cm, a value consistent with data obtained in the laser experiments. What is remarkable is that the gain/cm achieved by pumping the 575 nm band is approximately ten times greater in ß"-alumina than in a Nd:YAG laser pumped with a flashlamp. This behavior suggests that it should be possible with Nd(III) ß"-alumina crystals to achieve lasing in crystals of very small volume, a feature that is of considerable interest for a variety of applications.

Another potential device application for ß"-alumina is in devices for optical phase conjugation. The strong absorption coefficient at 575 nm can be easily saturated. This enables ß"-alumina to be considered a nonlinear-optical material in which the nonlinearity arises from saturated absorption. Using Nd(III)-exchanged ß"-alumina, Boyd et al. [47] have carried out a degenerate four-wave mixing experiment. A phase conjugate reflectivity of 0.3% was produced. In view of the ion exchange properties of ß"-alumina, it may be possible to obtain resonant transitions in other parts of the spectrum.

7.1.2 Er(III), Pr(III), Ce(III), Tb(III), and Mixed ß"-aluminas

Exploration of the spectroscopic characteristics of other lanthanide-exchanged ß"-aluminas, including those containing Er(III), Pr(III), and Ce(III), is just beginning. Er(III) is particularly interesting because it has hypersensitive transitions ($^4G_{11/2}$ and $^2H_{11/2}$), and it is interesting to know whether the oscillator strengths of these transitions are substantially stronger in ß"-alumina than in other oxide hosts. Initial results indicate that the oscillator strengths are high but not exceptional. The hypersensitive transitions of Er(III) ß"-alumina are three to four times greater than that of Er(III):$YAlO_3$ or Er(III):YAG , but comparable to those of Er(III):Y_2O_3 .

Table 3: Experimental Parameters for Nd(III) β"-Alumina

Sample Characteristics

Nd^{3+} Concentration	$5.7 \times 10^{20}/cm^3$
Sample Thickness	300 μm

Optical Properties

Measured Lifetime	420 μsec
Theoretical Lifetime	874 μsec
Effective Linewidth at 1.06 μm	256 cm^{-1}
Calculated Stimulated Emission Cross Section	2.08×10^{-20} cm^2
Measured Stimulated Emission Cross Section	2.1×10^{-20} $cm^2 \pm 0.5 \times 10^{-20}$ cm^2
Measured Power Gain	4.1/cm

Ce(III) is of considerable interest for optical applications because it has broad, strong absorption in the ultraviolet which results from the dipole-allowed, 4f→5d transition. Its absorption is compatible with the emission region for mercury and xenon flash lamps, and its emission is in the blue. Emission and excitation spectra for Ce(III) in ß"-alumina are shown in Figure 6.

One of the major potential applications for Ce(III) is as a sensitizer for the luminescence of trivalent lanthanide ions that have weakly-absorbing 4f→4f electronic transitions. The Ce(III)-Tb(III) energy transfer couple has been studied extensively because Ce(III) can improve the efficiency of the 540 nm luminescence of Tb(III) in phosphor applications [48]. In general, energy transfer in lanthanide-activated crystals is not achieved easily, because it is often limited by problems of concentration gradients, compositional inhomogeneities and lattice distortions. However, energy transfer in ß"-alumina with Ce(III) as a sensitizer was recently observed by Momoda et al. [49]. The systems investigated include Ce(III)-Nd(III), Ce(III)-Tb(III), and Ce(III)-Pr(III). To make these compositions, it was necessary to experiment with double-doping techniques by which Ce(III) and the desired lanthanide activator were both ion exchanged from the same molten salt. In all cases studied, the relative concentration of the ions in the melt determined the relative concentrations in the crystal. The Ce(III) emission was suitable for the various lanthanide activator ions, and energy transfer was observed in all three systems.

A clear indication of Ce(III)→Nd(III) energy transfer is shown in Figure 7. In this excitation spectrum (based on the detection of Nd(III) emission at 859 nm), samples doped only with Nd(III) show the presence of lines at 346, 355 and 428 nm which are characteristic of Nd(III) levels. In doubly-doped Ce(III)-Nd(III), the excitation spectrum displays the broad excitation band characteristic of Ce(III) (compare Fig. 7 with Fig. 6). This behavior demonstrates conclusively that Nd(III) emission in doubly-doped Ce(III)-Nd(III) ß"-alumina arises from absorption by the Ce(III) with subsequent energy transfer to Nd(III) ions. A series of measurements with varying Ce(III)/Nd(III) ratios has shown that a maximum efficiency of 90 % can be achieved for Ce(III)→Nd(III) energy transfer [49].

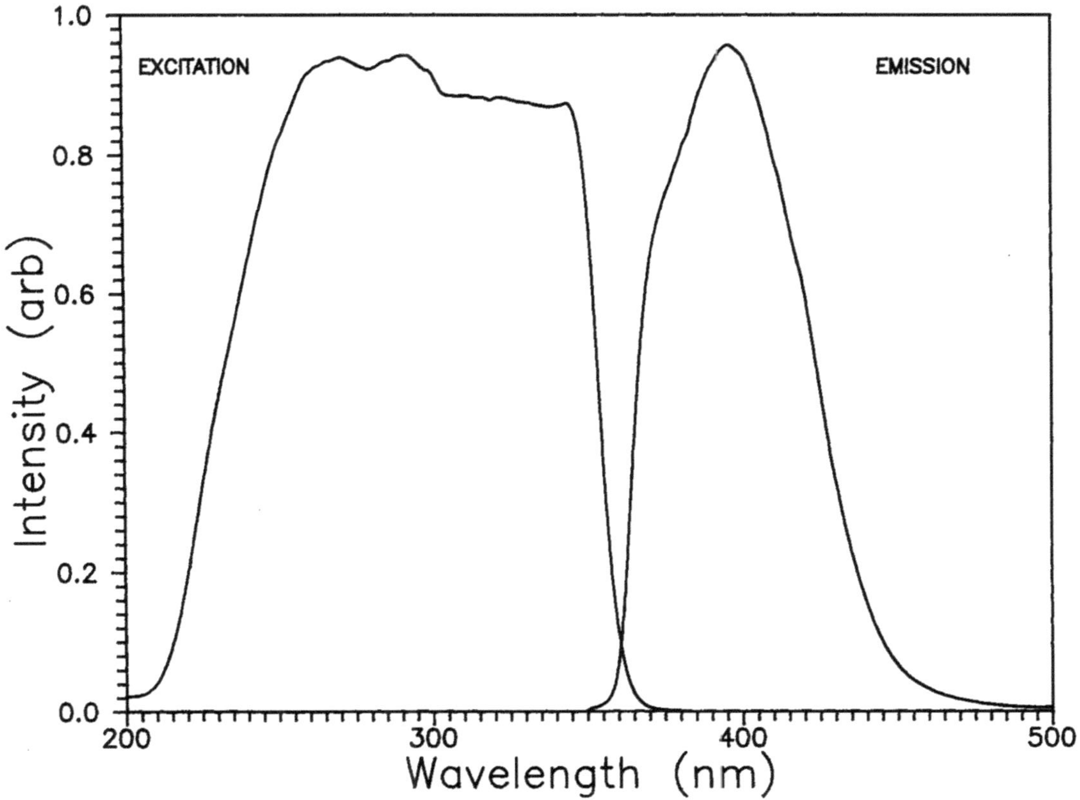

Figure 6. Excitation and emission spectra (room temperature) for Ce(III) exchanged ß"-alumina.

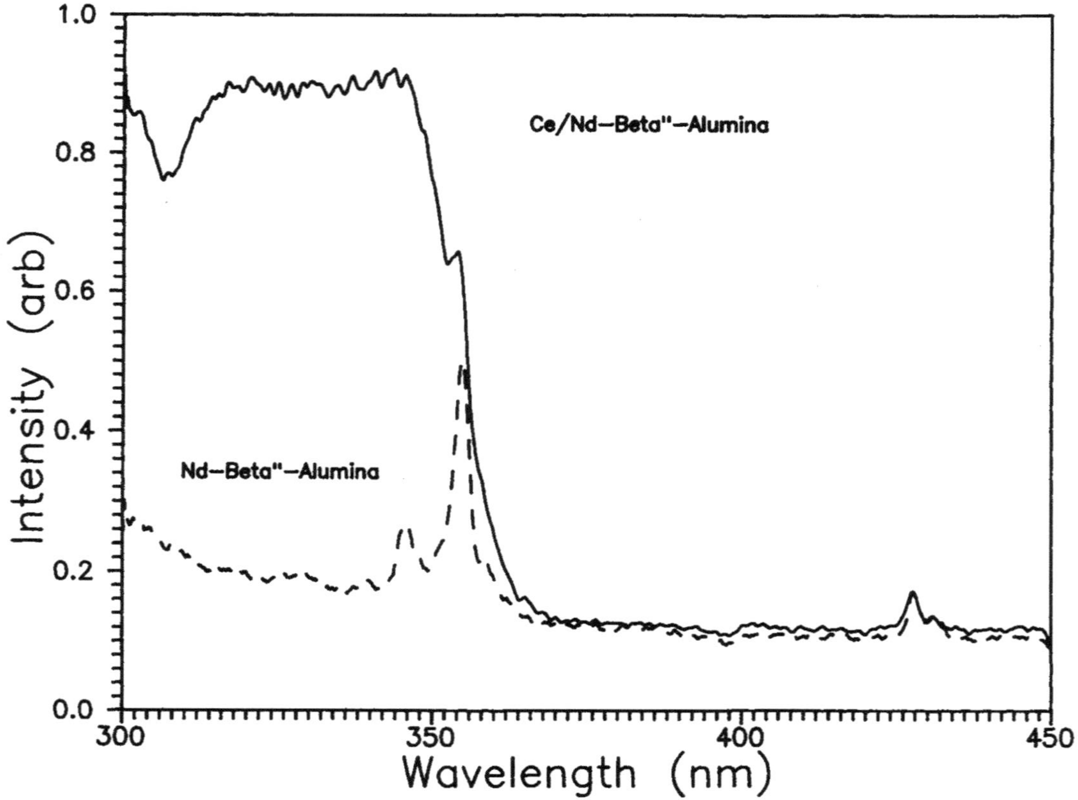

Figure 7. Excitation spectra for the 859 nm emission of Nd(III): Dashed line: ß"-alumina containing only Nd(III); Solid line: ß"-alumina containing mixture of Ce(III) and Nd(III).

7.1.3 Eu(III) ß"-Alumina

Eu(III) ß"-alumina has been the subject of a recent examination of the factors which influence Eu(III) bonding and spectroscopy in different Na(I)-Eu(III) ß"-alumina compositions [50]. Particular attention was paid to the evolution of the optical characteristics of the samples as a function of Eu(III) concentration and thermal history.

Two distinct types of fluorescence spectra were identified for Eu(III) ions in >98% exchanged Eu(III) ß"-alumina at 77K. The emission peaks which arise are the result of transitions from the $^5D_0$ excited state down to the ground state ^7F_J multiplet. Preliminary site-selective laser spectroscopy experiments at 10K have verified that Eu(III) appears to exist in at least three distinct but very similar sites in these highly-exchanged samples. Further analysis indicates that one site has (or is very close to having) inversion symmetry and another does not have inversion symmetry. These are presumably the mO and BR sites, respectively.

The effects of thermal history on the Eu(III) spectroscopy suggest that Eu(III) ions do not favor the BR site and only occupy it when trapped there by rapid cooling from high temperatures. At elevated temperatures, a disordered distribution of Eu(III) ions in the conduction planes is favored as the result of configurational entropy. If the material is cooled quickly, some of the Eu(III) ions are trapped in BR sites, which are not favored at lower temperatures. If the material is annealed at 350°C, however, these Eu(III) ions in BR sites relax into mO sites. Davies et al. [37] and Queenan and Davies [42] have shown that ordered and disordered structures of the mobile ions can be formed in the trivalent ß"-aluminas, and that the microstructures of the materials vary greatly with thermal history. It appears that the Eu(III) site occupations are also affected by thermal history. Similar effects in the Na-Eu(II) ß"-alumina system have also been reported [51, 52].

7.2 Transition-Metal Exchanged ß and ß"-Aluminas

7.2.1 Cr(III) ß"-Alumina

Studies of Cr(III) in both ß and ß"-alumina have been reported [53, 54, 55, 56]. In the work with ß-alumina, Cr(III) was present as a natural impurity in as-grown single crystals of Na(I) ß-alumina [53, 54]. Luminescence spectroscopy of Cr(III) was investigated for a series of samples containing Na(I), Li(I), K(I), Rb(I), Ag(I) and Tl(I) as the mobile ions. Fluorescence originated from the $^2E \rightarrow ^4A_2$ transition for Cr(III) in an octahedral site. Using lifetime measurements and time resolved spectroscopy, Mariotto et al. [53] were able to establish that Cr(III) ions were present in two non-equivalent sites within the spinel block: the Al(4) in the center of the spinel block and the Al(1) close to the conduction plane. In the Al(1) site, the Cr(III) is sensitive to the cation disorder in the conduction plane, so the splitting of the ^2E state is not resolved. The position and width of this line depend upon the nature of the ion in the conduction plane and its site occupancy [54].

In ß"-alumina, Cr(III) was introduced by ion exchange and was therefore located only in the conduction plane [55, 56]. Two prominent emission bands were observed. The band at 700 nm was assigned to the ^2E level while that at 820 nm was considered to come from the Stokes-shifted $^4T_2$ level [55]. Both bands exhibited inhomogeneous broadening, and lifetime measurements were consistent with the transition assignments. Powell et al. [55] ascribed the different bands to non-equivalent sites at which the crystal fields were influenced by local distributions of Na(I) ions and vacancies. They proposed that Na(I) motion, as indicated by the temperature dependence of fluorescence intensity and lifetime, was instrumental in making the lattice sites spectroscopically different.

7.2.2 Cu(I) ß"-Alumina

The spectroscopy of Cu(I) is particularly interesting because of its emission in the blue-green part of the spectrum. However, obtaining Cu(I) in a usable laser host material has been difficult, especially in oxides. The primary reason is that traditional growth processes invariably lead to the presence of Cu(II), which interferes with and usually quenches the Cu(I) luminescence. Using ion exchange processes, however, Barrie et al. [57] were able to control the valence of copper and to produce ß"-alumina crystals containing $\approx 10^{19}$ Cu(I)/cm^3. The resulting materials showed broad green luminescence peaked at 535 nm. This luminescence behaved as expected for a vibrationally-allowed transition that is dependent upon phonon coupling. Specific energy levels for the emission have yet to be assigned, in part because a large energy difference exists between excitation and emission. Thus, rather than emission arising solely from the excitation of Cu(I) ions, there is a distinct possibility that interactions may occur between Cu(I) ions in the conduction plane.

Another interesting and possibly very useful feature of Cu(I) ß"-alumina is that its emission can be tuned over much of the visible spectrum [57]. This is possible because the *c*-lattice parameter of ß"-alumina can be varied [4]. For example, a series of Cu(I)-doped samples have been prepared in which the Na(I) was replaced by a co-dopant. The purpose of the co-dopant was to expand or contract the ß"-alumina conduction plane around the Cu(I) ion. This results in a larger or smaller crystal field, and causes the emission to shift toward the blue or the red accordingly. What is more, single pass gain measurements at 514.5 and 632.8 nm have given values of about 0.1/cm. The implication is not only that Cu(I) ß"-alumina may lase, but also prove to be a laser host whose output is tunable.

7.2.3 Co(II) and Ni(II) ß"-Alumina

Co(II) and Ni(II) ß"-alumina are of interest as potential tunable laser materials with output in the infrared. The laser transitions for both ions, however, occur only in an octahedral crystal field [58]. Thus, it is important to identify the site occupancies for both Co(II) and Ni(II) in ß"-alumina. The spectroscopy of Co(II) has been investigated in some detail. Co(II) has been introduced into both ß and ß"-alumina by high temperature reaction [59, 60]. In both cases, optical absorption spectroscopy indicated that the Co(II) species appeared to occupy only tetrahedral sites in the spinel block. Recently, optical absorption and ESR measurements were made on ion exchanged Co(II) ß"-alumina, in which the Co(II) ions are located only in the conduction plane [61]. These spectra were also dominated by Co(II) in tetrahedral sites. The absorption bands and the resulting Dq/B value (0.51) were indicative of Co(II) in tetrahedral symmetry, while the ESR spectra and g-tensor components were consistent with Co(II) in a site of C_{3v} symmetry, i.e., a BR site. The situation for Ni(II) in ß"-alumina is not so well resolved [56]. The absorption spectrum of Ni(II) in ß"-alumina cannot fully be explained by a tetrahedral environment for Ni(II), and further study of this system is justified.

VIII. Conclusions

ß"-alumina is the only crystalline solid electrolyte known to the authors which conduct divalent and trivalent cations, as well as monovalent cations and various protonic species. It is nearly a universal solid electrolyte for cations and the host for a wide variety of fascinating solid state chemical phenomena: cation diffusion, electronic conductivity, defect formation, proton transport, ion pairing, mixed ion effects, formation of two-dimensional ionic 'alloys', order/disorder reactions, and many others.

Surely some of the most interesting ß"-aluminas are those that luminesce. Many different optically-active ions can be readily incorporated into the ß"-alumina structure by low-temperature ion exchange reactions. The facile ion exchange reactions of ß"-alumina make it an exceptionally versatile optical material in which the nature, concentration, and valence of the doping ions can be controlled simultaneously, as has been shown so elegantly for the copper and europium forms. In addition, ion exchange makes possible multiple-doping schemes which can be used both to modify the local crystal field around dopant ions and to bring about energy transfer among dopant ions.

The ß"-aluminas are not a single compound but a broad family of solid electrolytes. The Na(I) form was critical in launching the field of solid state ionics and has been central to the development of the Na/S battery. The authors hope, however, that this article has shown that the fortunes of the Na/S battery are only one aspect of the future of the ß"-aluminas. The unique ion exchange chemistry of the ß"-aluminas has yielded many new materials, the most interesting of which are those with potential for use in opto-electronic devices.

The field of solid state ionics has been traditionally concerned with electrolytes and electrochemistry. The ß"-aluminas expand the field into lasers and optical devices, technologies generally not associated with solid state ionics.

Nonetheless, these applications, like those of solid electrolyte batteries, take advantage of that very special property - fast ion transport in the solid state - that is the principal theme of solid state ionics.

Acknowledgments

Much of the work discussed in this chapter was supported by the U.S. Office of Naval Research. Additional support at the University of Pennsylvania from the NSF-MRL Program under grant no. DMR-8519059, and, at the University of Uppsala, from the Swedish Natural Science Research Council has been gratefully appreciated. We also are grateful for the the contributions of our students and colleagues: M. Aldén, A.J. Alfrey, J.D. Barrie, W. Carrillo-Cabrera, J.A. DeNuzzio, W.S. Frydrych, L.A. Momoda, P. Nelson, G.S. Rohrer, M.A. Saltzberg, O.M. Stafsudd and D.L. Yang.

References

1. Y.F.Y. Yao, and J.T. Kummer, J. Inorg. Nucl. Chem.., 29, 2453, (1967).
2. J.T. Kummer, Prog. in Solid State Chem.., 7, 141, (1972).
3. B. Dunn and G.C. Farrington, Mat. Res. Bull., 15 (1980) 1773
4. G.C. Farrington and B. Dunn, Solid State Ionics, 7 (1982) 267
5. G.C. Farrington, B. Dunn, and J.O. Thomas, Appl Phys.. A32 (1983) 159
6. Rankin and Merwin - see ref 4
7. G. Yamaguchi, Elect. Chem Soc. Japan. 11, 260, (1943).- Suzuki??
8. J. Théry and D. Briançon, Compt. Rend. 254, 2782, (1962)
9. N. Weber and A. Venero, presented at the 72nd annual meeting of the Am. Ceramic Soc., May 1970
10. M. Bettman and C.R. Peters, J. Phys. Chem.. 73(6), 1774, (1969)
11 DeVries R. and Roth, W. L., J. Solid State Chem. 20(2), 111, (1977)
12. Kennedy - review of beta - recent
13. K.G. Frase, J.O. Thomas, and G.C. Farrington, Solid State Ionics, 9&10, 307, (1983).
14. J.L. Briant and G.C. Farrington, J. Solid State Chem., 33, 385, (1980).
15. M. Aldén, Thesis, University of Uppsala, Uppsala, Sweden, 1985
16. M.A. Saltzberg, Thesis, University of Pennsylvania, Philadelphia, PA, 1988
17. B. Dunn, B.B. Schwarz, J.O. Thomas, and P.E.D. Morgan, Solid State Ionics, *in press.*
18. N.A. Toropov and M.M. Stukalova, Dokl. Akad. Nauk. SSSR. 24, 459, (1939).
19. S. Sattar, B. Ghosal, M.L. Underwood, H. Mertwoy, M.A. Saltzberg, W.S. Frydrych, G.S. Rohrer, and G.C. Farrington, J. Solid State Chem., 65 (1986) 231.

20. J. Tegenfeldt, M. Underwood, and G.C. Farrington, Solid State Ionics, 18/19 (1986) 668

21. D.L. Yang, B. Dunn, and P.E.D. Morgan, *submitted for publication*

22. B. Ghosal, E.A. Mangle, M.R. Topp, B. Dunn, and G.C. Farrington, Solid State Ionics, 9&10 (1983) 273.

23. G. Rohrer and G.C. Farrington, Solid State Ionics, *in press.*

24. W.L. Roth, F. Reidinger, and S. LaPlaca, in Superionic Conductors, W.L. Roth and G.D. Mahan, eds., (Plenum Press, New York),1976.

25. J.P. Boilot, G. Collin, P. Colomban, and R. Comes, Phys Rev B, 22(12), 5912, (1980).

26. R. Seevers, J. DeNuzzio, G.C. Farrington, and B. Dunn, J. Solid State Chem., 50 (1983) 146

27. J.O. Thomas, M. Aldén, and G.C. Farrington, Solid State Ionics, 9&10, 301, (1983).

28. J.O. Thomas, M. Aldén, G.J. McIntyre, and G.C. Farrington, Acta. Cryst., B40, 208, (1984).

29. J.P Boilot, P. Colomban, M.R. Lee, G. Collin, and R. Comes, Solid State Ionics, 9&10, 315, (1983).

30. G. Collin, J.P. Boilot, P. Colomban, and R. Comes, Solid State Ionics,18&19, 663, (1986).

31. W. Hayes, K. Kjaer, F.L. Pratt, and B. Schonfeld, J. Phys C, 18, L567,(1985).

32. W. Hayes, and F.L. Pratt, J. Phys. C, 19, 929,(1986).

33. W.S. Frydrych, Thesis, University of Pennsylvania, Philadelphia, PA, (1986).

34. W. Carrillo-Cabrera, J.O. Thomas, and G.C. Farrington, Solid State Ionics, 9&10, 245, (1983).

35. W. Carrillo-Cabrera, J.O. Thomas, and G.C. Farrington, Solid State Ionics, 18&19, 645, (1986).

36. W. Carrillo-Cabrera, J.O. Thomas and G.C. Farrington, Solid State Ionics, *in press.*

37. P.K. Davies, A. Petford, and M. O'Keeffe, Solid State Ionics, 18&19, 624, (1986).

38. F. Garzon and P.K. Davies, Solid State Ionics, *in press*

39. G.S. Rohrer and G.C. Farrington, *unpublished results*

40. M.A. Saltzberg, J.O. Thomas, and G.C. Farrington, *submitted for publication*

41. B. Dunn, D.L. Yang and D. Vivien, J. Solid State Chem., *in press*

42. R.B. Queenan and P. Davies, Solid State Ionics, *in press.*

43. M. Jansen, A.J. Alfrey, O.M. Stafsudd, D.L. Yang, B. Dunn and G.C. Farrington, Opt. Lett. 9, 119 (1984)

44. A.J. Alfrey, O.M. Stafsudd, D.L. Yang, B. Dunn, and L. Salmon, J. Chem. Phys., *in press*

45. A.J. Alfrey, O.M. Stafsudd, B. Dunn, D.L. Yang, L. Salmon and E.Y. Wong, *submitted for publication*

46. H. Dai, P. Nelson, and O.M. Stafsudd, *unpublished results*

47. R.W. Boyd, M.T. Gruneisen, P. Narum, D.J. Simkin, B. Dunn and D.L. Yang, Opt. Lett., 11, 162 (1986)

48. G. Blasse and A. Bril, J. Chem. Phys., 51, 3252 (1969)

49. L.A. Momoda, J.D. Barrie, B. Dunn and O.M. Stafsudd, *submitted for publication*

50. M.A. Saltzberg and G.C. Farrington, *submitted for publication*

51. M.A. Saltzberg, P.K. Davies and G.C. Farrington, Mat. Res. Bull., 21, 533 (1986).

52. M.A. Saltzberg, P.K. Davies, F.H. Garzon and G.C. Farrington, Solid State Ionics, in press.

53. G. Mariotto, M. Montagna and F. Rossi, J. Phys. C 19, 3029 (1986)

54. G. Mariotto, M. Montagna and F. Rossi, Solid State Ionics 18/19, 636 (1986)

55. R.C. Powell, G.E. Venikouas, L. Xi and J.B. Bates, J. Lumin. 37, 1 (1987)

56. J.D. Barrie, B. Dunn, O.M. Stafsudd and G.C. Farrington, Solid State Ionics, 18/19, 677 (1986)

57. J.D. Barrie, B. Dunn, O.M. Stafsudd and P. Nelson, J. Lumin., 37, 303 (1987)

58. P. Hammerling, A.B. Budgor and A. Pinto (eds), "Tunable Solid State Lasers", Springer series in Optical Sciences, Vol. 47 (Springer-Verlag, New York, 1985)

59. J.R. Akridge and J.H. Kennedy, J. Solid State Chem., 29, 68 (1979)

60. S. Chen, D.R. White, H. Sato, C.J. Sandburg and H.R. Harrison, Proc. Conf. on High Temperature Solid Oxide Electrolytes, F.J. Salzano, ed., (Brookhaven National Laboratory, Brookhaven, 1983) Vol. 2, p. 121

61. B. Dunn, J.D. Barrie and D. Vivien, J. Mater. Sci. Letts., 6, 679 (1987)

62. M. Aldén, J.O. Thomas, and G.C. Farrington, Acta Cryst., C40, 1763 (1984)

63. M. Aldén, J.O. Thomas, and G.C. Farrington, Acta Cryst., C41, 1700 (1985)

64. W. Carrillo-Cabrera, J.O. Thomas, and G.C. Farrington, Solid State Ionics, 17, 223 (1985)

FLUORIDE SOLID ELECTROLYTES: ION DYNAMICS

J. Schoonman
Delft University of Technology
Laboratory for Inorganic Chemistry
Julianalaan 136
2628 BL Delft
The Netherlands

D.R. Franceschetti
Department of Physics
Memphis State University
Memphis, Tennessee 38152
U.S.A.

ABSTRACT

Metal fluorides with the fluorite (CaF_2) structure are known to accept large amounts of LaF_3, rare earth (III) fluorides, ThF_4 and UF_4. The defect structures and dynamical behaviour of fluoride ions in these fluorite-type solid solutions have attracted considerable attention and the more recent developments are presented. Impedance spectroscopy has been used widely to study the mechanism underlying the bulk diffusion responsible for ionic conductivity. Interfacial and bulk impurity oxygen diffusion influences the small-signal a.c. response of fluorite-type solid solutions. Theoretical expressions are presented for the small-signal a.c. response. The defect structures of these types of solid solutions are dominated by defect association effects, extended defect clustering, ion size effects, and short range ordering. These phenomena often give rise to dielectric relaxation phenomena. The dynamical behaviour of bound fluoride ions has been studied by ionic thermocurrent (ITC) and a.c. dielectric loss (DL) techniques. Relaxation of macroscopic space charge, i.e. ionic conductivity, may contribute to the dielectric spectra. Useful information concerning diffusive motion can be obtained from the study of local motions of bound ionic defects.

INTRODUCTION

Metal fluorides with the fluorite structure exhibit anti-Frenkel disorder, i.e. anion vacancies and interstitials, and have promising solid electrolyte properties. Alkaline-earth fluorides and β-PbF_2 attain at high temperatures, starting roughly at 0.8 T_m where T_m is the melting point, ionic conductivity values that are typical for ionic melts. On going through this so-called Faraday transition high ionic conductivity values are due to fast fluoride ion conduction via vacancies and interstitials, and not to a liquid-like disorder in the anion array (1, 3).

In recent years the defect structure and the electrical properties of anion-excess fluorite-type solid solutions have engaged wide attention (1, 4-26). The isothermal conductivity of anion-excess solid solutions $M_{1-x}Me_xF_{2+x}$ and $M_{1-x}Me'_xF_{2+2x}$ (M = Ca, Sr, Ba, Pb, Me = rare earth, Me' = Zr, Th, U) increases more rapidly then linearly with solute content over appreciable ranges of x. At low and moderate temperatures the ionic conductivity of these solid solutions can easily be enhanced by eight orders of magnitude. As a consequence the temperature of the Faraday-transition at which the conductivity starts to level off to the values typical for molten salts are depressed substantially (18, 27, 28).

These fluoride ion solid electrolytes have been incorporated into electrochemical systems for thermodynamic studies, chemical sensing, and energy conversion and storage. Fluorite-type solid solutions have been utilized in thin-film galvanic cells (4, 29, 30). Although at ambient temperatures such all-solid state batteries can be applied for low-energy purposes, the performance characteristics improve substantially at elevated temperatures (31). Inevitable anode passivation can be depressed using alloy anodes (29, 30). Further improvements are to be expected if appropriate insertion electrodes become available.

Calcium fluoride has been conventionally used as a solid electrolyte to measure fluorine chemical potentials (32). However, its use as solid state membrane in oxygen and sulfur sensors has been described (33, 34). With the advance of microelectronic techniques in the manufacture of ion-selective electrodes, a growing demand for suitable solid-state contacts has emerged. Solid-state contacts are more compatible with microelectronic manufacturing than the conventional solution contacts. Such all-solid state fluoride electrodes are being developed.

Further research aimed towards optimization of all solid-state electrochemical systems is critically dependent on our insights into the mechanism of fast ionic conduction, the relations between chemical composition, defect structures, and optimized ionic conductivity, as well as interfacial polarization phenomena. Especially in these concentrated solid solutions the ionic conductivity is strongly influenced by defect association and clustering phenomena, ion size effects, short range ordering, and the formation of microdomains.
While the measurement of electrical conductivity represents a very sensitive technique to investigate ionic transport, other dynamic techniques such as nuclear magnetic resonance (NMR) relaxation, thermal depolarization - Ionic Thermal Current, ITC, and thermally stimulated (de)polarization current, TSP(D)C - and a.c. dielectric relaxation are currently employed to study diffusive and localized motions of point defects in the present solid solutions. In addition, these techniques provide information that contributes to unraveling the defect structure of these solid solutions.

It is the aim of the present review to high-light impedance spectroscopy, thermal depolarization phenomena, and mechanisms of ionic conductivity of these fluorite-type solid solutions.

SMALL-SIGNAL A.C. RESPONSE OF FLUORITES.

Small-signal a.c. response measurements can provide a great deal of insight into bulk and interfacial processes occurring in an electrode/electrolyte system. In a typical study the system impedance or admittance is measured over a wide range of frequencies, and an attempt is made to determine an equivalent linear electrical circuit, with a small number of circuit elements, which would accurately reproduce the observed response. The structure of the equivalent circuit frequently permits identification of the individual elements with such processes as bulk conduction, dielectric relaxation, double layer charging, and the adsorption of charged or neutral species. The dependence of the circuit elements on such parameters as temperature, pressure and impurity concentration provides further detailed information concerning the kinetics of the individual processes.

Attempts to derive system response functions and the corresponding equivalent circuits from a detailed microscopic model of the charge transport and interfacial processes for solid electrolytes date from the early 1950's (35, 36), and the most detailed treatment of the problem is to be found in the more recent work of Macdonald and coworkers (37-40). Fairly extensive studies of the small-signal a.c. response of a number of pure and doped fluorides have appeared, partly as a test of some of this theoretical work (9, 14, 18, 20, 41-44). The comparison of experiment and theory has led to some new theoretical developments (45, 46) which are outlined below.

Many fluorite-type fluorides are known to be subject to degradation when exposed to atmospheric oxygen, and the involvement of oxygen in ionic charge transport has been demonstrated by titration experiments using cells of the type Cu + $Cu_2O/CaF_2/Ni$ (47, 48). Considering the similar ionic radii of oxide and fluoride ions, 0.136 nm and 0.133 nm, respectively, and their identical electronic structure, it is not surprising that oxygen migrates readily via fluoride ion vacancies (41, 43, 48, 49). Oxygen may enter the fluorite lattice through reaction with atmospheric oxygen, with water vapour, or with oxygen adsorbed at an electrode. Possible chemical reactions of fluorite-type fluorides with the gas phase include

$$1/2O_2 + 2F_F^X \rightleftarrows V_F^{\cdot} + O_F' + F_2 \quad (1)$$

and

$$H_2O + 2F_F^X \rightleftarrows V_F^{\cdot} + O_F' + 2HF \quad (2)$$

while electrochemical redox reactions include

$$V_F^{\cdot} + O_{ads} + 2e' \rightleftarrows O_F' \quad (3)$$

$$V_i^X + O_{ads} + 2e' \rightleftarrows O_i'' \quad (4)$$

and

$$F_F^X + O_{ads} + e' \rightleftarrows O_F' + 1/2F_2 \quad (5)$$

The small-signal a.c. response of dilute and concentrated solid solutions has attracted widespread attention. Nominally pure β-PbF_2 has been studied in depth. Raistrick et al. (50) found that the small-signal a.c. response of polished, nominally pure β-PbF_2 crystals with sputtered Pt or Au electrodes in inert ambients could be described by an equivalent circuit with a single RC parallel combination in series with an element of impedance

$$Z = A\omega^{-\alpha} - iB\omega^{-\alpha} \quad (6)$$

The parameters A, B, and α in Eq. (6) cannot, however, be regarded as independent since the Kramers-Kronig relation leads to the constraint B = A.tan ($\alpha\pi/2$). With this restriction, Eq. (6) becomes the constant-phase-angle (CPA) element Q with impedance

$$Z_Q = K(i\omega)^{-\alpha} \quad (7)$$

with $A = K.\cos(\alpha\pi/2)$ and $B = K.\sin(\alpha\pi/2)$. The numerical data of Raistrick et al. (50) for Au electrodes can be fit to the equation $\log A = (1.46 \pm 0.21) \log B - (4.19 \pm 1.46)$, which is not consistent with this important constraint. The data of Jonscher and Réau (51) for pressed pellets of β-PbF_2 in contact with Au are concordant with this constraint.

Schoonman et al. (41) have examined the small-signal a.c. response of pure β-PbF_2 and of $Pb_{1-x}Bi_xF_{2+x}$ ($x \leq 0.05$) solid solutions. A sample impedance plot of $Pb_{1-x}Bi_xF_{2+x}$ in an oxygen containing ambient is shown in Figure 1.

The data were first analyzed with the equivalent circuit of Figure 2. This circuit, which includes a single diffusional impedance, Z_D, of the finite-length Warburg form

$$Z_C = Z_{Do}(i\omega H^2)^{-1/2}\tanh[(i\omega H^2)^{1/2}] \tag{8}$$

was derived by Franceschetti and Macdonald (54) for a solid electrolyte in which one of the mobile charge carrying species reacts at the electrode to form a neutral species which then diffuses into either the electrode or the electrolyte. Here $H = \delta(D)^{-1/2}$, where δ denotes a diffusion length, and D a diffusion constant. Although satisfactory fitting results were obtained with this circuit, more accurate fits were achieved when the finite length Warburg element was replaced with a CPA element, Eq. (7).

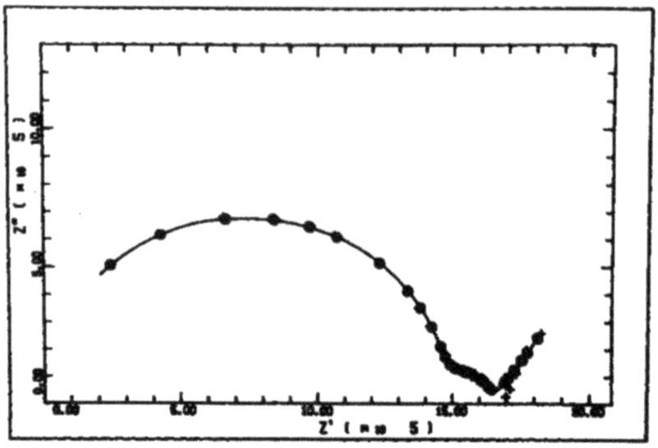

Figure 1. Impedance spectrum for the cell $Pt/Pb_{1-x}Bi_xF_{2+x}/Pt$ (x = 0.013) at 292 K. P_{O_2} = 10 Pa. Frequency range 1 Hz - 45 kHz. Cell constant 64.096 m^{-1}. + experimental data. o equivalent circuit fitting (see text).

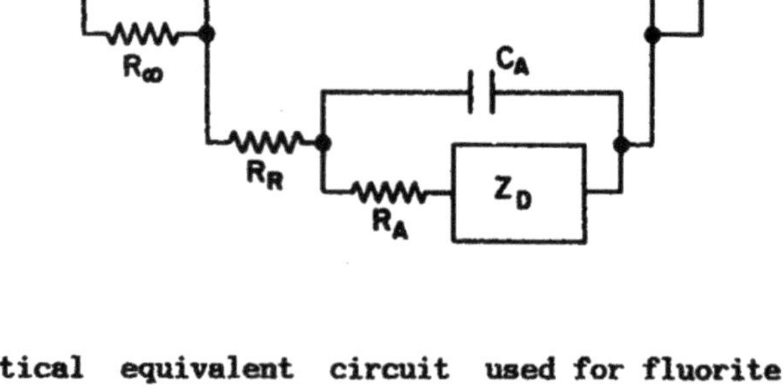

Figure 2. Theoretical equivalent circuit used for fluorite impedance data analysis. C_g = geometric capacitance, R_∞ = bulk resistance, C_R = (diffuse double layer capacitance - C_g), R_R = electrode reaction resistance, R_A, C_A = adsorption process elements, Z_D = diffusional impedance.

Consideration of the possible electrode mechanisms, Eqs. (3) - (5), indicates the need for a modification of the circuit of Figure 2 if it is to accurately describe the kinetics of electrode reactions involving impurity oxygen, since the flux of fluoride ions or vacancies towards the electrode would be dependent on two distinct oxygen transport processes, the transport of oxygen through the electrode, perhaps along the pores or internal surfaces of a sputtered or paint electrode, and the transport of oxygen through the crystal. The theoretical analysis of this case, as developed by one of the present authors (45), is based on the linearized electrode reaction rate equations.

$$J_p = k_1 p_F - k_2 n_{Ox} - k_e \Gamma_{Ox} \tag{9}$$

for the flux of vacancies out of the crystal towards the electrode, and

$$J_{Ox} = k_4 \Gamma_{Ox} - k_5 c_{Ox} \tag{10}$$

for the flux of oxygen atoms away from the crystal and into the electrode. These flux equations correspond most directly to the reaction of Eq. (3), but treatments based on Eqs. (4) or (5) should lead to formally similar results. Here p_F and n_{Ox} are the excess concentrations of fluoride ion vacancies and oxide ions, respectively, near the crystal surface, Γ_{Ox} is the concentration of adsorbed oxygen at the electrode/electrolyte interface, and c_{Ox} is the concentration of oxygen atoms inside the electrode near the crystal surface. It was further assumed that the motion of oxygen occurred by diffusion, both within the cyrstal (diffusion constant $D_{Ox,c}$, possibly describing the motion of neutral $(V_F O_F)^x$ pairs), and within the electrode (diffusion constant $D_{Ox,e}$).

The theoretical treatment leads to the equivalent circuit of Figure 3 for the interfacial impedance. The full circuit includes C_g, R_∞ and C_R (viz. Figure 2). The circuit parameters, which are associated with the electrode reaction and oxygen transport, are related to the rate constants, k_1, ..., k_5, the equilibrium concentration of fluoride ion vacancies P_{Fo} and the thickness l of the crystal and l_e of the electrode according to

$$R_R = kT/(4e^2 P_{Fo} k_1) \tag{11}$$

$$R_A = k_3 R_R / k_4 \tag{12}$$

$$C_A = (k_3 R_R)^{-1} \tag{13}$$

$$Z_{D,c} = Z_{EW,c} (i\omega)^{-1/2} \tanh[(i\omega H_c^2)^{1/2}] \tag{14}$$

and

$$Z_{D,e} = Z_{EW,e} (i\omega)^{-1/2} \tanh[(i\omega H_e^2)^{1/2}] \tag{15}$$

with

$$Z_{EW,c} = k_2 R_R (D_{Ox,c})^{-1/2}$$

$$Z_{EW,e} = (k_3 k_5 R_R / k_4)(D_{Ox,e})^{-1/2}$$

$$H_c = l(2D_{Ox,c})^{-1/2} \text{ and } H_e = l_e (D_{Ox,e})^{-1/2}.$$

The equivalent circuit of Figure 3 was used to analyze several sets of admittance data for the solid solutions $Pb_{1-x}Bi_xF_{2+x}$ with Pt paint electrodes over the frequency range 10^{-4} Hz to 50 kHz for temperatures between 290 K and 720 K. Arrhenius plots of the circuit parameters $Z_{EW,c}$, R_∞, $Z_{EW,e}$ and H_e^2 are given in Figures 4 and 5. These plots yielded activation enthalpies of 0.34, −0.05, 0.24, 0.35, 0.32 and 0.59 eV, respectively, for $Z_{EW,e}$, H_e^2, C_A, $Z_{EW,c}$, C_R and R_∞. Although it was possible to determine a value for H_e^2 at each temperature, the data did not extend to low enough frequencies to permit an accurate determination of H_c^2.

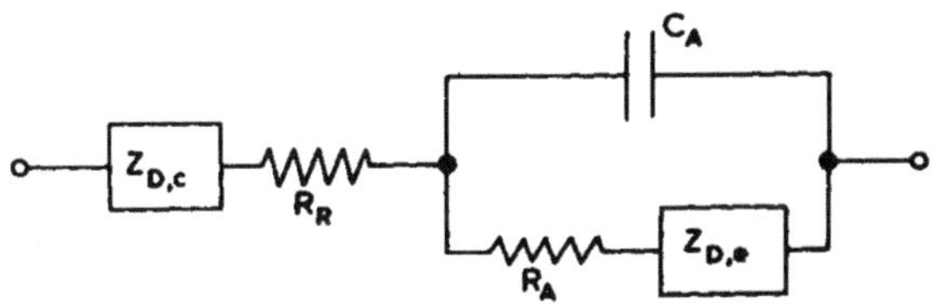

Figure 3. Theoretical equivalent circuit derived on the basis of reaction (3). $Z_{D,e}$ ≡ diffusion in electrode, $Z_{D,c}$ ≡ diffusion in crystal.

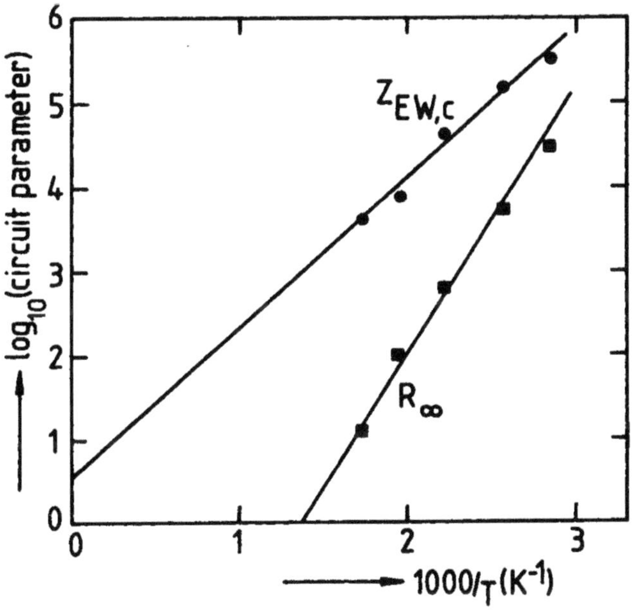

Figure 4. Temperature dependence of the circuit parameters $Z_{EW,c}$ and R_∞ for $Pb_{0.987}Bi_{0.013}F_{2.013}$ with paint electrodes.

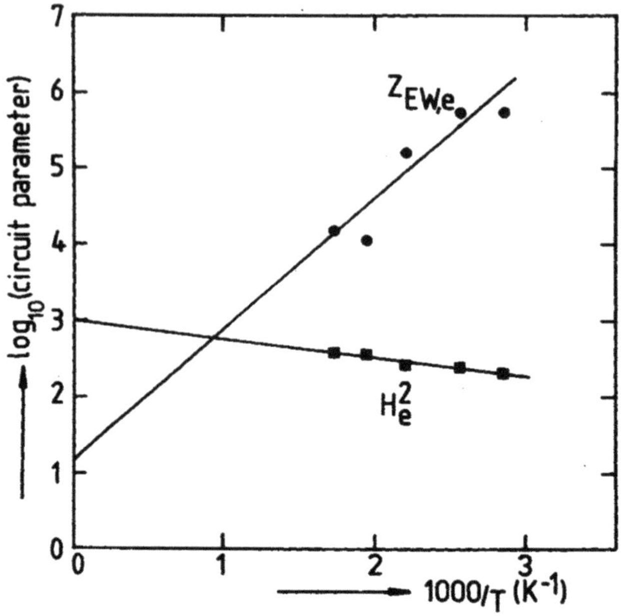

Figure 5. Temperature dependence of the circuit parameters $Z_{EW,e}$ and H_e^2 for $Pb_{0.987}Bi_{0.013}F_{2.013}$ with Pt paint electrodes.

In order to confirm the role of oxygen in the electrode mechanism, measurements were made at various oxygen partial pressures from 1.46 Pa to 1623 Pa. The circuit Warburg parameters were found to vary monotonically with P_{O_2} according to $Z_{EW,e} \propto P_{O_2}^{-1/8}$ and $Z_{EW,c} \propto P_{O_2}^{+1/16}$, while the adsorption related resistance displayed an approximate $P_{O_2}^{+1/4}$ dependence. The remaining parameters displayed little or no dependence on the oxygen pressure. Couturier et al. (43) have studied the reaction of thin β-PbF_2 films with oxygen. Besides oxygen analysis by nuclear methods and ESCA, they have analyzed the frequency dispersion using a simple parallel RC combination for the interface in series with the bulk resistance. The interfacial cpacitances thus obtained decreased with oxygen pressure and started to level off beyond partial pressures of 130 Pa. The bulk resistance decreased in oxygen. After storage of 1000 hr at room temperature the temperature dependence of the ionic conductivity (R_∞^{-1}) clearly revealed enhanced fluoride ion vacancy conduction, the enhancement being attributed to vacancies created via reaction (1).

Redox reaction (3) has recently been utilized by Chou and Rapp (48) to account for oxygen uptake by CaF_2. They have employed several equivalent circuits to model admittance spectra of CaF_2 with Pt foil contacts. For admittance spectra recorded in oxygen purged argon they have employed the equivalent circuit as introduced by Couturier et al. (43). The resistance in parallel to the interface capacitance is quite large indicating almost ideal blocking behaviour. Parts of admittance spectra recorded in oxygen ambient could be modelled with the Randles circuit with an ideal Warburg impedance Z_W. In the Randles circuit the interface capacitance is shunted by a series combination of a charge-transfer resistance and Z_W. As in aqueous electrolytes (53, 54) Chou and Rapp consider a field-free diffusion process of oxygen in CaF_2 in the presence of fluoride ion vacancies acting as the supporting electrolyte. Ideal Warburg response would lead to a 45° slope line in the low-frequency regime of an impedance spectrum, contrary to experimental data which indicate slopes of 48° and 55°. In addition, the impedance data below 100 Hz deviate from the straight line. A bending over toward the real axis in the impedance diagram seems to indicate that at lower frequencies the a.c. diffusion layer becomes comparable with the thickness of the electrolyte. Incidently, this bending over has also been observed for β-PbF_2 in small oxygen pressures. However, for this system the bending over starts below 25 mHz (41).

A recent application of the non-linear complex least squares minimization, as has been used in the aforementioned data analysis of $Pb_{1-x}Bi_xF_{2+x}$ (e.g. Figure 1), reveals that the Randles circuit cannot adequately describe the reported (48) admittance/impedance spectra of the system $Pt(O_2)/CaF_2$. Equivalent circuit fitting using the CPA impedance with for α values in between 1/2 and 1 yielded better fits to the data.

A Randles-type circuit in which both the interface capacitance C and Z_W have been replaced by a CPA impedance was found to adequately describe the small-signal a.c. response of the fluorite-type solid solutions $Ba_{1-x}La_xF_{2+x}$ (52). The CPA impedance replacing C is nearly capacitive [$\alpha \simeq 0.9$ (Eq.7)]. This deviation from an ideal capacitance has been attributed to current inhomogeneities at the surface, probably associated with surface roughness. Values for the exponent α for the CPA element replacing Z_W were found to lie in the range of 0.5 to 0.6. Since the CPA element is equivalent to a Warburg impedance when α = 0.5, this branche of the circuit is attributed to a diffusional process, quite probably involving impurity oxygen as was the case for the other fluorites. The Randles-type circuit can thus be viewed as related to those in Figures 2 and 3.

As can be seen from the above discussion of CaF_2, BaF_2, β-PbF_2 and their solid solutions, the interfacial phenomena of fluorite materials with impurity oxygen diffusion can be quite complex, and may require a rather complicated equivalent circuit for adequate representation. In order to obtain useful estimates of circuit parameters the availability of nonlinear complex least squares fitting is a prerequisite.

Although such conventional plots as parallel capacitance and conductance vs. frequency, and/or admittance or impedance spectra are useful in illustrating some aspects of the small-signal a.c. response of solids and liquids, they are two-dimensional representations of three-dimensional phenomena. Three dimensional perspective plotting and fitting of impedance and admittance data illustrates directly the interdependences between the three variabels: real part, imaginary part, and frequency. In addition, the three planar projections included show the associated three relationships between any two of these variables. Macdonald et al. (55, 56) have illustrated the usefulness of such plotting for synthetic data derived from given equivalent circuits, and for data obtained from measurements on lumped-constant circuits and on ionic conductors. An illustrative example for β-PbF_2 with Pt paint electrodes is given in Figure 6.

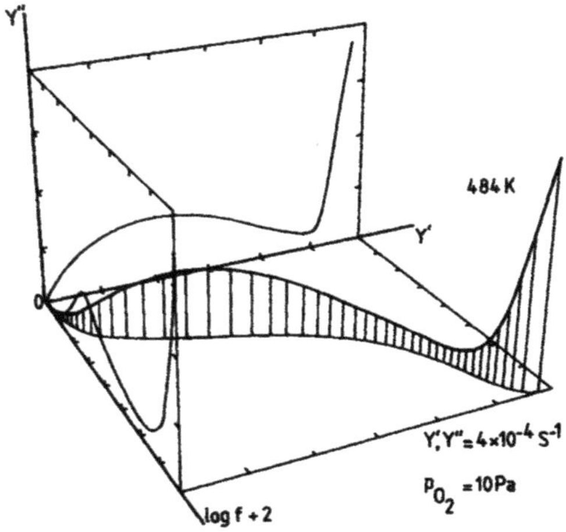

Figure 6. 3-D admittance spectrum of Pt/β-PbF_2/Pt at 484 K. Frequency range 10^{-2} Hz to 1 MHz.

Ouwerkerk and Schoonman (18, 57) have employed 3-D plotting, and equivalent circuit fitting to obtain accurate high-temperature bulk ionic conductivities of the fluorite-type solid solutions $Ba_{1-x}U_xF_{2+2x}$ ($x \leq 0.15$) and $Pb_{1-x}U_xF_{2+2x}$ ($x \leq 0.10$). While at elevated temperatures the low-frequency regimes of the admittance spectra indicate residual oxygen to take place in diffusion processes at the electrode-electrolyte interface, the high-temperature conductivity data were mainly used to obtain a relation between depressed Faraday transition temperatures and solute content.

Recently, Ivanov-Shits et al. (58) have studied the electrical properties of SrF_2- and BaF_2 based solid solutions with all rare-earth trifluorides using impedance spectroscopy. All the immittance spectra recorded in oxygen free ambients could be modelled with a simple equivalent circuit comprising $R_bQ_1pQ_2s$, where p and s stand for parallel and series, respectively. The bulk response, R_bQ_1p, represents the ionic resistance in parallel to bulk dielectric relaxation processes. Q_2 represents the interfacial electrode polarization phenomena. The bulk relaxation processes arise from extended defect association and clustering, characteristic for especially the concentrated fluorite-type solid solutions.

THERMAL DEPOLARIZATION PHENOMENA.

The Thermally Stimulated Depolarization Current (TSDC) technique, often referred to as Ionic Thermal Current (ITC) technique, has become widely used since its introduction by Bucci and Fieschi (59), and in recent years many studies have been reported, which describe TSDC results of the present fluorite-type solid solutions (22, 26, 60-69).

In TSDC the sample with bound charge is polarized at a temperature T_p sufficiently high so that the relaxation time T of the dipoles is short enough for the polarization associated with the dipoles to reach an equilibrium value P_o in a relatively short time. Then the sample is cooled down to low temperatures, and the external polarization field is switched off. Upon heating at a constant rate the reorienting dipoles cause depolarization currents, which can be analyzed to yield the number of polarizable species, their relaxation times, the reorientation energies, and their concentration, provided that the dipole moments are known. (61).
Bucci and Fieschi (59) have developed a set of equations describing the depolarization current due to N_d reorienting non-interacting dipoles per unit volume, each with a dipole moment μ, and reorientation energy E. The dipoles reorient with a unique characteristic relaxation time

$$\tau(T) = \tau_0 \exp(E/kT) \tag{16}$$

where τ_0 is the characteristic dipole relaxation time. If one has a partially oriented state at a temperature T where the sample has a polarization P(T), and is heated with a linear rate b in the absence of an electric field then the depolarization current density is given by

$$I(T) = P_0/\tau_0 \exp(-E/kT) \exp[-1/b\tau_0 \int_0^T \exp(-E/kT')dT'] \quad (17)$$

For the equilibrium polarization P_0 at the polarization temperature T_p the relation

$$P_0 = \frac{N_d\mu^2 E_p \alpha}{kT_p} \text{ for } kT_p >> \mu E_p \quad (18)$$

holds, where k is the Boltzmann constant, and α a geometrical factor. For the fluorites α is 1/3 (70). Eq. (17) represents an asymmetrical curve. The first exponential dominates in the low-temperature region, and is responsible for the initial increase of the current with temperature, while the second exponential, which dominates in the high-temperature region, slows gradually the current rise down, and subsequently depresses it rapidly. Lenting et al. (60), Laredo et al. (70), and Roos et al. (71) have observed that experimental TSDS spectra and the monoenergetic Bucci and Fieschi theory, i.e. Eq. (17) disagree systematically, a disagreement which is well outside the experimental uncertainty. Broadening of the TSDC relaxation peaks as a function of the dipole concentration has been observed by Shelly and Miller (72) in powdered samples of NaF-doped CaF_2, and is ascribed to dipole-dipole interactions. Likewise, it has been found that agreement between experimental and calculated TSDC spectra is greatly improved by assuming a Gaussian distribution of the reorientation energy to account for electrostatic interactions between the dipoles, and contributions from elastic lattice deformations (60).

A refined approximation to the TSDC equation, taking into account peak broadening by assuming a Gaussian distribution D(E) of width σ has been presented by Laredo et al. (70), and Roos et al. (71). If P(E,T)dE is the polarization of the sample at temperature T due to dipoles with reorientation energies between E and E + dE, than

$$P(E,T)dE = P(T)D(E)dE \quad (19)$$

Reorientation of these dipoles contributes to the current at temperature T

$$I(T,E)dE = \frac{P(E,0)}{\tau_0} dE \exp(-E/kT) \exp[\frac{-1}{b\tau_o} \int_0^T \exp(-E/kT')dT'] \quad (20)$$

The total current produced by the sample at temperature T can be found by integrating Eq. (20), using

$$P_p = \int_{-\infty}^{+\infty} P(E,0)dE \quad (21)$$

and a Gaussian energy distribution of the form

$$D(E) = \frac{1}{\sigma\sqrt{2\pi}} \exp[\frac{-(E-E_o)^2}{2\sigma^2}] \quad (22)$$

which yields

$$I(T) = \frac{P_o}{\tau_o\sigma\sqrt{2\pi}} \int_{-\infty}^{+\infty} \exp(-E/kT)\exp[\frac{-(E-E_o)^2}{2\sigma^2}]\exp[\frac{-1}{b\tau_o} \int_0^T \exp(-E/kT')dT']dE \quad (23)$$

Usually experimental TSDC spectra are fit either to a three-parameter TSDC equation without energy distribution [Eq. (17)], or to a four-parameter TSDC equation with energy distribution [Eq. (23)] (70, 71). It has been shown by Roos et al. (71) that in the latter case it is sufficient to limit the integration boundaries to $E_o \pm 5\sigma$, which considerably reduces the computing time.

In addition to dipolar relaxations the migration of free ionic charge carriers towards the blocking electrodes, i.e. space charge build-up, causes a large TSDC peak to appear. Its relaxation time is controlled by the defect conduction mechanism, and is given by

$$\tau = \frac{\varepsilon\varepsilon_0}{\sigma_i} \quad (24)$$

where σ_i is the ionic conductivity, ε_0 the permittivity of free space, ε the dielectric constant, while for the ionic conductivity the Arrhenius equation

$$\sigma_i = \frac{\sigma_o}{T} \exp (-\Delta H_\sigma/kT) \tag{25}$$

is used.

Here σ_o contains the concentration of the conducting defects, their charge, the jump distance, the attempt frequency and the migration entropy. ΔH_σ represents the migration enthalpy. One now obtains an expression for the temperature of the maximum of this space charge peak T_{max} by differentiating the conventional expression for the depolarization current [Eq. (17)], and inserting equations (24) and (25). The result is (64, 73),

$$\frac{\Delta H_\sigma}{kT_{max}} = \frac{\sigma_o}{b\varepsilon\varepsilon_o} \exp (- \frac{\Delta H_\sigma}{kT_{max}}) + 1 \tag{26}$$

This equation enables the calculation of ΔH_σ from T_{max} data at each composition. The calculated values can then be compared with the experimental values of ΔH_σ from ionic conductivity measurments at elevated temperatures.

Ionic conductivity measurements on fluorites are generally interpreted in terms of mobile fluoride ion vacancies and interstitial fluoride ions. These defects are thermally generated according to an anti-Frenkel mechanism. The fluoride interstitials migrate via an indirect or insterstitialcy mechanism.

Upon incorporating trivalent rare earth ions Me^{3+}, and tetravalent dopants Me^{4+} the extrinsic concentration of fluoride interstitials is increased.

$$MeF_3 \rightarrow Me_M^{\cdot} + 2F_F^{x} + 2F_i' \tag{27}$$

and

$$Me'F_4 \rightarrow (Me')_M^{\cdot\cdot} + 2F_F^{x} + 2F_i' \qquad (28)$$

The defect structure of dilute and concentrated solid solutions have been studied in great detail by many groups. For CaF_2, SrF_2 and BaF_2 there exists general consensus as to the presence of simple neutral nearest neighbour (nn, tetragonal symmetry C_{4v}) and next-nearest-neighbour (nnn, trigonal symmetry C_{3v}) associates $(Me_M F_i)^x$. Theoretical and experimental studies have revealed that the ratio of the nn to nnn associates strongly depends on the lattice parameter of the host lattice : nn associates appear to dominate in CaF_2, while nnn associates dominate in BaF_2. In SrF_2 the type of associate depends on the size of the rare earth ion, the nn associate being present in crystals doped with the light rare earths, and nnn associates in crystals doped with the heavy rare earth ions. The thermal depolarization phenomena of the dilute anion-excess solid solutions have been the subject of many investigations. Eq. (17) is used to fit the TSDC spectra in order to obtain the dipolar relaxation parameters. Usually resonable fits are obtained for low dopant levels. With increasing dopant levels line broadening occurs. Eq. (23) can then be used to fit the TSDC peak. A recent example from a TSDC study of solid solutions in the system BaF_2-UF_4 is presented in Figure 7.

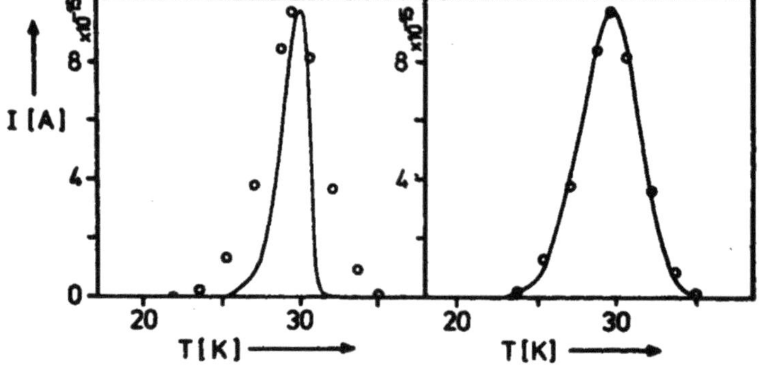

Figure 7. TSDC relaxation peak of $Ba_{0.97}U_{0.03}F_{2.06}$. (a) Computer fitting result using Eq. (17) and (b) using Eq. (23).

In Table I some reorientation energies E of nn and nnn associates are compared with the migration enthalphies of fluoride interstitials in dilute fluorite matrices.

Table I

Reorientation energies E_{nn} and E_{nnn}, and migration enthalpies $\Delta H_m(F_i')$ for some anion-excess solid solutions.

Solid Solution	E_{nn} (eV)	E_{nnn} (eV)	$\Delta H_m(F_i')$ (eV)
$Ca_{1-x}La_xF_{2+x}$	0.475	-	0.79
$Ba_{1-x}Y_xF_{2+x}$	0.45	0.56	0.714
$Ba_{1-x}La_xF_{2+x}$	0.391	0.544	0.714
$Ba_{1-x}U_xF_{2+x}$	0.38	0.51	0.714
$Sr_{1-x}La_xF_{2+x}$	0.444	-	0.74

It is generally agreed that the bound motion of fluoride interstitials can also be described by the interstitialcy conduction mechanism. The results in Table I then clearly reveal that the local motion of bound fluoride interstitals is greatly perturbed by the presence of aliovalent lattice disturbing defects. Insight into the dynamical behaviour of these bound ionic defects is extremely useful in understanding the mechanism of diffusive ionic motion in the more concentrated solid solutions. This will be delt with in the next section.

With increasing dopant content extensive defect clustering occurs. The structures of the different defect clusters as well as dipolar relaxations of bound fluoride interstitals in these clusters have been the subject of many investigations. For the more complex defect clusters reorientation energies have been found in the range 0.04 to 0.40 eV, i.e. well below the reorientation energy values of the nnn and nn dipoles given in Table I, as can be expected when more complex arrangements of dopant ions and fluoride interstitials are formed.

Usually TSDC measurements are performed starting at liquid nitrogen temperature (LNT). With regard to the numerical values of the reorientation energies of interstitials in extended defect clusters it is obvious that extended and polarizable defect clusters will reorient well below LNT. In fact, for $La_{1-x}Ba_xF_{2+x}$ (x = 0.133) a reorientation peak occurs at 21 K. This depolarization phenomenon has been ascribed to a cis→trans relaxation of the relaxed lattice fluoride ion above and below the plane of the well-known 222-cluster (17, 65). For the solid solutions $Ba_{1-x}U_xF_{2+2x}$ ($0.0001 \leq x \leq 0.02$) a reorientation peak shows up at 29.7 K. This relaxation can be ascribed to a 212-cluster (65). Because this cluster contains only one dopant ion it occurs if a neutral $(U_{Ba}^{\cdot}.2F_i^{\prime})^x$ associate is formed. Matar et al. (75) proposed for $Ca_{1-x}U_xF_{2+2x}$ solid solutions a defect cluster, comprising a dopant ion, three fluoride interstitials and one fluoride ion vacancy. This so-called trimer with a dipole moment in the [111] direction can be formed from (212)-clusters with a dipole moment in the [110], [101] and [011] direction in a single step as is shown in reverse in Figure 8 for [111]→[110] (65).

In an unpolarized $Ba_{1-x}U_xF_{2+2x}$ crystal the 212-clusters are oriented in twelve different directions, with the two fluoride interstitals in the plane of the cluster pointing at each of the twelve Ba^{2+} ions around a cluster U^{4+} ion. The dipole moments of the clusters will be in the [110], and all related directions, such as the [011] direction, the [$\bar{1}$01] direction, etc. To polarize a 212-cluster with its dipole moment in the [110] direction to a trimer with its dipole moment in the [111] direction takes three steps. The subsequent depolarization steps are depicted in Fig. 8. The last two steps (Figs. 8b and 8c) are in fact identical, and therefore they will have the same activation energy. The first step, from a trimer with its dipole moment in the [111] direction to a 212-cluster with its dipole moment in the [110] direction consists of only small and practically unhindered displacements of four fluoride ions. The activation energy for this reorientation is very low. Therefore, it does not show up as a distinct relaxation peak in the TSDC scan of a $Ba_{1-x}U_xF_{2+2x}$ sample, because this type of depolarization of the trimer will already take place when the electrical field is switched off at liquid helium temperature. It can therefore be concluded, that if the polarized state of a 212-cluster is a trimer, the activation energy for reorientation will be less than 0.04 eV.

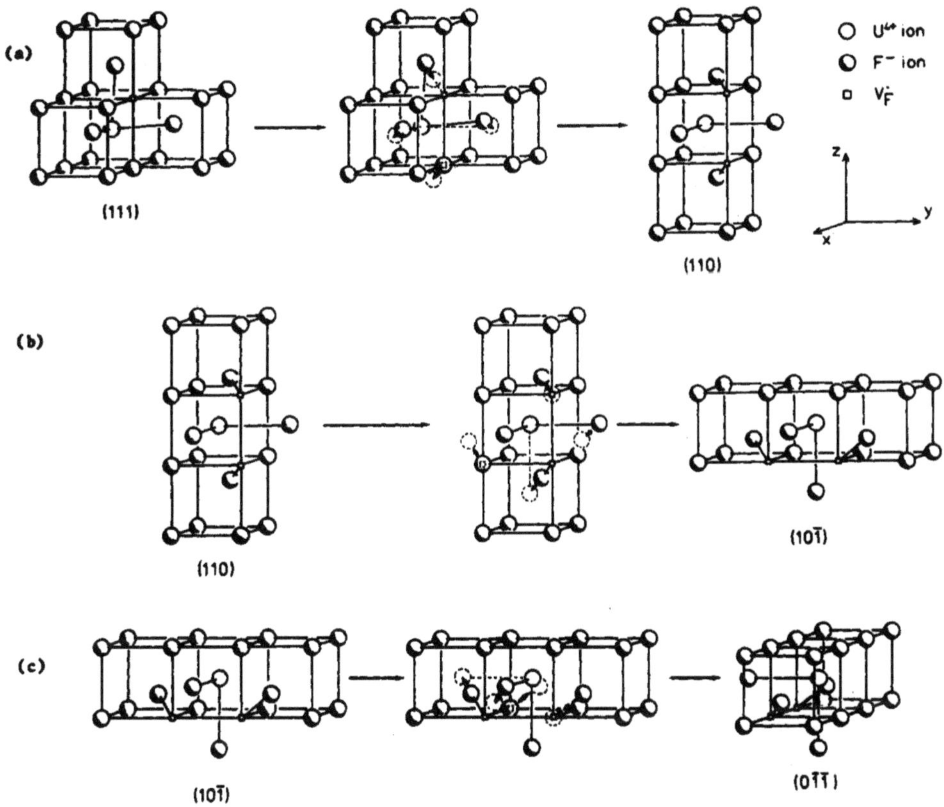

Figure 8. The reorientation of a trimer having its dipole moment in the [111] direction, to a 212-cluster with its dipole moment in the [110] direction.

The reorientation peak at 29.5 K for $Ba_{1-x}U_xF_{2+2x}$ solid solutions has a computed activation enthalpy of 0.09 eV, and is the most prominent feature in the TSDC scan, apart from the space-charge peak. Because the steps as depicted in Figs. 8b and 8c require the same activation enthalpy, they will both contribute to the depolarization current. Hence, the 29.5 K peak is assigned to this type of 212-cluster reorientations. In Fig. 9 a scheme of all possible polarization steps with $\Delta E = 0.09$ eV is pictured, along with the fractions of the total 212-cluster concentration involved in those steps.

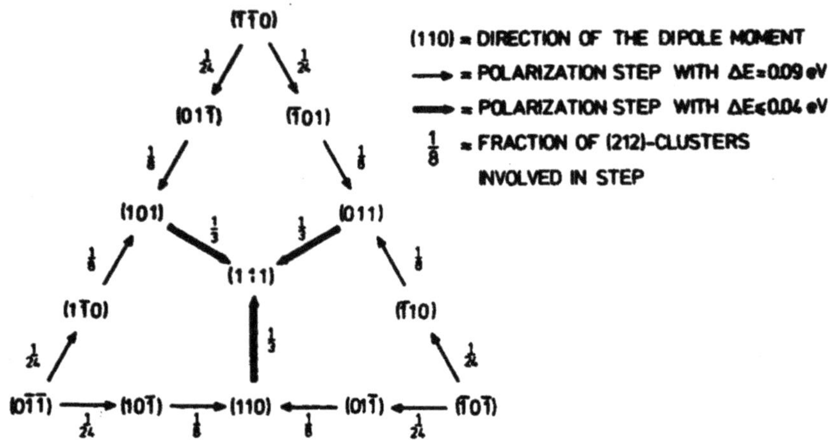

Figure 9. A schematic representation of the polarization steps for 212-clusters in $Ba_{1-x}U_xF_{2+2x}$

It is obvious that the unraveling of all possible dipolar-type relaxations in dilute and concentrated solid solutions indeed requires TSDS scans down to liquid helium temperature.

TSDC studies on a variety of fluorite-type solid solutions with defect clusters have been reported in the recent literature and reviewed by Schoonman (74).

IONIC CONDUCTIVITY

The electrical properties of dilute fluorite-type solid solutions have been studied for a long time (76, 77). The defect chemical parameters governing the thermal generation and migration of the anti-Frenkel defects have been documented in detail in the literature for the alkaline earth fluorides and cubic lead fluoride.

In recent years, particular attention has been focused on the electrical properties of the concentrated anion-excess solid solutions $M_{1-x}Me_xF_{2+x}$ and $M_{1-x}Me'_xF_{2+2x}$, because at low and moderate temperatures their ionic conductivities can be enhanced by eight orders of magnitude, despite the fact that extended defect clustering occurs (1, 4-26, 58, 78). Detailed studies of the concentrated solid solutions reveal that isothermal conductivities increase exponentially with solute content, while the conductivity activation enthalpy decreases linearly. Based on the trends given in Table I, these conductivity effects have been accounted for in the Enhanced Ionic Motion (EIM) model (5), which is based on the presence of a large number of lattice disturbing defects, giving rise to a broad distribution of interstitial defect jumps with slightly different activation enthalpies. This quasi-continuous distribution governs the conduction mechanism.

Extensive TSDC studies by Den Hartog et al. (26, 27, 79) have been explained in terms of a percolation model assuming that highly conductive inclusions are formed in the solid solution matrix. For high solute levels the conductive regions overlap, thus giving rise to the percolation phenomenon, responsible for ionic conduction. According to Den Hartog et al. extended defect clusters occur only in $Sr_{1-x}Me_xF_{2+x}$ solid solutions with Me = Er-Lu, and Y. For all other rare earth dopants simple dipoles $(Me_{Sr}F_i)^x$ constitute the disorder for x up to 0.4.

Recently, Ivanov-Shits et al. (44, 58, 78) have systematically studied the temperature and concentration dependence of the ionic conductivity of numerous single crystals of solid solutions based on SrF_2 and BaF_2 with all rare earth trifluorides. For solute contents $0 < x \leqslant 0.40$ the ion transport parameters are explained within the framework of the "defect regions" model (80). A defect region consists of a core comprising rare-earth dopant ions and immobilized fluoride ions, for instance a defect cluster, and of a periphery of distorted fluorite lattice in which mobile fluoride interstitals may be present, and which are bound to the core. Thus, the defect region has a composition and a local crystal structure, which differs from the fluorite matrix. Their number increases with increasing solute content and at a certain concentration x_p a spatial overlap of the defect regions occurs. The formation of a closed three-dimensional network of the defect regions corresponds to the percolation threshold, and is accompanied by changes in electrical, thermal and mechanical properties.

The defect regions formed in the solid solutions $M_{1-x}Me_xF_{2+x}$ and $M_{1-x}Me'_xF_{2+2x}$ will differ not only in their conduction values, but also in their geometric dimensions. If it is assumed that all defect regions in one system are identical and distributed statistically in the solid solution than the volume V of the defect regions can be estimated in terms of the percolation conductivity model using the relation (78)

$$V = \frac{N_d b_p a^3}{4x_p} \tag{29}$$

where N_d is the number of solute ions in the defect region, b_p the volume fraction of the defect regions in the solid solution, a the lattice parameter, and x_p the percolation threshold determined from conductivity data. Eq. (29) was deduced making usc of the fact that the number of formula units in the fluorite unit cell is 4 (78). From percolation theory it follows that b_p has a value of 0.25 to 0.30 (81).

For the dilute solid solutions $Sr_{1-x}Me_xF_{2+x}$ with Me = La, Ce, Pr, Nd ($x < x_p = 0.12-0.15$) Ivanov-Shits et al. (58) observed a pronounced increase of the isothermal ionic conductivity. In this region nearest neighbour associates $(Me_{Sr}F_i)^x$ are assumed to be present, while as a result of dipole-dipole interactions and variations in the lattice elastic deformation energy a quasi-continuous distribution of the reorientation enthalpy of these associates occurs. The conduction mechanism is based on reorienting associates.

For $x > x_p$ the following heterovalent substitution scheme has been proposed (58):

$$(Sr_4F_{23})^{15-} \rightarrow (Me_4F_{26})^{14-} + F_i^{1-}(ext) \quad (30)$$

The defect region is formed by the Me_4F_{26} cluster, while F_i' is assumed to be located in the periphery, and hence mobile. The concentration dependence of the ionic conductivity is then described by the EIM model (5):

$$\sigma T = \sigma_o \exp[-(\Delta\bar{H}_m - \frac{p^2}{4kT})/kT] \quad (31)$$

The conductivity activation enthalpy is given by

$$\Delta H_\sigma = \Delta\bar{H}_m - \frac{p^2}{4kT} \quad (32)$$

Here $\Delta\bar{H}_m$ is the average value of ΔH_σ, and p the width of the Gaussian distribution of interstitialcy jump enthalpies.

For the BaF_2-based solid solutions Ivanov-Shits et al. (58) presented a structural model of the ionic conductivity based on the presence of (A_6F_{36-38}) cuboctahedral clusters with A = Ba or Me. Such cuboctahedral clusters substitute Ba_6F_{32} groups in the BaF_2 matrix. This block isomorphism can be represented a priori by

$$(Ba_6F_{32})^{20-} \rightarrow (Ba_2Me_4F_{36})^{20-} \rightarrow$$

$$\rightarrow \begin{matrix}(BaMe_5F_{37})^{20-} \\ (BaMe_5F_{36})^{19-} + F_i^{1-}\ (add)\end{matrix} \rightarrow$$

$$\begin{matrix}(Me_6F_{36})^{18-} + 2\ F_i^{1-}\ (add) \\ (Me_6F_{37})^{19-} + F_i^{1-}\ (add) \\ (Me_6F_{38})^{20-}\end{matrix} \quad (33)$$

The additional fluoride ions F_i' (add) may be placed either inside the cuboctahedron or outside in the periphery as a mobile fluoride interstital, F_i' (ext). Using further refinements on this structural model Ivanov-Shits et al. (58) described the composition dependence of the ionic conductivity of BaF_2-based solid solutions for $x < x_p$. In addition, differences in ionic conductivities between SrF_2-based and BaF_2 based solid solutions could be accounted for with this structural model.

For $x > x_p$ the EIM model was used to account also for the composition dependence of the ionic conductivity of the BaF_2-based solid solutions. For solid solutions with tetravalent dopants, like e.g. $Ba_{1-x}U_xF_{2+2x}$, the cluster approach and the EIM model have been used also to account for the temperature and composition dependence of the isothermal ionic conductivity (19, 20, 22, 57, 65).

In addition to impedance spectroscopy, TSDC spectra can be used to obtain ionic conductivity parameters, because the usual high temperature peak is ascribed to the relaxation of macroscopic space charge, i.e. ionic conduction. Unlike for the dipole or cluster reorientation peaks, the position of this peak is strongly dependent on the solute content, and for $Ba_{1-x}La_xF_{2+x}$ and $Ba_{1-x}U_xF_{2+2x}$ solid solutions it has been shown that the lowering of the temperature of the maximum of the space charge peak (T_m) with increasing solute content is very similar to the composition dependence of ΔH_σ. As has been described in the previous section the maximum temperature T_m of this peak is related to the conduction enthalphy ΔH_σ of the fluoride interstitials via Eq. (26). In Table II some values of ΔH_σ (TSDC) calculated from the TSDC spectra with Eq. (26) are compared with ΔH_σ (cond) values derived from ionic conductivity measurements of $Ba_{1-x}U_xF_{2+2x}$ solid solutions.

Table II

ΔH_σ (TSDC) compared with ΔH_σ (cond) for $Ba_{1-x}U_xF_{2+2x}$ (65)

x	ΔH_σ(TSDC) (eV)	ΔH_σ(cond) (eV)
0.012	1.137	1.07
0.030	0.761	0.79
0.113	0.549	0.536
0.168	0.506	0.493

The agreement is very satisfactory, and indicates that ionic conduction processes can be studied down to low temperatures.

CONCLUDING REMARKS

Impedance spectroscopy in a wide range of frequencies enables the accurate determination of bulk conductivity values of fluorites, and to distinguish between impurity oxygen diffusion at the solid electrolyte-electrode interface and in the bulk of the fluorite solid electrolyte.

The study of localized ionic motions in dilute and concentrated fluorite-type solid solutions with the TSDC technique provides detailed information on the mechanism of ionic conductivity.

The transport-structure relations need to be studied down to 5K, if reorientation processes in clusters are to be unraveled.

The defect regions model together with the EIM model can account for the composition and temperature dependence of the ionic conductivity of dilute and concentrated fluorite-type solid solutions.

Acknowledgements

The authors are very grateful to Professors A.K. Ivanov-Shits and V.I. Simonov, and Drs. N.I. Sorokin, B.P. Sobolev, P.P. Fyodorov, B.A.

Maksimov, and A.M. Golubev of the Institute of Crystallography, Moscow, for the use of their structural and electrical studies on numerous fluorite-type solid solutions prior to publication.

REFERENCES

1. Schoonman, J., "Ionic Transport in Some Fluorite Structured Solid Solutions", in: "Fast Ion Transport in Solids", P. Vashishta, J.N. Mundy, G.K. Shenoy (eds.); North-Holland, New York, Amsterdam, Oxford; 1979, p. 643.

2. Schoonman, J., Solid State Ionics 1, 121 (1980).

3. Catlow, C.R.A., Comments Solid State Phys. 9, 157, (1980).

4. Rèau, J.M., Portier, J., "Fluorine Ion Conductors", in: "Solid Electrolytes", P. Hagenmuller, W. van Gool (eds.); Academic Press, New York, San Francisco, London; 1978, p. 313.

5. Wapenaar, K.E.D., Schoonman, J., J. Electrochem. Soc. 126, 667 (1979).

6. Rèau, J.M., Lucat, C., Portier, J., Hagemuller, P., Mat Res Bull, 13, 877 (1978).

7. Lucat, C., Portier, J., Rèau, J.M., Hagenmuller, P., Soubeyroux, J.L., J. Solid State Chem. 32, 279 (1980).

8. Soubeyroux, J.L., Rèau, J.M., Matar, S., Hagemuller, P., Lucat, C., Solid State Ionics 2, 103 (1982).

9. Wapenaar, K.E.D., Koesveld, J.L. van, Schoonman, J., Solid State Ionics 2, 145 (1981).

10. Fontanella, J.J., Wintersgill, M.C., Chadwick, A.V., Saghafian, R., Andeen, C.G., J. Phys. C. 14, 2451 (1981).

11. Fontanella, J.J., Wintersgill, M.C., Welcher, P.J., Chadwick, A.V., Andeen, C.G., Solid State Ionics 5, 585, (1981).

12. Darbon, P., Réau, J.M., Hagenmuller, P., Solid State Ionics, 2, 131; (1981).

13. Schoonman, J., Solid State Ionics 5, 71 (1981).

14. Wapenaar, K.E.D., Ionic Conduction and Dielectric Loss in Fluorite-Type Solid Solutions, Ph. D. Thesis; University of Utrecht, (1981).

15. Oberschmidt, J., Lazarus, D., Phys. Rev. B 21, 5823 (1980).

16. Wapenaar, K.E.D., Schoonman, J., J. Solid State Chem. 25, 31 (1978).

17. Kjems, J.K., Andersen, N.H., Schoonman, J., Clausen, K., Physica 120B, 357 (1983).

18. Ouwerkerk, M., Schoonman, J., in: "Solid State Chemistry 1982" Eds. R. Metselaar, H.J.M. Heijligers, J. Schoonman, Elsevier, Amsterdam/Oxford/ New York (1983), p. 247.

19. Ouwerkerk, M., Kelder, E.M., Schoonman, J., Miltenbuerg, J.C. van, Solid State Ionics 9210, 531 (1983).

20. Ouwerkerk, M., Schoonman, J., Solid State Ionics 12, 479 (1984).

21. Bendall, P.J., Catlow, C.R.A., Corish, J., Jacobs, P.W.M., J. Solid State Chem. 51, 159 (1984).

22. Ouwerkerk, M., Veldkamp, F.F., Andersen, N.H., Schoonman, J., Solid State Ionics 16, 125 (1985).

23. Ivanov-Shits, A.K., Sorokin, N.I., Fedorov, P.P., Sobolev, B.P., Sov. Phys. Solid State 25, 1007 (1983).

24. Fedorov, P.P., Turkina, T.M., Sobolev, B.P., Mariani, E., Svanter, M., Solid State Ionics 6, 331 (1982).

25. Aalders, A.F., On the Fluorine Ionic Motion in Some Solid Electrolytes. Ph.D. Thesis; University of Utrecht, (1985).

26. Dorenbos, P., Mechanism of Ionic Transport in Rare Earth Doped Alkaline Earth Fluorides. Ph.D. Thesis; University of Groningen, (1988).

27. Catlow, C.R.A., Comins, J.D., Germano, F.A., Hartley, R.T., Hayes, W., Phys. Letters 71A, 97 (1979).

28. Catlow, C.R.A., Comins, J.D., Germano, F.A., Hartley R.T., Hartley, W., Owen, I.B., J. Phys. C. 14, 329 (1981).

29. Schoonman, J., Wolfert, A., Solid State Ionics 3/4, 373 (1981).

30. Schoonman, J., Wolfert, A., J. Electrochem. Soc. 127, 1522 (1981).

31. Schoonman, J., Wapenaar, K.E.D., Oversluizen, G., Dirksen, G.J., J. Electrochem. Soc. 126, 709 (1979).

32. Egan, J.J., Heus, R.J., Z. Phys. Chem. 49, 38 (1966).

33. Ramanarayanan, T.A., Narulo, M.L., Worrell, W.L., J. Electrochem. Soc. 126, 1360 (1970).

34. Ramanarayanan, T.A., Worrell, W.L.,, J. Electrochem. Soc. 127, 1717 (1980).

35. Macdonald, J.R., Phys. Rev. 92, 4 (1953).

36. Friauf, R.J., J. Chem. Phys. 22, 1329 (1954).

37. Macdonald, J.R., J. Chem. Phys. 58, 4982 (1973).

38. Franceschetti, D.R., Macdonald, J.R, J. Electroanal Chem. 82, 271 (1977).

39. Macdonald, J.R, Franceschetti, D.R., J. Chem. Phys. 68, 1614 (1978).

40. Franceschetti, D.R., Macdonald, J.R, J. Electroanal. Chem. 101, 307 (1979).

41. Schoonman, J., Macdonald, J.R., Franceschetti, D.R., Solid State Ionics 3/4, 365 (1981).

42. Franceschetti, D.R., Schoonman, J., Macdonald, J.R., Solid State Ionics 5, 621 (1981).

43. Couturier, G., Danto, Y., Gibaud, R., Salardenne, J., Solid State Ionics 5, 621 (1981).

44. Ivanov-Shits, A.K., Melnikov, V.V., Sorokin, N.I., Elektrokhimiya 23, 766 (1987).

45. Franceschetti, D.R., Solid State Ionics 5, 613 (1981).

46. Macdonald, J.R. (ed.), "Impedance Spectroscopy". John Wiley & Sons, New York, Chichester, Brisbane, Toronto, Singapore (1987).

47. Chou, S.F., Rapp, R.A., in "High Temperature Metal Halide Chemistry", D.L. Hildebrand, D.D. Cubicciotti (eds.). Proc. vol. 78-1. The Electrochemical Society, Princeton, N.J. (1978) p. 392.

48. Chou, S.F., Rapp, R.A., J. Electrochem. Soc. 130, 506 (1983).

49. Levitskii, V.A., Hammou, A., Duclot, M., Deportes, C., J. Chim. Phys. 73, 305 (1976).

50. Raistrick, I.D., Ho, C., Hu, Y.W., Huggins, R.A., J. Electroanal. Chem. 77, 319 (1977).

51. Jonscher, A.K., Réau, J.M., J. Mat. Science 13, 563 (1978).

52. Wapenaar, K.E.D., Schoonman, J., Solid State Ionics 2, 253 (1981).

53. Rehbach, M., Sluyters, J.H., Recl. Trav. Chim. Pay Bas 79, 1092 (1960).

54. Sluyters-Rehbach, M., Sluyters, J.H., in "Electroanalytical Chemistry". A.J. Bard (Ed.) Vol. IV (1970) p. 1.

55. Macdonald, J.R., Schoonman, J., Lehnen, A.P., Solid State Ionics 5, 137 (1981).

56. Macdonald, J.R., Schoonman, J., Lehnen, A.P., J. Electroanal. Chem., 131, 77 (1982).

57. Ouwerkerk, M., The Ionic Conductivity and Defect Structure of Fluorite-Type Solid Solutions $Ba_{1-x}U_xF_{2+2x}$. Ph.D. Thesis. State University of Utrecht, (1986).

58. Ivanov-Shits, A.K., Sorokin, N.I., Fedorov, P.P., Sobolev, B.P., Solid State Ionics, in the press.

59. Bucci, C., Fieschi, R., Phys. Rev. Letters 12 16 (1964).

60. Lentink, B.P.M., Numan, J.A.J., Bijvank, E.J., Hartog, H.W. den, Phys. Rev. B14, 1811 (1976).

61. Laredo, E., Puma, M., Figueroa, D.R., Phys. Rev. B19, 2224 (1979).

62. Laredo, E., Puma, M., Suarez, N., Figueroa, D., Solid State Ionics 9&10 479 (1983).

63. Laredo, E., Suarez, N., Bello, A., Puma, M., Figueroa, D., Schoonman, J. Phys. Rev. B32, 8325 (1985).

64. Wapenaar, K.E.D., Koekkoek, H.G., Turnhout, J. van, Solid State Ionics 7, 225 (1982).

65. Ouwerkerk, M., Veldkamp, F.J., Schoonman, J., Solid State Ionics 22, 173 (1987).

66. Hartog, H.W. den, Langevoort, J.C., Phys. Rev. B24, 3547 (1981).

67. Meuldijk, J., Hartog, H.W. den, Phys. Rev. B28, 1036 (1983).

68. Meuldijk, J., Meulen, R. van der, Hartog, H.W. den, Phys. Rev. B29, 2153 (1984).

69. Hartog, H.W. den, Dorenbos, P., Postma, W.J., Phys. Rev. B34, 7496 (1986).

70. Laredo, E., Puma, M., Suarez, N., Figueroa, D.R., Phys. Rev. B23, 3009 (1981).

71. Roos, A., Buijs, M., Wapenaar, K.E.D., Schoonman, J., J. Phys. Chem. Solids 46, 645, 655 (1985).

72. Shelly, R.D., Miller, G.R., J. Solid State Chem. 1, 218 (1970).

73. Wang, D.Y., Nowick, A.S., Solid State Ionics 5, 551 (1981).

74. Schoonman, J., Revue chim. Minèrale 23, 647 (1986).

75. Matar, S.F., Rèau, J.M., Hagenmuller, P., Catlow, C.R.A., J. Solid State Chem. 52, 114 (1984).

76. Bollmann, W., Phys. Stat. Sol. (a) 18, 313 (1973).

77. Nagel, L.E., O'Keeffe, M., In "Fast Ion Transport in Solids". (W. van Gool, Ed.) North-Holland, Amsterdam (1973), p. 165.

78. Ivanov-Shits, A.K., Sorokin, N.I., Sobolev, B.P., Fedorov, P.P., Solid State Physics 28, 2552 (1986) (In Russian).

79. Meuldijk, J., Hartog, H.W. den, Phys. Rev. B27, 6376 (1983).

80. Ivanov-Shits, A.K., Sorokin, N.I., Sobolev, B.P., Fedorov, P.P., In "International Symposium on Systems with Fast Ionic Transport" (Dom techniky, Bratislava) (1985), p. 99.

81. Kirkpatrick, S., Rev. Mod. Phys. 45, 574 (1973).

OXYGEN ION CONDUCTORS

B.C.H. Steele
Department of Materials
Imperial College
London, U.K.

1. INTRODUCTION

Mass transport in metallic oxides has many technological implications and an extensive literature already exists on this subject. The present chapter on oxygen ion transport is principally concerned with the properties of metallic oxides used as solid state electrolytes (ionic conductors) or electrodes (mixed conductors) for high temperature electrochemical cells. This contribution is not intended to provide a comprehensive literature review but rather to summarise recent trends for both experimental results and theoretical interpretations of those data. Accordingly only developments since 1981 will be surveyed as the relevant chapters in the book, 'Non Stoichiometric Oxides' (1) provide competent analyses of our understanding of oxygen ion transport in metallic oxides upto to that date. After a brief survey of new data further attention will be given to three topics that were included in the 1981 book, applications of computer sumulation, experimental evidence for defect interactions and association, and influence of microstructure on anionic conductivity. Additional topics discussed in the present chapter include the interpretation of complex impedance measurements, sections on mixed anionic/electronic conduction and the role of protonic conductivity in oxides. A number of reviews relating to oxygen ion transport (2,3,4,5,6,7) have been published in the past seven years and the author has found these most useful in preparing the present chapter and it is recommended that these publications also be consulted for further information on oxygen ion transport.

2. EXPERIMENTAL TECHNIQUES

2.1. Impedance Spectroscopy

Complex impedance measurements are invariably used at present to determine oxygen ion conductivity values. This method was introduced during the 1970's and has the advantage of being able to separate the various contributions, bulk, grain boundary and electrode, to the total conductivity (Fig. 1). The technique has been comprehensively reviewed in a recent book (8) which includes useful summaries about experimental details including automation of data acquisation and manipulation. A well constructed apparatus should be able to provide conductivity data with accuracies of ± 10% over a wide range of impedances, frequencies and temperatures. The accuracy should usually be sufficient to enable investigators to distinguish between various theoretical interpretations, some of which are discussed in later sections.

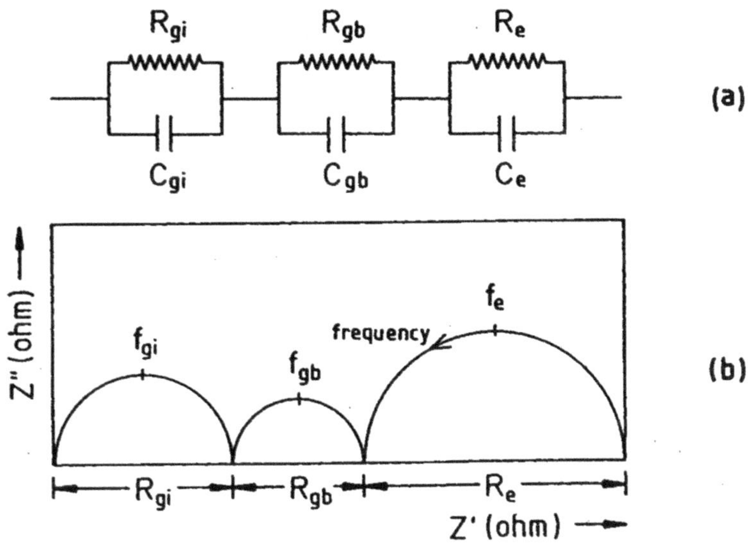

Fig. 1. Idealised equivalent circuit (a) and complex impedance diagram (b) for a two phase ceramic electrolyte.

2.2. Oxygen Diffusion Methods

In principle oxygen diffusion (D) measurements can provide information about anionic conductivity (σ) using the Nernst-Einstein equation,

$$\frac{\sigma}{D} = \frac{N q^2}{kT}, \qquad \text{............(i)}$$

where N is the number of anion sites per unit volume. As there is no long-lived radioactive isotope of oxygen the oxygen self-diffusion measurements are carried out using the stable isotopes ^{17}O and ^{18}O. Until recently the principal techniques involved measurement of the depletion of the isotope from an enriched gas (9) or isotopic exchange followed by nuclear reaction depth profiling (10). Compared to these methods the introduction of dedicated SIMS (secondary ion mass spectrometry) instruments to determine the actual isotopic profile in the solid has many advantages and can produce excellent data (11,12,13). The technique is particularly useful for measurements of oxygen ion conductivity in mixed anionic/electronic conductors and examples of these measurements are provided in section 4 of the present chapter.

Pulsed NMR (nuclear magnetic resonance) techniques (14) can also provide information on diffusion coefficients in fast ion conductors. The stable ^{17}O isotope possesses a spin of $^5/_2$ and an appropriate quadrupole moment and so in principle data relating to the transport properties of point defects can be derived from observations of the quadrupole interactions on NMR relaxation of ^{17}O. Fuda et al (15) have reported measurements of ^{17}O NMR spin-lattice relaxations in dilute $Ce(Y)O_{2-x}$ solid solutions. These investigators extracted values for the oxygen vacancy diffusion coefficient but the reported activation energy ($\sim$ 0.5 eV) is somewhat low compared to relevant activation energies (0.6 - 0.7 eV) obtained from isotopic exchange and ionic conductivity data. Interpretation of the complex NMR spectra produced by the ^{17}O nucleus is difficult and further investigations are probably required to improve the technique and deconvolution of the data.

An interesting technique, quasi elastic light scattering (QELS), has been applied to single crystals of yttria-stabilized zirconia (16). Oxygen ion vacancy diffusion coefficients were derived from an analysis of the temperature dependence of the scattering intensity at a frequency shift of 3GHz. The compositional dependence of the scattering intensity indicated that the concentration of the mobile charge carriers (anion vacancies) decreased as the yttria content increased which is in accordance with the ionic conductivity behaviour of concentrated zirconia-yttria solid solutions (see section 6.3). This technique should be generally applicable to those anionic electrolytes that can be prepared in single crystal form and further measurements should be made on solid electrolytes containing small concentrations of anion vacancies to facilitate interpretation of the data and to provide a more rigorous analysis of the technique.

Zirconia solid solutions have also been examined by IR (17) and Raman (18) spectroscopy but the phonon dispersion curves are very complicated for these highly defective solids and it is difficult to assign unequivocally the various modes to specific complex defect configurations.

The migration of oxygen ion defects (interstitials or vacancies) provides the mechanism for oxygen ion transport. These species can be trapped at associated defect complexes, and so information about the number and interaction energy of these complexes is important to interpret ionic transport data. An associated pair can act as both an electric and elastic dipole and the relevant theory has been discussed by Nowick (6). A variety of techniques have been used to probe an elastic and dielectric relaxations and the reader should consult the relevant literature (6, 19,20) for further information on these investigations including ultrasonic attenuation, and ionic thermocurrent techniques which can provide additional information about the details of the ionic transport mechanism.

Finally it should be noted that oxygen ion diffusion coefficients have been determined in $Zr(Y)O_{7-x}$ solid solutions using a gamma ray angular correlation technique (21) requiring a small concentration of ^{181}Ta nuclei on the Zr site.

3. OXYGEN ION CONDUCTIVITY DATA

3.1. General Comments

Anionic conductivity in a variety of oxide structures is surveyed in the present section. Most attention is given to oxides of the fluorite structure as the solid solutions of CeO_2 and ThO_2 have provided much of the experimental data for the development of theoretical models that can satisfactorily explain oxygen ion transport in the dilute solution range (< 2% anion vacancies). For the technologically important zirconia based solid electrolytes, as well as the doped hafnia systems, there exists no dilute fluorite solid solution range as approximately 8 mole % of dopant is required to stabilize the cubic fluorite structure. The behaviour of these concentrated solid solutions will be briefly discussed in section 6.3. Emphasis in this section, therefore, is given to examining how the theoretical models developed for dilute fluorite solid solutions can be extended to other structures including the pyrochlore, rare-earth and perovskite systems. Some consideration will also be given to preliminary results available for the metastable tetragonal zirconia dilute solutions. A selection of ionic conductivity data for the various structural systems is given in Fig. 2 and will be noted that the highly disorded δ-Bi_2O_3 material and associated solid solutions still exhibit the highest oxygen ion conductivity values followed by the fluorite CeO_2-Gd_2O_3 solid solutions. The dominant position occupied by these fluorite structures will be commented upon and the possibility of synthesising alternative superior oxygen ion conductors will be briefly discussed in section 3.7 dealing with miscellaneous oxygen ion conductors.

3.2. Ceria Solid Solutions.

As mentioned earlier these solid oxide electrolyte materials have often been used as model systems to examine theoretical concepts concerning defect associations and vacancy clustering. In contrast to zirconia, pure ceria adopts the cubic fluorite structure and so the effect of cation dopants can be studied from zero to large concentrations ($\sim$ 20 mole %). Moreover the radius of the Ce^{4+} ion is

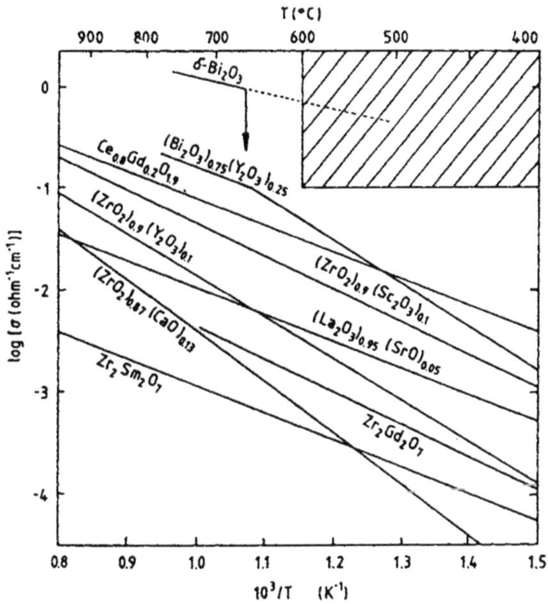

Fig. 2. Compilation of oxygen ion conductivity vs temperature for selected ceramic electrolytes.

such that a variety of dopant cations can be introduced having either larger or smaller radii. Based on the earlier comments of Nowick (22) and other investigators, Kilner and Steele (23) emphasised the importance of the defect pairs that are formed due to interaction between the anion vacancies, $V_O^{\bullet\bullet}$, and aliovalent cations, M'_{Ce} introduced by defect reactions of the type:

$$M_2O_3\ (CeO_2) \rightarrow 2M'_{Ce} + V_O^{\bullet\bullet} + 3O_O^X$$

The total activation energy (ΔH_T) for ionic conduction is thus given by the following expression:

$$\Delta H_T = \Delta H_m + \Delta H_a,$$

where ΔH_m is the enthalpy of migration of an oxygen ion and ΔH_a is association energy of the complex defect $[Me'_{Ce} - V_O^{\bullet\bullet}]^{\bullet}$. Whilst qualitative agreement was noted between the analytical expressions and experimental data it was still necessary to provide an explanation for the observation of Gerhart-Anderson and Nowick (24) that the association energy (ΔH_a) in CeO_2-M_2O_3 systems was dependent upon the particular dopant ion, M^{3+}. Kilner and Brook (25) suggested that the association energy incorporated a strain energy term due to the mis-match between host and dopant cation. These ideas received support from the calculations of defect binding energies by Butler et al (26). Experimental and calculated values are shown together in Fig. 3, and it can be seen that there is good quantitative agreement between the two sets of data. This model has now been widely used to optimise oxygen ion conductivity in a variety of oxide solid solutions and will be extensively involved in subsequent sections. Kilner (27) has further proposed that because the defect binding energy is dependent upon the relative radii of host and dopant, plots of lattice parameter versus concentration should yield relevant qualitative information about the expected strain component, and derived values for ThO_2, $BiO_{1.5}$, CeO_2 and ZrO_2 host lattices together with the trivalent rare earth elements are summarised in Fig. 4. Some interesting observations can be drawn from this plot:

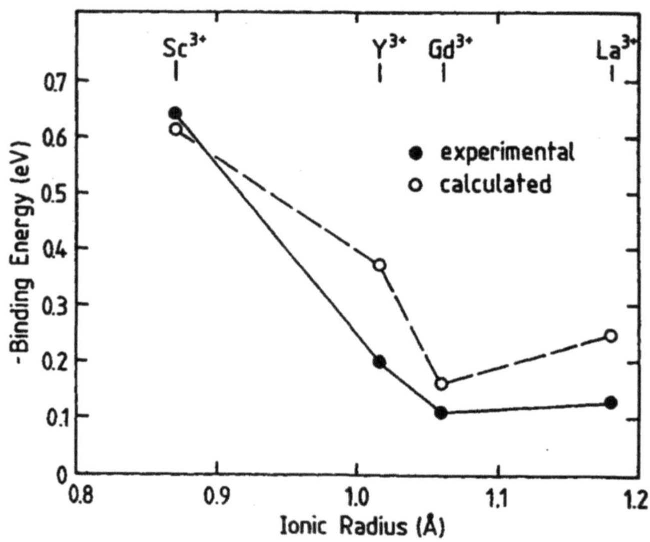

Fig. 3. The variation of the association enthalpy with ionic radius in trivalent doped CeO_2. From reference 3.

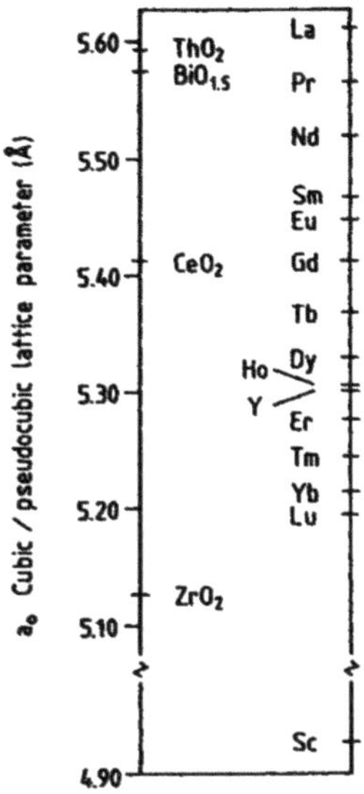

Fig. 4. Lattice parameter/ionic radii relationships for typical fluorite oxide hosts and trivalent dopants (from ref. 3).

(a) Th^{4+} (ThO_2) is, in general, too large a host cation; most dopants are smaller and thus high strain components are predicted because both calculation and experimental confirm that the association energy is much larger when the dopant is smaller than the host.

(b) Zr^{4+} is, in general, too small a host cation as most dopant cations will have larger radii.

Because the anion vacancies are strongly bound in the associated defect pairs the concentration of free vacancies available to contribute to ionic conductivity is relatively small at low temperatures as shown by the calculations of Kilner and Waters (28). The presence of the associated pairs influences how the pre-exponential factor (σo) and activation energies (ΔH_T) for conduction vary with dopant concentration as summarised in the following Table 1.

TABLE 1

	Pre-exponential Factor	Total Activation Energy
Divalent Dopants	Proportional to $(M_{Ce}'')^{\frac{1}{2}}$	$\Delta H_m + \Delta H_a/2$
Trivalent Dopants	Independant of (M_{Ce}')	$\Delta H_m + \Delta H_a$
ΔH_m, ΔH_a represent migration and association enthalpies, respectively.		

These conclusions have been derived by Kilner and Steele (23) and Nowick (6).

The preceding defect association model can thus explain satisfactorily the general features of ionic conductivity in fluorite oxide systems at least in the dilute solution regime. However some refinements of this model are required to interpret additional experimental data. For example, the relationships summarised in Table 1 indicate that the total activation energy should be independent of composition whereas in many systems it has been observed that this quantity initially decreases with increasing concentration. An explanation for this observation has been provided by Wang et al (29) who have suggested that coulombic interactions occur between the associated defects, $[V\ddot{o}^{\cdot} - M']^{\cdot}$, and the free dopant ions, M'. This model is developed (6) by supposing that the charged defects form a regular array of charges of alternating sign within the host lattice. The effect of this is to lower the association enthalpy term, ΔH_a, so that the conductivity is proportional to $(M_{Ce}')^{1/3}$ which is in good agreement for the data for dilute solutions of CeO_2-Y_2O_3.

Inspection of Fig. 3 reveals that the association energy for scandia doped ceria is very large. Introduction of this dopant, therefore, should have a scavenging effect on anion vacancies produced by other dopants. This scavenging effect has been demonstrated (24) by adding Y_2O_3 to produce vacancies in addition to those contributed by Sc_2O_3. For example with ½ % Y_2O_3 and 1 % Sc_2O_3 dopant additions the ionic conductivity value was about 30 times smaller than for ½% Y_2O_3 alone and ΔH_a was correspondingly increased.

In contrast the addition of two dopants can be controlled to optimise the oxygen ion conductivity in ternary systems. This is illustrated by the work of Zhen et al (30) for the system CeO_2-Gd_2O_3-Y_2O_3. The total dopant concentration was maintained at 10 mole % but the concentration of Y_2O_3 was varied between 0 and 10%. A minimum in the activation energy (0.82 eV) was noted for the composition $Ce_{0.9}Gd_{0.09}Y_{0.01}O_{1.95}$ which had a lattice parameter very close to that of the CeO_2 host lattice as summarised in Table 2.

TABLE 2.

Measured Lattice Parameters and Activation Energies of $Ce(GdY)O_{2-x}$.

Composition	Cell parameter (A°)	Activation Energy (eV)
CeO_2	5.412	-
$Ce_{0.9}Gd_{0.1}O_{1.95}$	5.423	0.88
$Ce_{0.9}Gd_{0.09}Y_{0.01}O_{1.95}$	5.416	0.82
$Ce_{0.9}Gd_{0.05}Y_{0.05}O_{1.95}$	5.407	0.92
$Ce_{0.9}Y_{0.1}O_{1.95}$	5.404	0.97

Because of the relatively high ionic conductivity of CeO_2 solid solutions (Fig. 2) there is still interest in using these materials in technological devices such as oxygen monitors and fuel cells. However the tendency of Ce^{4+} to be reduced to Ce^{3+} at low oxygen partial pressures according to the reaction:

$$O_O^X \rightarrow \tfrac{1}{2}O_2 + V_O^{\bullet\bullet} + 2e',$$

restricts the range of oxygen partial pressure over which the ionic transference number (t ion) remains close to unity. At 800°C this range can extend to 10^{-12} atm (0.64V) which could be adequate for selected applications (inclusing monitors) but not, unfortunately for fuel cell applications where open circuit voltages around 1 volt are typical. Recent determinations (31) of t ion in the CeO_2-CaO system provide a useful summary of the relevant experimental techniques and contains appropriate references to earlier investigations.

3.3. Zirconia Solid Solutions

As mentioned in section 3.1 approximately 8 mole % of dopant is required to stabilise cubic fluorite solid solutions of zirconia and the behaviour of these highly defective solid electrolytes is briefly discussed in section 7.3.

The dilute solution models developed for CeO_2 and ThO_2 can however be applied to metastable tetragonal zirconia materials. Advances in ceramic technology have provided fabrication routes to stabilise tetragonal zirconia solid solutions by preparing microstructures with small (< 0.5 μm) grain size. Originally these ceramics received much attention because of their excellent mechanical properties arising from transformation - toughening mechanisms. However in 1984, Bonanos et al (32) demonstrated that tetragonal $Zr(Y)O_{2-x}$ ceramic electrolytes exhibited higher anionic conductivities than cubic $Zr(Y)O_{2-x}$ samples below about 400°C. This combination of superior mechanical and electrical properties stimulated further investigations into the electrical behaviour of tetragonal ZrO_2 solid solution and data are now available for the systems $Zr(Y)O_{2-x}$, (33,34) $Zr(Er)O_{2-x}$, (35) $Zr(In)O_{2-x}$ (36). These data are summarised in Table 3 and Fig. 5.

Examination of Fig. 5 suggests that the conductivity of these tetragonal solid solutions is independant of dopant concentration which would be expected from the simple model of associated defect pairs summarised by Kilner and Steele (23). However it was mentioned in the previous section (3.2) that Nowick and co-workers had demonstrated that for dilute CeO_2-Y_2O_3 solid solutions the simple defect pair assoc-

TABLE 3.

Data for selected tetragonal zirconia solid solutions

Composition 3 mole %	Activation energy E_A(eV)	Pre-exponential $\log_{10} \sigma_o$	Radius Ratio $\frac{r(R^{3+})}{r(Zr^{4+})}$	Lattice Parameter Å a	c	Volume (Å)^3
Y_2O_3(33)	0.77	-	1.21	-	-	-
Y_2O_3(34)	0.98	6.1	1.21	5.09	5.17	133
Er_2O_3(35)	0.84	3.9	1.09	5.085	5.161	134
In_2O_3(36)	0.79	4.6	0.10	5.08	5.18	134

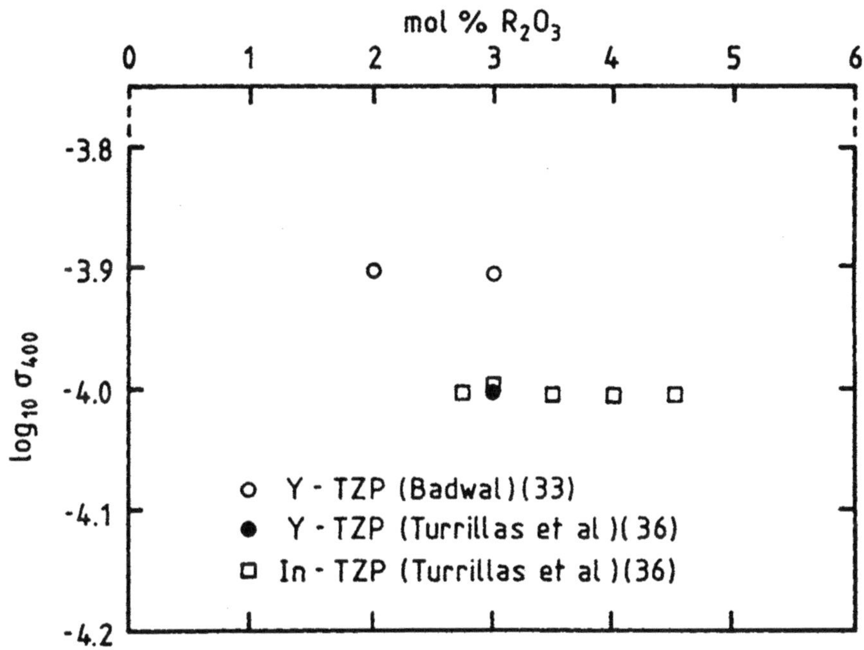

Fig. 5. Bulk conductivity at 400°C of selected tetragonal zirconia solid solutions as a function of dopant concentration.

iation model was inadequate and that the conductivity was proportional to (dopant concentration)$^{1/3}$. This model assumed an ordered array of $(Y_{Ce}' - V_O^{\bullet\bullet})^{\bullet}$ and Y_{Ce}' defects. It is possible that the defect arrangement in the zirconia tetragonal solid solutions is different and possible evidence for this is provided by the preliminary computer simulation studies of Cormack (37) which suggest that the stable defect arrangement in tetragonal zirconia electrolytes differs from that in the cubic zirconia solid solutions.

Attention is also drawn to the relatively low values of the pre-exponential term σ_0, for the dopants Er_2O_3 and In_2O_3. As outlined in section 7 the classical model for migration of free defects suggests a value for σ_0 around 10^6 and the relevant data for tetragonal $Zr(Y)O_{2-x}$ electrolytes is in accordance with this value. However further investigations are required on these tetragonal solid solutions before any definite conclusions can be made and it will be interesting to examine data for the associated ZrO_2-TiO_2-R_2O_3 and ZrO_2-CeO_2-R_2O_3 ternary tetragonal solid solutions.

3.4. Pyrochlore ($A_2B_2O_7$) Electrolytes.

Oxide pyrochlores have the general formula $A_2B_2O_7$. The structure is closely related to that of the fluorite and for the present purpose it can be visualised as a fluorite (M_4O_8) with $^1/_8$ of all oxygen sites vacant. However unlike defect fluorites of the same stoichiometry, in the perfect pyrochlore structure the oxygen and cation sub-lattices are both ordered. Intrinsic Frenkel disorder on the oxygen sub-lattice places some oxygens on the normally unoccupied sites and the vacancies thus created are responsible for the observed ionic conductivity. Reviews of anionic conductivity in the pyrochlores are available (3,38) and investigations on these systems have provided valuable insight into the influence of cation ordering on the magnitufe of the associated oxygen ion conductivity.

In general, the tendency to cation order depends upon the size difference between the dopant and host cation. For example, at the pyrochlore stoichiometry, $La_2Zr_2O_7$, when the rare earth ion (La^{3+}) becomes larger than Gd^{3+} (1.06 Å in 8 co-ordination) the ordered pyrochlore structure forms, but when the rare earth ion is smaller the disordered fluorite phase forms. The ordered phases can, however, disorder at higher temperatures and, for example, $Gd_2Zr_2O_7$ disorders at ∿ 1550°C. The disorder temperature increases as the size difference between the two cations becomes larger reflecting the increased stability of the ordered lattice.

Anion order is somewhat more difficult to define. Where the cation sizes are sufficiently different to provide a stable ordered cation lattice, the anion lattice will follow and order. This link between cation and anion order has been demonstrated by Rossell (39) for $Zr_3Er_4O_{12}$, by Faktor et al (40) for the system $Nd_2Zr_2O_7$ - $Nd_2Ce_2O_7$ and by Moon (41) for the system $Gd_2Ti_2O_7$ - $Gd_2Zr_2O_7$.

Van Dijk et al (42) have made a detailed study of the conductivity of $Gd_2Zr_2O_7$ and $Nd_2Zr_2O_7$ pyrochlores and their solid solutions. For both stoichiometric compositions they found a minimum in the activation energy (ΔE) and pre-exponential term (σo) versus composition curves. The results for $Gd_2Zr_2O_7$ are depicted in Figs. 6(a) and 6(b). For $Nd_2Zr_2O_7$, σ_0 is about three orders of magnitude lower than for $Gd_2Zr_2O_7$ and this causes a local minimum in the conductivity. For $Gd_2Zr_2O_7$ a local maximum is seen. Neutron diffraction studies(43), on $Nd_2Zr_2O_7$, showed that Frenkel disorder on the oxygen sublattice was always less than 5%, even at 600°C, and this accounts for the low value of σ_0 and also agrees with the Raman data (44) which showed greater oxygen disorder for $Gd_2Zr_2O_7$. Greater oxygen disorder creates more charge carriers and thus increases σ_0. In both systems E_{act} and σ_0 increase dramatically for the hyper- and hypostoichiometry compositions, consequently the conductivity decreases (42). No breaks in the E_{act}, σ_0 and conductivity curves are observed at the transition from pyrochlore to fluorite solid solution indicating the coherent intergrowth of the two phases (42). The influence of long-range order on E_{act} and σ_0, and so on the conductivity, was illustrated by Van Dijk et al (42).

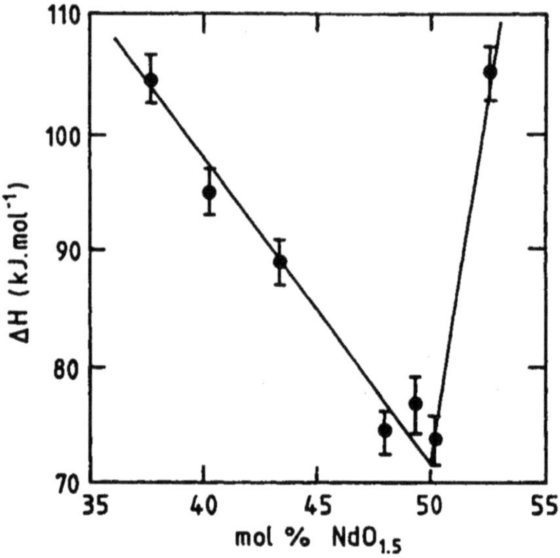

Fig. 6(a). The activation energy for ionic conductivity as a function of composition in the ZrO_2-$GdO_{1.5}$ system. From ref. 3.

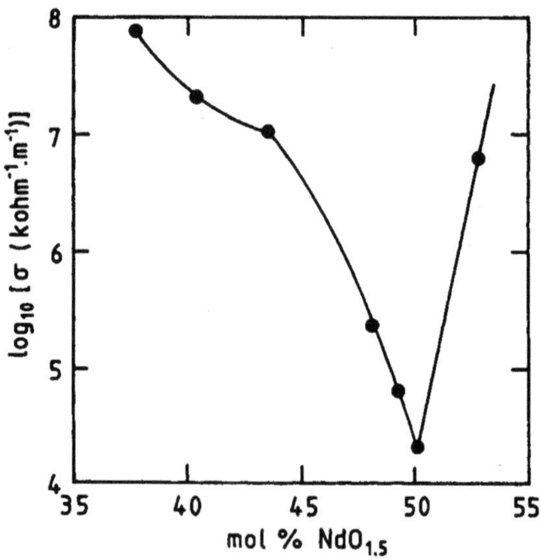

Fig. 6(b) The pre-exponential factor for ionic conductivity as a function of composition in the ZrO_2-$GDO_{1.5}$ system. From ref. 3.

They took an annealed single crystal of $Gd_{0.52}Zr_{0.48}O_{1.74}$, which showed pyrochlore x-ray reflections, and a quenched crystal of the same composition which showed only the fluorite reflections. The quenched crystal had a higher σ_0, and also a higher E_{act}, than did the annealed crystal. Overall, the annealed pyrochlore sample was the better conductor. The higher value of σ_0 for the fluorite is due to the increased number of charge carriers brought about by greater disorder on the oxygen sublattice. A detailed discussion of the probable pathways for oxygen migration in the ordered pyrochlore structure is provided by Van Dijk et al (38) and it appears that association effects and the generation of simple Frenkel defects only play a minor role in the perfectly ordered material. The complexity of the local arrangements in the pyrochlore and fluorite ZrO_2-$GdO_{1.5}$ system has been further explored by Vehara et al (45) using EXAFS techniques which provide evidence that the average co-ordination number of the Zr^{4+} ions decreases with increasing Gd^{3+} content. As the co-ordination number of the Gd^{3+} ions does not appear to vary with compositional changes it follows that the anion vacancies are located preferentially adjacent to the Zr^{4+} ions.

These studies of conductivity in the pyrochlores have shown that cation order reduces the activation energy; and that oxygen order, by reducing the concentration of charge carriers, lowers the pre-exponential σ_0, relative to a fluorite of the same stoichiometry. Order on the oxygen sublattice appears to be determined by the degree of order on the cation sublattice. Thus, a large difference in cation size will give a small activation energy but also a small σ_0; a small difference in size will give a higher σ_0, but also a higher activation energy. The activation energy does not appear to change dramatically as the degree of cation order decreases. For $Nd_2Zr_2O_7$ it is 0.77 eV and for $Gd_2Zr_2O_7$ it is 0.87 eV (46). The difference in σ_0, on the other hand is quite large. For $Nd_2Zr_2O_7$ σ_0 is $\sim 2 \times 10^4\ \Omega^{-1}m^{-1}K$ and for $Gd_2Zr_2O_7$ it is $\sim 1 \times 10^7\ \Omega^{-1}\ m^{-1}K$. These two competing factors would seem to preclude the development of useful oxygen-ion conductors in these systems as, ideally, full cation order and anion disorder would be required. The ionic conductivities of $Zr_2Gd_2O_7$ and $Zr_2Sn_2O_7$ are included in Fig. 2 for comparison.

Attempts to overcome these limitations by introducing lower valent cations on either the A sites, e.g. $Gd_{2-x}Ca_xZr_2O_{7-x/2}$ or B sites, $Gd_2Zr_{2-x}Sc_xO_{7-x/2}$, have also proved unsuccessful (47) as the conductivity decreased with increasing values of x. It should be noted though that Faktor (48) obtained enhanced conductivity with Ca doped $Nd_2Zr_2O_7$ samples.

3.5. Doped Rare-Earth Oxides

Structures of the rare earth oxides may be considered as being derived from an anion deficient fluorite structure (M_4X_6) and so might be expected to exhibit disorder on the oxygen sub-lattice. Moreover one of the M-O bond lengths is relatively long which suggests that the energetics of oxygen ion migration could be relatively low. Vanbaelinghem (49) and co-workers investigated oxygen ion conductivity in the system La_2O_3-SrO and indeed confirmed that this system had a reasonable ionic conductivity (see Fig. 2). Later studies by Zhen et al (50,51) were concerned with the effect of dopant cation radius and examined the electrical conductivity of La_2O_3 solid solutions doped with MgO, CuO, SrO and BaO. Maximum conductivity and minimum activation energy ($\sim$ 0.6eV) was obtained with the SrO dopant which is in accordance with the association energy for [$V_O^{\bullet\bullet}$ - M'] defect pairs being dependant upon the relative radius of the host ($\sim$ 1.16 Å for seven-co-ordination) and dopant ($\sim$ 1.18 Å) cation. The conductivity of the La_2O_3-SrO system was also optimised by varying the dopant level. A conductivity maximum of 10^{-2}S/cm at 800°C was obtained when the dopant composition attained about 12 mole %. As can be seen from Fig. 2 this value approaches the conductivity of cubic zirconia-yttria systems. Although the activation energy at this composition has increased to 0.78 eV this value is still lower than that reported for cubic zirconia-yttria ($\sim$ 8 mole %) electrolytes. This does suggest that the enthalpy of motion ΔH_m for oxygen ions in doped La_2O_3 is relatively low.

Ceramics containing La_2O_3 and SrO degrade when exposed to moist air due to the formation of hydrate/carbonate phases and so technological applications of these ceramics are unlikely. Ceramics based on Y_2O_3, however, are more stable and some information regarding anionic conductivity in these materials is available. Norby and Kofstad (52, 53) have investigated the electrical conductivity of pure Y_2O_3 as a function of oxygen partial pressure in wet and dry atmospheres and they concluded that interstial protons, $Hi^{\cdot}$, were the dominating positively charged defect upto about 1000°C in contrast to anion vacancies, $V\ddot{o}^{\cdot}$. The presence of significant protonic conductivity in oxides is an important observation and is discussed further in section 4.2. In the present context the same authors suggest that doping Y_2O_3 by lower valent cations (e.g. Ca^{2+}) to produce Ca_y' also leads to the incorporation of hydrogen to form compensating interstial protonic species, $Hi^{\cdot}$ rather than oxygen ion vacancies $V\ddot{o}^{\cdot}$. In this way they suggest that a proton conductor may have been obtained by Steele et al (54) during investigations on Y_2O_3 doped with 6 mole % of CaO. It should be possible to resolve these differences in interpretation by NMR. It should be noted, however, that the ionic conductivities measured by Steele et al (54) on doped Y_2O_3 and Sn_2O_3 do suggest a much lower ionic conductivity value than that reported by Zhen et al for La_2O_3 solid solutions although the earlier measurements of Steele et al did not separate the relative contributions of bulk and grain boundary resistivities.

3.6. Bi_2O_3 Based Electrolytes

It is generally accepted that δ-Bi_2O_3 has an anion deficient fluorite structure in which the bismuth ion occupies fcc sites. There is still controversy, however, about the arrangement of the oxygen ion sub-lattice in which formally (Bi_4O_6) 25% of the normal anion sites are vacant. The δ-phase of Bi_2O_3 is thermodynamically stable between 729°C and 824°C (melting point) and exhibits the highest value ($\sim$ $1 Scm^{-1}$) of oxygen ion conductivity so far reported (see Fig. 2) for a solid electrolyte. It is not surprising, therefore, that considerable effort has been expended in trying to understand the mechanism of ionic conduction in δ-Bi_2O_3 and also in the development of dopant stabilised low temperature forms of δ-Bi_2O_3.

A recent paper by Jacobs et al (55) summarises many of the structural models that have been proposed. Furthermore these authors conclude that computer simulation models favour anion vacancy ordering along the <111> direction as originally proposed by Sillen (56). However ordered anion vacancies can also occur in the <110> direction but the spatial extent of these ordered chains is limited because of interactions with nearby defects. Moreover the Bi^{3+} cations also undergo displacements from the fcc positions. Any proposed model for the structure of δ-Bi_2O_3 should also include an explanation of why the oxygen vacancies, which are formally uncharged, can result in high values for the oxygen ion conductivity. Kilner and Faktor (3) have suggested a domain boundary model in which conductivity results from a co-operative movement of domain walls rather than the migration of isolated vacancies. Although Jacob et al (57) examined the idea of charged boundaries for a variety of configurations they concluded that a domain mechanism of charge transport was unlikely. Instead these authors propose that the charged defects in δ-Bi_2O_3 are vacancies and pseudo-interstials (that is oxygen ion occupying normally-vacant sites in the <111> superlattice) associated with <110> defect chains. It is proposed that these defects have a high mobility due to two mechanisms of charge transport. One of these involves the transfer of charge carriers between collinear <110> defects; the other depends on the delocalisation of charge that accompanies the addition of a vacancy or a pseudo-interstial to a chain of <110> defects which is analogous to the Grotthuss model of proton conduction in water.

Although the details of charge transport in δ-Bi_2O_3 remain controversial even less is known about the various Bi_2O_3 solid solutions. Conductivity data for a selection of these materials are presented in Fig. 7. It is impossible to draw any general conclusions from these

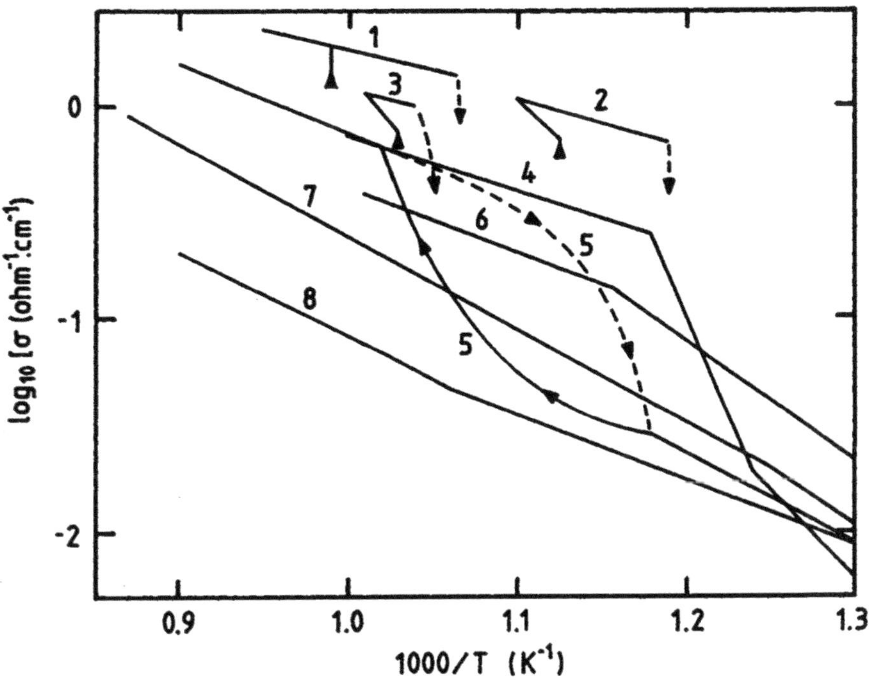

Fig. 7. The conductivities of various Bi_2O_3 based materials. From ref 3

data and the reader is referred to discussions of the specific systems in reviews by Kilner and Faktor (3) and by Boivin (58). It should be noted however, that several authors, for example, Laarif and Theobald (59), Infante et al (60) emphasise that the anion arrangement in defect cluster/chains is probably optimised to accommodate to the position of the lone pair orbital associated with the Bi^{3+} cation.

The incorporation of double dopants is also interesting as Meng et al (61) have noted that the δ-phase can be stabilized with lower total concentrations of two dopants compared to the situation for the simple binary systems.

Attention is also drawn to the high oxygen ion conductivities of $Bi_{24}Pb_5Ca_3O_{44}$(62) and $Bi_{0.571}Pb_{0.428}O_{1.285}$ (63). The latter material toughened with tetragonal zirconia particles, has been successfully employed as a membrane for oxygen separation. Although the oxygen ion conductivity of these Bi_2O_3 solid solutions is high technological applications of these materials are limited (64) due to the reduction of Bi_2O_3 when cathodic potentials below approximately 0.64 ($pO_2 \sim 10^{-13}$ atm at 600°C) are applied to the electrolyte. There are reports in the literature (65,66) that Bi_2O_3 based electrolytes are kinetically stable at lower oxygen partial pressures ($\sim pO_2^{-20}$). However these observations must be treated with caution as it is doubtful whether Bi_2O_3 systems could remain in the unreduced state for long periods at these low oxygen partial pressures.

Finally attention is drawn to investigations on oxygen ion conductivity on catalysts composed of complex mixtures of bismuth and molybdenum oxides. The Mars-van Krevelen mechanism (67) for the partial oxidation of hydrocarbons involces the lattice oxygen of the oxide catalyst. Oxygen abstracted from the catalyst surface is replenished by transport of oxygen through the bulk catalyst particle and so oxygen transport can influence the overall reaction kinetics. For example, Boon and Metselaar (68) have reported that $Bi_6Mo_2O_{15}$ is a pure oxygen ion conductor.

3.7. Perovskite and Related Oxides

The perovskite (ABO_3) structure provides interesting opportunities to investigate oxygen ion conductivity as the structure is very tolerant of changes in the ionic radii of the A and B sites and can also accommodate large concentrations of cation vacancies particularly on the A sites. Takahashi et al (69,70) initiated investigations on perovskite electrolytes and demonstrated that selected solid solutions could have reasonably high ionic conductivities (e.g. $\sim 5\times10^{-3}$S/cm at 800°C). A series of other perovskite oxides were examined by Brownall et al (1).

Preliminary theoretical and experimental investigations involving a group of perovskite oxides, $LaA\ell O_3$, (La represents rare earth cation) have been carried out by Kilner and Brook (25). For these oxides lattice simulation calculations suggested that ΔH_m (enthalpy of migration) was only weakly dependant upon the type of lanthanide cation. The same calculations also predicted a value for ΔH_m of 0.8 ± 0.2 eV which is significantly higher than the ΔH_m value (∿ 0.6 eV) determined for the fluorite oxides. A further prediction was that enthalpies of association ΔH_m would be strongly dependant upon dopant cation size with the probability of a minimum occurring when $r_{doping} \sim r_{host}$. This is in accordance with the situation prevailing for the fluorite and other host lattices mentioned in earlier sections. In a survey of the energetics of oxygen ion transport in a variety of host lattices Steele and Kilner (72) concluded that perovskite oxides did not appear, at present, to be promising candidate materials for electrolytes exhibiting fast oxygen ion transport. However Bonanos et al (137) have recently provided evidence for high oxygen ion conductivity in $BaCe_{0.85}Gd_{0.15}O_{2.925}$ (see Section 4.2).

Perovskite building blocks (ABO_3) feature in many complicated defect oxide structures such as cryolite related phases $(A_3B\ O_3)_n$. Reports in the Russian literature (73,74) suggested that oxides having the compositions $Sr_6Nb_2O_{11}$, $Ba_6Ta_2O_{11}$ exhibited ionic conductivities comparable to zirconia based electrolytes due to disorder on the anionic sub-lattice. However investigations by Turrillas et al (36) were unable to confirm these reports.

Another series of defect perovskite related structures $A_3B_2O_7$ have also been investigated, including $Sr_3Zr_2O_7$, $Sr_3Ti_2O_7$, $Ba_3Ti_2O_7$ (73). Again the measured conductivity values were inferior to $Zr(Y)O_{2-x}$ fluorite solid solutions. The high oxygen diffusivities reported for La_2CuO_4 which has the K_2NiF_4 structure ($LaO.\ LaCuO_3$) are discussed in section 4. Other compositions in this series merit further attention, although at present it is difficult to discern a pattern in the conductivity data for defect perovskite related structures.

3.8. Miscellaneous Electrolytes

The principal oxide structural types, fluorite, pyrochlore, rare-earth and perovskite, exhibiting fast oxygen ion conduction have been surveyed in previous sections. However a variety of other materials have also been investigated and demonstrated to exhibit oxygen ion conductivity although the magnitude of the ionic conductivities measured has so far not been sufficiently high to attract attention for technological applications. A selection of these materials are listed in the present section to illustrate the range of structural types that have been examined.

Anionic conduction in both the low temperature β or h- modification and high temperature α or H- form of Ta_2O_5 have been measured (75,76). The electrical conductivity of $BiVO_4$ samples having the scheelite structure has been measured (77), and also oxygen ion conductivity in mullite ($A\ell_6Si_2O_{13}$) ceramic components (78). Oxygen ion diffusion in beta alumina materials (79) has been determined, and electrical transport measurements (80) on a series of calcium and barium aluminates show evidence for anionic conduction in certain of the compounds studied particularly for the composition $Ca_{12}A\ell_{14}O_{33}$ (136). Finally it should be noted that phases having tetragonal symmetry such as $Ba_3Ge_{0.9}A\ell_{0.1}O_{4.95}$ have been investigated (36) and shown to have low values of ionic conductivity.

4. MIXED CONDUCTORS

4.1. Anionic and Electronic Conductors

Materials exhibiting mixed electronic and oxygen ion conductivity are of interest because of possible technological applications as membranes for separating oxygen from air and other gaseous mixtures. They also display attractive features when used as fuel cell electrodes in solid oxide fuel cell systems. In this second application the presence of significant oxygen ion fluxes in the porous oxide electrode can increase the effective area at which the relevant Faradaic reaction can occur thus reducing polarisation losses.

Solid state electrochemical techniques for determining oxygen ion conductivities in mixed conductors have often been reviewed (81,82,83, 84) but few data are available in the literature using these methods. An alternative approach is to determine the oxygen self diffusion coefficient using $^{18}O/^{16}O$ exchange measurement followed by analysis of the diffusion profile in the solid by means of dynamic secondary ion mass spectrometry (SIMS) as already mentioned in section 2.2. The Nernst-Einstein equation provides a link between the self-diffusion coefficient (D^*) and anionic conductivity (σ_i) so that ionic fluxes can be expressed in terms of either D^* or σ_i as indicated below:

$$\frac{\sigma_i}{D_i^*} = \frac{Nq^2}{kT} \qquad \ldots\ldots\ldots \alpha$$

$$j_{O2} = \frac{kT}{q^2} \cdot \frac{\sigma_i}{L} \int_{pO_2''}^{pO_2''} te \; d \ln pO_2, \qquad \ldots.\beta$$

This expression can be simplified by assuming that the oxide is highly disorded and a predominately electronic conductor (i.e. $\sigma_e \; \sigma_{ion}$), so that the electronic transference number (σ_e) is approximately unity and is independant of oxygen partial pressure. It follows that,

$$j_{O2} = \frac{kT}{q^2} \cdot \frac{\sigma_i}{L} \ln pO_2''/pO_2' \text{ ions/m}^2\text{/sec} \qquad \ldots\ldots\ldots\ldots \gamma$$

$$\text{or } j_{O2} = \frac{D^*N}{L} \ln pO_2''/pO_2' \text{ ions/m}^2\text{/s} \qquad \ldots\ldots\ldots\ldots \delta$$

$$\text{and } i = j\,q = \frac{kT}{q} \cdot \frac{\sigma_i}{L} \ln pO_2''/pO_2 \text{ A/m}^2\text{/s} \qquad \ldots\ldots\ldots\ldots \varepsilon$$

In the preceding expressions, N is the number of anion sites per unit volume, and L is the thickness of the oxide sample, with the other symbols having their usual meaning.

The Nernst-Einstein expression given in equation α is useful in that it does provide a means of comparing oxygen ion fluxes in electrolytic (te ∿ 0) and electronic (te ∿ 1) oxide materials. A selection of data is provided in Fig. 8 in which oxygen self diffusion coefficients (D_i^*) have been converted to anionic conductivities. The conversion used a temperature of 1000°K, q = 2e, and assumed a constant value for N of 10^{28} sites/m³,

$$\frac{\sigma_i}{D^*} \sim 7.4 \times 10^{10} \text{ (SI units)} = 7.4 \times 10^4 \text{ (cgs units)}$$

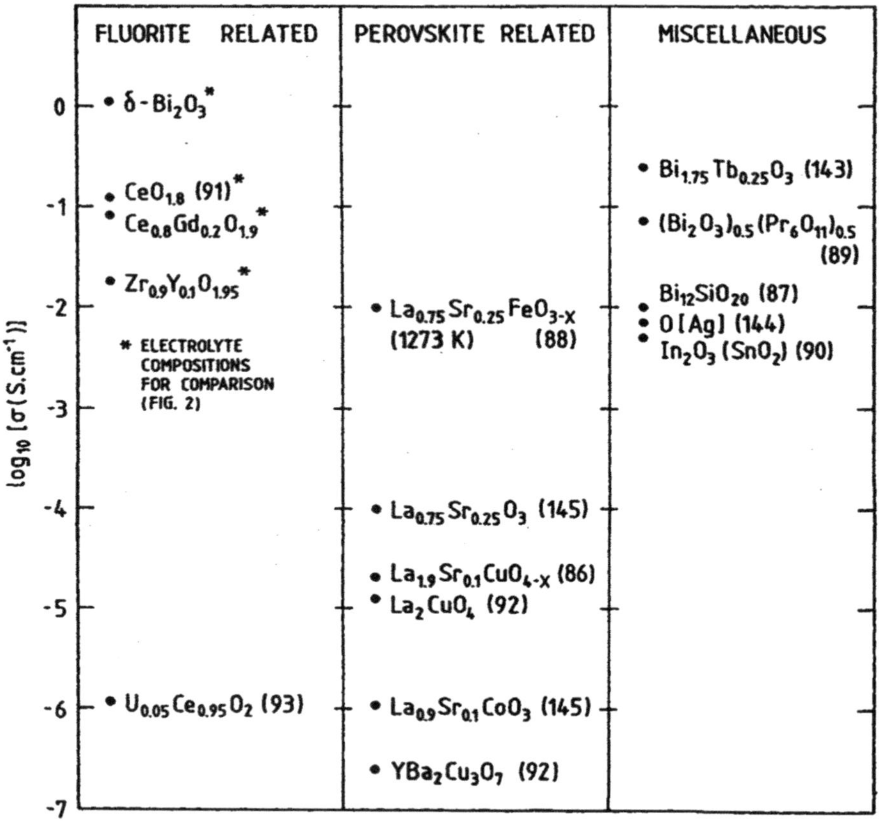

Fig. 8. Compilation of selected oxygen ion conductivities at 1000 K in predominately electronic conductors.

At present it is difficult to draw general conclusions about the data presented in Fig. 8 and so attention is drawn to specific materials. The $(Bi_2O_3)_{0.5}(Pr_6O_{11})_{0.5}$ composition has an oxygen ion conductivity comparable to the δ - Bi_2O_3 phase. Highly defective perovskites such as La(Sr)FeO_{3-x} and La(Sr)CoO_{3-x} exhibit anionic conductivity values approaching electrolyte compositions based on zirconia solid solutions. As expected disordered fluorite materials such as CeO_{2-x} exhibit high values for oxygen ion conductivity, and reports (85) that UO_2-Y_2O_3, UO_2-Sc_2O_3 fluorite solid solutions perform well as electrode materials is probably due to high oxygen ion fluxes in these materials although no data have been published. The perovskite related electronic superconductor materials $La_2Cu\ O_4$ (K_2NiF_4 structure) and $YBa_2Cu_3O_{7-x}$ also possess significant anionic conductivity values. There is evidence that defect clusters can form in these oxide superconductors and support for this view is provided by the data of Routbort et al (86) who have measured oxygen diffusion in $La_{2-x}Sr_xCuO_{4-y}$. For moderate values of x (0.1 - 0.2) the oxygen diffusion coefficient decreases with increasing Sr content due to strong Sr_{La}' - $V_O^{\bullet\bullet}$ interactions. The formation of strongly bound associated defects is to be expected from the models discussed in section 3.2. The only general trend is that high oxygen ion conductivity in the mixed conductors occurs in similar structures to those used for oxide electrolyte which is not of course surprising. Attention has already been drawn in section 3.6 to the importance of oxygen ion transport in partial oxidation catalysts which also exhibit mixed conduction.

4.2. Anionic and Protonic Conductors

It has been accepted for many years that small quantities of hydrogen can dissolve in oxides and influence the defect structure and defect dependant properties of many metal oxides. The early work of Hill (95), Shores and Rapp (96) provide appropriate examples, and further references are available in the publications of Norby and Kofstad (97). From the published information it would appear that hydrogen defects only have a significant influence on the total ionic conductivity in oxides with low intrinisic defect concentrations. For highly doped anionic conductors exhibiting high values of oxygen mobilities protonic conductivity is usually insignificant as demonstrated for $Zr(Y)O_{2-x}$ solid solutions by Wagner (98). It follows, therefore, that the series of papers published by Iwahara et al (99) have attracted much attention because these investigators have claimed high values for protonic conductivity in Yb doped $SrCeO_3$. Subsequent work by Mitsui et al (100), and Iwahara (101) suggested that rare-earth doped $BaCeO_3$ exhibited even higher conductivities. Further studies by Scherban et al (102) and by Bonanos et al (103) have provided additional data and confirmed that a significant protonic conductivity does occur in these doped perovskites, which is in accordance with diffusion studies of Ishigaki et al (104). However the relative values of protonic, oxygen ion, and electronic conductivity remain in dispute and indeed will be a function of the gaseous environment surrounding the samples. Bonanos et al (137), for example, have recently demonstrated that $BaCe_{0.85}Gd_{0.15}O_{2.925}$ exhibits values of oxygen ion conductivity greater than $Zr_{0.9}Y_{0.1}O_{1.95}$. Although when this material is incorporated into a H_2/O_2 fuel cell the water content of both anode and cathode compartments increases. Open circuit emf measurements on mixed conductors are not very helpful because the observed values reflect the relative exchange currents of the associated Faradaic reactions,

$$\tfrac{1}{2}\ O_2 + 2e \rightarrow O^{2-}$$

$$H_2 \rightarrow 2H^+ + 2e\ ,$$

and mixed potentials are probable (105). Moreover the samples used by many investigators were not very dense ($\sim$ 90% theoretical density) and so the role of protonic surface conductivity requires to be considered as emphasised in a recent note by Dryden (106).

A careful study by Lee et al (107) on single crystal $K\,Ta\,O_3$ indicated that for low concentrations of dopant ions (Fe^{3+}, Ni^{2+}, Cu^{2+}, etc) the amount of dissolved protons was determined by the equilibria,

$$Fe_2O_3\ (KTaO_3) \rightarrow 2Fe''_{Ta} + 3O_O^x + 2V_O^{\cdot\cdot}$$
$$H_2O(g) + V_O^{\cdot\cdot} \rightarrow 2H^{\cdot} + O_O^x$$

It is not immediately apparent, however, what factors govern the maximum solubility of hydrogen in heavily doped oxides. Examination of the thermodynamic data of Dickens and Wiseman (107a) for the hydrogen tungsten bronze, $H_{0.35}WO_3$,

i.e.
$$0.175\ H_2(g) + WO_3 \rightarrow H_{0.35}WO_3,$$
$$\Delta G° = -9{,}600 + 21.4\ T\ (J),$$

indicates that $H_{0.35}WO_3$ becomes unstable at 175°C when the external hydrogen partial pressure is 1 atm. These data suggest that the concentration of protonic charge carriers in heavily doped oxides is unlikely to exceed 1000 ppm (0.1 mole %) at temperatures approaching 1000°C. However the investigations of Ishigaki et al (103) suggest that deuterium can be present in Sr $Ce_{0.95}Yb_{0.05}O_{3-\delta}$ at concentrations upto 1 mole %. Clearly high concentrations of protonic charge carriers and/or high values for the protonic mobility are required to explain the protonic conductivity values (approx. 10^{-2}S/cm at 850°C) measured by Iwahara and co-workers for Sr $Ce_{0.95}Yb_{0.05}O_{3-\delta}$. Investigations of hydrogen transport in these interesting materials by NMR techniques should provide further information about the mechanism of protonic transport.

5. MICROSTRUCTURAL CONSIDERATIONS

The oxygen ion conductors discussed in the present chapter are usually examined as polycrystalline ceramic solids and so it is necessary to consider the role of grain boundaries and other microstructural features such as pores and second phase inclusions. As mentioned in the introduction impedance spectroscopy usually enables the bulk and grain boundary contributions to be separated provided the R-C time constant of the simple series equivalent circuit usually used to model ceramic electrolyte materials are sufficiently different.

It is usual to treat the ceramic microstructure as an array of cubic shaped grains separated by flat grain boundary regions as depicted in Fig. 9. This model often termed the brick layer model was introduced by Beekemans and Heyne (108) and its validity has been exam-

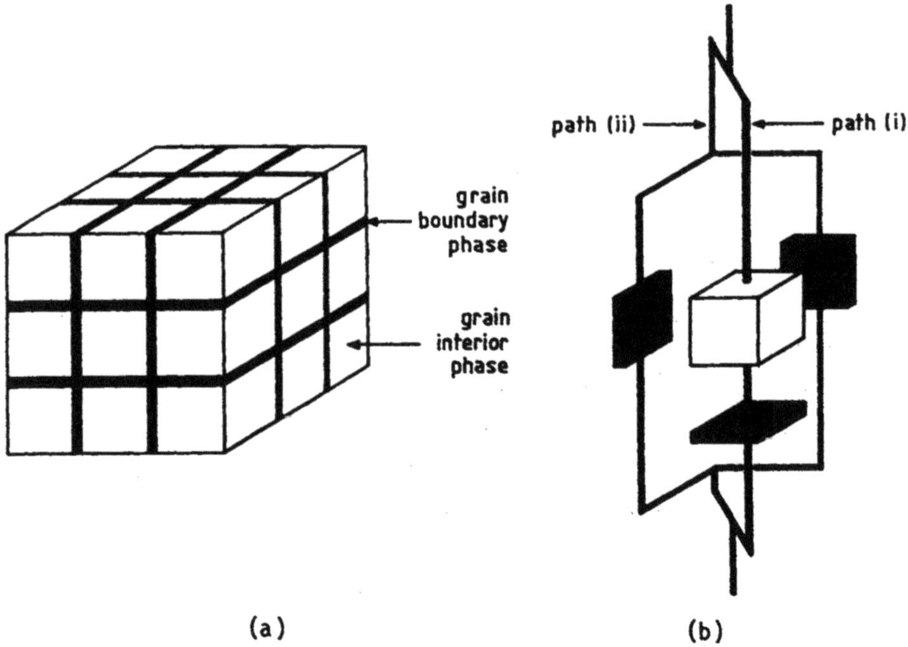

Fig. 9. Brick layer model for ceramic electrolyte samples. (a) Overall view. (b) Exploded view of single cell showing parallel electrical paths, (1) across grain boundaries, (2) along grain boundaries.

ined by Bonanos et al (109). For most situations it was concluded that the simple model was an adequate representation of ceramic electrolyte microstructure and information can be extracted about the electrical behaviour of both the bulk oxide grains and grain boundary phases. In this respect measurements on bicrystals (110, 111) of doped CeO_2 have been useful in that they tend to confirm the assumptions used for the polycrystalline ceramic materials.

Many engineering ceramics contain small quantities ($\sim$ 0.1 %) of aluminosilicates present either as impurities from the original raw materials or as additions to aid the sintering process. The presence of these alumino-silicate phases can have a deleterious effect on the overall D.C. conductivity. However, the electrical properties of these ceramic electrolytes can often be improved by the addition of small quantities of other oxides which can influence the distribution of the second phases. For example, it has been noted (112) that small additions of Al_2O_3 to ZrO_2 based ceramic electrolytes at levels above the solubility limit can reduce the grain boundary resistance. It has been suggested that this is accomplished by the scavenging of SiO_2 to form discrete non conducting particles of mullite ($Al_6Si_2O_{13}$). Subsequent grain boundary movement during grain growth can leave these particles in the interior of grains where their influence on the ionic conductivity is minimal. Further examples of the use of impedance spectroscopy to clarify microstructural features of zirconia based electrolytes are provided by Butler et al (113), and the review by Bonanos et al (109) demonstrates how the microstructure can be optimised for oxygen ion conductivity.

It has been observed that certain ceramic electrolytes particularly those based on Bi_2O_3 solid solutions do not exhibit a grain boundary response even though silica and alumina impurities are present in the initial powders used to fabricate the material. It is believed that Bi_2O_3 based electrolytes can dissolve the alumino-silicate impurities as bismuth oxide has a much greater solubility for Al_2O_3 and SiO_2 compared to the situation for zirconia based electrolytes.

It has been possible in the ceramic electrolyte systems so far discussed to provide interpretations of the grain boundary spectra in terms of the presence of second phases with electrical properties differing from those of the grain interiors. However even ultrapure zirconia based ceramics may still exhibit 'grain boundary' arcs and it is necessary to recognise that the electrical behaviour of grain-grain interfacial contact regions can differ from that of the interior even in the absence of second phases. For example, dopants such as Y_2O_3 may segregate to the interfacial region to provide a relatively high resistance layer. In addition ceramic grain boundaries will usually be charged producing either a depletion (or accumulation) space charge region which would be expected to exhibit electrical behaviour dissimilar to that of the bulk material. The influence of interfacial segregation and space charges has been discussed by Heyne (114), Burggraaf et al (115), Steele and Butler (116). It is anticipated that the electrical response of these space charge regions will be influenced by such processing variables as cooling rate which will determine the distribution of dopants and impurities and evidence for this effect has been provided by Leach et al (117) and El Barhmi (118). Second phases, segregation, and space charge regions can all be expected to be present in commercial ceramic electrolytes and so a complicated situation can often be produced which can only be deconvolated with difficulty.

Another type of microstructure developed in ceramic electrolytes is that of a dispersed second phase within the host matrix. The second phase can be introduced by additions or by solid state precipitation and is a feature of the transformation toughened ceramic electrolytes now being produced for a variety of technological applications. Excellent accounts of the mechanism of transformation toughening are given in the series of conferences on the Science and Technology of Zirconia (119) and it is noteworthy that tetragonal zirconia additions to Bi_2O_3 based electrolytes have also proved effective (63). The electrical response of two phase dispersions in zirconia based electrolytes has been discussed by Bonanos et al (109). Much attention has also been given to the role of dispersed second phase in 'composite' electrolytes which exhibit enhanced ionic conductivity (120). Much of the earlier work (121,122) was concerned with lithium ion conductors but more recently enhanced anionic conductivity has been observed with CaF_2 and $SrCl_2$ matrix composite conductors, and Gray (123) has measured enhanced anionic conductivity in tetragonal ZrO_2-CeO_2 composite electrolytes.

6. CONCLUDING REMARKS

6.1. General

Most technological applications of oxygen ion conductors require high specific conductivity values ($\sim 10^{-1}$S/cm) which in turn implies both a high concentration of charge carriers and a large mobility of these charge carriers. However our understanding of ionic conductivity in the high concentration range is still at a rudimentary level for most systems (e.g. beta-alumina, glasses, fluorite oxide solid solutions) and so it is customary to discuss the situation for low defect concentration levels and then to examine how far the models proposed for the dilute range can be extended to the high concentration range. This approach will be followed here and supplemented by a brief survey of some empirical relationships that have been derived to summarise the behaviour of selected oxygen ion conductors.

Finally the chapter will conclude with remarks relating to the likelihood of synthesising new materials with improved oxygen ion conductivity.

6.2. Dilute or Low Concentration Range

In the low concentration range it is possible to treat the defect equilibria equations in terms of concentrations instead of activities by assuming the various defects present obey Henry's Law. This simplification gives rise to the so-called 'mass action' equations which were originally used to interpret the transport properties of the alkali halides (124). As discussed in section 3.2. this approach has been very successful in interpreting oxygen ion conductivity in CeO_2-R_2O_3 solid solutions up to the concentration exhibiting the maximum conductivity which typically occurs at 1.5-2% anion vacancies (3-4 mole % R_2O_3). Interaction between theory and experiment has been most productive and strategies have been derived to optimise oxygen ion conductivity by minimising the elastic strain energy component of the dopant vacancy, $R_{Ce}^{'}$ - $V_O^{\bullet\bullet}$, association energy term. This can be accomplished when the mismatch between the effective host and dopant cation radius is a minimum.

Whilst this approach has undoubtedly been useful in providing a framework for the optimisation of oxygen ion conductors the model is over simplified and the details of the specific defect configuration also needs to be taken into account. For example, as discussed in section 3.2. Nowick has demonstrated that for dilute $Ce(Y)O_{2-x}$ electroytes the ionic conductivity should be proportional to (dopant concentration)$^{1}/_{3}$ and not be independant of dopant concentration as predicted by the simple associated defect pair complex model. At this stage it should also be noted that a model due to Brailsford (125) also predicts that the activation energy for conduction should be proportional to (dopant concentration)$^{1}/_{3}$. In addition this model indicates how the macroscopic dielectric constant increases with increasing anion vacancy concentration.

In contrast it appears from the results for tetragonal zirconia solid solutions discussed in section 3.3 that the simple model provides an adequate reprepresentation of the variation of conductivity with composition. That different electrolyte solid solutions require different detailed theoretical models is not surprising and this situation does emphasise the need to consider the chemistry involved in the local defect configurations. Some cations find it difficult to modify their local environment and adopt defect arrangements that allow the more stable co-ordination number to be retained (45,126). Defect structures in dilute zirconia solid solutions would not necessarily be expected to be the same in doped ceria electrolytes because whereas Ce^{3+} is a stable ion Zr^{3+} is very unstable, and the Zr^{4+} ion prefers to exist in structural arrangements with co-ordination numbers of 7 or 8. Mention has also been made in section 3.6 that the anion configurations in defect clusters associated with Bi_2O_3 solid solutions are probably optimised to accommodate to the position of the lone pair orbital associated with the Bi^{3+} cation. More emphasis on the chemistry of the ions involved, therefore, could be fruitful in the development of the details of the structural features influencing rapid oxygen ion migration.

6.3. Concentrated Solutions.

The results already summarised in earlier sections show that an ionic conductivity in fluorite oxide solid solutions goes through a maximum around 2% anion vacancies followed by a sharp decrease as the dopant concentration is increased. This strong decrease in conductivity is due to increasing values of the activation energy. Computer simulation calculations (127) are able to generate the general shape of the conductivity-dopant concentration curves, and this approach does emphasise that the drop in conductivity is associated with an increasing number of deep-trap sites at the higher concentrations which effectively immobilize the vacancies. Quantitative analysis of the situation will be difficult as it was necessary for Catlow and Parker (142) to assume a random dopant distribution to simplify the computing procedures, whereas in practice ordering is also common on the cation sublattice (6, 23).

Interpretation of ionic transport data in disordered solids also arises when considering the electrical response of the solid as a function of frequency. Interactions between defects will lead to a frequency dependent conductivity having the general form,

$$\sigma = \sigma_0 + j \varepsilon_0 \varepsilon^* \omega$$

where ε^* is the complex permittivity. The variation of conductivity with frequency can be due to a distribution of relaxation times, and various models have been proposed (8). The depression of the semicircles (Fig. 1) often encountered in impedance spectroscopic measurements on ceramic electrolytes is often interpreted in terms of a distribution of relaxation times. New approaches are being developed to account for the effect of disorder on transport process and a useful summary is provided by Macdonald et al (8). In a disordered material there will be a distribution of energy barriers and so there will be a distribution of waiting or residence times for a particle on a site. The distribution function $\psi(t)$ for the residence time has been discussed by Scher and Lax (128), and the associated expressions have been used by Abelard and Baumard (129) to interpret the conductivity-frequency response of zirconia solid solutions. An interesting extension of the dispersive transport model has been proposed by Glarum (130), who suggested that relaxation at a given site can only occur when approached by a mobile charge carrier. This concept has recently been applied by Elliott (131) to ionic conduction in glasses and it would be interesting to examine whether the same approach would be fruitful for interpreting the impedance spectra of concentrated solutions of fluorite oxide solid solutions.

So far the discussion has been concerned with oxygen ion electrolyte materials. It should be emphasised, however, that the general concepts probably also apply to mixed anionic/electronic conductors. Reference (section 4.1) has already been made to oxygen ion conductivity in the superconducting material $La_{2-x}Sr_xCuO_4$ in which addition of strontium produces a rapid decrease in oxygen ion diffusion values due to trapping of oxygen vacancies within $Sr_{La}^{'}$ - $V_O^{\bullet\bullet}$ defect associates.

Recent studies on ionic conductivity in the concentrated solution region, e.g. ZrO_2-Sc_2O_3 (132), ZrO_2-Yb_2O_3-Y_2O_3 (133), HfO_2-Er_2O_3 (134) confirm the overall trends reported by earlier workers but provide no additional insight into the detailed mechanisms. The data of Badwal (132) for the ZrO_2-Sc_2O_3 system suggests an associated defect pair energy of 0.5 eV but the break in the conductivity-temperature plot at 850°K is indicative of a lower value for the association energy (28). Corman and Stubican (133) demonstrate that the conductivity of $Zr_{0.9}Y_{0.1}O_{1.95}$ gradually increases as Yb_2O_3 is substituted for Y_2O_3. This observation is in accordance with the model discussed in section 3.2. as the strain energy term is expected to be smaller as the radius of Yb^{3+} is closer to the radius of Zr^{4+} than Y^{3+}. Similar comments also apply to the results of Pascual et al (135) for the system $Zr_{0.9}Y_{0.05}Er_{0.05}O_{1.95}$.

6.4. Activation Energy - Pre-Exponential Relationships.

Hohnke (138) has drawn attention to the linear relationship between $\ln\sigma_o(A)$ and ΔE for the system $Zr(In)O_{2-x}$ where σ_o and ΔE are the pre-expontial term and activation energy, respectively, for ionic conductivity. This is an example of the empirical Meyer-Neldel (M-N) rule, $\ln A = a(\Delta E) + b$, which has been discussed by Dosdale and Brook (139, 140). More recently Nowick et al (141) have examined the physical basis for the M-N rule in terms of classical models (124) for lightly doped alkali halides and proposed a modified M-N rule,

$$\ln(A/A_{ref}) = \Delta E/kT_o = (\Delta Ha - \Delta Hm)/kT_o \qquad \ldots\ldots 2$$

where ΔHa is the association energy for defect pairs, ΔHm is the enthalpy of motion, T_O is the intersection point between stage II (free carriers) and stage III (associated defects) and A ref is given by the usual expression,

$$A\ ref = (Z\ \lambda^2 e^2\ V_0\ k/6)\ \omega_0\ c\ (1-c).$$

Equation 8 differs from the original M-N rule because A ref contains the factor C(1-c) which depends upon the concentration of the sample. In practice, however, the quantity log C(1-c) does not change appreciably over a series of compositions, so that the departure from the original M-N rule can be expected to be small.

A selection of data are presented in Fig. 10. Reasonable values of T_O = 930K and ΔHm = 0.77 eV are obtained for $Zr(In)O_{2-x}$, indicating the plausibility of the concept of a transition between stage III and stage II around 930°K. The $Zr(Y)O_{2-x}$ series also conforms to the M-N rule even though the data are for different structures, tetragonal, tetragonal + cubic, cubic. A similar situation prevails for the La_2O_3-SrO compositions.

It would appear, therefore, that the M-N rule does provide a useful way of examining the ionic conductivity behaviour exhibited by a family of oxygen ion electrolytes. Attention can then be focussed upon departures from the simple relationship so that initially the experimental results can be checked, and if verified, then alternative interpretations must be provided which do not involve the somewhat restrictive assumptions implied in the M-N rule that ΔHm and σ_O are constant over appreciable changes of composition.

6.5. Future Developments

The resurgence of interest in solid oxide fuel cells and the observation of rapid oxygen ion transport in high T_C superconductors is likely to stimulate both experimental and theoretical investigations of oxygen ion migration in solids. It is difficult to predict whether significant advances will be made in the near future but the high oxygen ion conductivities reported for Ba $Ce_{0.85}Gd_{0.15}O_{2.925}$, and the

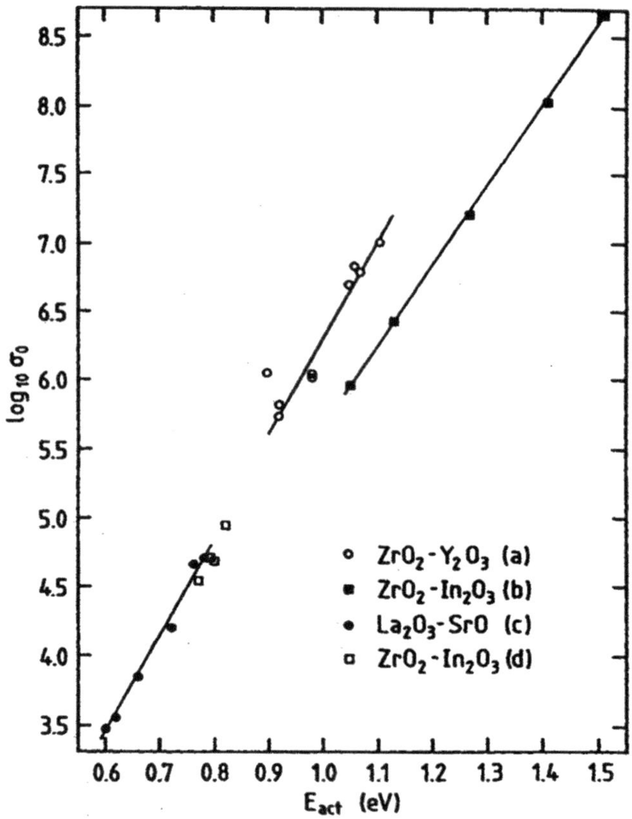

Fig. 10. Compilation of pre-exponential factors vs. activation energies for selected ceramic electrolytes.

$La_{2-x}Sr_xCu\,O_{4-y}$ family, do suggest that additional interesting materials will continue to be discovered. As emphasised in section 3.3, it is likely that in the search for new materials more emphasis will be given to chemical aspects of the ions involved so that the local configurations can be optimised to provide low energy pathways for the migrating oxygen ions.

7. ACKNOWLEDGEMENTS

The author wishes to thank past and present members of the Wolfson Unit for Solid State Ionics at Imperial College for many useful discussions which have contributed to the preparation of the present chapter the final version of which was completed on 15th April 1988.

8. REFERENCES.

1. 'Nonstoichiometric Oxides' ed. O. Toft Sorensen. (Academic Press, 1981).
2. Proceedings of Conf. on High Temperature Solid Oxide Electrolytes Aug. 16-17, 1983. Compiled by F.J. Salzano. Publication BNL-51728 (Brookhaven Nat. Lab. New York, 1983).
3. J.A. Kilner and J.D. Faktor, p.347 in 'Progress in Solid Electrolytes' Eds. T.A. Wheat, A. Ahmad and A.K. Kuriakose. Publication ERP/MSL 83-94 (TR) (Energy, Mines and Resources, Ottawa, Canada, 1983).
4. Solid Electrolytes ed. E.C. Subbarao (Plenum, 1980).
5. E.C. Subbarao and H.S. Maiti. Solid State Ionics 11, 317, 1984.
6. A.S. Nowick p.143 in 'Diffusion in Crystalline Solids'. eds. G.E. Murch and A.S. Nowick. (Academic Press, 1984).
7. H. Matzke. Adv. in Ceram. 17, 1, 1986.
8. Impedance Spectroscopy. ed. J.R. MacDonald (John Wiley, 1987).
9. B.C.H. Steele and J.M. Floyd. Proc. Brit. Ceram. Soc. 19, 55, 1971.
10. D.J. Derry, D.G. Lees and J.M. Calvert. Proc. Brit. Ceram. Soc. 19, 77, 1971.
11. J.A. Kilner, B.C.H. Steele and L. Ilkov. Solid State Ionics 12, 89, 1984.
12. C. Monty. Radiat Eff. 74, 29, 1983.
13. J.A. Kilner. Mat. Sci. Forum. 7, 205, 1986.
14. D. Brinkman. p.1 in 'Progress in Solid Electrolytes'. Eds. T.A. Wheat, A. Ahmad and A.K. Kuriakose. Publication ERP/MSL 83-94 (TR) (Energy, Mines and Resources, Ottawa, Canada, 1983).
15. K. Fuda, K. Kishio, S. Yamauchi, K. Fueki and Y. Onoda. J. Phys. Chem. Solids. 46, 1141, 1985.
16. T. Suemoto and M. Ishigame. Solid State Ionics 21, 225, 1986.
17. M. Ishigame, E. Yoshida, S. Shin and Y. Shimada. Solid State Ionics 20, 105, 1986.

18. M. Ishigam and E. Yoshida. Solid State Ionics 23, 211, 1987.
19. M. Weller and H. Schubert. J. Am. Ceram. Soc. 69, 573, 1986.
20. D.P. Almond and A.R. West. Solid State Ionics 26, 1988.(To be published).
21. A. Baudry, P. Boyer and A.L.de Oliveira. J. Phys. Chem. Solids. 43 871, 1982.
22. A.S. Nowick. Comments Solid State Phys. 9, 85, 1979.
23. J.A. Kilner and B.C.H. Steele. p.233 in Nonstoichiometric Oxides. ed. O. Toft Sorensen. (Academic Press, 1981).
24. R. Gerhart-Anderson and A.S. Nowick. Solid State Ionics 5, 547, 1981.
25. J.A. Kilner and R.J. Brook. Solid State Ionics 6, 237, 1982.
26. V. Butler, C.R.A. Catlow, B.E.F. Fender and J.H. Harding. Solid State Ionics 8, 109, 1983.
27. J.A. Kilner. Solid State Ionics 8, 201, 1983.
28. J.A. Kilner and C.D. Waters. Solid State Ionics 6, 253, 1982.
29. D.Y. Wang, D.S. Park, J. Griffith and A.S. Nowick. Solid State Ionics 2, 95, 1981.
30. Y.S. Zhen, S.J. Milne and R.J. Brook. Sci. of Ceramics 14, 1987 in press.
31. H. Arai, T. Kunisaki, Y. Shimiza and T. Seiyama. Solid State Ionics 20, 241, 1986.
32. N. Bonanos, R.K. Slotwinski, B.C.H. Steele and E.P. Butler. J. Mat. Sci. Lett. 3, 245-248, 1984.
33. S.P.S. Badwal. J. Mat. Sci. Lett. 6, 1419, 1987.
34. A. Sellars and B.C.H. Steele. Centre for Advanced Ceramics. Imperial College, London, U.K. Private Communication.
35. J.R. Jurado, C. Moure and P. Duran. p.305 in Transport-Structure Relations in Fast Ion and Mixed Conductors. Eds. F.W. Poulsen, N.H. Andersen, K. Clausen, S.Skaarup and O.T. Sorensen (Risø, Denmark, 1985).
36. X. Turrillas, A.P. Sellars and B.C.H. Steele. Abs. B2-5 in 6th Intl. Conf. on Solid State Ionics Sept. 6-11, 1987. Garmisch-Partenkirchen, Germany. To be published in Solid State Ionics.

37. A.N. Cormack. p.15 Transport-Structure Relations in Fast Ion and Mixed Conductors. Eds. F.W. Poulsen, N.H. Andersen, K. Clausen, S. Skaarup and O.T. Sorensen (Risø, Denmark, 1985).
38. M.P. Van Dijk, A.T. Burggraaf, A.N. Cormack and C.R.A. Catlow. Solid State Ionics 17, 159, 1985.
39. H.J. Rossell. Adv. in Ceramics 3, 47, 1981.
40. J.D. Faktor, J.A. Kilner and B.C.H. Steele. Stud. Inorg. Chem. 3, 207, 1983.
41. P.K. Moon and H.L. Tuller. Abs. B2-6 in 6th Intl.Conf. on Solid State Ionics. Sept. 6-11, 1987. Garmisch-Partenkirchem, Germany. To be published in Solid State Ionics.
42. T. Van Dijk, K.J. de Vries and A.J. Burggraaf. Phys. Stat. Sol. (a), 58, 115, 1980.
43. T. Van Dijk, R.B. Helmhaldt and A.J. Burggraaf. Phys. Stat. Sol. (b) 101, 765, 1980.
44. B.E. Schestz and W.B. White. J. Am. Ceram. Soc. 62, 468, 1979.
45. T. Uehara, K. Koto, S. Emura and F. Kanamaru. Solid State Ionics 23, 331, 1987.
46. A.J. Burggraaf, T. Van Dijk and M.J. Verkerk. Solid State Ionics, 5, 519, 1981.
47. T. Fournier, J.Y. Nots, J. Muller and J.C. Joubert. Solid State Ionics. 15, 71, 1985.
48. J.D. Faktor. 'A Study of Oxygen Ion Conductivity in the System $Nd_2Zr_2O_7$-$Nd_2Ce_2O_7$'. Ph.D. Thesis, University of London, 1984.
49. F. Vanbaelinghem. J. Appl. Electrochem. 6, 67, 1976.
50. Y.S. Zhen, S.J. Milne and R.J. Brook. p.1989 in High Tech. Ceramics ed. P. Vincenzini (Elsevier 1987).
51. Y.S. Zhen, S.J. Milne and R.J. Brook. Proc. Brit. Ceram. Soc. to be published.
52. T. Norby and P. Kofstad. Solid State Ionics 20, 169, 1986.
53. T. Norby and P. Kofstad. J. Am. Ceram. Soc. 69, 784, 1986.
54. B.C.H. Steele, B.E. Powell and P.M.R. Moody. Proc. Brit. Ceram. Soc. 10, 87, 1968.

55. P.W.M. Jacobs and D.A. MacDonaill. Solid State Ionics 23, 279, 1987.
56. L.G. Sillen. Ark. Kemi. Mineral. Geol. 12A(18), 1, 1937
57. P.W.M. Jacobs and D.A. MacDonaill. Solid State Ionics 23, 295, 1987.
58. J.C. Boivin and D.J. Thomas. Solid State Ionics 5, 523, 1981.
59. A. Laarif and F. Theobald. Solid State Ionics 21, 183, 1986.
60. C.E. Infante, C. Gronemeyer and F. Li. Solid State Ionics 25, 63, 1987
61. G. Meng, C. Chen, P.Yang and D. Dingkun. Abs. P1-17 in 6th Intl. Conf. on Solid State Ionics. To be published in Solid State Ionics.
62. W.L. Cai, I.D. Raistrick and R.A. Huggins. Abs. 169. J. Electrochem. Soc. 131, 295C, 1984.
63. M. Dumelie, G. Nowogrocki and J.C. Boivin. Abs. P1-45 Proc. of 6th Intl. Conf. on Solid State Ionics. To be published in Solid State Ionics.
64. T. Takahashi, T. Esaka and H. Iwahara. J. Appl. Electrochem. 7, 299, 1977.
65. C. Wang, X. Xu and B. Li. Solid State Ionics. 13, 135, 1984.
66. J.R. Durado, C. Moure, N. Valverde and P. Duran. Abs. P1-44 Proc. of 6th Intl. Conf. on Solid State Ionics. To be published in Solid State Ionics.
67. Chemistry of Catalytic Processes, B.C. Gates, J.R. Katzer and G.C.A. Schuit. (McGraw Hill, 1979).
68. L. Boon and R. Metselaar. Solid State Ionics 16, 201, 1985.
69. T. Takahashi and H. Iwahara. Denki Kagaku 35, 433, 1967.
70. T. Takahashi. p.1011 in 'Physics of Electrolytes' Vol. 2, ed. J. Hladik (Academic Press, 1972).
71. K.W. Browall, O. Muller and R.H. Doremus. Met. Res. Bull. 11, 1475 1976.
72. B.C.H. Steele and J.A. Kilner. p.136 in Proc. of Conf. on High Temperature Solid Oxide Electrolytes Aug. 16-17, 1983. Compiled by F.J. Salzano. Pub. BNL-51728 (Brookhaven Nat. Lab. New York, 1983).

73. V.I. Spibsin, Russ. J. Inorg. Chem. 27, 464, 1982.

74. A. Ya Neyman, A.L. Podkorytov and V.M. Zhukovsky. Abs. OC17-P2/32 4th Intl. Conf. on Solid State Ionics, Grenoble, July 4-8, 1983.

75. R. Bredesen and P. Kofstad. Solid State Ionics. To be published.

76. A.E. McHale and H.L. Tuller. Solid State Ionics, 5, 515, 1981.

77. T. Lu and B.C.H. Steele. Solid State Ionics 21, 339, 1986.

78. G.Y. Meng and R.A. Huggins. Solid State Ionics 11, 271, 1984.

79. A.E. McHale, F.A. Kilner and B.C.H. Steele. p.217 in 'Transport in Non Stoichiometric Compounds' Eds. G. Simkovich and V.S. Stubican (Plenum 1985).

80. R. Metselaar and A.M. Hoefsloot. Solid State Ionics 24, 305, 1987.

81. C. Wagner. Progr. Solid State Chem. 10, 3, 1975.

82. L. Heyne, p.169 in 'Solid Electrolytes' ed. S. Geller. (Springer-Verlag, 1977).

83. W. Weppner and R.A. Huggins. Ann. Rev. Mater. Sci. 8, 269, 1978.

84. G.J. Dudley and B.C.H. Steele. J. Solid State Chem. 31, 233, 1980.

85. S.P.S. Badwal, M.J. Bannister and M.J. Murray. J. Electroanal. 168, 363, 1984.

86. J.L. Routbort, S.J. Rothman, B.K. Flandermeyer, L.J. Nowicki and J.E. Baker. J. Mater. Res. 3, 116, 1988.

87. M. Miyayami, Y. Suenaga and H. Yanagida. J. Mat. Sci. 18, 3023, 1983.

88. T. Kenjo, T. Maruyama, T. Yoshida, Y. Kouno and Y. Saito. Report of Res. Lab. of Eng. Mats. Tokyo Inst. of Tech. Nagatsuta, Yokahama 227 11, 17, 1986.

89. T. Esaka, H. Iwahara and K. Kunieda. J. Appl. Electrochem. 12, 235 1982.

90. J. Berger, I. Riess and D.S. Tannhauser. Solid State Ionics, 15, 225, 1985.

91. B.C.H. Steele and J.M. Floyd. Proc. Brit. Ceram. Soc. 19, 55,1971.

92. J.A. Kilner. Dept. of Mats. Imperial College. London. Private Communication.

93. H.L. Tuller, J.A. Kilner, A.E. McHale and B.C.H. Steele. p.315 in Reactivity of Solids. Eds. P. Barrett and L.C. Dufour. (Elsevier 1985).

94. F.R. Van Buren, G.H.J. Broers, A.J. Bouman and C. Boesveld. J. Electroanal Chem. Interfacial Electrochem. 88, 353, 1978

95. G.J. Hill. Brit. J. Appl. Phys. Ser. 2 1, 1151, 1968.

96. D.A. Shores and R.A. Rapp. J. Electrochem. Soc. 119, 300, 1972.

97. T. Norby and P. Kofstad. J. Physique Cl, Suppl. 2, 47, C-849, 1986.

98. C. Wagner. Ber. Bunsenges. Phys. Chem. 72, 778, 1968.

99. H. Iwahara, T. Esaka, H. Uchida and N. Maeda. Solid State Ionics 3/4, 359, 1981; 11, 117, 1983; 18/19, 1003, 1986.

100. A. Mitsui, M. Miyayama and H. Yanagida. Solid State Ionics 22, 213 1987.

101. H. Iwahara, H. Uchida, K. One and K. Ogaki. J. Electrochem. Soc. 135, 529, 1988.

102. T. Scherban, W.K. Lee and A.S. Nowick. Abs. B6-3, p.171 6th Intl. Conf. on Solid State Ionics, Sept. 6-11th 1987, Garmisch Partenkirchen. Germany.

103. N. Bonanos, B. Ellis and M.N. Mahmood. Abs. B6-2 p. 170 ibid.

104. T. Ishigaki, S. Yamaudi, K. Kishio and K. Fuecki. Solid State Ionics, 21, 239, 1986.

105. B.C.H. Steele. Solid State Ionics 12, 391, 1984.

106. J.S. Dryden. Solid State Ionics 23, 319, 1987.

107. W.K. Lee, A.S. Nowick and L.A. Bootner. Solid State Ionics 18/19, 989, 1986.

107a P.G.Dickens and P.J. Wiseman. p.211, Vol. 10, Solid State Chem. Inorganic Chem. Series Two. Intl. Review of Sci. (Butterworths, 1975).

108. N.M. Beckemans and L. Heyne. Electrochim. Acta. 21, 303, 1976.

109. N. Bonanos, B.C.H. Steele and E.P. Butler. p.191 in 'Impedance Spectroscopy'. Ed. J.R. MacDonald (John Wiley, 1987).

110. A.L. Dragoo, C.K. Chiang and A.D. Franklin. Solid State Ionics 7, 249, 1982.
111. K. El Adham and A. Hammou. Solid State Ionics 9 & 10, 905, 1983.
112. J. Drennan and E.P. Butler. J. Am. Ceram. Soc. 65, C-194, 1982.
113. E.P. Butler, R.K. Slotwinski, N. Bonanos, J. Drennan and B.C.H. Steele. p.572 in Sci. and Tech. of Zirconia II. Eds. N. Claussen, M. Ruhle and A.H. Heuer. Advances in Ceramics. Vol.12 (Am. Ceram. Soc. 1984).
114. L. Heyne p.425 in 'Mass Transport in Solids'. eds. F. Beniere and C.R.A. Catlow. (Plenum Press, 1983).
115. A.J. Burggraaf, M. van Hemert, D. Scholten and A.J.A. Winnubst. p.797 Mat. Sci. Monograph. 28B (Reactivity of Solids) 1985.
116. B.C.H. Steele and E.P. Butler. Brit. Ceram. Proc. 36, 45, 1985.
117. C. Leach, P. Tanev. and B.C.H. Steele. J. Mater. Sci. Lett. 5, 893 1986
118. A. El Barhmi, E.J.L. Schouler, A. Hammou and M. Kleitz. Abs. p1-39 6th Int. Conf. on Solid State Ionics, Garmisch-Partenkirchen, Sept. 6-11, 1987.
119. Science and Technology of Zirconia Vol. 3, Vol. 12, (Am Ceram. Soc. 1981 and 1984.
120. J. Maier. J. Electrochem. Soc. 6, 1524, 1987.
121. S. Fujitsu, M. Miyayama, K. Koumoto, H. Yanagida and T. Kanozawa. J. Mat. Sci. 20, 2103, 1985.
122. S. Fujitsu, K.Koumoto and H. Yanagida. Solid State Ionics 18 & 19 1146, 1986.
123. D.J. Gray. Dept. of Materials, Imperial College. Private Communication.
124. L.W. Barr and A.B. Lidiard. p.152 in Vol. 10 of 'Physical Chemistry' Eds. H. Eyring, D. Henderson and W. Jost. (Academic 1970).
125. A.D. Brailsford. Solid State Ionics 13, 87, 1984.
126. C.R.A. Catlow, A.V. Chadwick, G.N. Greaves and L.M. Moroney. J. Am. Ceram. Soc. 69, 272, 1986.
127. C.R.A. Catlow. p.405 in 'Mass Transport in Solids'. Eds. F. Beniere and C.R.A. Catlow (Plenum 1983).

128. H. Scher and M. Lax. Phys. Rev. B7, 4491, 1973.
129. P. Abelard and J.F. Baumard. Phys. Rev. B26, 1005, 1982.
130. S.H. Glarum, J. Chem. Phys. 33, 639, 1960.
131. S.R. Elliott. Solid State Ionics. To be published.
132. S.P.S. Badwal. J Mat. Sci. 18, 3117, 1983.
133. G.S. Corman and V.S. Stubican. J. Am. Ceram. Soc. 68, 174, 1985.
134. G.W. Jordan and M.F. Berard. J. Am. Ceram. Soc. 68, 464, 1985.
135. C. Pascual, J.R. Jurado and P. Duran. J. Mat. Sci. 18, 1315, 1983.
136. M. Lacerda, J.T.S. Irvine, F.P. Glasser and A.R. West. Nature, 332, 525, 1988.
137. N. Bonanos, B. Ellis and M.N. Mahmood. Solid State Ionics. To be published.
138. D.K. Hohnke. J. Phys. Chem. Solids. 41, 777, 1980.
139. T. Dosdale and R.J. Brook. J. Mat. Sci. 13, 167, 1978.
140. T. Dosdale and R.J. Brook. Solid State Ionics. 8, 297, 1983.
141. A.S. Nowick, W.K. Lee and H. Jaim. Abs. C5-6 in Proc. of 6th Intl. Conf. on Solid State Ionics. To be published in Solid State Ionics.
142. C.R.A. Catlow and S.C. Parker. Chap. 15 in 'Computer Simulation of Solids'. Eds. C.R.A. Catlow and W.C. Mackrodt. (Springer-Verlag. 1982).
143. P. Shuk, S. Jakobs and H.H. Mobius. Z. Anorg. Allg. Chem. 524, 144, 1985.
144. T.A. Ramanayaran and R.A. Rapp. Met. Trans. 3, 3239, 1972.
145. T. Ishigaki, S. Yamauchi, K. Kishio, J. Mizusaki and K. Fueki. J. Solid State Chem. 73, 179, 1988.

INTERCALATION COMPOUNDS:
METAL IONS IN CHALCOGENIDE AND OXIDE HOSTS

Keld West

Institute of Physical Chemistry,
The Technical University of Denmark, DK2800 Lyngby, Denmark.

1. INTRODUCTION

Intercalation is the reversible insertion of mobile *guest* species into a rigid *host* lattice containing a contiguous network of empty lattice sites of appropriate sizes. Normally the main factors controlling the rate of solid-state reactions is the nucleation and growth of product phases and transport of reactants in the solid to the reaction site. As the activation energy barriers for these processes usually are high, most solid-state reactions and transformations proceed at low rates at room temperature. There are, however, groups of solid-state reactions that are known to be less restricted kinetically: *Topotactic* reactions maintain a close structural relationship between the existing phase and the product, and nucleation and growth of product phases are here greately facilitated. An outstanding example is the austenite-martensite transition in steel, a topotactic phase transformation where growth of the product phases occur at velocities approaching the speed of sound[1]. Phases containing a connected system of empty lattice positions may exhibit unusually high diffusion rates for species fitting into these sites, as seen in *fast ion conductors*.

In intercalation reactions both of these favourable kinetic conditions are present. During the intercalation process the host lattice may undergo changes in structural dimensions to accomodate the guest, but the integrity of the host lattice is preserved. Originally the term intercalation referred to the insertion of an extra day in the calender, such as the 29th of February in leap years, and it is in solid-state chemistry traditionally used in connection with lamellar or layered host materials as shown in fig. 1. Some authors prefer the more general term *insertion* in connection with other types of hosts, but the two term are here used interchangeably.

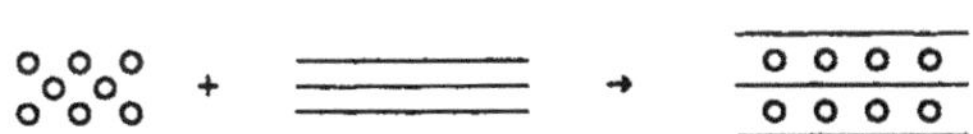

Fig. 1. Schematic representation of intercalation reaction in a layered host material.

Intercalation chemistry has a long history. According to Weiss[2] one of the key processes in the original manufacture of china in the 6th and 7th century was modification of the mechanical properties of kaolinite through the intercalation of urea or formamide. The first tungsten bronzes were described in 1823 by Wöhler[3], and graphite intercalation compounds in 1841 by Schauffautl[4]. It was, however, only when structural investigations became possible that the common reaction principle leading to the formation of these compounds became evident. In a systematic x-ray study of a series of intensely colored sodium tungsten bronzes, Hägg discovered in 1935[5] that these substances actually belonged to a continuous solid solution series: Na_xWO_3 $0.32 \leq x \leq 0.93$, all having the same cubic W-O-lattice. The large stoichiometric range seen here is typical of intercalation systems.

Today a wide variety of intercalation compounds is known. The guests include both molecular species and ions. Insertion of ions require the simultaneous accomodation of compensating charges by the host matrix, either delocalized in its band structure or localized by reduction/oxidation of discrete atomic or molecular entities of the host. In both cases the electronic conductivity of the host is an important factor in determining the kinetics of the intercalation reaction. Lithium and sodium intercalation has, however, been demonstrated in NASICON-type titanium phosphates[6], which would normally be regarded as electronic insulators. Also in the case of molecular intercalates a partial charge transfer between guest and host is often seen.

A wide range of host structures has been investigated, ranging from three-dimensional framework structures with empty lattice positions forming channels or interconnected cavities, over two-dimensional layer structures, where the insertion takes place in van der Waals gaps between layers, to one-dimensional chain structures or nearly amorphous materials where the inserted species occupy positions between the structural elements, fig. 2.

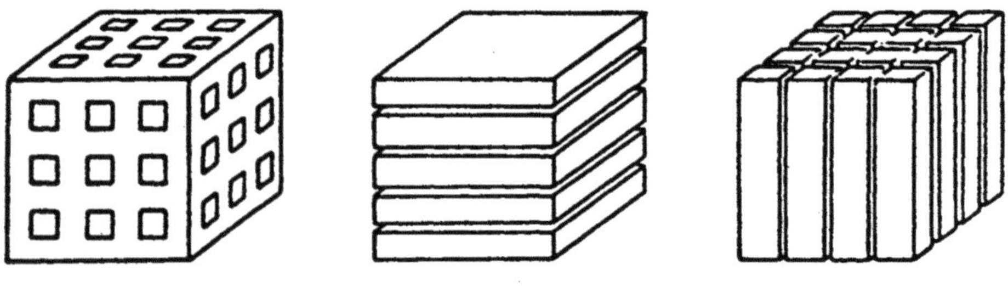

Fig. 2. Schematic representation of host structures with different dimensionality[7].

The *layered* materials have proven to be particulary versatile host materials, as they can adapt to guest species of very different size by changes in interlayer spacing. In this way tantalum disulphide, TaS_2, with an original interlayer spacing of 6.1 Å forms complexes with long-chain amines such as octadecylamine where the interlayer spacing is increased up to 57 Å[8], as well as with small ions as lithium resulting in only a small increase to 6.4 Å[9].

The high degree of dimensional stability that is required of the host lattice demands strong, directional bonding. This is only obtained in compounds with an appreciable degree of covalency as metal oxides or chalcogenides (sulphides, selenides and tellurides) or in carbon based materials as graphite or conjugated polymers. The fractional covalent character of a single bond A-B can be estimated from the electronegativities of species A and B (X_A and X_B)[10]:

$$f_c = 1-f_i = \exp[-\frac{1}{4}(X_A - X_B)^2] \qquad (1)$$

where f_i is called the ionicity of the bond.

As the intercalation process can be performed at temperatures where the host lattice is kinetically stable against transformations that would involve bond breaking, intercalation reactions may lead to metastable products. In this way intercalation reactions constitute a new synthetic route to phases that cannot be prepared by conventional high temperature techniques. This preparation method has been coined *chimie douce* or *soft chemistry*. An example of a product synthesized this way is the layered vanadium disulphide, VS_2, a metastable analog to hexagonal titanium disulphide, TiS_2, which was prepared by extraction of lithium ions from the stable compound $LiVS_2$ using iodine as oxidation agent[11]:

$$LiVS_2 + \frac{1}{2}I_2 \rightarrow LiI + VS_2 \qquad (2)$$

In the same way a cubic modification of TiS_2 has been prepared by oxidation of the high temperature stable spinel phase $CuTi_2S_4$ with ferric chloride or bromine in acetonitrile at room temperature[13,14]:

$$CuTi_2S_4 + \frac{1}{2}Br_2 \rightarrow CuBr + 2TiS_2 \qquad (3)$$

In these reactions ions are extracted using a chemical oxidation agent, a convenient technique for making larger quantities of a definite composition. Murphy[12] has proposed a series of suitable agents with different redox potentials that can be used for these reactions. A reagent often used to prepare fully lithiated compounds is n-buthyl lithium (n-BuLi)[15].

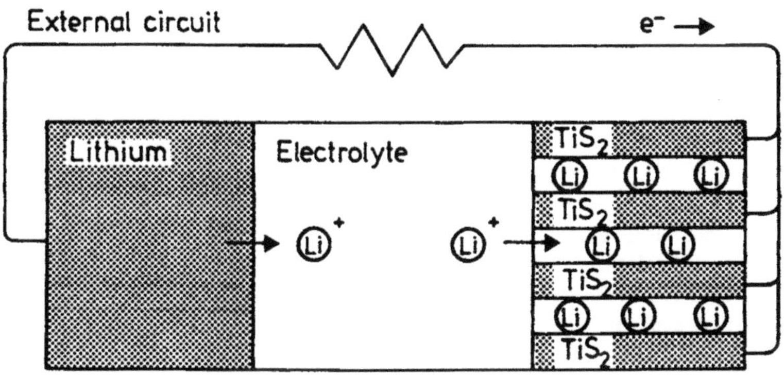

Fig. 3. Schematic representation of electrochemical cell for the intercalation of lithium ions in the layered host TiS_2.

Alternatively, intercalation reactions involving ionic intercalates can be readily controlled and studied by electrochemical techniques. In these cases the host material is used as one of the electrodes in an electrochemical cell. As the other electrode a suitable source or sink of the intercalate is used. For alkali metal intercalates, this electrode will typically be the free alkali metal, fig. 3. Under equilibrium conditions, the open circuit voltage, *emf* (ϵ) of such a cell will be determined by the difference in free energy of the metal in metallic state and in intercalated state ("the intercalation affinity": $-\Delta G_I$):

$$\epsilon = \frac{-\Delta G_I}{nF} \tag{4}$$

where F is Faradays constant and n is the charge of the metal ions in the electrolyte.

Intercalation reactions have many scientific and technological applications, including the possibility of "tuning" electrical or mechanical materials properties, the development of new catalysts and sensors, and the use as electrode materials in secondary (rechargeable) batteries. In this area the intercalation materials have paved the way for a new battery concept: *the all solid-state battery.*

In the present paper the main emphasis will be on ionic intercalates of transition metal oxides and chalcogenides. It is not the intent to present an exhaustive review of the field which is large and rapidly growing, but rather to exemplify important and general features of intercalation systems. The reader is referred to recent reviews[16-22] and monographs[23-24] for more comprehensive treatments and for information on the important subjects intercalation in graphite, organic metals and clays.

2. HOST MATERIALS AND STRUCTURES

The properties of host materials result from the interplay of chemical and structural factors. As already mentioned, a certain degree of covalency is necessary to impart to the host structure the required rigidity. The carbon based materials such as graphite or the conjugated polymers are the most covalent host materials. These materials are *amphoteric*, *i.e.* they are able to intercalate both anions and cations. The oxides and chalcogenides of transition metals are partially ionic in character. Due to differences in electronegativity ($X_O = 3.5$, $X_S = 2.5$, $X_{Se} = 2.4$[25]), the oxides will be the most ionic, whereas the sulphides and selenides have sufficiently small net charges on the chalcogen atoms to make van der Waals bonded layer structures possible. Due to coulombic interactions only cations and electron donors (Lewis bases) can be intercalated in these materials. Recently it has been reported[26] that indium selenide, InSe, in addition to various cations also is able to intercalate iodide ions. The amphoteric behaviour is consistent with the highly covalent bonding in this material.

The most important of the structural factors is the number and size of sites available for intercalation, and the features of the electronic band structure determined by the coordination symmetry of the transition metals. The guest-host interactions are, however, influenced by the host structure in many other ways, some of which are exemplified in the following.

2.1 Three Dimensional Host Structures

Host structures strongly bound in three dimensions exhibit a high degree of specificity with respect to the species inserted. This is due to the spatial restrictions set by the geometry of the diffusion path in the host. The "bottlenecks" determine a maximum size of the intercalant, but it also appears that ions too small to fit the channels will experience a restricted mobility as they tend to stick to the channel walls. In the molybdenum cluster compounds - "Chevrel phases" - the ions are inserted in intersecting channels between Mo_6X_8 clusters, and an unusually wide variety of metal atoms can be accomodated.

2.1.1 Bronzes. In analogy with the bronze colored $NaWO_3$, the term "bronze"[27-30] is used for ternary metal oxides of the general formula $M_xT_yO_z$, where T is a transition metal capable of being reduced, and M is another metal or hydrogen. The composition parameter x varies in accordance with the reduction of T, whereas y and z are constants. A further property which is normally only stated implicitly, is that the bronzes are *thermodynamically stable* over a certain range of compositions x. As the composition changes, variations in colour and electrical properties of the bronze are observed. Most bronzes have a three dimensional framework structure, and though the composition of the host corresponds to a binary oxide, the framework found in the bronze is not necessarily stable in the unintercalated state. The formation of these compounds can be regarded as crystallisation of a binary oxide in the presence of ions too large to find suitable interstitial sites, therefore an entirely different structure is adopted. After the host has been formed around the guest, it is possible to adjust the composition in certain intervals especially in the case of the smaller cations (H, Li, Na, Ag, Cu), whereas the ionic mobility in bronzes of larger cations is severely limited by intercavity bottlenecks. Titanium, vanadium, molybdenum and tungsten bronzes have been extensively studied as these phases exhibit a series of unusual physical and structural properties, and as they can be prepared with varying concentration of valence electrons due to their wide stoichiometric ranges. Some features of this class of materials will be illustrated using the vanadium pentoxide bronzes as examples:

The vanadium bronzes of composition $M_xV_2O_5$ crystallize in different structures depending on the concentration and size of the inserted species M. Some of these structures are shown in fig. 4, and their composition intervals are given in table 1. It is noteworthy that virtually identical α and β structures are formed with all the listed cationic intercalants.

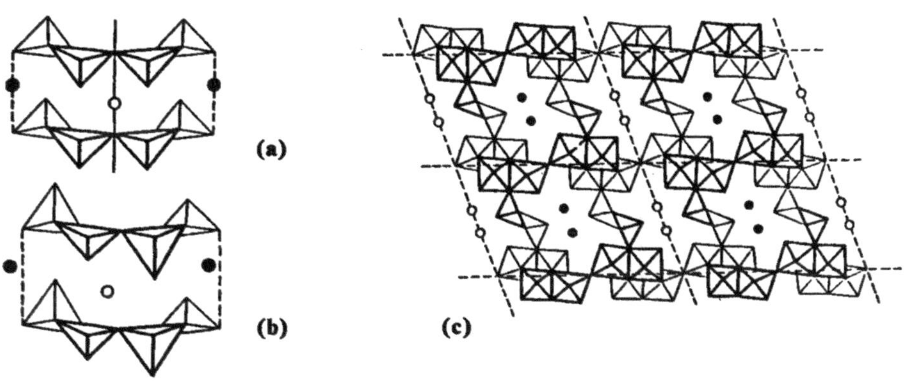

Fig. 4. Structure of vanadium pentoxide bronzes: (a) α-phase, (b) α'-phase, (c) β-phase. Circles mark possible sites of intercalated ions.

The α-phase is structurally closely related to V_2O_5 with layers made up of distorted VO_5 bipyramids forming double zig-zag chains and cross-linked through shared corners. With only a weak V--O interlayer bond this material has both been classified as three-dimensional and two-dimensional. Although the multiplicity of the interlayer sites normally occupied by the inserted ions corresponds to a limiting composition MV_2O_5, the α-phase is normally only formed with small concentrations of intercalant, the concentration range decreasing with increasing ionic radius of M^+. A similar phase, α', with decreased symmetry is however formed with high concentrations of sodium. In the α'-phase there are two types of vanadium sites due to the presence of approximately equal numbers of V^{+4} and V^{+5} atoms.

Table 1.
Composition intervals of $M_xV_2O_5$ bronzes.

Intercalant	Phase	Composition range		Ref
H	α	$0 \leq x \leq 3.8$	*)	31
Li	α	$0 \leq x \leq 0.04$		32
	α	$0 \leq x \leq 2$	*)	33,34
	β	$0.22 \leq x \leq 0.37$		32
	β'	$0.44 \leq x \leq 0.49$		32
	β+β'	$0.22 \leq x \leq 0.62$		35
	β	$0 \leq x \leq 1.33$	*)	36
	γ	$0.88 \leq x \leq 1$		35
	γ	$0.5 \leq x \leq 1$	*)	37
	δ	$0.9 \leq x \leq 1$	*)	38
Na	α	$0 \leq x \leq 0.02$		35
	β	$0.22 \leq x \leq 0.40$		35
	β'	$0.65 \leq x \leq 0.69$		35
	α'	$0.70 \leq x \leq 1.00$		39
	η	$1.28 \leq x \leq 1.45$		39
	κ	$1.68 \leq x \leq 1.82$		39
	-	$0.30 < x < 2.00$	*)	40
Cu	α	$0 \leq x \leq 0.02$		35
	β+β'	$0.26 \leq x \leq 0.64$		35
	ε	$0.85 \leq x \leq 1$		35
Ag	α	$0 \leq x \leq 0.01$		35
	β+β'	$0.29 \leq x \leq 0.41$		35
	δ	$0.67 \leq x \leq 0.86$		35

*) Composition range of metastable low temperature phases.

At room temperature insertion of up to 3.8 H^+ [22] and 3.30 Li^+ [34] per formula unit in α-V_2O_5 has been reported. At small degrees of intercalation, $x = 0.33$ in $H_xV_2O_5$[52] and $x < 0.9$ in $Li_xV_2O_5$, intercalation causes only small changes in the host lattice. In the interval $0.9 < x < 1$ a new room temperature phase (δ) is formed in the lithium system. The α-δ phase transformation is accompanied by a decrease in symmetry like the α-α' transition in the sodium system, but structural investigations have revealed that this transition is more extensive, involving relative displacement and puckering of the layers[53]. On the insertion of more than 3.0 H^+ or 2.0 Li^+ the oxides become amorphous as they loose coherence in the direction perpendicular to the layers. The nature of the sites occupied by lithium in the composition interval $1 < x < 2$ is not known. Attempts to intercalate sodium in α-V_2O_5 at 80°C initially leads to the formation of α'-NaV_2O_5. Upon further sodium insertion, the host undergoes an irreversible transformation to a phase capable of highly reversible sodium insertion over a wide composition interval[40] with a stoichiometric energy density of 430 Wh/kg, see fig. 5.

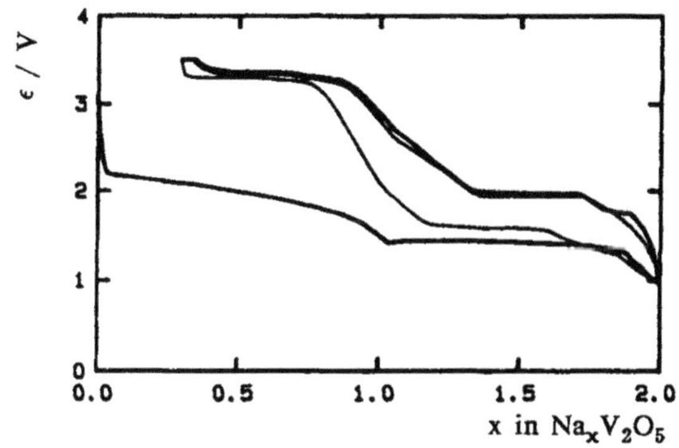

Fig. 5. Constant current current cycling of Na | $NaClO_4$-PEO | α-$Na_xV_2O_5$ cell (80°C). Current density 25 μA/cm². First (heavy line) and second (thin line) cycles are shown[40].

The β-structure is made up of double zig-zag ribbons of VO_6 octahedra joined by corner sharing and further linked by additional chains of doubled VO_5 bipyramids. The structure can be regarded as a tunnel structure, where the inserted ions reside at sites in two rows along the tunnels. At high concentrations of intercalant a cooperative displacement of inserted ions into alternative tunnel sites takes place, forming the β'-phase of theoretical limiting composition $M_{2/3}V_2O_5$. At room temperature the entire composition interval from $x = 0$ to $x = 2/3$ can be covered by electrochemical intercalation in β-$Li_xV_2O_5$. The β-β' phase transformation is highly reversible and proceeds at a high rate, as it does not require structural rearrangements of the host lattice. On further lithium intercalation, two new phases are obtained in the composition interval up to x=1.33[36]. The nature of these phases has not been determined, but the ready reversibility of the phase transitions at room temperature suggest that they must be closely related to the β- and β'-phases. Attempts to perform a similar expansion the composition interval of β-$Na_xV_2O_5$ at 80°C resulted in loss of crystallinity of the host when the high temperature stability limit ($x = 0.40$) was exceeded[40].

At high contents of intercalate, the bronze structures formed are dependent on the inserted ion. The γ-phase formed in the lithium system is a layered structure quite different from the α-phase. At room temperature up to 0.5 lithium can be extracted from γ-LiV_2O_5, but attempts to increase the lithium content beyond $x = 1.05$ lead to disproportionation of the bronze into V_2O_3 and $LiVO_3$[37].

2.1.2 Framework structures: Lithium insertion in vanadium oxides. The vanadium oxides that have been shown to undergo lithium insertion are characterized by a high degree of localization of the valence electrons in contrast *e.g.* to the cubic tungsten bronze which is best described in terms of a model involving a degenerate free electron gas[41]. The individual vanadium atoms can thus be assigned a formal valence or oxidation state, and oxides with all oxidation states between V^{+2} and V^{+5} are known. Apparently the energy differences between oxidation states V^{+3}, V^{+4}, and V^{+5} are relatively small, as evidenced by the wealth of oxides with two separate valence states in the same phase, in some cases even coordinated to the same number of oxygens. Also a phase with simultaneous occurence of V^{+3}, V^{+4}, and V^{+5}, $(Fe,V)_{18}O_{35}$ is known[42].

In order to facilitate lithium ion transport, continuous diffusion paths must be present in the host lattice. It is therefore not surprising that all the vanadium oxides showing lithium insertion are open structures with a relatively low density (Table 2) compared to *e.g.* the rutile modification of VO_2 (ρ = 4.67 g/cm^3) or to the Magnèli phases V_nO_{2n-1} (n = 2: ρ = 4.84 g/cm^3, n = 3: ρ = 4.76 g/cm^3) into which lithium ions are not inserted in appreciable amounts[43].

Table 2
Physical properties of Lithium-Vanadium-Oxides.

Substance	Density (g/cm^3)	Composition range		Energy density Wh/kg	Wh/ℓ
$Li_{1+x}V_3O_8$	3.49 (x=0.5)	$0.12 < x < 2$	(25 °C)	440	1240
		$0 < x < 3$	(120 °C)	650	1650
α-$Li_xV_2O_5$	3.40 (x=0.0)	$0 < x < 1$	(25 °C)	460	1300
		$0 < x < 2$	(120 °C)	750	1840
		$0 < x < 3$	(120 °C)		**)
β-$Li_xV_2O_5$	3.65 (x=0.3)	$0 < x < 1.33$	(25 °C)	590	1680
$Li_xV_2O_5$ (Xerogel)	-	$0 < x < 1.1$	(120 °C)	430	
$Li_xV_5O_{12}$	-	$0.73 < x < 4.45$	(n-BuLi)*)		
$Li_xV_3O_7$	3.58 (x=0.0)	$0 < x < 4$	(n-BuLi)*)		
$Li_xV_4O_9$	-	$0 < x < 4$	(n-BuLi)*)		
$Li_xV_5O_{11}$	-	$0 < x < 5$	(n-BuLi)*)		
$Li_xV_6O_{13}$	3.92 (x=0.0)	$0 < x < 8$	(25 °C)	890	2150
$Li_xVO_2(B)$	4.05 (x=0.0)	$0 < x < 0.6$	(25 °C)	390	1200
		$0 < x < 0.82$	(120 °C)	550	1560
$Li_{1-x}VO_2$	4.40 (x=0.0)	$0 < x < 1$	(120 °C)	820	2310**)

*) In these materials, the composition interval cited is determined from n-BuLi intercalation. Cyclic voltammetry has shown, that Li-insertion is reversible in the major part of this interval.

**) The host structure of these materials is not kinetically stable in the composition interval cited. Thus, the stoichiometric energy density given in the table can not be utilized in cycling.

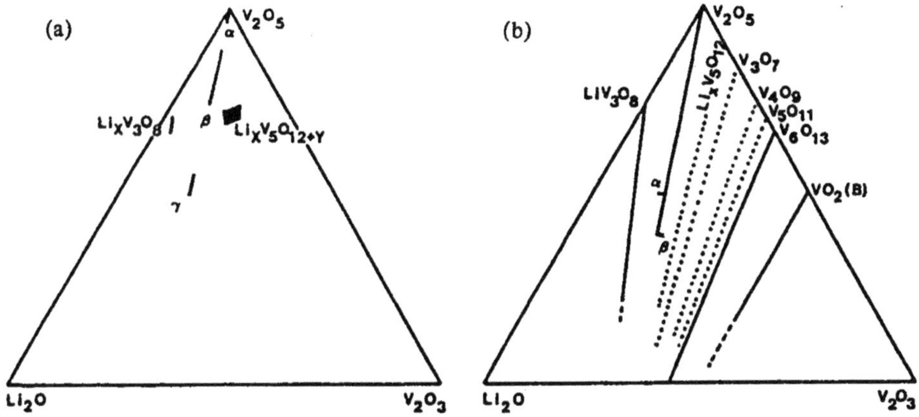

Fig. 6: Phase diagrams for lithium-vanadium(III, IV, V) oxides. **(a)** Thermodynamically stable bronzes. **(b)** Composition interval of lithium insertion compounds at ambient temperature. Solid lines are reversible composition intervals demonstrated by electrochemical methods at 25°C, broken lines indicate additional capacity at 120°C. Dotted lines indicate composition intervals spanned by n-BuLi[49].

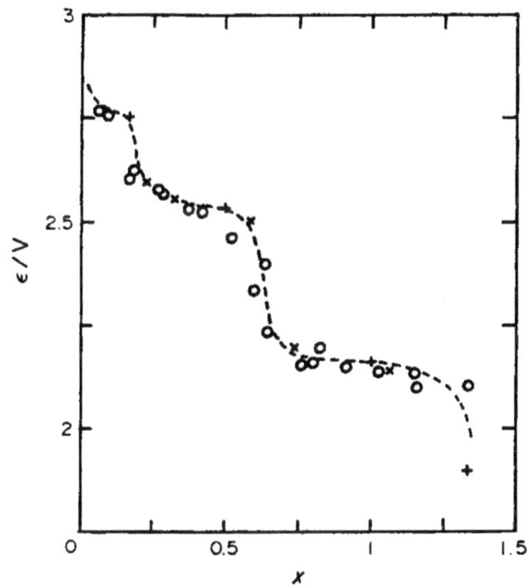

Fig. 7. Open circuit voltage of Li | $Li_xV_6O_{12.78}$ versus degree of intercalation, x[50].

The thermodynamic stability limits of the lithium-vanadium bronzes are indicated on fig. 6(a). As shown in the previous section, these limits can often be exceeded at ambient temperature without changes of the host structure. Several vanadium oxides, that do not form stable bronze phases, may accomodate large amounts of lithium or other small cations anyway at temperatures where the oxide matrix is kinetically stable against structural rearrangements, i.e. below 200-300°C, forming metastable insertion compounds. The composition intervals for some of these compounds are indicated on fig. 6(b). Many of the vanadium oxides have been studied with regard to their application as the active component of rechargeable lithium cells, and calculated stoichiometric energy densities are collected in table 2. Most of these open host frameworks have rather large unit cells with more than one type of site for the inserted ions. As these sites do not have the same affinity for inserted ions, the emf vs composition curve will exhibit two or more steps, reflecting the sequential filling of inequivalent host sites, as seen in the Li/V_6O_{13} system, fig. 7.Due to the unavailability of single crystals, information on the actual sites occupied by lithium in the thermodynamically unstable lithium inserted vanadium oxides, usually can not be obtained from x-ray diffraction techniques. At the same time the large ratio between incoherent and coherent scattering by vanadium atoms limits the usefulness of neutron diffraction. In these cases one therefore has to rely heavily on inferences from model substances and analogies with the stable bronze phases.

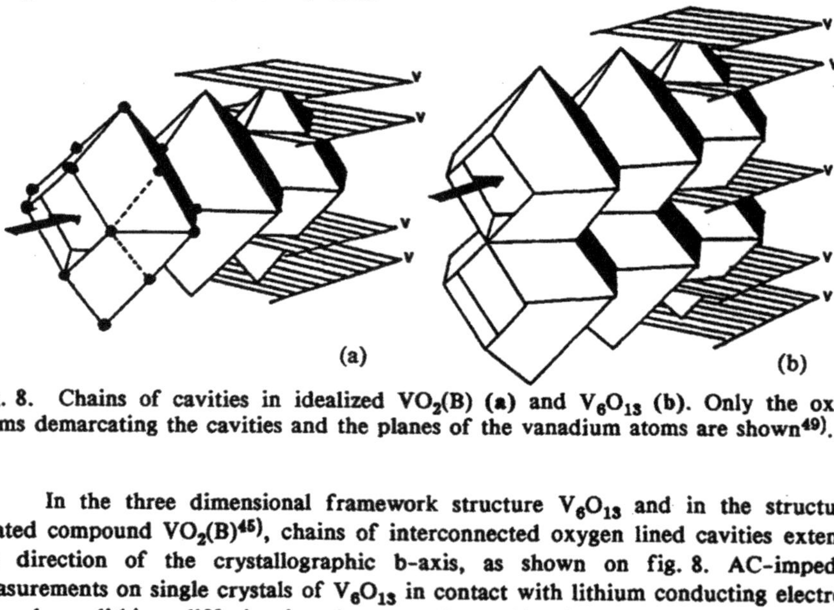

Fig. 8. Chains of cavities in idealized VO_2(B) (a) and V_6O_{13} (b). Only the oxygen atoms demarcating the cavities and the planes of the vanadium atoms are shown[49].

In the three dimensional framework structure V_6O_{13} and in the structurally related compound VO_2(B)[45], chains of interconnected oxygen lined cavities extend in the direction of the crystallographic b-axis, as shown on fig. 8. AC-impedance measurements on single crystals of V_6O_{13} in contact with lithium conducting electrolyte have shown lithium diffusion in other crystallographic directions to be negligible[44]. It is thus assumed that the lithium ions inserted into V_6O_{13} and VO_2(B) are confined to these channels. Cava *et al*[46] have used neutron diffraction powder profile analysis to determine the structure of lithium inserted FeV_3O_8. FeV_3O_8 is a structural analog to VO_2(B), and by reaction with n-BuLi it takes up 2 lithium per formula unit, and the same number per cavity. These lithium were shown to occupy five-coordinated square pyramidal sites, forming opposite apexes of the cavities in fig. 8(a). This coordination is resemblant of the "M1 sites" in β-V_2O_5[47].

There are six square pyramidal sites per double cavity in V_6O_{13} (see fig. 8(b)) and one double cavity per formula unit. V_6O_{13} can accomodate 8 Li per formula unit, so other types of sites must be present in this structure. As coordination to oxygen is a critical factor in the stabilization of inserted lithium, these sites are most likely to be the square faces joining the cavities, with the possibility of coordination to 4 coplanar oxygens, a coordination also seen in $Li_{0.36}WO_3$[48]. This leads to a total of 9 lithium sites per formula unit. However, from intercalation experiments on a series of V_6O_{13+y} ($0.22 < y < 0.11$) it was concluded that vanadium cannot be reduced beyond V^{+3} without destroying the host lattice[50]. The maximum lithium uptake is thus limited by the number of "electronic sites".

Potential vs composition curves measured during electrochemical cycling of $Li_xV_6O_{13}$ and $VO_2(B)$ are shown on fig. 9. It is seen that the structural similarity between V_6O_{13} and $VO_2(B)$ is reflected in the similarity between the voltage curves of these materials. On this basis the following model for the sequential filling of the different sites in V_6O_{13} has been proposed[51]: The first plateau at 2.75 V, not present in $VO_2(B)$, indicates that a completed structure is present after insertion of 1 lithium per double cavity. The square planar site joining the two parts of the double cavity is the only site with this multiplicity, and it is situated as far as possible from the densely vanadium populated "VO_2-layers". It is thus assumed that the first lithium entering the structure occupies this site, and that further lithium insertion leads to occupation of the two opposite square pyramidal sites, displacing lithium from the square planar site into one of these sites too. This leads to a completed structure containing 2 Li per cavity as in $VO_2(B)$. Further lithium insertion results in an abrupt expansion of the unit cell[50], to accomodate a more dense population of the available lithium sites.

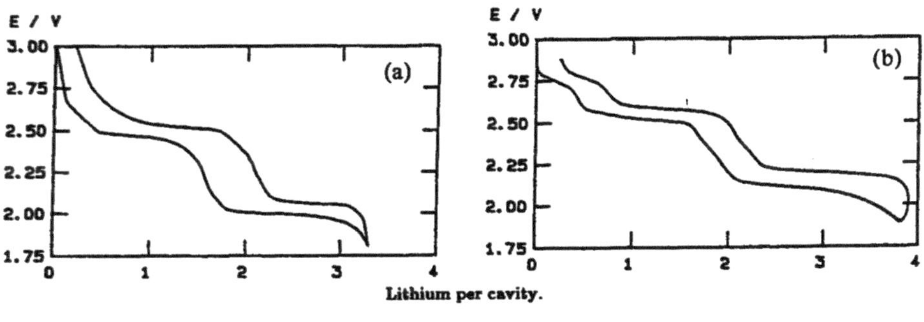

Fig. 9. Potential vs composition calculated as lithium per cavity. **(a)** $Li/Li_xVO_2(B)$ in $LiCF_3SO_3/PEO$ (120°C), constant current = 5 $\mu A/cm^2$ (4 formula units per cavity). **(b)** $Li/Li_xV_6O_{13}$ in $LiAsF_6/PC$ (25°C), calculated from voltammogram, sweep-rate = 5 $\mu V/s$ (2 cavities per formula unit).

Emf data for a number of the lithium-vanadium oxides are collected in fig. 10 as a function of the average oxidation number of V. The emf curves for the kinetically stable compounds fall in a relatively narrow region, with some dependence on the oxidation number. This demonstrates, that the magnitude of the emf is largely governed by the "chemistry" of the Li-V-O-system, whereas the structure of the host lattice determines the magnitude of the composition interval and the details in the shape of the emf curve. Based on fig. 10 and assuming that the entire interval from V(5) to V(3) can be utilized for Li insertion, the upper limit of the reversible stoichiometric energy density obtainable within the Li-V-O-system can be estimated to 1200 Wh/kg.

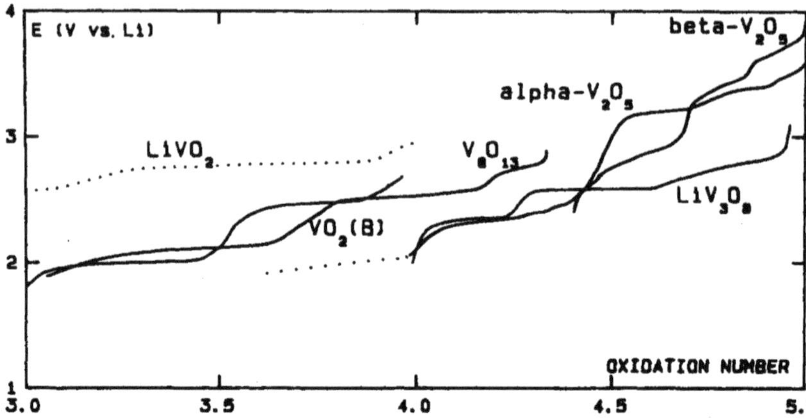

Fig. 10. Emf-curves for lithium insertion in various lithium vanadium oxides (at 25°C or 120°C) as function of the mean oxidation state of vanadium. Dotted lines indicate that these reactions lead to changes in the host lattice[33].

The electronic conductivity (σ_e) measured on compressed powders of a number of Li inserted vanadium oxides is shown on fig. 11. The measured conductivities were not corrected for porosity or intergrain resistances, but a general trend is clear from the results: The electronic conductivity decreases by several orders of magnitude upon Li insertion, except at oxidation states close to V^{+5} where an initial increase is seen. The decrease in σ_e shows that the electrons transferred to the oxide during intercalation are localized at discrete vanadium atoms, and is perhaps a consequence of the increasing distance between these atoms. Because of the low electronic conductivity of the fully lithiated oxides, conductive material must be added to electrodes with vanadium oxides as active electrode material.

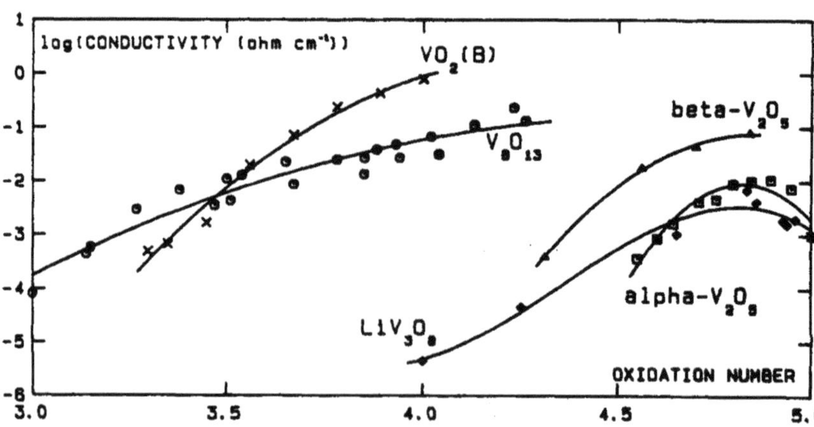

Fig. 11. Electronic conductivity (σ_e) of compressed powders (at 70 MPa) of various lithium vanadium oxides[33].

2.1.3 Molybdenum cluster compounds. The molybdenum cluster compounds - or "Chevrel phases" $M_xMo_6X_8$ (X = S, Se) - are formed with a wide range of main group and transition metal guests, table 3. The structure can be considered as a three dimensional array of units consisting of a Mo_6 octahedron surrounded by a X_8 cube, fig. 12. Six of the chalcogens form bridges between adjacent units, being coordinated to three Mo in one unit and one Mo in the neighbouring unit. The two remaining chalcogens can be fully or partially substituted by halogen atoms, thereby increasing the valence electron concentration on the Mo_6 unit[54]. In the interconnected framework of Mo_6X_8 units, a three-dimensional network of intersecting channels is formed, in which the M atoms reside.

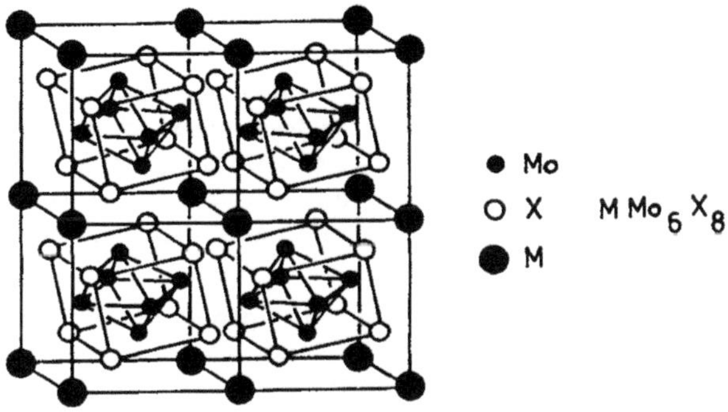

Fig. 12. Structure of molybdenum cluster compound $M_xMo_6X_8$.

Table 3.
Elements reported as ternary elements (M) in Chevrel phases $M_xMo_6X_8$, X = S, Se[55].

H													
Li													
Na	*Mg*											*Al*	
K	*Ca*	*Sc*				**Mn**	**Fe**	**Co**	**Ni**	**Cu**	**Zn**		
Rb	*Sr*	*Y*							*Pd*	**Ag**	**Cd**	*In*	*Sn*
	Ba	*La*									*Hg*	*Tl*	*Pb*
Ce	*Pr*	*Nd*	*Sm*	*Eu*	*Gd*	*Tb*	*Dy*	*Ho*	*Er*	*Tm*	*Yb*	*Lu*	
Th		*U*	*Np*	*Pu*	*Am*								

M : Elements forming intercalation compounds $M_xMo_6X_8$ with strongly bound or nearly immobile intercalant.
M : Elements forming intercalation compounds $M_xMo_6X_8$ with room temperature mobility of intercalant.

There are two classes of Chevrel phases[56]. In compounds of the first type the ternary element M is a small cation like Li^+, Cu^+ or Zn^{2+} distributed over two sets of six-fold lattice sites. These sites are located slightly off the origin of the unit cell ("inner sites") and in the tunnels joining these sites ("outer sites"). For steric reasons[57,58] only part of these sites can be occupied simultaneously, but it appears that the number of guest ions is limited by the number of electrons that can be accepted by the host. From band structure calculations it is concluded, that the Mo_6 unit has 24 bonding electronic levels separated from higher levels by a 0.8 eV gap[59]. If it is assumed that insertion of M only adds electrons to an otherwise unchanged band of the Mo_6 unit (*the rigid band model*) a transfer of maximum four electrons per formula unit is thus allowed in $M_xMo_6X_8$, and two in the fully halogen substituted compounds. These compounds often exhibit wide homogeneity ranges, and due to the large number of equivalent sites high ionic mobilities are encountered[60,61]. Apparantly there is a critical upper limit for the radius, r_i, of the inserted ion as sodium (r_i = 0.95 Å) can be inserted in both the sulphide and the selenide while silver (r_i = 1.26 Å) can only be inserted in the selenide and potassium (r_i = 1.33 Å) in neither[20]. This type of Chevrel phases is the only class of intercalation compounds where room temperature mobility of divalent transition metal ions has been unambigously demonstrated. Chevrel phases of the second kind, e.g. $PbMo_6X_8$ and $SnMo_6X_8$, are formed with larger or more charged ions, situated at the origin of the unit cell. In this case the guest ions are essentially immobile at ambient temperature, and only narrow composition intervals around $M_1Mo_6X_8$ are encountered.

The variations of voltage with composition for Li | $Li_xMo_6S_8$ and Zn | $Zn_xMo_6S_8$ are shown in fig. 13. In most $M_xMo_6S_8$ phases with small ions a sharp potential step is seen at x = 1, propably due to a steric blocking of further insertion into the inner sites. Lithium insertion into Mo_6S_8 has proven to be very reversible with no qualitative change in the voltage curves during more than a thousand deep cycles[62]. Li | $Li_xMo_6S_8$ has thus attracted interest as electrode couple for secondary batteries, despite its rather low energy density (>260 Wh/kg, 1050 Wh/ℓ).

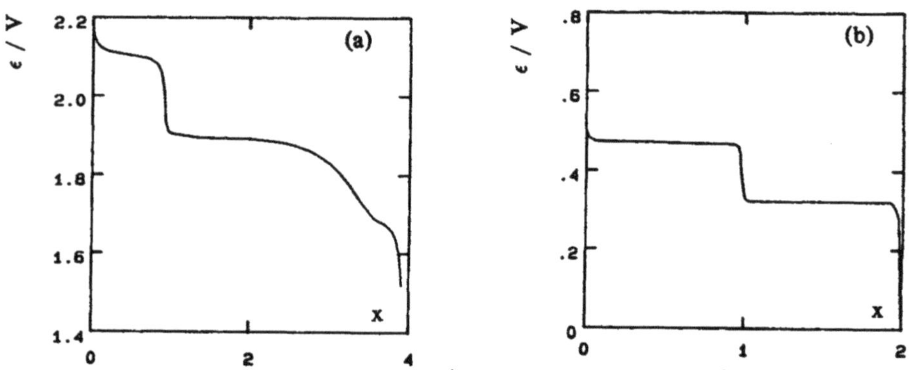

Fig. 13. Potential vs composition for M | $M_xMo_6X_8$ cells. (a) Li | $Li_xMo_6Se_8$[63], (b) Zn | $Zn_xMo_6X_8$[61].

2.2 Layered Host Materials

When the dimensionality of the host structure is reduced, it becomes possible to deform the lattice extensively during the intercalation reaction without rearranging bonds in the host matrix. Low-dimensional host lattices are thus capable of intercalating species of widely varying size.

The layered host materials are sandwich structures of layers internally bound by strong, predominantly covalent forces. The "vacant" planes between sandwiches are called *galleries* or *van der Waals gaps*, indicating that the layers are held together only by van der Waals forces, allthough weak covalent and ionic contributions also can be present[65]. The weakness of the interlayer forces offers the possibility of introducing foreign atoms or molecules into the van der Waals gap.

The layered intercalation hosts can be classified into three general categories according to the rigidity of their layers with respect to distortions perpendicular to the layer planes[66]. The three classes are exemplified on fig. 14.

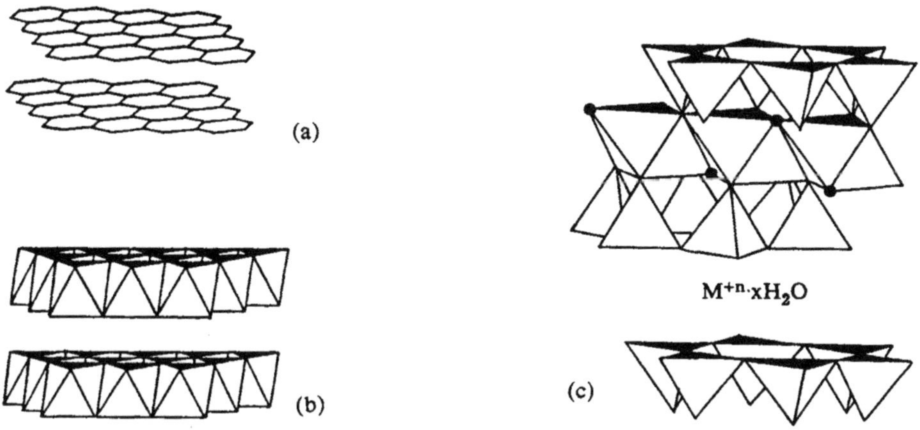

Fig. 14. Examples of layered host structures. **(a)** The hexagonal arrays of carbon atoms in graphite. **(b)** The octahedral coordination polyhedra of 1T transition metal dichalcogenides. **(c)** Idealized structure of 2:1 sheet silicate (smectite) showing the oxygen framework of the layers. Possible hydroxyl sites indicated as filled circles.

Class I is constituted by graphite, where the layers are atomic monolayers, fig. 14(a). The screening of the coulomb repulsion between the inserted species is modest, and strong guest-guest and guest-host interactions lead to intercalation products with fixed compositions, where only a well defined fraction of the available sites is occupied. The limiting compositions for the alkali metal intercalates are : MC_6 (M: Li, Na) or MC_8 (M: K, Rb, Cs). Due to flexibility of the layers, strong deformations are seen at the boundaries between intercalated and unintercalated regions of the crystal. In order to minimize the associated "boundary energy" these regions tend to coalesce and due to interlayer interactions form *stages* - one dimensional superlattices, in which domains with one monolayer of guest species are separated by n host layers, giving the stage n compound, as illustrated on fig. 15.In class II the layers are made up of a single sheet of coordination polyhedra, fig. 14(b). This class includes the transition metal dichalcogenides (TMD's), the layered transition metal oxyhalides and the transition metal phosphorous trichalcogenides. In these compounds the screening of the repulsion between the intercalates is efficient, and continuous broad intercalation ranges with a limiting composition corresponding to complete occupation of one type of the available sites are generally seen. Staging is only seen in few of these compounds, and mostly in cases where the intercalation reaction involves a sliding of the layers relative to each other. Fisher and Kim[67] have determined the average decay length of layer distortions around lithium intercalates in graphite and titanium disulphide, TiS_2, to be 5 and 7 Å, respectively. This illustrates that both class I and II have flexible layers, but with different rigidity.

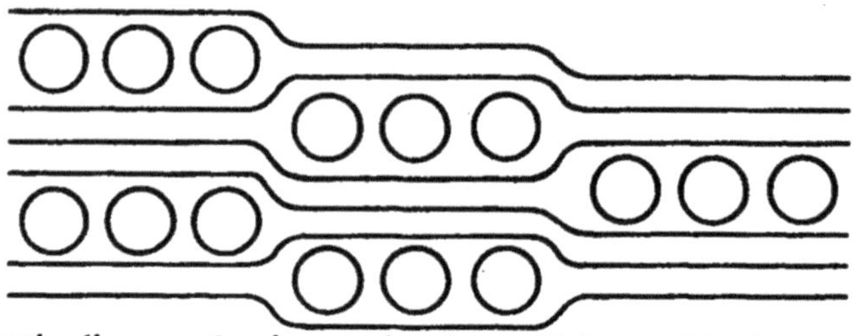

Fig. 15. Schematic diagram showing staging (stage 3) in graphite intercalation compounds.

In the host structures of class III each layer is formed from multiple sheets of coordination polyhedra. This class covers the phyllosilicate or clay minerals, where the layers are build up as various combinations of tetrahedral and octahedral sheets. Silicon ions are found in the tetrahedral sites, whereas the octahedral positions are occupied by aluminium or magnesium ions. Homo- and/or aliovalent substitution in both the anionic and the cationic networks yield a multitude of minerals with either neutral or negatively charged layers. The negative charge is compensated by intercalated cations located in the interlamellar space. As these minerals are not subject to redox reactions, the charge on the layers remains fixed, but some groups of sheet silicates (especially *smectites*, fig. 14(c)) readily undergo ion exchange as well as intercalation of a number of neutral molecules. The smectites are able to take up more than a single molecular layer in each gallery, in contrast to the intercalation in graphite where each monolayer of intercalant is separated by one or more host layers. Intercalation of multiple layers of solvent, which also can be seen in some cases with class II hosts, can ultimately lead to *exfoliation*, when the layers become detached forming colloidal suspensions. Due to the rigidity of the host layers of class III the intercalation reaction often proceeds discontinuously as whole monolayers are absorbed in a single step. For the same reason the interlayer spacing of smectites can be preserved - even when the swelling solvent is removed - by *pillaring*, where large, thermally stable cations props the layers apart[68]. The decay length of layer distortions in these materials is measured in tens of ångström.

The continuous monophasic intercalation of ions in layered hosts is thus only observed with class II materials. Some pertinent features of these systems are exemplified in the following:

2.2.1 The layered transition metal dichalcogenides. Intercalation in layered transition metal dichalcogenides (TMD's) was first reported in 1959 by Rüdorff and Sick[64], reacting alkali metals dissolved in liquid ammonia with MoS_2 and WS_2. Since then a rapidly growing number of guest-host combinations have been studied. In table 4 the elements known to form layered dichalcogenides and to form intercalates with these compounds are collected. It is however only the intercalation compounds with group I and Ib elements that conform with our previous requirement of room temperature mobility of the intercalant. Most of the transition metal and post transition metal intercalants form partially covalent bonds with the hosts.

Table 4.
Elements known to form layered transition metal dichalcogenide intercalation compounds, M_xTX_2.

H														
Li														
Na												*Al*	*Si*	
K	*Ca*		Ti	V	Cr	*Mn*	Fe	*Co*	*Ni*	**Cu**	*Zn*	*Ga*	*Ge*	
Rb	*Sr*		Zr	Nb	Mo					**Ag**	*Cd*	*In*	*Sn*	
Cs	*Ba*	*Eu,Yb*	Hf	Ta	W						*Hg*	*Tl*	*Pb*	*Bi*

T : Transition metals forming layered dichalcogenides TX_2. The first row transition metals are also known as guest species.
M : Elements forming intercalation compounds M_xTX_2 with strongly bound or nearly immobile intercalant.
M : Elements forming solvent free intercalation compounds M_xTX_2 with room temperature mobility of intercalant.

The structure of the TMD's consists of sandwiches of a sheet of transition metal atoms (T) between two close packed layers of chalcogen atoms (X). Depending on the positioning of these sheets relatively to each other, two different coordinations around the metal atoms are obtained: either trigonal prismatic or slightly distorted octahedral (trigonal antiprismatic symmetry) as shown on fig 16.

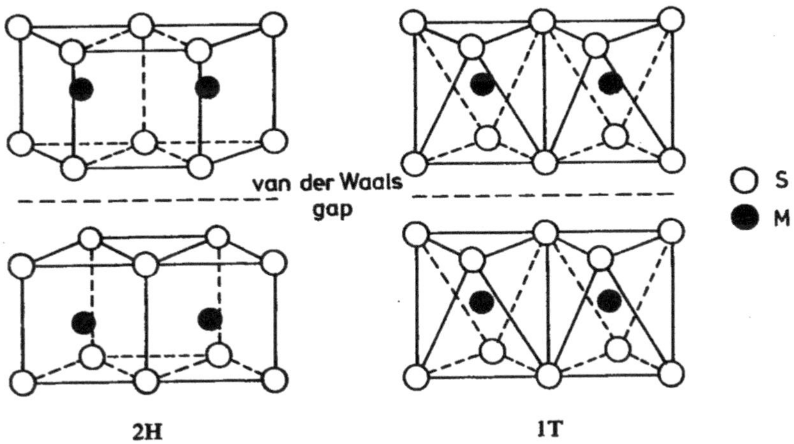

Fig. 16. Trigonal prismatic coordination (TP) and octahedral (O) coordination of transition metal in TX_2. Close packing of these layers leads to the 2H and 1T polytypes, respectively.

Furthermore, adjacent sandwiches may be packed in a number of different ways, leading to a rich variety of possible structures (*polytypes*), differing only in stacking sequence. During the formation of intercalation compounds, the structure may be altered by displacement of layers relatively to each other. The coordination around the transition metal is largely determined by the ionicity of the metal-chalcogen bond. The strongest covalent bond can be formed in the trigonal prismatic coordination, whereas the close packed anion array of the octahedral coordination is preferred if the bonds are predominantly ionic or if the metal ion is very small compared to the chalcogens. The same type of arguments can be used for the coordination of intercalated ions, and Rouxel[69] and Hibma[65] have constructed ionicity-structure diagrams that unambiguously relates the coordination to a size factor and to the charge on the chalcogens. The general trend is that small ions and/or high charges on the chalcogen sheets favour the octahedral coordination whereas the trigonal prismatic coordination is seen with larger ions. These relations are illustrated by alkali metal intercalation into TiS_2, fig. 17:

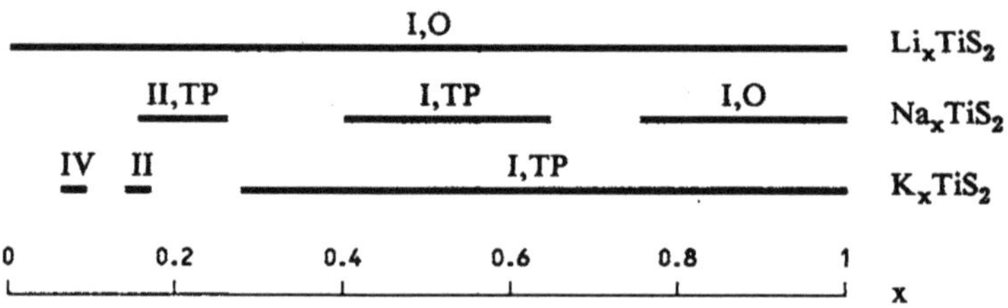

Fig. 17. Phase widths of alkali metal intercalates in titanium disulphide: Li_xTiS_2, Na_xTiS_2[70] and K_xTiS_2[71]. The staging is indicated by **I**, **II** and **IV**, and the coordination is indicated by **O**: distorted octahedral (trigonal antiprismatic symmetry) and **TP**: trigonal prismatic.

The small lithium ion (r_i = 0.68 Å) fits into the octahedral site in the van der Waals gap with nearly no distortion, and the octahedral coordination is maintained through the whole composition range $0 \leq x \leq 1$. Intercalation of the larger sodium ion (r_i = 0.98 Å) induces a shift of the layers relatively to each other, yielding the trigonal prismatic coordination of sodium ions. As this process requires a relatively large amount of energy, initially a stage 2 phase is formed. With high degrees of intercalation, the charge on the layers becomes so high that the octahedral coordination is stabilized. For small degrees of intercalation of potassium ions (r_i = 1.33 Å) both a stage 4 and a stage 2 phase have been identified[71]. The potassium ion is too large for an octahedral phase to be formed.

The stacking sequence interferes with intercalation reactions in several ways: The coordination around the intercalant influences the kinetics of intercalation, as diffusion of larger ions may be severely restricted in octahedrally coordinated layers, where the diffusion path passes through tetrahedral sites. Also phase shifts involving relative displacement of layers may be kinetically slow processes. Repeated intercalation/extraction of Na in TiS_2 is thus only possible at practical rates in the composition interval corresponding to the **I,TP** phase of fig. 17[72]. The coordination around the transition metal influences the electronic properties of the host material. This is exemplified in TaS_2 existing in both a 1T (transition metal in octahedral coordination) and a 2H (transition metal in trigonal prismatic coordination) modification, see fig. 16. The 1T modification is stable above 750°C, and it is obtained by quenching from the reaction temperature. In this modification periodic lattice distortions associated with the electronic instability known as charge density waves (CDW's) are present to temperatures above 250°C. Under certain conditions low-dimensional conductors are able to minimize their electronic band energy by undergoing a Peierls transition to a distorted state with a spatially fluctuating charge density. At low temperatures the CDW's may be in registry with the crystal lattice (commensurate), whereas incommensurate lattice distortions occur at higher temperatures. The 2H modification of TaS_2 is much more stable towards lattice distortions and only undergo CDW transitions well below room temperature. The two TaS_2 modifications also behave differently as host structures for lithium intercalation as seen from the emf vs composition curves shown in fig. 18.

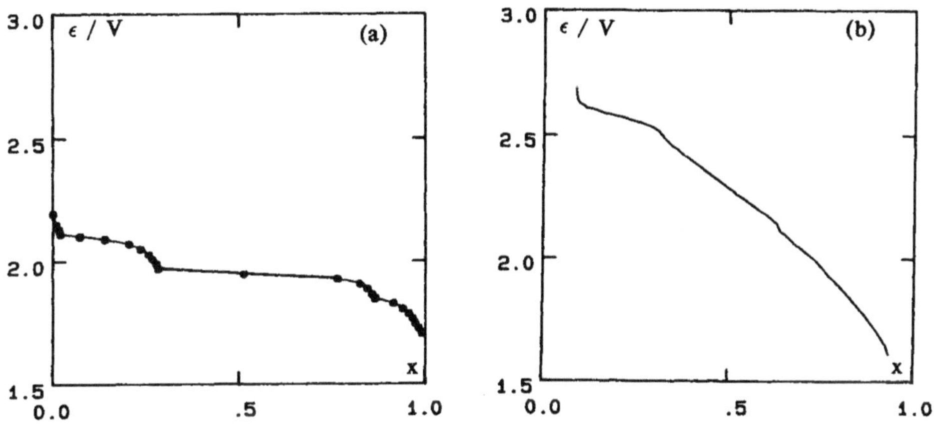

Fig. 18. Potential as function of compostion parameter x for $Li|Li_xTaS_2$ at room temperature. (a) Modification with Ta in octahedral configuration ($1T-TaS_2$)[73]. (b) Modification with Ta in trigonal prismatic configuration ($2H-TaS_2$)[74].

In both materials lithium ions are inserted into octahedral sites of nearly equal size. The higher initial voltage seen in the $2H-Li_xTaS_2$ system shows that this modification takes up lithium ions more readily, despite the more ionic character of $1T-Li_xTaS_2$. These facts indicate that electronic factors in this case are determining the intercalation affinity. Whereas the $2H-Li_xTaS_2$ system shows subtle features at $x = 1/3$ and $x = 2/3$ indicating the formation of ordered structures at these compositions[74], the $1T-Li_xTaS_2$ system undergoes a phase transition starting at $x = 1/4$. It has been suggested that the presence of CDW's shows that the host is susceptible to distortions, and that intercalation induces the lattice distortion responsible for the phase shift[75]. The TMD which has attracted most interest as host for lithium intercalation is TiS_2 (1T modification) which has fast intercalation kinetics and high stability of the host lattice that can withstand several hundreds of deep intercalation/extraction cycles without deterioration. The stoichiometric energy density calculated from the emf vs composition curve on fig. 19 is 480 Wh/kg or 1190 Wh/ℓ.

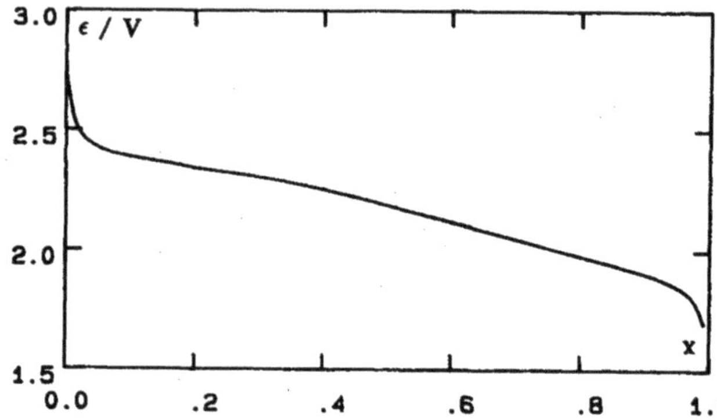

Fig. 19. Potential as function of composition parameter x for Li | Li_xTiS_2 at room temperature[76].

Hydrogen intercalation is influenced by the stacking sequence in an unusual way[22]. In compounds with the 2H structure (H_xNbS_2 and H_xTaS_2) nearly all hydrogen atoms are located within the TX_2 sandwich in the same plane as the metal atoms. A hydrogen atom is thus closely surrounded by three metal atoms, and is immobile on the NMR time scale. Only a small fraction of the hydrogen atoms ($x < 0.1$) enters the lattice as mobile protons, presumeably residing in the van der Waals gap. In compounds with the 1T structure (H_xTiS_2 and H_xVS_2), however, all the intercalated hydrogen atoms are mobile and situated in the van der Waals gaps, clustered at adjacent tetrahedral sites.

2.2.2 Transition metal phosphorous trichalcogenides. The structure of the transition metal phosphorous trichalcogenides[78] (TPX_3, X = S, Se) can be regarded as a TMD in which one third of the metal sites have been substituted with phosphorous pairs, P_2, giving the formula $T_{2/3}(P_2)_{1/3}X_2$. The intralayer bonding is however more involved than suggested by this analogy, with predominantly ionic bonding between T^{2+} cations and $(P_2X_6)^{4-}$ anions. The $(P_2X_6)^{4-}$ anion consists of two flat PS_3 pyramids linked through a phosphorous-phosphorous bond, the length of which is adapted to the cation size[79]. TPX_3 compounds are formed with T = Mn, Fe, Co, Ni, Zn, and Cd, but compounds with an additional cation or vacancies can be formed with V, Cr, Cu, and In. The phosphorous trichalcogenides thus constitute a very rich class of intercalation host materials. In accordance with the ionic character of the layers, most TPX_3 compounds undergo cation substitution reactions (ion exchange) as well as intercalation reactions at room temperature. A quite original combination of these reactions has been reported[80]:

$$2xM^{+} + y(\mathrm{solv}) + TPX_3 \rightarrow T_{1-x}M_{2x}(\mathrm{solv})_yP_3 + xT^{2+} \qquad (5)$$

where M^+ is a large alkali cation (K^+, Rb^+, Cs^+ or NH_4^+) or an organometallic species. After reaction, the solvated cations are found in the van der Waals gap, leading to an increased interlayer separation. Pure ionic intercalation has been demonstrated with $NiPS_3$, $FePS_3$, $FePSe_3$, $CoPS_3$, $MnPS_3$, and $MnPSe_3$[78] in reaction with n-BuLi. The size of the van der Waals gaps expressed as sulphur-sulphur distance is fairly constantly 3.24 Å in the TPS_3 compounds. This is much larger than in the TMD's and it does not change upon lithium intercalation. The lithium uptake is found to be reversible up to the composition $Li_{1.45}TPS_3$ - close to the theoretical maximum $Li_{1.5}TPS_3$ where all the octahedral sites in the van der Waals gaps are filled. Further reaction with n-BuLi is possible, but is propably associated with an irreversible destruction of the host. The chemical shift of the ^7Li NMR signal indicates that lithium is completely ionized throughout the composition range $0 < x < 1.45$ in Li_xNiPS_3[81]. The transferred electron is considered to be accomodated at host levels of mainly transition metal d character[78], yielding the following redox scheme:

$$xLi + NiPS_3 \rightarrow (Li^+)_x(Ni^{2+})_{1-x/2}(Ni^0)_xPS_3 \qquad (6)$$

involving the total reduction of part of the nickel ions. Electrochemical intercalation of lithium into $NiPS_3$ has been studied as a possible alternative to TMD's as electrode material for secondary and primary batteries[83-85]. The discharge behaviour is reported to be dependent on the type of electrolyte used, and an unusually strong polarisation leading to irreversible capacity loss is observed with higher current densities. The quasi equilibrium voltage-composition curve shown on fig. 20 indicates that two sites are involved in the intercalation reaction in the reversible range. These are propably the *2d* and *4h* octahedral sites of space group *C2/m*. At high degrees of intercalation an irreversible reaction takes place leading to destruction of the host lattice and the formation of Li_2S and non-miscible Ni/P phases[84]. The stoichiometric energy density calculated from fig. 20 ($0 \le x \le 1.5$) is ≈500 Wh/kg (1300 Wh/ℓ). Due to the poor electronic conductivity of Li_xNiPS_3 a conductive compound must be added to electrodes using this material, but the lack of volume expansion on Li intercalation might compensate for this by increasing electrode cycling performance. So far only little has been reported on the reversibility of the $Li/NiPS_3$ system.

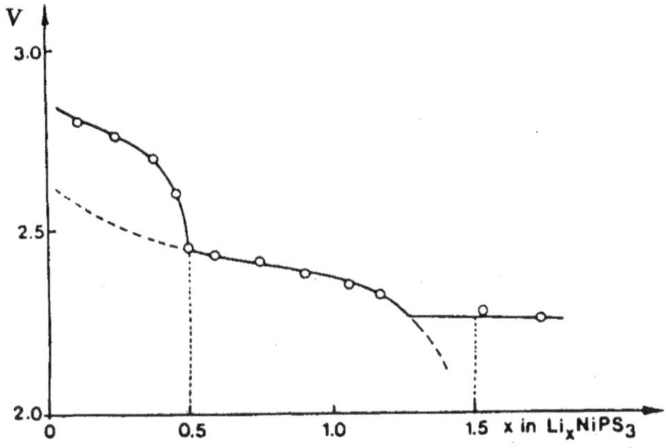

Fig. 20. Variation of open circuit potential of the cell: Li | $LiClO_4$-PC | Li_xNiPS_3[82].

2.2.3 Layered oxides Layered oxide hosts analogous to the TMD's are not stable due to strong repulsive coulombic interactions between essentially ionically bound oxygen atoms. Stoichiometric MTO_2 oxides, where M is an alkali metal, are known as structural analogues of the fully intercalated TMD's, with octahedral coordination of the transition matel (T) and either octahedral or trigonal prismatic coordination of the alkali ion. By soft chemical or electrochemical methods the alkali ions can to a certain extent be extracted from these oxides, demonstrating that these ions are mobile as in the TMD's. For sodium containing oxides only limited composition intervals ($0.75-0.8 < x < 1$) have been obtained[86)], whereas nearly all lithium can be extracted from $LiVO_2$[87)] and $LiCoO_2$[88)]. In Li_xVO_2 the electrochemical lithium extraction occurs at a higher potential level than seen for other lithium-vanadium oxides within this range of oxidation numbers, showing that lithium is strongly bound in this compound, fig. 10. Structural investigations[89)] have demonstrated that when x is decreased below 0.3, vanadium ions migrate to vacant positions in the lithium layer. In Li_xCoO_2 too, lithium extraction proceeds at very high potentials. Compositions down to $x = 0.07$ have been reported[88)], accompanied by a structural breakdown when $x < 0.3$.

Another type of layered oxide host lattices is exemplified by the $M_{1+x}V_3O_8$ (M = Li, Na) compounds. In these materials alkali metal ions are situated between puckered layers of VO_6-octahedra, leaving 1 alkali metal ion per formula unit to stabilize the structure of the completely oxidized compounds. Reversible lithium and sodium insertion have been demonstrated up to the limiting compositions $Li_4V_3O_8$ and $Na_3V_3O_8$[33,40,90)] with good capacity retention when used in rechargeable electrochemical cells.

2.3 One-Dimensional and Amorphous Hosts

The structures of one-dimensional solids are made up of chains of coordination polyhedra extending in one crystal direction. The intrachain bonding is generally much stronger than the bonding between individual chains. As a consequence, these materials are susceptible to extensive lattice defects involving improper alignment of the individual structure elements. There is no clear boundary between the crystalline one-dimensional materials with long-range ordering of the chains and the amorphous materials in which the interchain coordination is relaxed and where long-range order is lost. As host materials for intercalation these structures are characterized by a large number of sites which are easily adapted to the size of intercalants. Due to the structural instability, reactions between guests and host will often involve extensive structural rearrangements of the host around the guest species. In this case the reactions are therefore better characterized merely as topotactic.

The transition metal trichalcogenides TX_3 are crystalline one-dimensional compounds made up of chains of face sharing trigonal TX_6 prisms. The individual strings are linked by loose T--X bonds. Two chalcogens in the basal plane of the prism are closer together, showing that dichalcogen bonds X-X are formed. TiS_3 and $NbSe_3$ both react with 3 lithium per formula unit[18,91,92)]. Insertion of the first two Li result in oxidtion of the dichalcogenide bond. In TiS_3 this leads to a shift in the Ti coordination to the more stable octahedral configuration, while Nb remains trigonally prismatic coordinated. In $NbSe_3$ all Li can be extracted, but subsequent reintercalation proceeds at a higher voltage, indicating some distortion of the host[91,93)]. Only one Li can be extracted from the rearranged Li_3TiS_3[18,92,94]. Cointercalation of solvent leads to deterioration of both materials upon extended cycling in organic electrolytes. At 280°C two Na ions can be intercalated in TiS_3, of these only one is reversible[95].

Molybdenum trisulphide, MoS_3, is amorphous to x-rays, but modern spectroscopic techniques have shown its local structure to resemble $NbSe_3$, but with Mo-Mo pairing along the chains[96-99]. MoS_3 can insert up to four alkali metal atoms per formula unit by reaction with n-BuLi or with sodium or potassium napthalide[100]. The disulphide bonds are attacked during the insertion reaction, but electrochemical cycling of MoS_3 in lithium cells[101] shows that of the four lithium initially intercalated per formula unit the three can be cycled reversibly with a stoichiometric energy density of 767 Wh/kg. Electrochemical cycling of sodium in MoS_3 is less reversible. In the first discharge 1.5 - 2.5 Na/MoS_3 is inserted, depending on the details of the synthesis of the host material[100,102]. With both ions the discharge curves are sloping smoothly, showing that host sites are filled without ordering of the intercalants.

By polycondensation of decavanadic acid a gel of highly polymerized vanadium oxide can be prepared ($M_w > 10^6$)[103]. After careful dehydration of this gel an amorphous form of $V_2O_5 \cdot nH_2O$ ($n < 0.1$) is obtained. The structure of the oxide consists of ribbons of corner and edge sharing VO_6 octahedra, between which ions and solvent molecules can be inserted. By discharge in an electrochemical cell, this compound accepts 1 Li/V_2O_5 with a smooth and featureless discharge curve without any of the inflexions seen for crystalline α-$Li_xV_2O_5$, fig. 10. Surprisingly, this material shows good capacity retention upon cycling[49].

3. THERMODYNAMICS OF INTERCALATION

In this section the factors influencing the variation of the equilibrium voltage, ϵ, of electrochemical cells with intercalation electrodes will be discussed. Generally, ϵ will depend on the concentration, x, of intercalant in the electrode(s). This variation is often discussed in terms of $\partial x/\partial \epsilon$ curves where the fine structure of the potential variation become evident. $\partial x/\partial \epsilon$ curves are either computed by numeric differentiation of equilibrium voltage vs composition curves, or directly from linear sweep voltammograms at low sweep-rates, fig. 21. The cell considered will be of the type shown on fig. 3, with a negative electrode consisting of pure intercalant, M, and a positive intercalation electrode, H. The electrode reactions can be written in differential form as:

$$\delta M \rightarrow \delta M^+ + \delta e^- \qquad (7)$$

$$\delta M^+ + \delta e^- + M_xH \rightarrow M_{x+\delta}H \qquad (8)$$

and the electrode potentials will be given by the expressions:

$$(\pi_M - \phi) = (\mu_{M^+} - \mu_{M(M)})/F \qquad (9)$$

$$(\pi_H - \phi) = (\mu_{M^+} - \mu_{M(H)})/F \qquad (10)$$

where π is the Fermi potential of the electrode phase ($\pi = -\overline{\mu}_{e^-}/F$), ϕ the Galvani potential of the electrolyte phase and (H) and (M) indicate the host lattice and the pure intercalant, respectively. Under equilibrium conditions there will be no gradients in electrolyte potential or electrolyte concentration, and the equilibrium voltage or emf, ϵ, is given by:

$$\epsilon = (\pi_H - \pi_M) = (\mu_{M(M)} - \mu_{M(H)})/F \tag{11}$$

Whereas the chemical potential of M in the pure state, $\mu_{M(M)}$, is independent on the reaction parameter x, $\mu_{M(H)}$ will generally vary with the degree of intercalation, resulting in a composition dependence of ϵ. Assuming that the host matrix is unchanged during the intercalation reaction, the chemical potential of the intercalant, $\mu_{M(H)}$, can be written as the sum of three terms:

$$\mu_{M(H)} = \mu^{\ominus}_{M(H)} - TS_c + \mu^{E}_{M(H)} \tag{12}$$

a standard term accounting for the constant affinity between guest and host matrix, a configurational molar entropy term accounting for the distribution of the guest over the available sites, and finaly an excess interaction term accounting for the guest-guest interactions.

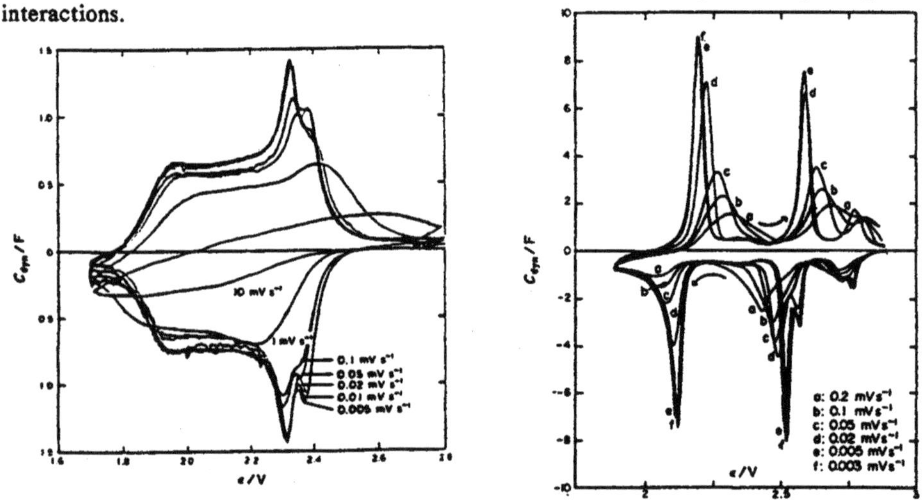

Fig. 21. Single sweep voltammograms of (a) $Li|Li_xTi_{1.056}S_2$[76] and (b) $Li|Li_xV_6O_{12.86}$[50]. The current (i) is normalized as dynamic capacity ($C_{dyn} = i/\alpha$, α = sweep-rate). Sweep-rates are indicated in the figure.

As long as there is a close spatial correlation between the inserted ion and its corresponding electron, and as only one type of sites for the intercalant is considered, the configurational entropy term will be given by:

$$S_c = -R \ln\left[\frac{x}{x_{max} - x}\right] \tag{13}$$

where x_{max} is the limiting value of the composition parameter x in M_xH. Equation (13) stresses the analogy between intercalation and adsorption: In both cases the reactant is distributed over a limited number of sites, and very similar thermodynamic models can be used for these systems. In case the compensating electrons are distributed independently of the inserted ions, an additional electronic term must be included in the configurational entropy. A detailed knowledge of the band structure of the host is needed to determine this term, but here only two limiting cases will be considered: If the host is metallic, and remains metallic throughout the entire composition interval, the electronic contribution will be small. If the host is a semiconductor where electrons during intercalation enters the empty conduction band, the electronic contribution to the configurational entropy will be of the same form as eq. (13), possibly with a different value for x_{max}. Often the entropy terms are only dominating the emf expression near the limiting compositions $x = 0$ and $x = x_{max}$, and in most cases a single term as eq. (13) will be a good approximation, taking x_{max} as the limiting number of ionic or electronic sites. In cases where these numbers are equal or with semiconductors both terms should be included.

The main contribution to the interaction term is the screened coulombic repulsion between intercalated ions. Taking only nearest neighbour interactions into account[104] or using mean field theory[105] this term will show a linear dependence on the intercalation parameter x, giving the following expression for the chemical potential of the intercalant:

$$\mu_{M(H)} = \mu^{\ominus}_{M(H)} + RT \ln\left[\frac{x}{x_{max} - x}\right] + Ux \tag{14}$$

or using eq. (11):

$$\epsilon = \epsilon^{\ominus} + \frac{RT}{F} \ln\left[\frac{x_{max} - x}{x}\right] - \frac{U}{F} x \tag{15}$$

ϵ vs x curves calculated from this expression are shown on fig. 22 together with corresponding $\partial x/\partial \epsilon$ vs ϵ curves. It is seen that smooth, symmetrical emf curves and bell shaped $\partial x/\partial \epsilon$ curves are predicted by this model. Equation (11) has been found to describe the emf vs composition behaviour of cubic Li_xTiS_2 ($0 < x < 1$, $U/F = 0.423$ V)[14] and $Li_xMo_6S_8$ ($0 < x < 1$, $U/F = -0.0904$ V)[63] with good approximation.

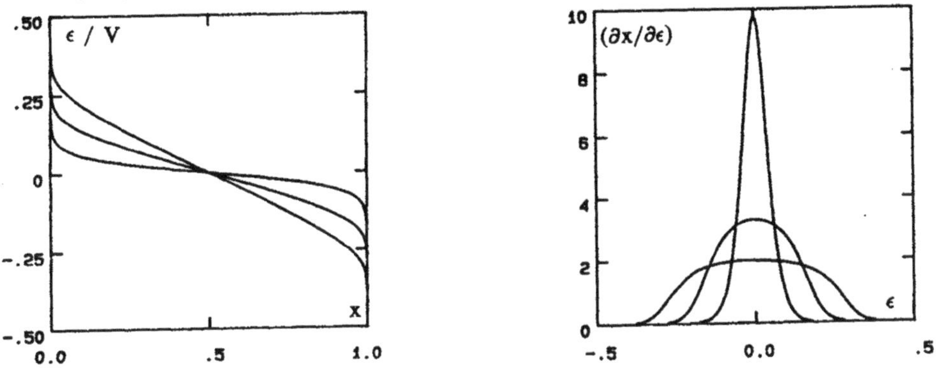

Fig. 22. Compositon variation of cell voltage (ϵ) as given by equation (11). (a) ϵ vs x for different values of the interaction parameter U/F as indicated. (b) Corresponding differential capacity curves ($-\partial x/\partial \epsilon$ vs ϵ).

Several intercalation systems, however, require more elaborate models. Jacobsen[106] calculated the screened electrostatic interaction between point charges confined in layers of a homogeneous dielectric separated between conducting planes. This model was found to account quantitatively for the overall interaction between lithium ions in hexagonal Li_xTiS_2, but excluded the possibility of local ordering of intercalated ions. Details in the $\partial x/\partial \epsilon$-curve indicate that some kind of ordering indeed does occur[107]. Lattice gas models[108], space renormalisation group and Monte Carlo techniques[105] have been used to describe the ordering of lithium in Li_xTiS_2 (see McKinnon and Haering[19] for a review of these models). It appears that none of the proposed models are able to fully account for both the thermodynamic and the structural data[109-112] accumulated on this subject.

In the expressions derived above it was assumed that the lattice was invariant during the intercalation process, and consequently that the affinity between guest and host was independent on composition. In practice the host is always to some degree perturbed by intercalation, and it may be necessary to include a term to account for the changes in guest-host interactions. Whereas guest-guest interactions and the entropy term are always "repulsive" and tend to distribute the intercalated species evenly over the available sites, guest-host interactions may result in condensation of intercalant in domains of high concentration in equilibrium with diluted regions. Phase separations of this type are seen when the energy gained by a cooperative lattice distortion is large enough to overcome the electrostatic repulsion between the participating ions. As long as the changes in guest-host interactions can be approximated by a linear term in the concentration of intercalant, the ϵ-x expression will retain the form of eq. (11). With this extension both positive and negative values of the interaction parameter U are possible. By inclusion of non-linear interaction terms the particle-hole symmetry is lost, and emf curves not symmetrical around $x = 0.5$ can be described. A model of this type has ben proposed to account for the elastic energy associated with intercalation into a layered host lattice[113].

Many host materials offer more than one type of sites for the intercalant, and in these materials the emf vs composition curve will exhibit two or more steps, reflecting the sequential filling of inequivalent host sites. Also ordering of the intercalants, and break-down of the rigid band model may invalidate the simplistic models outlined above for the composition dependence of the emf.

4. APPLICATIONS

Apart from the many scientifically interesting properties of intercalation compounds, these materials have also found a range of practical applications, of which some are briefly mentioned in the following.

The unique properties of intercalation compounds can be exploited in several ways in the field of catalysis[68,114]. Due to the inherent reversibility of intercalation reactions, these materials can be used as sources or sinks of molecular species or ions at controlled chemical activities as well as they can be used as mediators of redox reactions. In this way hydrogen bronzes like $H_xV_2O_5$ or H_xMoO_3 have been showed to catalyze hydrogenation of alkenes[115]. Intercalation compounds can also be used as medium for the catalysed reaction. Instead of promoting reactions at a solid/gas or solid/liquid interface as in conventional heterogeneous catalysis, this type of catalysts act upon intercalated reactants. The host lattice presents a specific chemical and physical environment where favourable transition states can be achieved. With this type of catalysis high selectivity and stereospecificity can be obtained. The high degree of control that can be imposed on the chemical and physical environment of the intramolecular sites of the sheet silicates or zeolites used for this purpose, allows ample room for tailoring of new catalysts.

Several intercalation compounds exhibit *electrochromism*, a reversible change in the optical properties of the host upon ion insertion. Such compounds can be used in *passive displays*, i.e. display devices modulating ambient light[116]. Displays based on proton or alkali metal insertion into materials like tungsten trioxide or iridium oxide have excellent visual contrast with wide viewing angle and good temperature stability. In some of the systems satisfactory switching times and cycle lives have been demonstrated, allthough corrosion problems still persist.

The use as reversible battery electrodes has prompted most of the research in intercalation systems. In batteries of the type shown on fig. 3 no new compounds or phases are created during cell operation, and the electrolyte is only required as an ionically conducting electronic barrier separating the electrodes. This is in contrast to for instance the lead/acid battery in which the electrolyte participates in the cell reaction and the discharge involves the formation of a new, incompatible solid phase ($PbSO_4$). The more simple cell reaction in intercalation systems imply that highly reversible battery systems may be attainable. The use of the highly electropositive alkali metals as intercalated species in these batteries makes high energy densities possible. Several battery systems using lithium and liquid electrolyte are on the treshold of commercialization: These cells are using MoS_2[117] or TiS_2[118] as active material in the positive electrodes.

In systems using liquid electrolyte there is a possibility of solvent cointercalation leading to electrode deterioration, and of parasitic reactions between cell components and the electrolyte. These problems are avoided by using a solid electrolyte - the *all solid state battery* concept as first proposed by Hever[119]. The generally inferior interfacial contact properties of solid phases may be offset by the higher structural stability and the possibility of operation at higher temperatures (up to ≈170°C). Several types of solid ionic conductors have been employed in all solid state battery designs with intercalation electrodes. Of these systems can be mentioned: Polymer electrolytes based on poly(ethylene oxide) in contact with V_6O_{13}[120], glass electrolytes in contact with TiS_2[121] (this battery is being developed at Eveready Corp. - at first only as a primary cell, but the performance as secondary system is also being investigated), and crystalline Li_3N in contact with TiS_2[122]. An "all-plastic" battery based on poly(vinylidene flouride) electrolyte and poly(acetylene) as positive electrode has also been considered[123]. Lately it has been demonstrated that solid state batteries with sodium instead of lithium are feasible[40], and it is likely that other systems will be developed in the future, for instance based upon hydrogen intercalation using the newly developed polymeric proton conductors[124].

5. REFERENCES

1. West, A.R., "Solid State Chemistry", p. 448, J. Wiley & Sons (1984).
2. Weiss, A., *Angew. Chem.*, 75 755 (1963).
3. Wöhler, F., *Ann. Chim. Phys.* 43 29 (1823) in Dickens, P.G. and Whittingham, M.S., *Quart. Rev.* 22 30 (1968).

4. Schauffautl, P., *J. Prakt. Chem.* **21** 155 (1841) in Whittingham, M.S. and Jacobson, A.J. (eds.), "Intercalation Chemistry", Academic Press (1982).
5. Hägg, G., *Z. Physik. Chem.* **B29** 192 (1935).
6. Delmas, C., Nadiri, A. and Soubeyroux, J.L., extended abstract *6th Int. Conf. Solid State Ionics* 107 (1987).
7. Schöllhorn, R., *Angw. Chem.* **92** 1015 (1980).
8. Gamble, F.R., Osiechi, J.H., Cais, M., Pisharody, R., DiSalvo, F.J., and Greballe, T.H., *Science* **174** 493 (1971).
9. Omloo, W.P.F.A.M. and Jellinek, F., *J. Less-Common Metals* **20** 121 (1970).
10. Pauling, L., "The Nature of the Chemical Bond", Cornell University Press (1945).
11. Murphy, D.W., Cross, C., DiSalvo, F.J., and Waszcak, J.V., *Inorg. Chem.* **16** 3027 (1977).
12. Murphy, D.W., Christian, P.A., *Science* **205** 651 (1979).
13. Schöllhorn, R. and Payer, A., *Angew. Chem.* **97** 57 (1985).
14. Sinha, S. and Murphy, D.W., *Solid State Ionics* **20** 81 (1986)
15. Murphy, D.W., Di Salvo, F.J., Hull, G.W., and Waszczak, J.V., *Inorg. Chem.* **15** 17 (1976).
16. Ebert, L.B., *Ann. Rev. Mater. Sci.* **6** 181 (1976)
17. Whittingham, M.S., *Progr. Solid State Chem.* **12** 41 (1978).
18. Abraham, K.M., *J. Power Sources* **7** 1 (1981/82).
19. McKinnon, W.R., Haering, R.R., in White, R.E., Bockris, J.O'.M., and Conway, B.E. (eds.), "Modern Aspects of Electrochemistry" **15** 235, Plenum Press (1983).
20. Schöllhorn, R. in Atwood, J.L., Davies, J.E.D. and MacNicol, D.D. (eds.) "Inclusion Compounds" 1 247, Academic Press (1984).
21. Rouxel, J. and Brec, R., *Ann. Rev. Mater. Sci.* **16** 137 (1986).
22. Ritter, C., *Z. Phys. Chem.* **151** 51 (1987).
23. Whittingham, M.S. and Jacobson, A.J. (eds.), "Intercalation Chemistry", Academic Press (1982).
24. Dresselhaus, M.S. (ed.), "Intercalation in Layered Materials", NATO ASI series, **B 148** , Plenum Press (1986).
25. Wells, A.F., *Structural Inorganic Chemistry*, Fifth Edition, Clarendon Press (1984).
26. Kovalyuk, Z.D., Lukyanyuk, V.K., Pyrlya, M.N., and Seredyuk, A.I., *Ukr. fiz. Zh.* **31** 754 (1986).
27. Wadsley, A.D., *Rev. Pure Appl. Chem.* **5** 165 (1955).
28. Dickens, P.G. and Whittingham, M.S., *Quart. Rev.* **22** 30 (1968).
29. Whittingham, M.S. and Huggins, R.A., *National Bureau of Standards Special Publications* **364** 51 (1972).
30. Hagenmuller, P., in "Comprehensive Inorganic Chemistry", **4** 54, Pergamon (1973).
31. Tinet, D. and Fripat, J.J., *Rev. Chim. Minér.* **19** 612 (1982).
32. Galy, J., Darriet, J., and Hagenmuller, P., *Rev. Chim. Minér.* **8** 509 (1971).
33. West, K., Zachau-Christiansen, B., Østergård, M.J.L., and Jacobsen, T., *J. Power Sources* **20** 165 (1987).
34. Wiesener, K., Schneider, W., Ilic, D., Steger, E., Hallmeier, K.H., and Brackmann, E., *J. Power Sources* **20** 157 (1987).

35. Hagenmuller, P., Galy, J., Pouchard, M., and Casalot, A., *Mat. Res. Bull.* **1** 45 (1966).
36. Zachau-Christiansen, B., West, K., and Jacobsen, T., *Solid State Ionics* **9&10** 399 (1983).
37. Liaw, B., Raistrick, I.D. and Huggins, R.A., *Solid State Ionics* **18&19** 828 (1986).
38. Murphy, D.W., Christian, P.A., DiSalvo, F.J., and Waszcak, J.V., *Inorg. Chem.* **18** 2800 (1979).
39. Pouchard, M., Casalot, A., Galy, J., and Hagenmuller, P., *Bull. Soc. Chim. Fr.* 4343 (1967).
40. West, K., Zachau-Christiansen, B., Jacobsen, T., and Skaarup, S., extended abstract *6th Int. Conf. Solid State Ionics* 372 (1987).
41. Raistrick, I., *Rev. Chim. Minér.* **21** 456 (1984).
42. Grey, I.E., Anne, M., Collomb, A., Muller, J., and Marezio, M., *J. Solid State Chem.* **37** 219 (1981).
43. DiSalvo, F.J., Murphy, D.W., and Waszczak, J.V., *Synth. Metals* **1** 29 (1979/80).
44. Jacobsen, T., West, K., Zachau-Christiansen, B., and Atlung, S., *Electrochim. Acta* **30** 1205 (1985).
45. Zachau-Christiansen, B., West, K., and Jacobsen, T., *Mat. Res. Bull.* **20** 485 (1985).
46. Cava, R.J., Santoro, A., Murphy, D.W., Zahurak, S., and Roth, R.S., *J. Solid State Chem.* **48** 309 (1983).
47. Galy, J., Darriet, J., Casalot, A., and Goodenough, J.B., *J. Solid State Chem.* **1** 339 (1970).
48. Wiseman, P.J. and Dickens, P., *J. Solid State Chem.* **17** 91 (1976).
49. West, K., Zachau-Christiansen, B., Landeira, M.J., and Jacobsen, T. in Poulsen, F.W., Hessel Andersen, N., Clausen, K., Skaarup, S., and Toft Sørensen, O. (eds.), *Transport-Structure Relations in Fast Ion and Mixed Conductors*, 265, Risø National Laboratory (1985).
50. West, K., Zachau-Christiansen, B., and Jacobsen, T., *Electrochim. Acta* **28** 1829 (1983).
51. West, K., Zachau-Christiansen, B., Jacobsen, T., and Atlung, S., *J. Power Sources* **14** 235 (1985).
52. Dickens, P.G., Chippindale, A.M., Hibble, S.J., and Lancaster, P., *Mat. Res. Bull.* **19** 319 (1984).
53. Cava, R.S., Santoro, A., Murphy, D.W., Zahurak, S.M., Flemming, R.M., Marsh, P, and Roth, R.S., *J. Solid State Chem.* **65** 63 (1986).
54. ter Haar, L.W. and DiSalvo, F.J., *Solid State Comm.* **60** 305 (1986).
55. Tarascon, J.M., DiSalvo, F.J., Murphy, D.W., Hull, G., and Waszczak, J.V., *Phys. Rev. B.* **29** 172 (1984).
56. Fischer, Ø., *J. Appl. Phys.* **16** 1 (1978).
57. Cava, R.S., Santoro, A. and Tarascon, J.M., *Solid State Chem.* **54** 193 (1984).
58. McKinnon, W.R. and Dahn, J.R., *J. Phys. C* **19** 5121 (1986).
59. Hughbanks, T. and Hoffmann, R., *J. Am. Chem. Soc.* **105** 1150 (1983).
60. Schramm, W., Gocke, E., and Schöllhorn, R., *Mat. Res. Bull.* **21** 929 (1986).
61. Gocke, E., Schramm, W., Dolscheid, P., and Schöllhorn, R., *J. Solid State Chem.* **70** 71 (1987).
62. Mulhern, P.J. and Haering, R.R., *Can. J. Phys.* **62** 527 (1984).
63. Coleman, S.T., McKinnon, W.R., and Dahn, J.R., *Phys. Rev. B* **29** 4147 (1984).

64. Rüdorff, W, and Sick, H.H., *Angew. Chem.* **71** 127 (1959).
65. Hibma, T., in Whittingham, M.S. and Jacobson, A.J. (eds.), "Intercalation Chemistry", 285, Academic Press (1982).
66. Solin, S.A., *J. Molec. Catal.* **27** 293 (1984).
67. Fischer, J.E. and Kim, H.J., *Phys. Rev. B* **35** 3295 (1987).
68. Pinnavaia, T.J., *Science* **220** 365 (1983).
69. Rouxel, J., *J. Solid State Chem.* **17** 223 (1976).
70. Whangbo, M.-H., Rouxel, J. and Trichet, L., *Inorg. Chem.* **24** 1824 (1984).
71. LeBlanc, A., Danot, M., Trichet, L., and Pouxel, J., *Mat. Res. Bull.* **9** 191 (1974).
72. Jacobson, A.J., *Solid State Ionics* **5** 65 (1981).
73. Thompson, A.H., *Physica* **99B** 100 (1980).
74. McKinnon, W.R., and Dahn, J.R., *Solid State Comm.* **48** 43 (1983).
75. Thompson, A.H. and DiSalvo, F.J., in Whittingham, M.S. and Jacobson, A.J. (eds.), "Intercalation Chemistry", 573, Academic Press (1982).
76. West, K.,Jacobsen, T., Zachau-Christiansen, B., and Atlung, S., *Electrochim. Acta* **28** 97 (1983).
77. Ritter, C., Müller-Warmuth, W. and Schöllhorn, R., *Solid State Ionics* **20** 283 (1986).
78. Brec, R., *Solid State Ionics* **22** 3 (1986).
79. Brec, R., Ouvrard, G. and Rouxel, J., *Mat. Res. Bull.* **20** 1257 (1985).
80. Clement, R., *J. Chem. Soc. Chem. Comm.* 647 (1980).
81. Le Méhauté, A.,Ouvrard, G., Brec, R., and Rouxel, J., *Mat. Res. Bull.* **12** 1191 (1977).
82. Le Méhauté, A., *C. R. Acad. Sc. Paris* **287** 309 (1978).
83. Thompson, A.H. and Whittingham, M.S., *Mat. Res. Bull.* **12** 741 (1977).
84. Le Méhauté, A., and Perche, P., *Power Sources* 7 623 (1979).
85. Foot, P.J.S., Katz, T., Patel, S.N., Nevett, B.A., Piercy, A.R., and Balchin, A.A., *Phys. Stat. Sol.* **100** 11 (1987).
86. Delmas, C. in Dresselhaus, M.D. (ed.), "Intercalation in Layered Materials", NATO ASI series, **B 148** 155, Plenum Press (1986).
87. Vidyasagar, K. and Gopalakrishnan, J., *J. Solid State Chem.* **42** 217 (1982).
88. Mizushima, K., Jones, P.C., Wiseman, P.J., and Goodenough, J.B., *Mat. Res. Bull.* **15** 783 (1980).
89. de Piciotto, L.A., Thackeray, M.M., David, W.I.F., Bruce, P.G., and Goodenough, J.B., *Mat. Res. Bull.* **15** 783 (1980).
90. Pistoia, G., Panero, S., Tocci, M., Moshtev, R.V., and Manev, V., *Solid State Ionics* **13** 311 (1984).
91. Murphy, D.W. and Trumbore, F.A., *J. Electrochem. Soc.* **123** 960 (1976).
92. Whittingham, M.S., *J. Electroanal. Chem.* **118** 229 (1981).
93. Carides, J.N. and Murphy, D.W., *J. Electrochem. Soc.* **124** 1309 (1977).
94. Hollek, G.L. and Driscoll, J.R., *Electrochim. Acta* **22** 647 (1977).
95. Zanini, M., *J. Electrochem. Soc.* **132** 588 (1985).
96. Liang, K.S., Cramer, S.P., Johnston, D.C., Chang, C.H., deNeufville, J.P., and Chianelli, R.R., *J. Non-Cryst. Solids* **42** 345 (1980).
97. Chang, C.H. and Chan, S.S., *J. Catal.* **72** 139 (1981).
98. Murugesan, T. and Gopalakrishnan, J., *Proc. Ind. Acad. Sci.* **91** 7 (1982).
99. Cramer, S.P., Liang, K.S., Jacobson, A.J., Chang, C.H. and Chianelli, R.R., *Inorg. Chem.* **23** 1215 (1984).

100. Jacobson, A.J., Chianelli, R.R., Rich, S.M., and Whittingham, M.S., *Mat.Res. Bull.* **14** 1437 (1979).
101. Auborn, J.J., Barberio, Y.L., Hanson, K.J., Schleich, D.M., and Martin, M.J., *J. Electrochem. Soc.* **134** 580 (1987).
102. West, K., Zachau-Christiansen, B., Jacobsen, T., and Atlung, S., *J. Electrochem. Soc.* **132** 3061 (1985).
103. Livage, J., *J. Solid State Chem.* **64** 322 (1986).
104. Crandall, R.S., Wojtowicz, P.J., and Faughnan, B.W., *Solid State Comm.* **18** 1409 (1976).
105. Berlinsky, A.J., Unruh, W.G., McKinnon, W.R., and Haering, R.R., *Solid State Chem.* **31** 135 (1977).
106. Jacobsen, T., West, K., and Atlung, S., *Electrochim. Acta* **27** 1007 (1982).
107. Thompson, A.H., *J. Electrochem. Soc.* **126** 608 (1979). See also Jacobsen, T., West, K., and Atlung, S., *J. Electrochem. Soc.* **126** 2169 (1979).
108. Osorio, R. and Falicov, L.M., *J. Phys. Chem. Solids* **43** 73 (1982).
109. Hibma, T., *J. Solid State Chem.* **34** 97 (1980).
110. Dahn, J.R., McKinnon, W.R., Haering, R.R., Buyers, W.J.L, and Powell, B.M., *Can. J. Phys.* **58** 2 (1980).
111. Dahn, J.R. and Haering, R.R., *Solid State Comm.* **40** 245 (1981).
112. Kleinberg, R.L., Silbernagel, B.G., and Thompson, A.H., *Solid State Comm.* **41** 401 (1982).
113. Dahn, J.R., Dahn, D.C., and Haering, R.R, *Solid State Comm.* **42** 179 (1983).
114. Boersma, M.A.M., *Cat. Rev. -Sci. Eng.*, **10** 243 (1974).
115. Marcq, J.P., Poncelet, G., and Fripiat, J.J., *J. Catal.* **87** 339 (1984).
116. Beni, G. and Shay, J.L., *Advances in Image Pickup and Display* **5** 83 (1982).
117. Brandt, K., *J. Power Sources* **18** 117 (1986).
118. Shen, D.H., Subba Rao, S., Yen, S.P.S., and Somoano, R.B., *J. Power Sources* **18** 127 (1986).
119. Hever, K.O., *J. Electrochem. Soc.* **115** 830 (1968).
120. Gauthier, M., Fauteux, D., Vassort, G., Bélanger, A., Duval, M., Ricoux, P., Chabagno, J.-M., Muller, D., Rigaud, P., Armand, M., and Deroo, D., *J. Electrochem. Soc.* **132** 1333 (1986).
121. Akridge, J.R. and Vourlis, H., *Solid State Ionics* **18&19** 1082 (1986).
122. Knutz, B. and Skaarup, S., *Solid State Ionics* **18&19** 783 (1986).
123. Nagamoto, T., Ichikawa, C., and Omoto, O., *J. Electrochem. Soc.* **134** 305 (1987).
124. Donoso, P., Gorecki, W., Berthier, C., Defendini, F., and Armand, M.B., extended abstract *6th Int. Conf. Solid State Ionics* 302 (1987).

MIXED IONIC ELECTRONIC CONDUCTORS

I. Riess and D.S. Tannhauser

Physics Department, Technion IIT.

Haifa, 32000, Israel

CONTENTS

1. INTRODUCTION

This article deals with effects which are caused by the fact that in a material we have both ionic and electronic conductivity.

Naturally in any material we will have in principle non-zero ionic and non-zero electronic conductivity (σ_i, σ_{el}). However, often the ratio between these two is extreme and then for many properties, such as total conductivity, the smaller one does not influence the result of an experiment.

The ratio between the ionic and electronic conductivities is often given by the transport numbers t_i and t_{el}. These are defined by

$$t_i = \frac{\sigma_i}{\sigma_i + \sigma_{el}} \tag{1}$$

$$t_{el} = \frac{\sigma_{el}}{\sigma_i + \sigma_{el}} \tag{2}$$

Since the conductivity is the product of concentration and mobility the values of t_i, t_{el} will change with temperature and can also be manipulated by controlling the concentration of electronic charge carriers or ionic defects through doping or through deviation from stoichiometry of the material.

Books and reviews on solid electrolytes usually contain information also on mixed ionic electronic conductors (MIEC's). Some of the relevant books are by Hauffe[1)], Kroeger[2)], Rickert[3)] and Schmalzried[4)]. Review papers were written among others by Wagner[5–6)] and Heyne[7)] (on polarization techniques), Tuller[8)] (on mixed conducting oxides), Weppner and Huggins[9)], and Dudley and Steele[10)] (on some of the experimental methods to investigate MIEC's).

Since steady state polarization methods and most of the time dependent methods using blocking electrodes have been reviewed before, we shall expand in this review on recent analytic studies of the steady state concerning the transport properties of a MIEC, the distribution of charge carriers, possible p-n and p-i-n quasi junctions in MIEC and analysis of the performance of a fuel cell based on a MIEC.

In order not to get into complications with charge carriers with a charge larger than one we shall limit our general equations to the hypothetical MIEC X^+Y^- and shall assume that the X^+ ions are mobile, the Y^- ions are immobile and the electronic current is carried by electrons. If we need to introduce explicit ionic defects we shall usually assume that the ionic current is carried by X^+ interstitials. However examples will also be given from materials for which the charges of the ions and/or the signs of ions and electronic carriers are different from the above. The details of the translation of the equations for these cases will then often be left to the reader. (When charged vacancies are the mobile species then the chemical potential μ_X in the equations should be replaced by μ_V.)

Many of the materials we deal with have a very high concentraton of defects. While it is customary to refer only to the electrons in the conduction band and not to all the electrons of the crystal when talking for example about mobility of electrons the convention for ionic conductors is less clear. We shall therefore either regard all the ions of one kind as charge carriers, or when talking about a small concentration of defects carrying the charge, shall limit our discussion to interstitials or vacancies. We shall use the Kroeger-Vink notation for defects. A list of symbols appears at the end of the article.

2. EXAMPLES OF MIXED CONDUCTORS

Co-existence of electronic and ionic conduction can occur under various conditions. In β -$Ag_{2+\delta}S$[11] (existing for T > 177 C) the silver ions propagate through interstitial sites[12] and conduct current through the material by moving from site to site. $Ag_{2+\delta}S$ is a very good ionic conductor with an ionic conductivity $\sigma_i = 3.8\Omega^{-1}$ cm^{-1} at 200 C[13] - practically independent of δ, the deviation from stoichiometry. The activation energy of the ionic conductivity is only 0.05 eV. β -$Ag_{2+\delta}S$ in equilibrium with liquid sulfur is close to stoichiometry[11]. Because of a moderate energy gap of 0.4 eV[14] it exhibits some electronic conductivity[15] carried both by electrons and holes.

The stoichiometry of $Ag_{2+\delta}S$ can be changed[15] by inserting silver up to $\delta = 2.5 \times 10^{-3}$. The additional silver enters into interstitial sites and acts as a donor according to the reaction

$$Ag \leftrightarrow Ag^{+} + e'$$

so that the concentration of quasi free electrons in the conduction band increases drastically. A non-metal to metal transition occurs at $\delta = 0.15 \times 10^{-3}$.[16] The molar concentration of these electrons n_e (given by δ) is small compared to that of the mobile ions, N_i (which is practically 2, since all silver ions take part in the conduction process). Since, however, the electron mobility is much higher than that of the ions, and $\sigma_e \sim 1000\ \Omega^{-1}$cm^{-1} at 200 C and $\delta = 2.5 \times 10^{-3}$, the nonstoichiometric solid is an electronic conductor even though $n_e << N_i$. In this example the defects contributing to electronic conduction are native ones, namely additional silver ions in interstitial sites.

We consider now another example of an undoped semiconductor in which ionic and electronic conduction occurs due to change of stoichiometry but for which n_e and N_i are comparable. Stoichiometric CeO_2 which has been studied extensively near 1000 K, is an insulator with respect to electrons (or holes) having a gap of 6 eV, [17] and also for ions. In reduced ceria, CeO_{2-x}, the oxygen vacancies contribute both to oxygen ion and to electron conduction[18]. The vacant sites with concentration N_i enable the motion of oxygen ions at elevated temperatures ($\sim$ 1000 K). In CeO_2 two electrons are contributed by the cations (Ce^{4+}) to each oxygen atom. When an oxygen atom is missing, these electrons reside on the vacancy. The electrons are not tightly bound to the oxygen vacancies and at $\sim$ 1000 K are excited to the 4f narrow band (3.5 eV above the valence band)[17] in which conduction occurs by small polaron hopping.[19,20]. As each vacancy can donate one or two electrons we have

$$n_e = \alpha N_i$$

where α is of order unity.

As a third example we take CeO_2 doped with CaO in which the motion of ions is enabled by vacancies. Ca^{++} enters CeO_2 substitutionally introducing a charged oxygen vacancy $V_O^{\cdot\cdot}$. The material is an oxygen ion conductor with a negligible electronic conductivity since the charged vacancies are compensated by electrons tightly bound to the Ca_{Ce} impurities. Electronic conduction can be introduced by reducing this material[21]. The additional vacancies formed act as donors in the manner mentioned in the previous example. By controlling the level of reduction one can tune the ionic transference number from $t_i \sim 1$ for a low level of reduction to $t_i << 1$ for a high level.

Yttria or calcia stabilized ZrO_2 (YSZ, CSZ) are similar to CaO doped CeO_2[22,23]. When stoichiometric with respect to oxygen, $Zr_{1-x}Y_xO_{2-x/2}$ is a good oxygen ionic conductor at $\sim$1300 K with $t_i \sim 1$, $t_e << 1$. However, when YSZ is reduced electron conduction appears in addition to the ionic one[22-25]. YSZ can also be made a p- type semiconductor by introducing an excess of oxygen into the lattice at elevated pressure[25]. The excess oxygen enters into vacant site and accepts two electrons

$$\frac{1}{2}O_2 + V_O^{\cdot\cdot} \rightarrow 2h^{\cdot}$$

In the last two examples doping was used to introduce the vacancies which enable ionic conduction. We turn now to two examples in which doping is used in the traditional way to introduce electronic conduction. In CeO_2 doped with UO_2 [26] the uranium is ionized to form U_{Ce} and electrons are introduced into the conduction band. The material is therefore an electronic conductor. In addition some extra oxygen can enter interstitial sites in the lattice and become ionized. These ionized interstitials are somewhat mobile, and one obtains a mixed ionic-electronic conductor.

Li-doped Si or Ge are different examples of MIEC's as the host lattice is covalently bonded. The mobile species are the dopant ions, Li, which act also as the donors that contribute to the electronic conductivity[27].

It is customary to limit the use of the name MIEC to those compounds in which the ionic conductivity is comparable to the electronic one and they are not too small (σ_i,$\sigma_e \geq 10^{-5}\Omega^{-1}\text{cm}^{-1}$), even though in some processes the minority charge carrier plays an important role. Thus for example MnO_{1+x} is an electronic conductor in which holes or electrons dominate the conductivity, depending on the oxygen pressure[28,29]. A change of stoichiometry in this material involves motion of manganese ions through the rigid oxygen sublattice. Since macroscopically only neutral atoms can move in such a process the ions are coupled to electrons or holes and we get ambipolar diffusion i.e. a mixed conductivity process. This will be treated in detail below. Other examples of semiconducting oxides are given in refs. (8,30-32).

We have seen that electronic and ionic conductivity can be introduced in three ways: a) Intrinsically combined with thermal excitation, b) By deviation from stoichiometry, and c) By doping. This is summarized in Table 1, which also includes the examples discussed before.

3. MEASUREMENT OF CONDUCTIVITY

In a solid conducting both electrons and ions one would like to determine both σ_e, and σ_i. The total conductivity σ_t is then

$$\sigma_t = \sigma_e + \sigma_i \tag{3}$$

If holes also contribute to the conductivity the total electronic conductivity is

$$\sigma_{el} = \sigma_e + \sigma_h \tag{4}$$

TABLE 1

Case	Conductivity	Cause of Conductivity: Intrinsic	Deviation from Stoichiometry	Doping	Example
a	Electronic		x		Ag_2S,
	Ionic	x			$YBa_2Cu_3O_7$*
b	Electronic		x		CeO_{2-x}
	Ionic		x		
c	Electronic		x		Doped CeO_2
	Ionic			x	
d	Electronic			x	CeO_2+UO_2
	Ionic		x		
e	Electronic			x	Li doped Si
	Ionic			x	
f	Electronic	x			Stoichiometric Cu_2S[a]
	Ionic	x			
g	Electronic	x			H^+ in Pd
	Ionic			x	
h	Electronic	x			MnO_{1+x}**
	Ionic		x		
i	Electronic			x	***
	Ionic	x			

* Chemical stoichiometry in $YBa_2Cu_3O_x$ requires $x = 6.5$. Therefore $x = 7$ implies a deviation from stoichiometry.

** For nearly stoichiometric MnO_{1+x} with $n \sim p$.

*** Any intrinsic ionic conductor into which electronic conductivity is introduced by doping.

(a) Wagner, J.B. and Wagner, C., J. Chem. Phys. 26, 1602 (1957).

We shall assume here that $\sigma_e >> \sigma_h$.

In many solids ions of one kind only contribute to the ionic conductivity. (However other cases may also occur, e.g. in KCL both K^+ and Cl^- are mobile and $\sigma_{K^+}/\sigma_{Cl^-} \simeq 6$).[33]

In a mixed conductor interacting with a non-uniform chemical surrounding which is subjected to an applied electric field, gradients in the concentration of electrons and ions may form. The conductivities and transference numbers then become local properties and one sometimes refers to average quantity $\bar{\sigma}_e, \bar{\sigma}_i$ and $\bar{t}_e, \bar{t}_i$ which will be defined below.

Before discussing the interpretation of conductivity measurements done on mixed ionic-electronic conductors, which may be quite involved, we consider first the direct determination of the average ionic transference number of the solid using the EMF method. We consider a mixed conductor conducting X^+ ions which interacts at its electrodes with two surrounding I, II in which the chemical potentials of the X atoms are μ_X^I and μ_X^{II}. The open circuit EMF, V_{oc}, is then given by the Nernst equation multiplied by $\bar{t}_i$ of the sample

$$V_{oc} = \bar{t}_i \frac{\mu_X^{II} - \mu_X^I}{q} \tag{5}$$

where we have assumed that the electrodes are reversible. In terms of activities a_X this can be written

$$V_{oc} = \bar{t}_i \frac{kT}{q} \ell n \frac{a_X^{II}}{a_X^I} \tag{6}$$

where k is the Boltzmann constant.

For example, to determine $\bar{t}_i$ of δ-Bi_2O_3, which is an oxygen conductor, a pellet of the material was placed between two gases with different oxygen partial pressures, and $\bar{t}_i$ was found to be greater than 0.95[34].

An error in this kind of measurement is often caused by a temperature difference between the two sides of the sample, this generates a thermal EMF which may be as large as 0.5 mV/C. If a small $\Delta\mu_X$ is impressed on the sample so that $\Delta\mu_X/q \leq 10$ mV then the sample stays almost uniform and the measured $\bar{t}_i$ reflects to a good approximation the local t_i. However $\bar{t}_i$ may be in error because of the thermal EMF. If $\Delta\mu_X$ is such that a signal of more than 100 mV is generated, the accuracy of the measurement is better but $\bar{t}_i$ is an average value over the non-uniform sample.

We have assumed that the electrodes are reversible. Electrodes are reversible with respect to a charge carrier when a current of this carrier does not entail a difference of its electrochemical potential between the two sides of the electrode.

A thin porous metal electrode with negligible resistance for electrons, exposed to X in the gas phase, is in effect also reversible for X^+ ions across this electrode: we have

$$\Delta\tilde{\mu}_e = 0 \tag{7}$$

and

$$\Delta\mu_X = 0 \tag{8}$$

therefore also

$$\Delta\tilde{\mu}_{X^+} = \Delta\mu_X - \Delta\tilde{\mu}_e = 0 \tag{9}$$

If an electrode on a MIEC is not reversible for either ions or electrons, the open circuit voltage will differ from that given by Eqs.(5,6), even though no current flows. The reason for this is that in a mixed conductor only the total current can be made to vanish in the open circuit condition,

$$j_e + j_i = 0 \tag{10}$$

The partial electronic or ionic current densities j_e respectively j_i will not vanish and therefore $\Delta\mu_X = \Delta\tilde{\mu}_e + \Delta\tilde{\mu}_X$ across the electrode will differ from zero[35]. Therefore the chemical potential of X at the surface of the mixed conductor just under the electrode will differ from that in the surrounding and the voltage measured between the two electrodes will differ from that given by the values of μ_X at the boundary surfaces of the mixed conductors.

Four point conductivity measurements on mixed conductors are subject to the same problem, i.e. probes which are not reversible can falsify the measurement. In a rectangular sample of a normal conductor the current flows everywhere in the same direction, say x (see Fig. 1).

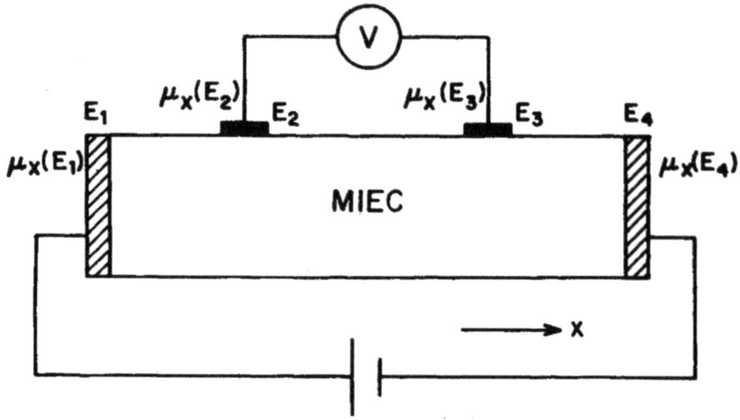

Fig. 1. Four point conductivity measurement configuration. This cannot be assumed to be a linear configuration when a mixed conductor is being tested.

However, in a mixed conductor with voltage probes attached to it, the interaction of the surrounding with the MIEC at the voltage probes results in partial currents j_e, j_i towards the voltage probes, i.e. not in the x direction. The field distribution ceases to be one dimensional and becomes 2 or 3 dimensional and an analytic analysis becomes difficult. It is then hard, for example, to extract the conductivities from the measured data. One can use the classical one dimensional analysis if one of the following conditions hold:

a) The surrounding is uniform and at least the current carrying electrodes are reversible. In this case the partial currents through the probes vanish, and one then measures the total conductivity.

b) Three blocking electrodes are used to ensure that only one type of charge carriers can flow, in this case the conductivity of the non blocked carrier is measured. One electrode should be left non-blocking, this permits the sample to interact with the surrounding so that the composition near this electrode is defined by the surrounding. If all electrodes were blocking the composition of the sample would not be well defined and depend on its history. The blocking can be either for ions or for electrons. The first can be done e.g. by using an electron conducting metallic layer. If the surrounding interacting with the mixed conductor is a gas one also has to make sure that the electrode blocks the gas phase. Otherwise, the gas can be ionized at the three phase boundary gas-metal-MIEC. For example, for oxygen the reaction O_2 (gas) + 4e (electrode) $\rightarrow 2O^{--}$ (MIEC) can take place and the electrode will not be blocking for O^{--} ions. Electrons can be blocked using a solid electrolyte that conducts the X^+ ions only.

A two dimensional configuration with four point electrodes for measuring the conductivity of a planar (singly connected) sample of an electronic conductor was suggested by van der Pauw[36]. Riess and Tannhauser[37] have shown that the steady state transport equations can be solved analytically also for a mixed conductor in a planar configuration with four reversible point electrodes exposed to a different surrounding at each of the electrodes. The analysis is valid for small variations in the charge carrier concentration imposed by a small difference $\Delta\mu_X$ in the chemical surrounding i.e.

$$| \mu_X(E_i) - \mu_X(E_j) | = | \Delta\mu_{X,ij} | << kT \tag{11}$$

Here $\mu_X(E_i)$ is the chemical potential of X at the surface of the MIEC just under electrode E_i (see Fig. 2) and indices i,j = 1,2,3,4 refer to the electrode number.

In Fig. 2 the current carrying electrodes are E_1 and E_4 and the voltage probes are E_2 and E_3. The current is therefore denoted by I_{14} and the voltage by V_{32}. By interchanging the role of electrodes E_2 and E_4 the current becomes I_{12} and the voltage V_{34}. The four measured quantities I_{12}, V_{34}, I_{14}, V_{32} are related by [37]

$$\begin{aligned} &exp\{-\pi d\sigma_t \mid V_{32} + \frac{\sigma_i}{\sigma_t}\frac{\Delta\mu_{X,32}}{q} \mid / \mid I_{14} \mid\} \\ &+exp\{-\pi d\sigma_t \mid V_{34} + \frac{\sigma_i}{\sigma_t}\frac{\Delta\mu_{X,34}}{q} \mid / \mid I_{12} \mid\} = 1 \end{aligned} \tag{12}$$

where d is the sample thickness. It should be noted that it is permitted to apply different chemical potentials $\mu_X(E_i)$ at the electrodes.

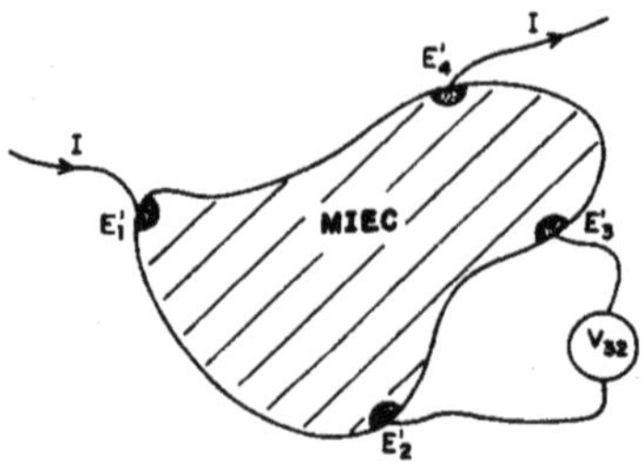

Fig. 2 Planar singly connected sample with four electrodes

If the surrounding is uniform then $\Delta\mu_{X,32} = \Delta\mu_{X,34} = 0$ and Eq. (12) reduces to an equation for σ_t alone. Once σ_t is determined one can change the surrounding to induce a known difference of e.g., $\Delta\mu_{X,32}$ and repeat the two I, V measurements. Using Eq. (12) σ_i can now be determined.

For a circular sample with equally spaced electrodes in a uniform surrounding Eq. (12) yields the original equation of van der Pauw[36)]

$$\sigma_t = \frac{\ell n 2}{2\pi d}\left(\left|\frac{I_{14}}{V_{32}}\right| + \left|\frac{I_{12}}{V_{34}}\right|\right) \tag{13}$$

We had to assume reversible electrodes otherwise one cannot relate the applied values of $\mu_X(E_i)$ and the measured voltage V_{ij} to the values actually on the mixed conductor which are referred to in the equations. If this assumption fails one has to eliminate one component of the current by using (three) blocking electrodes. The experimental setup and its interpretation then become identical to the usual van-der-Pauw method. The partial conductivity measured is that of the charge carriers not blocked. In this case one has also to limit the applied voltage to eliminate polarization leading to large changes in the sample composition. Thus to ensure that $\Delta\mu_X$ on the sample stays small (Eq. (11)) one has to add the constraint

$$| qV | < kT \tag{14}$$

4. TIME DEPENDENT TRANSPORT IN MIEC

4.1 Introduction

Measurements in the non steady state may yield information regarding diffusion coefficients, while in the steady state one measures conductivities which reflect a product of diffusion coefficient and charge concentration. Due to local neutrality conditions the motions of ions and electrons are coupled and one measures the chemical diffusion coefficient $\tilde{D}$. If one kind of charge carrier exists as a small minority, $\tilde{D}$ reduces to the component diffusion coefficient of the minority carrier D_e or D_X.

The solution of the transport equation has been carried out under various restrictive conditions always including the use of blocking electrodes and probes either for electrons or ions or combinations of the two kinds. Yokota[38)] discussed depolarization for small signals for which the spatial variations δn in charge concentration, n, are small: $\delta n << n$. This was used to determine the chemical diffusion coefficient $\tilde{D}_{Ag^+}$ in Ag_2Te. Yokota and Miyatani[39)] used the same approach to discuss ac measurements on Cu_2S and Ag_2Te and determined $\tilde{D}_{Cu^+}$ and $\tilde{D}_{Ag^+}$. An extension of Yokota's analysis to larger applied fields and hence larger values of the ratio $\delta n/n$ under various conditions of $\Delta\mu_X$ and inclusion of diffusion under a temperature gradient and a magnetic field, have been given recently by Miyatani and applied to Cu_2S and Ag_2Te[40-42)]. A pulse method[43,44)] whereby silver is injected into a long bar shaped sample of Ag_2Te and is detected by an Ag/AgI and Pt collector was recently employed to measure $\tilde{D}_{Ag^+}$ in Ag_2S.

A galvanostatic intermittent titration technique in which a pulse of constant ionic current is injected through a solid electrolyete into a MIEC has been discussed by Weppner and Huggins[45)]. The diffusion of the ions into the MIEC is determined from the measurement of the EMF established across the solid electrolyte with the MIEC serving as one of its electrodes. This and related methods are reviewed by Weppner and Huggins [9)].

Chebotin et al.[46] discuss the time dependence of the build up of polarization on oxides with blocking cathodes in the Hebb-Wagner polarization method.

Weppner presented a method to measure the component diffusion coefficient of electrons D_e as the minority carriers in a MIEC and applied it[25] to stabilized ZrO_2. In this method the sample is polarized and the decay of the polarization in an open circuit experiment is followed by measuring the time dependence of the cell EMF.

When ions are the minority carriers $\tilde{D}$ depends on D_{X^+}. Jin and Rosso[47] measured the component diffusion coefficient of Cu^+ in the semiconductor Cu_3VS_4 by a three probe polarization measurement.

4.2 Chemical diffusion

We have a process of chemical diffusion when the chemical composition is not in equilibrium and the boundary conditions are such that no electric current is passing through the MIEC. The situation when these boundary conditions are relaxed will be treated in Section 5. We shall derive a relation between the chemical and the tracer diffusion coefficient. We take as a simple model the system XY and assume that the electronic current is carried by the electrons and that there is no interaction between the defects except for the coulomb interaction which leads to charge neutrality.

The particle currents (No. of particles per m^2 and second) of ions and electrons are

$$J_X = -[X^+]b_X \frac{\partial \mu_{X^+}}{\partial x} - q[X^+]b_X \frac{\partial \phi}{\partial x} \tag{15}$$

$$J_e = -nb_e \frac{\partial \mu_e}{\partial x} + qnb_e \frac{\partial \phi}{\partial x} \tag{16}$$

where b_i are mechanical mobilities ($ms^{-1}Nt^{-1}$). We require because of charge neutrality that the total electric current is zero, i.e. $J_X - J_e = 0$, use conductivities defined by

$$\sigma_i = q^2 b_X [X^+] \tag{17}$$

$$\sigma_e = q^2 b_e n \tag{18}$$

and get the following expression for the electric field

$$\frac{\partial\phi}{\partial x} = \frac{\frac{\sigma_e}{q}\frac{\partial\mu_e}{\partial x} - \frac{\sigma_i}{q}\frac{\partial\mu_{X^+}}{\partial x}}{\sigma_i + \sigma_e} \tag{19}$$

We introduce this into Eq. (15) and get after some algebra

$$J_X = -\frac{\sigma_i\sigma_e}{q^2(\sigma_i + \sigma_e)}\left(\frac{\partial\mu_{X^+}}{\partial x} + \frac{\partial\mu_e}{\partial x}\right) \tag{20}$$

Since the introduction into the crystal of an ion and an electron is equivalent to introducing an X^{atom} we use $\mu_X = \mu_{X^+} + \mu_e$. We also use the electronic transport number

$$t_e = \frac{\sigma_e}{\sigma_i + \sigma_e} \tag{21}$$

and transform Eq. (20) to

$$J_X = -t_e[X^+]b_X\frac{\partial\mu_X}{\partial x} \tag{22}$$

Since $[X^+] = [X^{atom}]$ and $b_X = D_X^{Tr}/fkT$, where D_X^{Tr}/f is calculated from the conductivity via the Einstein relation and f is the correlation factor for diffusion of ions, usually of order one, we get

$$J_X = -t_e\frac{D_X^{Tr}}{fkT}\frac{\partial\mu_X}{\partial ln[X^{atom}]}\frac{\partial[X^{atom}]}{\partial x} \tag{23}$$

so that the chemical diffusion coefficient becomes

$$\tilde{D} = t_e\frac{D_X^{Tr}}{fkT}\frac{\partial\mu_X}{\partial ln[X^{atom}]} \tag{24}$$

Often it is illuminating to write this in terms of the activity a of the atoms which is defined through $\mu_X = kTlna_X$ and is for non-interacting particles at low concentration proportional to the concentration. We then get

$$\tilde{D} = t_e\frac{D_X^{Tr}}{f}\frac{\partial lna_X}{\partial ln[X^{atom}]} \tag{25}$$

We see that $\tilde{D}$ differs considerably from D^{Tr}. A slight generalization of our derivation shows that Eq. (25) is valid for any combination of multivalent ions, holes and electrons.

Eq. (20) is symmetric with respect to ions and electrons. It therefore leads also to an alternative form of Eq. (25) namely

$$\tilde{D} = t_i D_e \frac{n}{[X^{atom}]} \frac{\partial ln a_X}{\partial ln [X^{atom}]} \tag{26}$$

Note that $D_e n/[X^{atom}]$ is an effective diffusion constant the electrons calculated from σ_e assuming, as we did for ions, that there is one electron per atomic site.

Weppner and Huggins[45] define an "enhancement factor" for diffusion called W (in honor of C. Wagner)

$$W = t_e \frac{\partial ln a_X}{\partial ln [X^{atom}]} \tag{27}$$

which measures the ratio of the chemical diffusion constant and D^{Tr}/f. For example for CoO it is easy to show that[48]

$$\tilde{D} = \frac{1}{f} D^{Tr}_{Co}(z+1) \frac{[Co^{atom}]}{[V^{z'}_{Co}]}$$

where z is the effective charge of the mobile vacancies. We see that the enhancement factor equals the reciprocal of the relative vacancy concentration times $(z+1)$. If diffusion was dominated by neutral vacancies we would have $(z+1)=1$.

The former expression for $\tilde{D}$ is more intuitive if one remembers that $D^{Tr} = f D_V [V^{z'}_{Co}]/[Co^{atom}]$. It reads then

$$\tilde{D} = (z+1) D_V$$

and means that $\tilde{D}$ equals D_V enhanced by a factor $(z+1)$, depending on the ionization state of the vacancies. For neutral vacancies we would have $\tilde{D} = D_V$.

For stabilized zirconia one gets[48]

$$\tilde{D} = D_{el}$$

where $D_{el} = D_e$ or D_h depending on whether $\sigma_e >> \sigma_h$ or vice versa.

We see that in stabilized zirconia (and similarly in any conductor with $t_e << 1$)$\tilde{D}$ is determined only by the diffusion coefficient of the electronic charge carriers and has nothing to do with the tracer diffusion coefficient of the ions.

5. TRANSPORT IN MIXED CONDUCTORS IN THE STEADY STATE

5.1 General

In this chapter we discuss the steady state solution of the transport equations for ions and electrons for a MIEC with a non-uniform surrounding and with current flowing. We shall obtain first the $I - V$ relations and then for three models also the distribution of charge carriers. The calculation refers to locations inside the mixed conductor. We shall, however, assume that the electrodes are reversible so that the results refer also to the voltages and chemical potentials at the electrodes applied to the mixed conductor. The configuration considered is linear, as shown in Fig. 3 and the temperature uniform. The driving forces are due to the applied voltage V and the difference in the chemical surrounding $\mu_X^0 - \mu_X^L$.

The equations governing charge transport are:

$$j_X = -\frac{\sigma_X}{q} \nabla \tilde{\mu}_{X^+}, \quad with\ \sigma_X = q\nu_X[X^+] \tag{28}$$

$$j_e = \frac{\sigma_e}{q} \nabla \tilde{\mu}_e, \quad with\ \sigma_e = q\nu_e n \tag{29}$$

$$j_h = -\frac{\sigma_h}{q} \nabla \tilde{\mu}_h, \quad with\ \sigma_h = q\nu_h p \tag{30}$$

Cross coefficients representing direct coupling between fluxes are omitted.[49,50)]. The boundary conditions are given by (μ_X^0, μ_X^L), the chemical potentials of X in the (nonuniform) surrounding interacting with the MIEC, and by the applied voltage.

In the steady state the total electric current density as well as the ionic current density are uniform throughout the MIEC since the sample is not

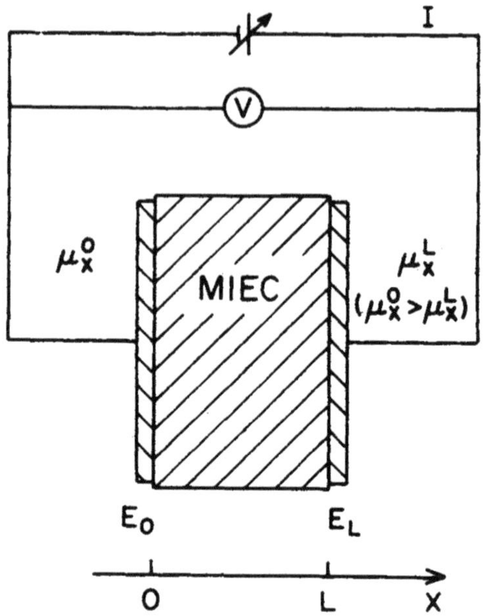

Fig. 3. Schematic representation of a MIEC connected to an external voltage V. E_0, E_L are reversible electrode. μ_X^0, μ_X^L are the chemical potentials of X in the surroundings near E_0, E_L resprectively.

being charged and no material is accumulated or depleted. Hence also the electronic current density $j_{el} = j_e + j_h$ is uniform. This enables the integration of the transport equations yielding the following $I - V$ relations:[51,52]
For the ionic current,

$$I_X = \frac{V_{th} - V}{R_X}, \tag{31}$$

where

$$I_X \equiv Sj_X \tag{32}$$

$$R_X \equiv \int_0^L \frac{dx}{S\sigma_i} \tag{33}$$

$$-qV_{th} \equiv \mu_X^L - \mu_X^0 \tag{34}$$

and S is the cross section of the MIEC perpendicular to the current direction. For the total electronic current

$$I_{el} = -\frac{V}{R_{el}} \tag{35}$$

where

$$I_{el} \equiv Sj_{el} = S(j_e + j_h) \tag{36}$$

and

$$R_{el} \equiv \int_0^L \frac{dx}{S(\sigma_e + \sigma_h)} \tag{37}$$

If one electronic species, say the electrons, predominates ($\sigma_e >> \sigma_h$) then R_{el} reduces to R_e where

$$R_e = \int_0^L \frac{dx}{S\sigma_e}$$

Eqs. (31),(35) may look deceptively simple, as if the system could be represented by an equivalent circuit with an Emf of V_{th} having an internal resistance R_X (in series) and an electronic shunt R_{el} in parallel to both the Emf source and R_X. However, R_e (or R_{el}), and R_X are not simple resistors. They are functions of the applied voltage and the $I-V$ relations therefore do not follow Ohm's law.

The total current, I, through the MIEC is the sum of the electronic and ionic ones. From Eqs. (31), (35)

$$I = I_{el} + I_X = \frac{V_{th} - V}{R_X} - \frac{V}{R_{el}} \tag{38}$$

Under open circuit (oc) conditions the total current I vanishes, but

$$I_{el} = -I_X \neq 0 \tag{39}$$

so that there is a leakage current of material through the MIEC in the open circuit conditions. The open circuit voltage is

$$V_{oc} = \frac{R_{el}}{R_X + R_{el}} V_{th} \equiv \bar{t}_i V_{th} \leq V_{th} \tag{40}$$

where R_{el}, R_X are evaluated at $V = V_{th}$. $\bar{t}_i$ is identified as the average ionic transport number. (See Eq.(5)). The open circuit voltage is close to the maximum possible value V_{th} for $R_X << R_{el}$ as is the case in solid electrolytes, but is much smaller than V_{th} for $R_X >> R_{el}$ i.e. in predominantly electronic conductors.

When the chemical "driving force" vanishes, $\mu_X^L = \mu_X^0$, $V_{th} = 0$ (Eq. (34)), and the $I - V$ relation reduces (from Eq. (38)) to

$$I = -\left(\frac{1}{R_X} + \frac{1}{R_{el}}\right) V \tag{41}$$

In order to get the dependence of R_X, R_e, and R_{el} on V, μ_X^0, μ_X^L, and T and to obtain a detailed $I - V$ relation, one has to discuss specific models.

5.2 MIEC with a high and uniform concentration of mobile Ions

A high concentration of mobile ions is a necessary condition for a solid to be a solid electrolyte. The discussion here refers therefore to solid electrolytes but is applicable also to those solids in which the electronic species, though a minority, are the dominant current carriers due to their very high mobility. We shall use as an example doped Ceria (D - CeO_2) exposed to oxygen on one side and a reducing atmosphere on the other. As model material we take now $X^{+}_{1+x}Y^{-}_{1-x}Z^{2-}_{x}$, where the divalent dopant Z substituting for Y serves to introduce a large fraction of X interstitials into the material.

The mobile ionic species in D-CeO_2 are oxygen vacancies $V_O^{\cdot\cdot}$ introduced by a lower valent dopant. Their concentration is high, of the order of a few mole % . In a reducing atmosphere some additional oxygen vacancies (and conduction electrons) are formed, but we shall only be interested in conditions where this additional concentration of vacancies can be neglected in a first approximation. The concentration of $V_O^{\cdot\cdot}$ is thus uniform under all relevant conditions. This corresponds in our model material to

$$\frac{\partial \mu_{X^+}}{\partial x} = 0 \tag{42}$$

The conduction electrons introduced in the reduction process may have a dramatic effect on the conductivity due to their high mobility. For D-CeO_2 the electron mobility is about 20 times larger than that of the ions[53]. Hence the local t_i on the reduced side of the sample (the side with large μ_X in Fig. 3), may be considerably less than unity.

To proceed we rewrite Eqs. (28), (29) for the one-dimensional configuration and use "ideal gas" statistics for the low concentration electrons. The electron chemical potential is then

$$\mu_e = \mu_e^0 + kT\ell n n \tag{43}$$

We use also the relations

$$\tilde{\mu}_{X^+} = \mu_{X^+} + q\phi \tag{44}$$

$$\tilde{\mu}_e = \mu_e - q\phi \tag{45}$$

$$\mu_{X^+} = \mu_{X^+} + \mu_e = \tilde{\mu}_{X^+} + \tilde{\mu}_e \tag{46}$$

where ϕ is the electric potential in the MIEC. This yields

$$j_X = -q\nu_X[X^+]\frac{\partial \phi}{\partial x} \tag{47}$$

$$j_e = kT\nu_e \frac{\partial n}{\partial x} - q\nu_e n \frac{\partial \phi}{\partial x} \tag{48}$$

where we have used also Eq. (43). j_X and j_e are uniform in the steady state. Hence, from Eq. (47) the electric field $\epsilon = -\partial\phi/\partial x$ is constant. This result of the model enables the integration of the transport equation (48), in which only n is a function of location x. A boundary condition is determined from $\mu_X^L - \mu_X^0$ and Eqs. (34), (42), (43) and (46). This yields

$$n_0/n_L = exp(q\beta V_{th}), \quad \beta = 1/kT \tag{49}$$

where n_0, n_L are the concentrations of electrons at $X = 0$ and $X = L$ respectively. Integration of Eq. (48) yields the distribution of electrons, $n(x)$

$$n(x) = n_0[1 - \frac{1 - exp(-\beta q V_{th})}{1 - exp(-\beta q(V_{th} - V))}(1 - exp\frac{-\beta q(V_{th} - V)x}{L})] \tag{50}$$

We bear in mind that due to local neutrality condition the electrons concentration reflects the concentration of excess X^+ created by the larger μ_X, and therefore also the local deviation from stoichiometry. We find that the magnitude and shape of $n(x)$ depends on the applied voltage V (Eq. (50)). This is demonstrated in Fig. 4. The change of $n(x)$ with V means that the applied voltage affects the local and overall deviation from stoichiometry. For a forward bias ($V > 0$) the width of the defect rich layer increases, and for $V > 2V_{th}$ $n(x)$ is almost uniform with $n(x) \simeq n_0$ throughout the sample. On the other hand for a reverse bias the width decreases and for $V < -V_{th}$ $n(x) \simeq n_L$ in most of the sample. For $V = V_{th}$, n is a linear function of x, as was also verified experimentally on Cu_2S.[54] The overall deviation from stoichiometry, δ, is the integral over $n(x)$,

$$\begin{aligned} \delta &= \frac{V_m}{N_A} n_0[1 + (1 - exp(-\beta q V_{th})) \\ &\times (\frac{1}{\beta q(V_{th} - V)} - \frac{1}{1 - exp(-\beta q(V_{th} - V))})] \end{aligned} \tag{51}$$

where V_m is the molar volume and N_A is Avogadro's number. The dependence of δ on V is shown in Fig. 5. δ is seen to be most sensitive to changes in V for $V \sim V_{th}$, i.e. when the applied voltage is close to the cell EMF.

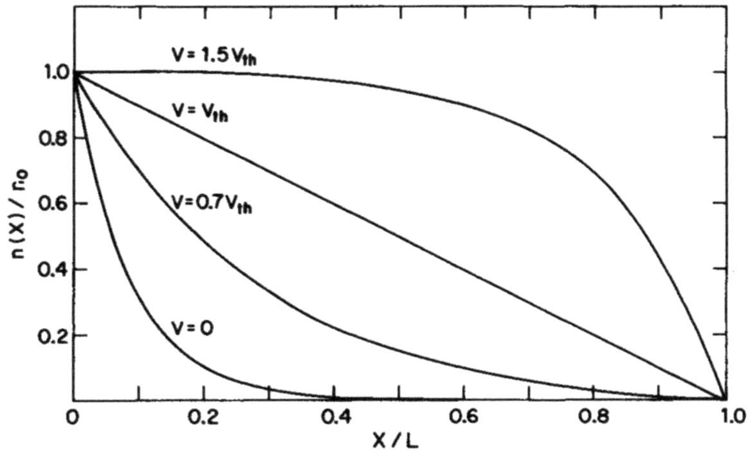

Fig. 4. $n(x)$ vs. x for different values of V for $V_{th} = 1$ V, $T = 1000$ K ($n_L/n_0 = 10^{-5}$), $t_i^L \sim 1$.

The asymptotic values of the electronic conductivity and current density are for $V \geq 2V_{th}$ and $\beta qV >> 1$

$$\sigma_e^0 \approx qn_0\nu_e, \tag{52}$$

$$j_e \approx \sigma_e^0 \frac{V_{th} - V}{L}, \tag{53}$$

and for $-V \geq V_{th}, \quad -\beta qV >> 1$

$$\sigma_e^L \approx qn_L\nu_e, \tag{54}$$

$$j_e \approx \sigma_e^L \frac{V_{th} - V}{L}, \tag{55}$$

In the $I - V$ relations the electronic current is asymptotically linear in V both for $V > 2V_{th}$ and for $-V > V_{th}$, but the slope may differ by a few orders of magnitude between the two limits. This is shown in Fig. 6.

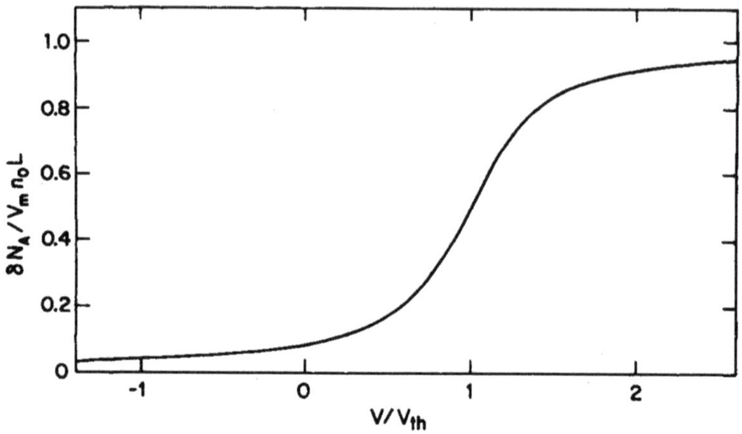

Fig. 5. δ vs. V (for $V_{th} = 1$ V, $T = 1000$K, $t_i^L \sim 1$).

The solution of j_e for arbitrary V is

$$j_e = -\sigma_e^L \frac{V_{th} - V}{L} \times \frac{e^{\beta q V} - 1}{1 - e^{-\beta q (V_{th} - V)}} \tag{56}$$

The ionic current depends linearily on the applied voltage.

$$j_X = \sigma_i \frac{V_{th} - V}{L} \tag{57}$$

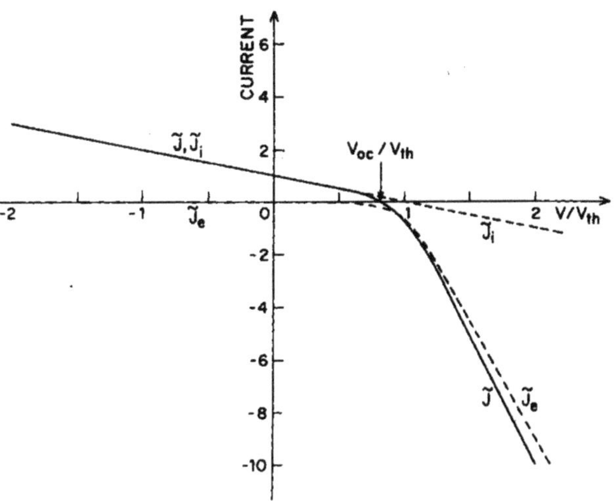

Fig. 6. Electronic, ionic and total current vs. voltage (for $V_{th} = 1$ V, $T = 1000$ K, $t_i^L \sim 1$).

5.3 p-n and p-i-n quasi-junctions

We consider now another case, a mixed ionic electronic conductor subjected as in Fig. 3 to different surroundings, with μ_X^0 sufficiently high to turn the MIEC near $x=0$ into n-type material, and μ_X^L sufficiently low to turn the MIEC near $x = L$ into p-type material. This forms a p-n or p-i-n quasi junction. The boundary values (n_0, n_L and p_0, p_L) depend again on the local equilibrium established with the surroundings, but the charge distribution inside the MIEC depends, within the limits of the boundary values, on the applied voltage. The possibility of both n-type and p-type behaviour in the same MIEC is not only hypothetical; this has been found in e.g. stabilized in ZrO_2[25,55], Cu_2O[56], Ta_2O_5[57], TiO_2[58] and NiO[59].

The p-n or p-i-n junction formed is different from a regular p-n junction in which the ionic defects are frozen in and only electrons and holes are in thermal equilibrium. In the MIEC the defects are mobile and will move because of the "driving force" $\mu_X^L - \mu_X^0$, even when $V = 0$. No space charge is formed in the junction since the mobility of the ions allows local neutrality to be maintained, though one side is n-type and the other p-type. A space charge will, however, be present if it is possible to quench the MIEC and freeze in the ionic defects, while keeping the electrons and holes in their respective conduction bands. In this case, one can "shape" the p, n and ionic defect distributions at elevated temperature by a proper choice of the applied voltage, μ_X^L, μ_X^0, and T and then quench the sample. The low temperature solid will be a regular p-n or p-i-n junction the structure of which has been controlled in situ at elevated temperature.

Quantitative analysis[52] of the high temperature behavior requires the solution of Eqs. (28)-(30) taking into consideration electron-hole interaction

$$e' + h^{\cdot} \leftrightarrow 0 \tag{58}$$

which implies

$$\mu_e + \mu_h = \tilde{\mu}_e + \tilde{\mu}_h = 0 \tag{59}$$

The distribution of $n(x)$, $p(x)$ in the MIEC for different applied voltages is shown in Fig. 7. For large forward bias ($V > 2V_{th}$ and $\beta qV >> 1$) the transition region between n and p regions is narrow. The electronic conductivity of the whole MIEC is high. For large reverse bias ($V \leq -V_{th}$ and $-\beta qV >> 1$) the n and p regions are narrow with a wide intrinsic region between them. The electronic conductivity is low, determined by the large intrinsic region. This leads to the following asymptotic $I - V$ relations for the electronic current

$$j_{el} \equiv qn_0\nu_{el}\frac{V_{th} - V}{L}, \quad V \geq 2V_{th}, \quad \beta qV >> 1 \tag{60}$$

and

$$j_{el} \approx 2qn_i\nu_{el}\frac{V_{th} - V}{L}, \quad -V \geq V_{th}, \quad -\beta qV >> 1 \tag{61}$$

where ν_{el} is the mobility of electrons and holes (assumed here to be equal) and n_i is the concentration of electrons and holes in the intrinsic region. The ionic current follows Eq. (57) and is linear in V.

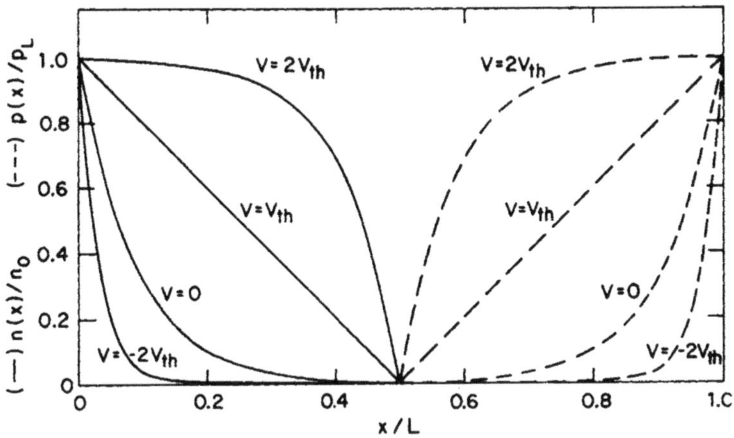

Fig. 7. $n(x)$ and $p(x)$ vs. x for different values of V (for $V_{th} = 1$ V, T=1000 K, $n_L/n_0 = 10^{-5}$, $p_0 = n_L$, $p_L = n_0, \nu_e = \nu_h$).

The asymmetry in the $I - V$ relation of the electronic current is shown in Fig. 8. In contrast to rectification by a normal p-n junction, j_{el} here depends on V linearly, not exponentially. The asymmetry is not due to a space charge, but due to a difference in composition induced by the applied voltage (see Fig. 7).

5.4 Equal Concentration of Mobile Ions and Conduction Electrons

In the previous two section we discussed models with $[X^+] >> n, p$. As mentioned in the introduction to this chapter one is interested also in understanding a model with equal concentrations of mobile ionic defects and, say, electrons in which $[X^+] = n >> p$,, which is relevant e.g. to reduced ceria CeO_{2-x}, and Li doped Ge.

Solving[51] the transport equations (28), (29) with the interstitial ions concentration $[X^+] = n$ yields an electron distribution $n(x)$ (and defect distribution $[X^+]$) changing linearly between the boundary value n_0, n_L.

$$[X^+] = n(x) = n_0 - (n_0 - n_L)x/L \tag{62}$$

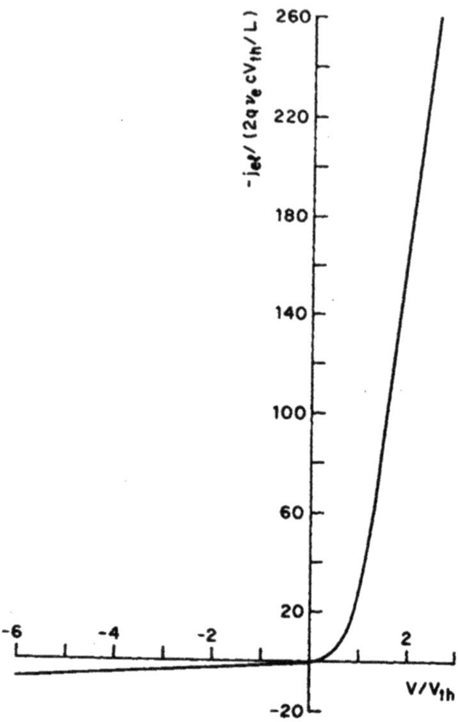

Fig. 8. Electronic current j_{el} vs. V (for conditions as in Fig. 7).

These distributions and therefore also the total deviation from stoichiometry δ are independent of the applied voltage V. The $I-V$ relations are:

$$j_e = -2\nu_e kT\frac{(n_0 - n_L)V}{LV_{th}} \tag{63}$$

and

$$j_X = 2\nu_i kT\frac{(n_0 - n_L)(V_{th} - V)}{LV_{th}} \tag{64}$$

i.e. the currents are linear in V. One expects the electron current to predominate due to the higher mobility of the electrons.

The ionic transport number is independent of x, it equals

$$t_i = \frac{\nu_X}{\nu_X + \nu_e} \tag{65}$$

5.5 Transport in Semiconducting and Metallic MIEC

If a MIEC is predominantly an electronic conductor, $\sigma_e >> \sigma_i$, it is regarded as a semiconductor or metal exhibiting also some ionic diffusion. Due to the high electrical conductivity $\Delta\tilde{\mu}_e$ under open circuit conditions is small: $\Delta\tilde{\mu}_e << \Delta\tilde{\mu}_{X^+}$ and the electric field $-\partial\phi/\partial x$ is therefore given by:

$$q\frac{\partial\phi}{\partial x} = \frac{\partial\mu_e}{\partial x} \tag{66}$$

In this special case the driving force for ionic motion equals the gradient of chemical potential of the atomic species. Schmalzried[60)] has reviewed the theory and experiments on MIEC oxides of this type.

6. APPLICATION TO FUEL CELLS BASED ON MIEC

Mixed-ionic-electronic-conductors have been considered as selective permeable membrane for ionic or atomic species, as electrodes in electrochemical devices, and as catalysts in which both the electrons and ionic species participate in the catalytic process. In other applications the composition of the MIEC is changed as in "smart windows", which change their optical absorption by an electric signal, variable resistance devices, conlometers, integrators analogue memories and timers. MIEC's have also been used in measurements of thermodynamic properties. All these applications have been reviewed in the past and we refer the reader to the literature[61)].

We shall discuss here the application of MIEC's to fuel cells. In the past MIEC's were ruled out for this application because of the electronic internal leak current. However, experiments and analysis by Takahashi et al.[62)],Choudhury and Patterson[63)], Tannhauser[53)], Ross and Benjamin[64)], and Riess[65] have shown that under proper conditions a MIEC can be used in a fuel cell. We shall present here the analytic solution[65] of this problem.

A fuel cell converts the chemical energy of the oxidation reaction of the fuel directly into electrical energy. A solid state (high temperature) fuel cell is based on an oxygen conducting solid electrolyte, e.g. stabilized (S-)ZrO_2, which at elevated temperatures is a fast anion conductor. The configuration is shown schematically in Fig. 3.

Energy waste is determined in part by heat loss due to the internal resistance of the electrolyte. It is therefore desired to minimize R_i. There are not many fast oxygen conductors that have a lower resistivity and can replace S- ZrO_2. One candidate is CeO_2 doped e.g. with 10 mole % Gd_2O_3. The ionic conductivity of this oxide at $\sim$ 800 C is three times higher than that of yttria S-ZrO_2 .[66,67]. However, D-CeO_2 has the disadvantage that it looses oxygen at the high temperatures and low oxygen partial pressures ($P(O_2) \leq 10^{-16}$ atm) that prevail on the fuel side of the cell. Bi_2O_3 and many doped Bi_2O_3 compounds have higher ionic conductivities than D-CeO_2 but were found to be unstable under the reducing conditions prevailing on the fuel side. However recent reports suggest that Er_2O_3 or Y_2O_3 doped Bi_2O_3 properly prepared may after all be stable under $P(O_2) \approx 10^{-16}$ atm at 700 C[68].

In this section we shall discuss the power output and efficiency of a fuel cell based on a MIEC with quantitative examples calculated for D-CeO_2 at 1000 K. We shall show that this cell will operate with less energy loss than one based on S-ZrO_2 (in which the electronic conductivity can be neglected) with equal geometry and temperature.

The model applicable to D-CeO_2 is that of a MIEC with a high concentration of mobile ionic species and a low concentration of mobile electronic charge carriers, discussed in Section 5.2. We use the solution of the transport equations given there to discuss the fuel cell performance. When the cell is connected to a load its voltage is within the range $0 < V < V_{oc}$. The open circuit voltage, V_{oc}, is

$$V_{oc} = V_{th} - \frac{kT}{q} \ell n \frac{\sigma_e^0 + \sigma_i}{\sigma_e^0 exp(-\beta q V_{th}) + \sigma_i} = \tag{67}$$

$$= V_{th} - \frac{kT}{q} \ell n \frac{t_i^L}{t_i^0} \equiv \bar{t}_i V_{th},$$

where indices 0 and L refer to the boundary regions at 0 and L. If the MIEC were a predominantly electronic conductor throughout its entire width, so that $\sigma_e^0 >> \sigma_i$ and $\sigma_e^0 exp(-\beta q V_{th}) = \sigma_e^L >> \sigma_i$, then $V_{oc} = 0$, as one would expect. For D-CeO_2, $t_i^L \sim 1$. Even if t_i^0 is as low as 0.1 the average ionic transference number, $\bar{t}_i$, is still quite high: $\bar{t}_i = 0.8$ at 1000 K.

Power output: The power output is

$$P = IV \tag{68}$$

where

$$I = S\frac{V_{th} - V}{L}[\sigma_i - \sigma_e^L \frac{exp(\beta q V) - 1}{1 - exp(-\beta q(V_{th} - V))}]. \tag{69}$$

One should operate the fuel cell far from the short circuit conditions where $V \rightarrow 0, P \rightarrow 0$ and from open circuit conditions, where $I = 0$, $P = 0$ but $I_i \neq 0$. Fuel is being consumed in both limiting conditions. Maximum power is expected to be delivered for $V \sim 0.5V_{th}$. The actual value of V that maximises the power can be calculated [65] from Eqs. (68,69). The maximum for D-CeO_2 (with $t_i^0 = 0.1$) is found at $V = 0.48V_{th}$, the maximum power is $P_{max} = 0.97 \times (0.25V_{th}^2/R_i)$, where R_i is the resistance of the cell to ionic current, $R_i = L/S\sigma_i$. P_{max} is only 3% lower than one would find, if the electronic conductivity could be neglected and $t_i^0 = t_i^L = 1$.

Efficiency: The efficiency of energy conversion at the maximum power is 0.46, i.e. 8% less than the efficiency of 0.50 that could be obtained, if $t_i^0 = t_i^L = 1$, as in S-ZrO_2. This loss is due to the electronic leak. Yet one has to bear in mind that R_i in D-CeO_2 is much lower (roughly by a factor 3) than in S-ZrO_2 of equal temperature , therefore the available maximum power $\approx 0.25V_{th}^2/R_i$ is almost 3 times larger for equal V_{th}.

We can conclude that a MIEC with a low t_i on the fuel side can still be of interest for use in a fuel cell.

REFERENCES

1. Hauffe, K., Reaktionen in und an Festen Stoffen, 2nd ed. Springer-Verlag, 1966.

2. Kroeger, F.A., The Chemistry of Imperfect Crystals, 2nd revised edition, Vol.I, II, III, North-Holland Pub. Co., 1974.

3. Rickert, H., Electrochemistry of Solids, an Introduction, Springer- Verlag, 1982.

4. Schmalzried, H., Solid State Reactions, Verlag-Chemie and Academic Press, 1974.

5. Wagner, C., in Proc. Int. Comm. Electrochem. Thermodyn. Kinetics (CITEC) 7, 361 (1957).

6. Wagner, C., Progress in Solid State Chemistry 10, 3 (1975).

7. Heyne, L., in: Solid Electrolytes, S. Geller Ed., Springer-Verlag, 1977, p.169.

8. Tuller, H.L., in: Nonstoichiometric Oxides, O.T. Sorensen Ed. 1981, p.271.

9. Weppner, W. and Huggins, R.A., Ann. Rev. Mater. Sci. 8, 269 (1978).

10. Dudley, G.J. and Steele, B.C.H., J. Solid State Chem. 31 , 233 (1980).

11. Wagner, C., J. Chem. Phys. 21, 1819 (1953).

12. Cava, R.J., Reidinger, F., Wuensch, B.J., J. Solid State Chem., 31, 69 (1980).

13. Rickert, H., Z. Phys. Chem. NF 23 , 355 (1960).

14. Rickert, H., Sattle, V. and Wedde, Ch., Z. Phys. Chem. NF 98, 339 (1975).

15. Sohege, J. and Funke, K., Ber. Bunsenges. Phys. Chem. 88, 657 (1984).

16. Shukla, A.K. and Schmalzried H., Z. Phys. Chem. NF 118, 59 (1979).

17. Wuilloud, E., Delley, B., Schneider, W.D. and Baer, Y., Phys. Rev. Letters 53, 202 (1984).

18. Tuller, H.L. and Nowick, A.S., J. Electrochem. Soc. 126, 209 (1979).

19. Tuller, H.L. and Nowick, A.S., J. Phys. Chem. Solids 38, 859 (1977).

20. Naik, I.K. and Tien T.Y., J. Phys. Chem. Solids 39, 311 (1978).

21. Garnier, J.E., Blumenthal, R.N., Panlener, R.J. and Sharma, R.K., J. Phys. Chem. Solids 37, 369 (1976). Tuller, H.L. and Nowick, A.S., J. Electrochem. Soc. 122, 255 (1975).

22. Schmalzried, H., Z. Elektrochemie 66, 572 (1962).

23. Burke, L.D., Rickert, H. and Steiner, R., Z. Phys. Chem. NF 74, 146 (1971).

24. Inouye, M., Iwase, M. and Mori, T., Transactions ISIJ 21, 54 (1981).

25. Weppner, W., Z. Naturforsch. 31a, 1336 (1976).

26. Tuller, H.L. and Stratton, T.S. in Proc. 3rd Int. Conf. on Transport in Nonstoichiometric Compounds, Penn. State Univ., PA, USA, June 10- 16, 1984.

27. Knoll, G.F., Radiation Detection and Measurement, John Wiley and sons, 1979, p. 414.

28. Mitoff, S., J. Chem. Phys. 31, 1261 (1959).

29. Hed, A.Z. and Tannhauser, D.S., J. Chem. Phys. 47, 2090 (1967).

30. Rudolph, J., Z. Naturforsch. 14a, 727 (1959).

31. Becker, J.H. and Frederikse, H.P.R., J. Applied Phys. 33(suppl.), 447 (1962).

32. Subba Rao, G.V., Ramdas, S., Mehrotra, P.N. and Rao, C.N.R., J. Solid State Chem. 2, 377 (1970).

33. Hauffe, K., Reaktionen in und an Festen Stoffen, 2nd ed., Springer-Verlag, 1966, pp. 65, 88.

34. Shuk, P. and Moebius, H.H., Z. Phys. Chem., Leipzig 266, 9 (1985).

35. Wang, D.Y. and Nowick A.S., J. Electrochem. Soc. 127, 113 (1980).

36. van-der Pauw, L.J., Phillips. Res. Rept. 13, 1 (1958).

37. Riess, I. and Tannhauser, D.S., Solid State Ionics 7, 307 (1982).

38. Yokota, I., J. Phys. Soc. Japan 16, 2213 (1961)

39. Yokota, I. and Miyatani, S., Japan J. Applied. Phys. 1, 144 (1962).

40. Miyatani, S., J.Phys. Soc. Japan 50, 1595 (1981).

41. Mitayani, S., J. Phys. Soc. Japan 53, 4284 (1984).

42. Miyatani, S., J. Phys. Soc. Japan 54, 639 (1985).

43. Miyatani, S. and Tabuchi A., J. Phys. Soc. Japan 50, 2050 (1981).

44. Yokota, I. and Miyatani, S., Solid State Ionics 3/4, 17 (1981).

45. Weppner, W. and Huggins, R.A., J. Electrochem. Soc. 124, 1569 (1977).

46. Chebotin, V.N., Brainin, M.I., Soloveva, L.M., Pakhnutov, I.A. and Lukach, Y.S., Soviet Electrochem. 22, 137 (1986).

47. Jin, Y. and Rosso, M., Solid State Ionics, 26, 237 (1988).

48. Tannhauser, D.S., J. Solid State Chem. 55, 40 (1984).

49. Miyatani, S., Solid State Comm. 38, 257 (1981).

50. Rickert, H., Electrochemistry of Solids, an Introduction, Springer-Verlag 1982, pp. 87, 217.

51. Riess, I., J. Phys. Chem. Solids 47, 129 (1986).

52. Riess, I., Phys. Rev. B35, 5740 (1987).

53. Tannhauser, D.S., J. Electrochem. Soc. 125, 1277 (1978).

54. Allen, L.H. and Bucks, E., J. Appl. Phys. 56, 327 (1984).

55. Weppner, W., J. Solid State Chem. 20, 305 (1977).

56. Hauffe, K., Reaktionen in und an Festen Stoffen, 2nd ed. Springer-Verlag 1966, p. 180.

57. Sasaki, Y., J. Phys. Chem. Solids 13, 177 (1960).

58. Huber, F., J. Electrochem. Soc. 110, 846 (1963).

59. Indira, K.S.. Rangarajan, S.K. and Doss, K.S.G., J. Electroanal. Chem. 21, 49 (1969).

60. Schmalzried, H. Reactivity of Solids 1, 117 (1986).

61. Huggins, R.A., Solid State Ionics 5, 15 (1980).

62. Takahashi, T., Ito, K. and Iwahara, H., J. Intl. Piles Combust. 3, 42 (1965).

63. Choudhary, N.S. and Patterson, J.W., J. Electrochem. Soc. 118, 1398 (1971).

64. Ross, B.N. and Benjamin, T.G., J. Power Sources 1, 311 (1976/77).

65. Riess, I., J. Electrochem. Soc. 128, 2077 (1981).

66. Obayashi, H. and Kudo, T., in: Solid State Chemistry and Energy Conversion, Goodenough, J.B. and Whittingham, M.S. Eds., Am. Chem. Soc., p. 316 (1977).

67. Riess, I., Braunshtein, D. and Tannhauser, D.S., J. Am. Ceram. Soc. 64, 479 (1981).

68. Jurado, J.R., Moure, C., Valverde, N. and Duran, P., 6th Intl. Conf. on Solid State Ionics, Garmisch, Federal Republic of Germany, September 6-11,1987.

b	mechanical mobility
D_e	diffusion coefficient of electrons
D_V	diffusion coefficient of vacancies
D_X	diffusion coefficient of ions X
$\tilde{D}$	chemical diffusion coefficient
D^{Tr}	tracer diffusion coefficient
E_i	electrode number i
f	correlation factor for diffusion
I	current
j_e	electronic current density
j_i, j_X	ionic curent density
J_e, J_X	electronic and ionic particle current densities
MIEC	mixed ionic electronic conductor
n, n_e	concentration of electrons
N_i	concentration of ions
N_A	Avogadro's number
p	concentration of holes
P	fuel cell power
q	electronic charge
t_e, t_{el}	electronic tranport number
t_i	ionic transport number
V	voltage
$V_{Co}^{\cdot}$	singly ionized cobalt vacancies etc.
V_m	molar volume
δ	deviation from stoichiometry
ϕ	electric potential
μ	chemical potential
$\tilde{\mu}$	electrochemical potential
ν	electrical mobility
σ_e, σ_{el}	electronic conductivity
σ_i	ionic conductivity
[ ]	concentration

APPLICATIONS OF SOLID ELECTROLYTES IN GALVANIC SENSORS

K.T. JACOB and TOM MATHEWS

Department of Metallurgy
Indian Institute of Science
Bangalore-560012, India

ABSTRACT

The concepts and theoretical origins of conduction domains for solid electrolytes and electrode polarization are outlined briefly. The point electrode made of the solid electrolyte material is useful for deflecting the semipermeability flux away from the electrode. The emf of galvanic sensors consisting of two solid electrolytes in intimate contact with each other and in which transport occurs by a common ion is reviewed. The voltage of such cells depends on the chemical potential of the active species at the interface between the two electrolytes, which can be evaluated from the transport properties of electrolytes using a numerical procedure. The factors governing the speed of response of solid electrolyte gas sensors are analyzed. A rigorous expression for the emf of non-isothermal galvanic sensors and the criterion for the design of temperature compensated reference electrodes for non-isothermal galvanic sensors are outlined. Non-isothermal sensors are useful for the continuous monitoring of concentrations or chemical potentials in reactive systems at high temperatures. The principles of operation of galvanic sensors for oxygen, sulphur, oxides of sulphur (SO_x,x=2,3), carbon, oxides of carbon (CO_x,x=1,2), oxides of nitrogen (NO_x,x=1,2) and silicon are discussed. The use of auxiliary electrodes in galvanic sensors to expand the detection capability of known solid electrolytes to a large number of species is explained with reference to sensors for sulphur and oxides of sulphur (SO_x,x=2,3). Finally the cause of the common errors in galvanic measurements and test for the correct functioning of galvanic sensors is given.

INTRODUCTION

Electrochemical gas sensing devices based on solid electrolytes can be potentiometric, amperometric or coulometric. In potentiometric sensors the emf of the galvanic cell is measured which is a function of the logarithm of the ratio P_1/P_2, where P_1 and P_2 are the partial pressures of the active component of the gas at the two electrodes. In amperometric sensors the limiting current, i_{lim}, through the solid electrolyte is measured for a preset value of applied voltage. The limiting current is proportional to the partial pressure of the active component in the gas[1]. One of the major advantages of potentiometric sensors over amperometric sensors is that the former have a logarithmic response and can measure gas concentration over an extremely wide range. Amperometric devices display a linear response. Unlike potentiometric oxygen sensors, amperometric oxygen sensors can only be used where the oxygen-reductant (or air/fuel) ratio is lean of stoichiometry. Additional modes of deterioration such as blocking of pores in the diffusion barrier by dust in the sample gas and structural changes of the porous ceramic, occur in amperometric devices. In coulometric sensors the active component of the gas is quantitatively removed from a chamber by electrolysis and the partial pressure determined from the charge passed. When almost all the active component of the gas is pumped out the current is discontinued and a period follows during which the active component diffuses into the chamber from the surrounding gas. The response is slow because of the time taken for equilibration. This chapter is primarily concerned with the application of solid electrolytes to galvanic sensors.

Galvanic sensors with solid electrolytes can be used for rapid and direct measurements of partial pressures in gases and of activities or concentration in liquids and melts. With optimized design of electrodes the signal is obtained almost instantaneously. Commercial devices for measuring the oxygen content of liquid melts and highly purified inert gases are now available. Further interesting applications lie in the field of medicine for studying the oxygen content of exhaled air, in industry, for example, in the analysis of exhaust gases from furnaces and combustion engines, and the use of gas sensors for automated process control ('robotics'), especially in view of the fast progress in microelectronics. The classical (mostly spectroscopic) methods of gas analysis are far from being suitable for these new types of application. One of the advantages of galvanic sensors is that the output is an electrical potential that can be readily used to actuate a control circuit.

The development of electrochemical sensors has rapidly gained momentum, as may be recognized by a fast increase in the number of publications over the last few years. Several excellent review articles cover the recent developments[2-7]. The present chapter will focus on some critical properties of solid electrolyte and electrode, bielectrolyte cell, response time of sensors, new developments in non-isothermal sensors, systematic principles governing functioning of different gas sensors and tests for correct functioning of galvanic sensors. Recent advances in enhancing the conductivity of solid electrolytes by dispersion of inert submicron second phase particles and by combined homo-hetro doping are discussed elsewhere in this publication.

1. CONDUCTION DOMAIN FOR SOLID ELECTROLYTES

The use of solid electrolytes in galvanic sensors is based on two salient features: (i) The open circuit emf E, across a solid electrolyte is a measure of the chemical potential difference, and (ii) the charge passed through the solid electrolyte is a measure of ion transport. A solid electrolyte is defined as a compound having ionic transference number, $t_{ion} > 0.99$. Solid electrolytes exhibit predominant ionic conductivity only over a limited range of temperature and chemical potential. Thus electrolytic conduction domain is an important factor limiting the application of solid electrolytes in galvanic sensors.

Theory

Wagner's phenomenological theory of mixed conduction gives

the relation[8)]

$$E = -\frac{1}{2Z_x F}\int_{\mu'_{x_2}}^{\mu''_{x_2}} t_{ion}\, d\mu_{x_2}$$

$$= -\frac{1}{Z_M F}\int_{\mu'_M}^{\mu''_M} t_{ion}\, d\,\mu_M \qquad ..(1)$$

where M and X denote the metal and nonmetal components of the M_xX_b mixed conductor respectively. The double and single primes refer to the electrodes contacting the opposite ends of the solid electrolyte, μ is the chemical potential, Z is valence and F is the Faraday constant. For a solid electrolyte with $t_{ion} > 0.99$ Eq.(1) can be simplified to,

$$E = -(\mu''_{x_2} - \mu'_{x_2})/2Z_xF = -(\mu''_M - \mu'_M)/Z_MF \qquad ..(2)$$

In general t_{ion} is a complex function of chemical potential as well as temperature.

At elevated temperatures solid electrolytes exhibit an ionic conductivity independant of the partial pressure, P_{x_2}, of the gaseous X_2 dimer, but their electron and hole conductivities, σ_n and σ_p, are found to be proportional to $P_{x_2}^{-1/x}$ and $P_{x_2}^{1/x}$, respectively, at constant temperature. The value of x is determined by the predominant defect structure of the solid electrolyte. For calcia stabilized zirconia the value of x is four. At constant P_{x_2} all the three conductivities follow an Arrhenius-type temperature dependance.

It is possible to write the total conductivity, σ_T, of a solid electrolyte as follows

$$\sigma_T = \sigma_{ion} + \sigma_p + \sigma_n \qquad ..(3)$$

where

$$\sigma_{ion} = \sigma^{o}_{ion} \exp\left(-\frac{Q_{ion}}{RT}\right) \qquad ..(4)$$

$$\sigma_{p} = \sigma^{o} P^{1/x}_{x_2} \exp\left(-\frac{Q_{p}}{RT}\right) \qquad ..(5)$$

$$\sigma_{n} = \sigma^{o}_{n} P^{-1/x}_{x_2} \exp\left(-\frac{Q_{n}}{RT}\right) \qquad ..(6)$$

Q_{ion}, Q_p and Q_n are the activation energies of ionic, hole and electron conduction. The quantities σ^{o}_{ion}, σ^{o}_{p}, σ^{o}_{n}, x, Q_{ion}, Q_p and Q_n are all independant of P_{x_2} and T to a first approximation.

A schematic 3-D representation of log σ surfaces in the log P_{x_2} and 1/T space is shown in Fig. 1. The electron, hole and ionic conductivities appear as planes whose intersections represent boundaries between ionic and electronic conductivities. In region A of figure 1, σ_p is much greater than either σ_{ion} or σ_n, so that σ_p and σ_T are virtually identical. For the same reason log σ_T surface conforms to the log σ_n plane in region C, whereas it conforms to the log σ_{ion} plane in region-D.

Domain boundaries separate the regions where each mechanism of conduction dominates. Analytical expressions for the different domain boundaries are[9)]

$$\log P_{pn} = -\frac{x}{2}\left(\frac{Q_n - Q_p}{2.303\,R}\right)\frac{1}{T} + \frac{x}{2}\log\frac{\sigma^{o}_{n}}{\sigma^{o}_{p}} \qquad ..(7)$$

$$\log P_{p} = -x\left(\frac{Q_{ion} - Q_p}{2.303\,R}\right)\frac{1}{T} + x\log\frac{\sigma^{o}_{ion}}{\sigma^{o}_{p}} \qquad ..(8)$$

$$\log P_{n} = x\left(\frac{Q_{ion} - Q_n}{2.303\,R}\right)\frac{1}{T} - x\log\frac{\sigma^{o}_{ion}}{\sigma^{o}_{n}} \qquad ..(9)$$

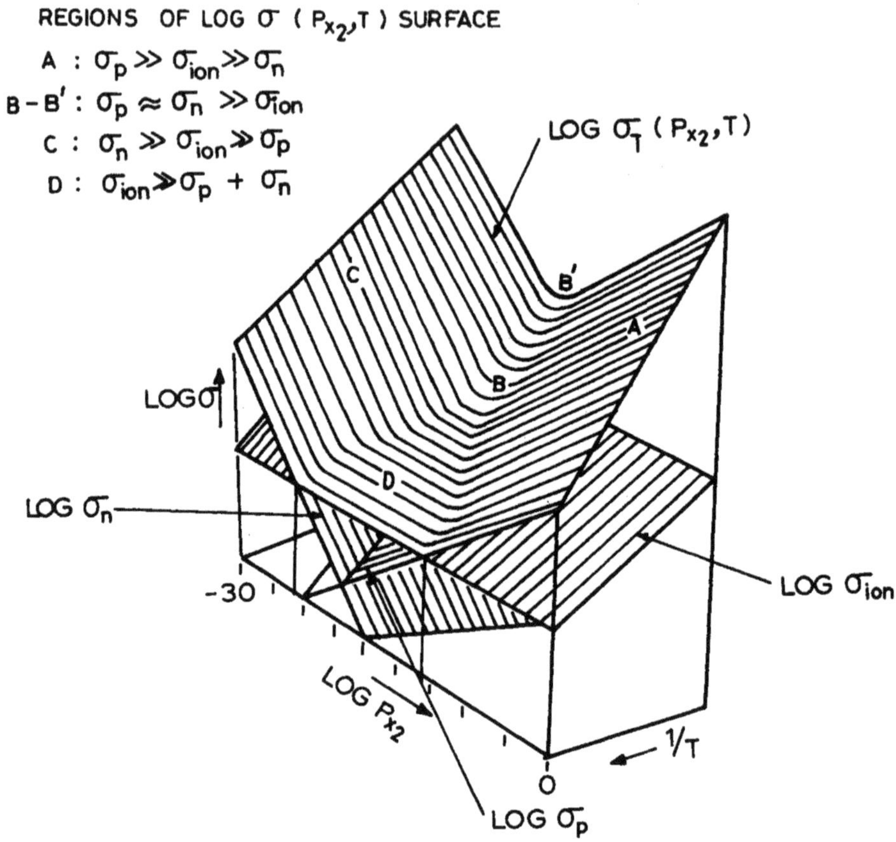

FIG.1: SCHEMATIC REPRESENTATION OF LOG σ SURFACES OVER LOG P_{x_2}, 1/T SPACE FOR $\sigma = \sigma_{ion}, \sigma_p, \sigma_n$ AND σ_{total} - AFTER PATTERSON[9] (COURTESY ELECTROCHEMICAL SOCIETY INC.)

where

P_{pn} is the value of P_{x_2} at which $\sigma_p = \sigma_n$,

P_p is the value of P_{x_2} at which $\sigma_p = \sigma_{ion}$, and

P_n is the value of P_{x_2} at which $\sigma_n = \sigma_{ion}$.

For sensor applications the ionic conductivity should be at least two orders of magnitude higher than either electronic or hole conductivity. The electrolytic domain boundaries may be deduced by equating σ_{ion} to $100\,\sigma_p$ and $100\,\sigma_n$. In $\log P_{x_2} - \frac{1}{T}$ space the equations for electrolytic domain boundaries are[10]

$$\log P_{x_2} = -x\left(\frac{Q_{ion} - Q_p}{2.303\,R}\right)\frac{1}{T} + x \log \frac{\sigma^o_{ion}}{100\,\sigma^o_p} \quad ..(10)$$

$$\log P_{x_2} = x\left(\frac{Q_{ion} - Q_n}{2.303\,R}\right)\frac{1}{T} - x \log \frac{\sigma^o_{ion}}{100\,\sigma^o_n} \quad ..(11)$$

Any application requiring electrolytic behaviour must correspond to $\log P_{X_2} - \frac{1}{T}$ conditions which lie between the lines corresponding to these two equations. The conduction domain map is essential for the assessment of the useful application conditions for a given solid electrolyte.

It is important to note that even within the electrolytic conduction domain, electron and hole conductivities have finite values. The residual electron or hole conductivity gives rise to electrochemical permeability of the migrating species. The flux flows from the electrode having a higher chemical potential to the one having a lower potential. The permeability is restricted to the mobile ion, and is therefore referred to as semipermeability of the solid electrolyte. In the absence of residual electron or hole conductivity, a flux cannot be sustained through the solid electrolyte. Ionic transport would then given rise to a space charge which will oppose further movement of ions. However, if electronic defects are mobile, a flux can be sustained by coupled transport of ions and electronic defects. The semipermeability is minimum in the middle of the electrolytic domain ($\sigma_p = \sigma_n$), where the sum of electron and hole conductivities exhibit a minimum.

2. ELECTRODE POLARIZATION

One of the important factors which results in incorrect values for composition of the active species in a galvanic sensor is electrode polarization. Polarization is caused by the physical or electrochemical permeability of the solid electrolyte. While physical permeability can be avoided by using a dense form of the solid electrolyte free from interconnecting porosity, electrochemical permeability results from intrinsic properties of the material. Because of the small but finite electronic conduction present in the solid electrolyte a charge-compensated ionic flux occurs, resulting in a mass transport of the active species across the electrolyte from the electrode of higher chemical potential to the one having a lower chemical potential. This phenomenon known as the electrochemical permeability or semipermeability will cause electrode polarization, especially when the sample electrode has a low thermodynamic capacity for the conducting species[11].

When gas electrodes are used presence of polarization manifests as flow rate dependance of emf. Although polarization can be reduced by increasing the flow rate, a practical limit is set by the cooling effect of the gas. Assymetric gas flows at the two electrodes can result in differential cooling and nonisothermal contributions to cell emf which are difficult to correct. A design solution to the polarization problem is the use of point electrode[12], shown in Fig. 2. The Pt electrode actually measures the chemical potential of oxygen in a microsystem formed at the triple phase region between the electrode, electrolyte and gas. The oxygen flux through the electrolyte due to the electrochemical semipermeability may alter the equilibrium oxygen chemical potential in the microsystem, if the exchange kinetics between the gas phase and the microsystem does not keep pace with the semipermeability flux[12]. However, this flux can be dissipated at the tip of the point electrode made of CSZ, so that, at the point of contact between the Pt lead and point electrode, no effect of this flux will be felt and the measured composition corresponds to that

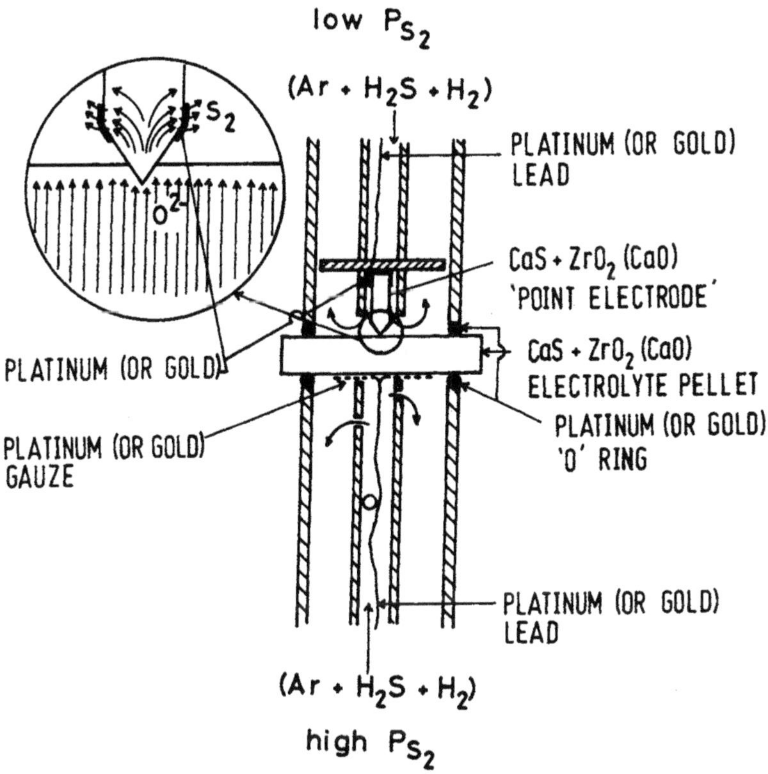

Fig. 2 : A schematic representation of the CaS. (CaO) ZrO_2 point electrode for deflecting the semipermeability flux from the measuring metal electrode-modified version[13)] of the system introduced by Fouletier et al[12]].

of the bulk gas. Jacob et al[13] have shown that the performance of the point electrode can be enhanced by applying a porous coating of a catalyst such as Pt at the tip of the point electrode made from the solid electrolyte material. The presence of the catalyst facilitates the dissipation of the semipermeability flux near the tip of the point electrode. Another way of minimizing semipermeability or its effect is by the choice of suitable reference system. The reference system consists of either a gas phase of suitable composition or a mixture of condensed phases which will generate the required chemical potential of the active species at the electrode. The semipermeability flux of the active species from the electrode of higher chemical potential to the one of lower chemical potential can be reduced by choosing a reference system with chemical potential close to that of the sample system. In gas reference systems a flowing gas stream is preferred over a static gas phase to minimize polarization effects arising from the semipermeability of the solid electrolyte since the flux arriving at the electrode can be carried away by the flowing gas. In oxygen sensors, air is a commonly used reference gas as it is easily available and has constant composition. For accurate measurements air should be dried and correction for barometric changes applied. Inert gas + oxygen mixtures are not suitable for $P_{O_2} < 10^2$ Pa because of the low thermodynamic capacity of the mixture[11]. Mixtures of $CO+CO_2$ and H_2+H_2O provide good buffered systems when the gas ratio ranges from 10^{-2} to 10^2. The usefulness of the H_2+H_2O mixtures is limited by severe thermal segregation in the gas phase and the elaborate procedures necessary to minimise it.

The use of flowing gas is inconvenient for sensors that must operate inside high temperature reactors and in remote locations. For such applications sealed condensed phase mixtures, inert to the solid electrolyte, are attractive. Thus in an oxygen sensor, a two phase metal/metal oxide mixture or two coexisting oxide phases of a binary system can be used. Mixtures like Ni+NiO and $Cr+Cr_2O_3$ are easily polarizable, especially when the oxygen content of the sample system is higher than that established by these mixtures. In such case an excess of the component that is being consumed due to the flux of oxygen is taken to minimize polarization[14]. The use of graded composition electrodes to minimize polarization without significantly affecting the buffer capacity of the reference system has been successfully used by the authors. The overvoltage, η, at a two phase $M+M_aO_b$ electrode, based on the kinetic model of Worrell and Isokoe[15], is given by;

$$\eta = \frac{RT\,(\Delta Y)\,I}{8.358\,F^2 ACD}$$

where C and D are the solubility and diffusivity of oxygen in the metal component of the two phase electrodes respectively. I is the current in amperes, A the effective interfacial area between the metal component and the electrolyte and ΔY is 1/3 the diameter of the metal particles. The overvoltage can therefore be minimized by

a) selecting a two phase electrode system with a high solubility-diffusivity product for oxygen in the metal,
b) increasing the fraction of the metal phase in the electrodes, and
c) using fine metal particles.

3. BIELECTROLYTE CELLS

Since each solid electrolyte has a limited electrolytic domain, it is useful to explore the possibilities of extending the electrolytic domain by combining two materials in series. Tretyakov and Muan[16] suggested that a bielectrolyte cell consisting of CSZ and YDT would be used to measure oxygen partial pressures over a wider range than would be possible with either electrolyte. Rapp and Shores[17] have given a general analysis for the interpretation of the voltage of a bielectrolyte cell. For the cell arrangement

$$\mu'_{x_2},\ \text{Pt}|\ \text{Electrolyte-A}|\ \text{Electrolyte-B}|\ \text{Pt},\ \mu''_{x_2}$$

$$E = -\frac{1}{2F}\int_{\mu'_{x_2}}^{\mu^{I}_{x_2}} \frac{t_i^A}{Z_x}\,d\mu_{x_2} - \frac{1}{2F}\int_{\mu^{I}_{x_2}}^{\mu''_{x_2}} \frac{t_i}{Z_x}\,d\mu_{x_2} \qquad ..(12)$$

where $\mu^{I}_{X_2}$ is the chemical potential of the species X_2 at the interface between the two electrolytes, t_i^A and t_i^B are the transport numbers of the conducting species in the electrolyte A and B respectively and Z_x the charge of the conducting ions. The interfacial potential, $\mu^{I}_{X_2}$, is established exclusively by the ionic and electronic transport through the two electrolytes and is unaffected by the surrounding environment, if the two electrolytes A and B are in perfect contact. The cell voltage E is the sum of the voltages across the electrolytes A and B and is given by Eq.(12). For two electrolytes having anions as the mobile ionic species, the steady-state rate of ionic and electronic transport in the one electrolyte will be equivalent and equal to those in the other electrolyte if the interface is not a source or sink for ions or electrons. Thus for a linear cell geometry [12)]

$$L_A^{-1} \int_{\mu'_{X_2}}^{\mu^{I}_{X_2}} t_2^A t_3^A \sigma^A d\mu_{X_2} = L_B^{-1} \int_{\mu^{I}_{X_2}}^{\mu''_{X_2}} t_2^B t_3^B \sigma^B d\mu_{X_2} \quad ..(13)$$

where L_A and L_B are the thickness of the electrolytes A and B, of unit area of cross section, respectively; t_2 and t_3 are the transport numbers of ions and electrons and σ^A and σ^B are the total conductivities of the electrolyte A and B respectively. Interfacial chemical potential, $\mu^{I}_{X_2}$, is the only unknown parameter in this expression. This equation may be solved numerically to obtain $\mu^{I}_{X_2}$. Then Eq.(12) can be evaluated to obtain the voltage of the bielectrolyte cell.

4. RESPONSE TIME

The response time is defined as the time taken by an electrochemical concentration cell to achieve a specified percent of its maximum open circuit voltage, when the concentration of the active component at one of the electrodes is suddenly changed. When the active component forms one of the constituents of the electrolyte material, variations in the concentration of the active species results in a certain finite compositional changes in the electrolyte. These compositional changes give rise to small deviations from the exact stoichiometric ratio. During a transition from one concentration to another, an amount of material equivalent to these stoichiometric changes must be either supplied to or removed from the solid electrolyte. The amount of active material involved in the change from one state to another and the density of the material flux with which the required amount is supplied determine the speed of the response of the sensor.

The concentration profile of the active component with time through the electrolyte, when there is a step change in the partial pressure of the gas at one of the electrodes, has been computed numerically[18], both for pure solid electrolytes and solid electrolytes containing an intergrain phase with mixed conductor properties. For a pure electrolyte the emf developed after a step change is constant and has the value corresponding to the Nernst equation right from the start. The reason is that t_{ion} in the equation

$$E = \frac{1}{2Z_x F} \int_{\mu'_{x_2}}^{\mu''_{x_2}} t_{ion} \, d\mu_{x_2} \qquad ..(14)$$

is unity and can be taken out of the integral. Consequently E becomes independant of the μ -distribution and is determined by the boundary value only, provided the material flux due to the change in stoichiometry of the electrolyte does not disturb the chemical potential at the boundary. This condition is achieved where the electrodes are perfectly reversible. Polarizable electrodes generate overvoltages as the result of the flux, which are large at short times.

The presence of a second phase exhibiting electrical properties different from those of the main material would alter the concentration profiles. Here also the emf starts with the ideal Nernst value. But when the diffusion front reaches the second phase and a gradient is established across the second phase, the latter begins to contribute to the emf and deviations occur because the value of t_{ion} in the second phase is lower than unity. A slow response time may also be observed in cases where a second phase is present at the electrode/electrolyte boundary. In such cases deviations from the theoretical value occur from the start[18].

The response time is also affected by the adsorption of the active species at the electrode-electrolyte interface and the rate of charge transfer to the adsorbed species[19], which results in the formation of a double layer at the interface. This changes the defect structure without a net change in stoichiometry. The rate of this reorganization can be correlated with the conductivity of the electrode zone. Winnubst et al[19] studied the effect of the electrode materials Pt and Au on response time of zirconia based solid electrolytes and found that gold electrodes result in shorter response time. Fouletier et al[20] have shown that Ag electrodes respond faster than Pt electrodes in oxygen probe for $1 > P_{O_2} > 0.39$.

Concentration steps from low to high gas partial pressure produced faster response than high to low concentration steps[19-21]. This can be understood as a bulk gas/boundary layer exchange process where the gas partial pressure is the relaxing variable. Assymmetry arises from the logarithmic dependence of emf on gas partial pressure. Response curves following concentration steps involving reacting gases differ from those where reaction is not involved. With reactive gases slower response may occur following either a high-to-low or a low-to-high change in gas partial pressure. Preequilibration of the gas with a secondary catalyst produces faster responses as compared to non-equilibrated mixtures[22].

The sensor emf as a function of time after a step change in P_{X_2}, can be described by the empirical relation[21]

$$E_{(t)} - E_{(o)} = (E_{(\infty)} - E_{(o)})\exp(-t/\tau)^{1/2} \qquad ..(15)$$

$E_{(t)}$, $E_{(o)}$, $E_{(\infty)}$ are sensor voltages at time t=t, t=0 and t=∞ respectively. The parameter τ is a characteristic time constant which exhibits an Arrhenius-type relationship. For non-reactive gas mixtures,

$$\tau = A(P_M)^{-1/2} \exp(E_{act}/RT) \qquad ..(16)$$

$$P_M = \{ P_{initial} + P_{final} \} /2 \qquad ..(17)$$

The partial pressure dependance of response time has not been clearly established for reactive gas mixtures.

In well buffered systems such as $CO+CO_2$ and H_2+H_2O response times are of the order of a few minutes even though oxygen partial pressures are very low[23)]. This is because of the high exchange current density associated with reactions,

$$CO_2 + V_O^{\bullet\bullet} + 2e' \rightarrow CO + O_o^x$$

$$H_2O + V_O^{\bullet\bullet} + 2e' \rightarrow H_2 + O_o^x$$

involving oxidized and reduced species. Polarization effect due to semipermeability flux are also small when buffered gas mixtures are used because the oxygen flux arriving at the electrode rapidly reacts with the reduced species in the gas without affecting the gas composition significantly. The amount of oxygen transported by the semipermeability of the electrolyte is insignificant when compared with concentrations of oxidised and reduced species in a well buffered gas.

5. NON-ISOTHERMAL SOLID STATE GALVANIC SENSORS

Isothermal solid state galvanic sensors have been extensively used for the determination of chemical potential or concentration of chemical species in industrial melts at high temperatures. At temperatures in excess of 1800K and in corrosive melts the chemical reactivity of the melt often destroys the ionic conducting properties of the solid electrolyte and thus degrades sensor response. A design solution to this problem consists in continuously introducing a rod of solid electrolyte into the melt at a rate commensurate with the reaction rate. Thus a functioning solid electrolyte element is always in contact with the melt and the sensor can be used in a continuous mode. In this proposed 'feeder design' of the sensor the reference electrode is placed at the other end of the solid electrolyte rod. The temperature of the reference electrode is thus much lower than that of the melt, which minimizes the chemical reaction between the solid electrolyte and the electrode. In this design, the galvanic sensor is non-isothermal. Theoretical equation for the emf of a non-isothermal galvanic sensor has been derived[24].

The main disadvantage of such non-isothermal sensors is the need to measure the temperature of both the electrodes. In the 'feeder design' the temperature of the reference electrode will usually vary with time; the temperature increasing as the reference electrode moves towards the melt. It is therefore useful to design temperature compensated reference electrodes for non-isothermal galvanic sensors. Theory of design of temperature compensated reference electrodes and experimental verification of the theory using fully stabilized $(CaO)ZrO_2$ as solid electrolytes is discussed by Jacob and Ramasesha[25] in detail.

EMF of Non-isothermal Galvanic Sensors

An expression for the emf of a non-isothermal cell can be derived from the phenomenological expressions of irreversible thermodynamics. It has been shown that for an oxide solid electrolyte[24],

$$E_{\text{non-isothermal cell}} = \frac{1}{4F}(\Delta\mu''_{O_2} - \Delta\mu'_{O_2}) + \frac{1}{4F}({}^{o}\mu''_{O_2} - {}^{o}\mu'_{O_2})$$

$$+ \frac{1}{2F}\left[\bar{S}_1 (T'' - T') + Q_1^* \ln \frac{T''}{T'}\right]_{SE}$$

$$- \frac{1}{F}\left[\bar{S}_2 (T'' - T') + Q_2^* \ln \frac{T''}{T'}\right]_{Pt} \qquad ..(18)$$

where $\Delta\mu_{O_2} = \mu_{O_2} - {}^{o}\mu_{O_2} = RT \ln P_{O_2}$.

The subscripts 1 and 2 stand for ions and electrons respectively, and T' and T" are the temperatures at the two electrodes. Eq.(18) contains terms such as $\bar{S}_i$ and Q_i^*, the partial entropy and the heat of transport of charged species respectively, which are difficult to calculate or measure directly.

To overcome this the emf of a thermocell in pure oxygen at unit activity is measured

$$E^{\text{Pure } O_2}_{\text{Thermocell}} = \frac{1}{4F}({}^{o}\mu''_{O_2} - {}^{o}\mu'_{O_2}) + \frac{1}{2F}(\bar{S}_1 (T''-T') + Q_1^* \ln \frac{T''}{T'})_{S.E.}$$

$$- \frac{1}{F}(\bar{S}_2 (T''-T') + Q_2^* \ln \frac{T''}{T'})_{Pt} \quad ..(19)$$

The emf of a non-isothermal cell can therefore be expressed as the emf of a thermocell operating under pure oxygen at 1 atm. pressure in the same temperature gradient and a contribution arising from the chemical potential gradient across the electrolyte.

$$E_{\text{non-isothermal cell}} = E^{\text{Pure } O_2}_{\text{Thermocell}} + \frac{1}{4F}(\Delta\mu''_{O_2} - \Delta\mu'_{O_2})$$

$$= E^{\text{Pure } O_2}_{\text{Thermocell}} + \frac{R}{4F}(T'' \ln P''_{O_2} - T' \ln P'_{O_2}) \quad ..(20)$$

When T" = T' (i.e. for an isothermal cell), the first term of the right of Eq.(20) is zero and the cell emf reduces to the familiar Nernst equation,

$$E_{\text{cell}} = \frac{RT}{4F} \ln \frac{P''_{O_2}}{P'_{O_2}} \quad ..(21)$$

Similarly for a solid electrolyte in which the cation is the migrating species, the emf of the non-isothermal cell is

$$E_{\text{non-isothermal}} = E^{\text{Pure M}}_{\text{Thermocell}} - \frac{1}{Z_1 F}(\Delta\mu''_M - \Delta\mu'_M) \quad ..(22)$$

where $E^{\text{Pure M}}_{\text{Thermocell}}$ is the emf of a thermocell operating in the same temperature gradient as the non-isothermal galvanic cell, but with the metal M at unit activity at each electrode. The term $\Delta\mu_M$ is equal to $RT \ln a_M$:

a_M is the activity of the metallic species M.

$$E^{\text{Pure M/O}_2}_{\text{Thermocell}} = \int_{T'}^{T''} \xi dT = \xi (T''-T') \qquad ..(23)$$

where ξ is the Seebeck coefficient or thermoelectric power of the metal ion or oxygen ion conducting solid electrolyte respectively.

Temperature Compensated Reference Electrode

If an electrode can be designed which reduces the Seebeck coefficient of the solid electrolyte to zero then the emf developed in the non-isothermal galvanic sensor would be same as that of the isothermal cell operating under the same activities of the active species. The emf of a nonisothermal oxygen sensor can be written as

$$E_{\text{non-iso}} = \xi(T''-T') + \frac{1}{4F}(\Delta\overline{H}''_{O_2} - T''\Delta\overline{S}''_{O_2} - \Delta\overline{H}'_{O_2} + T'\Delta\overline{S}'_{O_2}) \qquad ..(24)$$

where $T'' = T' + \Delta T$ and $\Delta\overline{H}''_{O_2} - T''\Delta\overline{S}''_{O_2} = \Delta\mu''_{O_2}(T'')$.

Hence

$$E_{\text{non-iso}} = \xi \Delta T + \frac{1}{4F}(\Delta\mu''_{O_2}(T') - \Delta\mu'_{O_2}(T') - \Delta T\, \Delta S''_{O_2}) \qquad ..(25)$$

If $\frac{1}{4F}\Delta\overline{S}''_{O_2} = \xi$, then the emf of the non-isothermal cell is identical with that of an isothermal cell with identical electrodes at T". Hence if we choose a reference electrode which satisfies the condition, $\xi = \frac{1}{4F}\Delta\overline{S}''_{O_2}$, then the sensor, can be used without precise knowledge of the temperature of the cold limb. Relative partial molar entropies, $\Delta\overline{S}_{O_2}/4F$, of some metal/metal-oxide reference electrodes are given in Table I. The values for Ni+NiO, $Cr+Cr_2O_3$ and $Mo+MoO_2$ are very close to the value of $\xi^{\text{Pure } O_2}$ (-0.423 mV/K) at a temperature of 1200K for fully stabilized zirconia. Hence they may be used in commercial applications for approximate temperature compensation. To obtain exact compensation it is necessary to take either the metal or

TABLE I

Relative partial molar entropies of some metal + metal oxide reference electrodes

Electrode	$\Delta \bar{S}_{O_2}/4F$ (mV.K^{-1}) 1200K	1875K
$Cu + Cu_2O$	-0.36902	-
Ni + NiO	-0.44107	-
Co + CoO	-0.36241	-
Fe + FeO	-0.33235	-
Mn + MnO	-0.38699	-
$Cr + Cr_2O_3$	-0.43780	-0.4420
$Mo + MoO_2$	-0.44185	-0.41338

oxide into solid solution and thereby alter the value of $\Delta\bar{S}_{O_2}$. For example it has been shown[25] that a solid solution $Ni_{0.65}Mg_{0.35}O^2$ in equilibrium with Ni provides exact compensation in nonisothermal oxygen sensors for liquid copper.

6. SENSORS FOR SPECIFIC ELEMENTS AND COMPOUNDS

Oxygen Sensors

Subsequent to the first application of calcia stabilized zirconia in an oxygen sensor by Weissbart and Ruka[26], a number of studies have been conducted to evaluate the performance of the sensor with platinum electrodes[27-30]. At the operating temperatures greater than 873K, the response time of the sensor is of the order of few milliseconds for $P_{O_2} > 10^2$ Pascals. For lower pressures the response time is larger. Recently ruthenium oxide electrodes have been found to give rapid response at temperatures as low as 500K. The emf of an oxygen sensor based on oxide solid electrolytes with $t_{ion} > 0.99$ is given by

$$E = -\frac{RT}{2Z_oF} \ln \frac{P''_{O_2}}{P'_{O_2}} \qquad ..(26)$$

where P''_{O_2} is the partial pressure of oxygen in the test gas (unknown) and P'_{O_2} is the partial pressure in the reference gas (known). The gauge can also analyse CO/CO_2 and H_2/H_2O mixtures with platinum electrodes.

As temperature decreases the cell resistance increases. Doped ceria, which exhibits higher conduction than zirconia and thoria based electrolytes[31,32], can be used in an oxygen sensor at lower temperatures for any given choice of electrode. It has been found that solid electrolytes such as CaF_2[33] and a two-phase $\alpha + \beta$-alumina[34] can be used to measure extremely low oxygen potentials and their use in oxygen sensors can extend the detection limits to lower pressures than possible with oxide electrolytes with fluorite structure. The oxygen partial pressure response of a CaF_2 cell such as

$$Pt,\ Ni+NiO|\ CaF_2|\ Cu+Cu_2O,\ Pt \qquad (\ I\ )$$

is due to the presence of fine dispersion of CaO in CaF_2[35]. The CaO forms as a result of reaction of CaF_2 with trace moisture in the ambient atmosphere at high temperature. Calcia and CaF_2 do not exhibit significant mutual solubility and are present at unit activity.

The electrode reactions can be represented as

$$CaF_2 + \frac{1}{2} O_2' + 2e^- \rightleftarrows CaO + 2F^- \qquad ..(27)$$

$$CaO + 2F^- \rightleftarrows CaF_2 + \frac{1}{2} O_2'' + 2e^- \qquad ..(28)$$

The open circuit emf of the cell-I is therefore given by Eq.(26).

Oxides of the type AO where the transference number of the cation, $t_{A^{2+}}$, is unity and AB_2O_4, where the sum of the transference numbers of cations ($t_{A^{2+}} + t_{B^{3+}}$) is unity, can also be used as solid electrolytes in oxygen sensors. This follows from the fact that the changes in the chemical potential of each element is related to the chemical potentials of other elements through the Gibbs-Duhem equation. The cell arrangement for an oxygen sensor where the solid electrolyte is a stoichiometric ternary compound with both cations mobile can be represented as,

$$Pt, P'_{O_2} |\ AB_2O_4 \|\ AB_2O_4 \|\ AB_2O_4 |\ P''_{O_2},\ Pt \qquad (\ II\ )$$

The emf of such a cell is given by,

$$E = \frac{RT}{nF} \ln \frac{P''_{O_2}}{P'_{O_2}} \qquad ..(29)$$

provided there is no activity gradient of component oxides AO and B_2O_3 across the solid electrolyte. If the left hand electrode is saturated with respect to AO and the right hand electrode with respect to B_2O_3 the cell can be represented as,

$$\mathrm{Pt, P_{O_2}} \left| \begin{matrix} AB_2O_4 \\ + \\ AO \end{matrix} \right| AB_2O_4 \left\| \begin{matrix} AB_2O_4 \\ + \\ B_2O_3 \end{matrix} \right| \mathrm{P''_{O_2}} \quad \mathrm{Pt} \qquad (\mathrm{III})$$

The emf is given by[36)]

$$E = \frac{RT}{4F} \ln \frac{P''_{O_2}}{P'_{O_2}} + \frac{1}{4F}\left(\frac{2}{3} t_{B^{3+}} - 2t_{A^{2+}}\right) \Delta G^o_{(AB_2O_4)} \qquad ..(30)$$

where $\Delta G^o_{AB_2O_4}$ is the standard free energy of formation of AB_2O_4. The positive sign between the two terms of Eq.(30) is to be replaced by a negative one, if AO is thought to be precipitated in the right layer and B_2O_3 in the left layer of the electrolyte.

Oxygen probes for liquid metals and alloys

The oxygen content in copper during wire bar casting has to be precisely controlled to obtain the desired quality. Oxygen probes based on stabilized zirconia have been successfully used for this purpose[37-39)]. Oxygen concentrations in the range 0.0010 wt. pct. to 0.2500 wt. pct. have been measured[39)]. Since alloying elements influence the activity coefficient of oxygen in copper, these effects must be considered if the emf is to be related to oxygen concentration. In steel making low oxygen contents are essential both for ingot and continuously cast steel and the solid electrolyte measuring technique has been tried for in situ measurements[40-46)]. The oxygen potential of liquid steel containing approximately 200 ppm dissolved oxygen is close to the lower electrolytic conduction boundary for calcia stabilized zirconia; so the expression for the sensor emf must take into account $t_{O^{2-}} < 1$. The emf is given by,

$$E = \frac{RT}{F} \ln \frac{P_n^{1/4} + P^{1/4}_{O_2\,(ref)}}{P_n^{1/4} + P^{1/4}_{O_2(melt)}} \qquad ..(31)$$

The partial pressure of oxygen over the melt is given by

$$P_{O_2}^{1/2} = \frac{\text{Wt. percent of O x } f_O}{K} \quad ..(32)$$

where K is the equilibrium constant for the reaction

$$\frac{1}{2}\ O_2 \rightarrow \underline{O}\ (\text{wt \%, Fe})$$

With increasing electronic conductivity significant polarization of the electrodes occur at steel making temperature resulting in erroneous values. The problem, however, can be minimized by using reference electrodes such as, $Cr+Cr_2O_3$, with oxygen potential close to that in the melt. Electrolytic conduction boundary for yittria doped thoria (YDT) is lower than for stabilized zirconia. Hence thoria based electrolytes are superior at low oxygen concentration. However, because of the radio-activity of thorium, YDT has not found application in steel industry. Unlike conventional chemical methods of analysis, the galvanic cell technique estimates the activity of dissolved oxygen and not the total oxygen content, including oxygen present in the form of oxides. This is very useful for computing the amount of deoxidizers required for achieving a specified oxygen content.

The oxygen potential of a steel melt can be related to the carbon potential when the partial pressure of CO is well defined. Therefore sensors can be designed for monitoring carbon based on oxide solid electrolytes provided a capillary connection is given to maintain a fixed partial pressure of CO over the melt.

An alternate technique for measuring oxygen activity is by voltametry[47,48]. Etsell and Flungas[47] have indicated that electrochemical removal of oxygen from copper and silver alloys using cells based on CSZ gives limiting currents controlled by oxygen diffusion and convection in the liquid metal. The limiting currents are proportional to the oxygen activity.

In addition to use in in-situ measurements, solid state sensors can form an integral part of anlaytical instruments. Thus the liquid metal sample containing carbon on combustion in a stream of oxygen produces CO_2, the concentration of which can be measured down the stream with the cell arrangement[49)],

$$\text{Pt, air} | \text{CSZ} | CO_2 + O_2, \text{Pt.}$$

The carbon content of the metal sample is related to the change in emf of the cell as

$$\text{Wt. percent of C} = \frac{4800\, n_{O_2} F}{RTW} \int \Delta E\, dt \qquad ..(33)$$

where n_{O_2} is the flow rate of pure oxygen at the inlet of the combustion furnace, W is the sample weight, ΔE the change in emf and t is the time.

The corrosive action of liquid sodium, used as the coolant in the liquid metal fast breeder reactor, on stainless steel depends on the dissolved oxygen content. Thus monitoring the oxygen content of liquid sodium is important for the smooth functioning of the reactor. An electrochemical oxygen meter using ThO_2-15% Y_2O_3 solid electrolyte and a Sn/SnO_2 reference couple has been used[50)].

Lambda sensors for automobiles

The operation of automobile engines at the stoichiometric air/fuel ratio,A/F, i.e. the one that contains just enough oxygen to convert all the hydrocarbons of the fuel to CO_2 and H_2O, improves the engine efficiency as well as reduces the content of toxic gases like CO and NO_x in the exhaust. A λ-sensor based on CSZ is used as a feed back controller to maintain the A/F ratio at the stoichiometric point. The parameter λ is defined as

$$\lambda = \frac{(A/F)_{actual}}{(A/F)_{stoich.}} \qquad ..(34)$$

As the A/F ratio becomes large ($\lambda > 1$) the exhaust gas has a high P_{O_2} and the emf decreases. For small A/F ratios ($\lambda < 1$) the insufficiency of oxygen causes the formation of the reducing species CO and H_2 which establishes a low P_{O_2} and the emf rises to values above 1 V. In passing from a reducing gas with $\lambda = 0.999$ to an oxidising gas with $\lambda = 1.001$, there is a change in P_{O_2} by a factor of 10^{15} at 973K[51]. Thus in the vicinity of the stoichiometric point, the sensor exhibits steep variation of emf of the order of 700 mV[52]. The voltage switching is used as a feed back to an electronic closed loop circuit[53,54].

In the lean fuel region, the sensitivity of the sensor to variations in P_{O_2} is poor, because sensor output voltage varies as $\ln P_{O_2}$. Greater sensitivity may be achieved using a linear sensor.

Linear sensors

The device is used to pump gas X_2 from the cathode to the anode, the electrode processes being

$$X_2 + 2V_X^{\bullet\bullet} + 4e^- \underset{\text{anode}}{\overset{\text{cathode}}{\rightleftarrows}} 2X_X \qquad ..(35)$$

where X_X is divalent. A porous barrier is fixed in front of the cathode to restrict the transport of X_2 to the electrode. A sufficient voltage is applied across the electrolyte to pump the gaseous species X_2 from the cathode to anode. When the P_{X_2} at the cathode is reduced to a value close to zero the limiting condition is attained. The current flowing, i_{lim}, is controlled by the rate of diffusion of X_2 through the porous barrier according to Fick's first law,

$$i_{lim} = nFD(Q/L)C_{X_2} \qquad ..(36)$$

where n is the number of electrons transferred per molecule of X_2 and D is the diffusion coefficient, Q the sum of the cross sections of the pores of effective length L and C_{X_2} is the concentration of X_2 in the sample gas. Eq.(36) shows that i_{lim} is proportional to the gas concentration.

The amperometric oxygen sensor can only be used in the lean fuel region. This is a serious limitation and arises because potentials are essentially referred to that of the anode in the sample gas. On passing through the stoichiometric point the anode potential changes by more than 500mV.

Typically the amperometric sensor is operated at 0.8V. On passing through stoichiometry to $\lambda < 1$ the overvoltage applied to the cathode is then $\sim 1.3V$ with respect to an air reference and this is now sufficiently cathodic to reduce CO_2 and H_2O resulting in a current flow[55]. This problem can be solved by using pump gauge devices. A reference gas of fixed composition is provided adjacent to the anode on a single two-electrode amperometric sensor. Traversal of stoichiometry then results in reversal of current flow so that regions $\lambda < 1$ and $\lambda > 1$ are clearly identified. Application of a fixed gaseous reference to the gauge of the pump-gauge provides a stable reference as negligible current need be carried by it[56]. A disadvantage of this arrangement is that a connection and suitable seal to the device must be provided to supply reference gas.

Sensors For Sulphur Bearing Gas Species

A number of solid electrolytes with or without sulphur as a primary component have been incorporated in sensors for sulphur. The literature in this field has recently been reviewed[57]. Despite the concerted efforts of many laboratories during the last two decades to find suitable sulphide electrolytes analogous to $CaO\text{-}ZrO_2$ or $Y_2O_3\text{-}ThO_2$ in oxygen sensors, no acceptable material has been identified in which the ionic transport number is higher than 0.99 over a large range of sulphur potentials and temperatures. This is because the electronic bandgaps in the sulphides are generally lower than the corresponding oxides[58]. Further they are unstable in oxidising atmospheres. Sulphur containing industrial atmospheres are either reducing with respect to oxygen as in coal gasifiers or oxidising as in combustion processes. Hence the choice of solid electrolytes for sensors of sulphur depends on the oxygen potential or the ratio of oxygen to sulphur potential in the gas phase.

Sensors for sulphur bearing gas species use three types of electrolytes; (i) sulphates (ii) sulphides and iii) a non-sulphide/sulphate electrolyte with an auxiliary sulphide/sulphate electrode. Sensors using either sulphide electrolytes or non-sulphide electrolyte with auxiliary sulphide electrodes are used in atmospheres of low oxygen potential where sulphur is usually present in the elemental form, S_2, or as H_2S. Sulphate electrolytes are used in atmospheres of high oxygen potential where sulphur will be predominantly present as SO_x(x=2,3).

Sensors for sulphur in low oxygen potential environments

Sulphides can be used as solid electrolytes or auxiliary electrodes at low oxygen potentials such as those encountered in coal gasification and liquifaction plants and in hot metal from iron blast furnaces. Mixtures of a metal and its sulphide can be used as a reference electrode. The choice of metal+sulphide mixture depends on the sulphur potentials of the ambient gas. The reference couple is usually chosen so that the cell emf has a small value. Gaseous reference electrode containing H_2 and H_2S mixture, although extensively used in laboratory studies because they provide primary calibration, are not suitable in practical sensors in view of the elaborate precautions necessary to minimize the thermal segregation.

Cells with sulphide electrolytes

The use of CaS doped with higher valent transition metal sulphides as an electrolyte has been explored by a number of investigators[59-62]. These studies indicate that CaS doped with Y_2S_3 is an ionic conductor with mobile Ca^{2+} ions for sulphur pressures $< 10^{-1}$Pa between 1043K and 1273K. At higher sulphur pressures hole conduction is observed[59,60]. In sharp contrast CaS doped with La_2S_3 has been reported to exhibit electronic conductivity at $P_{S_2} < 10^{-1}$Pa and ionic conductivity above this pressure over the temperature range of 973K to 1173K[63]. If the properties of La_2S_3 doped CaS are confirmed, then bielectrolyte cells using doped CaS can be employed to measure sulphur potentials over a wider range by suitable choice of the dopant.

Sensors using solid electrolytes not containing sulphur

Solid electrolytes CaF_2, β-alumina, Nasicon, α-AgI and (CaO)ZrO_2 have been used in sensors to measure sulphur potential with sulphide auxiliary electrodes.

Sensors for H_2S or S_2 based on CaF_2

The electrodes are designed in such a way to convert the sulphur potential in the gas into an equivalent fluorine potential. Measurements on the cell[64)]

$$Pt|Ar'+H_2S'+H_2'|\ CaS+CaF_2+(Pt)\|\ CaF_2\|\ (Pt)+CaF_2+CaS|\ H_2''+H_2S''+Ar''|Pt \quad \textbf{(IV)}$$

have shown that the emf of the cell is directly related through the Nernst equation to the difference of sulphur potentials established at the two $Ar+H_2+H_2S|$ electrode interfaces. The response time varied from approximately 9h at 900K to 2.5h at 1225K.

The emf developed across the CaF_2 electrolyte is fundamentally related to the partial pressure of fluorine across the electrolyte provided the transport number is $\geqslant 0.99$.

$$E = \frac{\Delta\mu^R_{F_2} \quad \Delta\mu^T_{F_2}}{nF} = \frac{RT}{nF} \ln \frac{P^R_{F_2}}{P^T_F} \qquad ..(37)$$

where F is the Faraday constant n is the number of electrons (=2) involved in the transfer of one molecule of fluorine gas between the two electrodes, R is the gas constant, T the absolute temperature, $P^R_{F_2}$ and $P^T_{F_2}$ are the fluorine partial pressures at the reference and test electrodes respectively. The gas phase containing $Ar+H_2S+H_2$ establishes a partial pressure of sulphur over the electrodes depending on the ratio of H_2 to H_2S by virture of the reaction

$$H_2(g) + \frac{1}{2} S_2(g) \rightarrow H_2S(g) \qquad ..(38)$$

$$\frac{1}{2} \Delta\mu_{S_2} = RT \ln P^{1/2}_{S_2} = \Delta G^o_{38} + RT \ln(P_{H_2S}/P_{H_2}) \qquad ..(39)$$

where ΔG^o_{38} is the change in the standard Gibbs' energy for reaction (38). The sulphur potential established by the gas phase, when in contact with the electrode pellet consisting of CaS and CaF_2 fixes a fluorine potential according to the exchange reacton,

$$CaF_2(s) + \frac{1}{2}\ S_2(g) \rightarrow CaS_{(s)} + F_2\ (g) \quad ..(40)$$

$$\Delta\mu_{F_2} = \frac{1}{2}\ \Delta\mu_{S_2} - \Delta G^o_{40} \quad ..(41)$$

provided the activities of CaF_2 and CaS are unity. Combining equations (37) and (41) we get

$$E = \frac{\Delta\mu^R_{S_2} - \Delta\mu^T_{S_2}}{2\ nF} = \frac{RT}{2nF}\ \ln\frac{P^R_{S_2}}{P^T_{S_2}} \quad ..(42)$$

Eq.(42) may also be expressed in terms of the ratio of H_2S to H_2 as

$$E = \frac{RT}{nF}\ \ln\frac{(P_{H_2S}/P_{H_2})^R}{(P_{H_2S}/P_{H_2})^T} \quad ..(43)$$

$$\frac{dE}{dT} = \frac{R}{n\ F}\ \ln\frac{(P_{H_2S}/P_{H_2})^R}{(P_{H_2S}/P_{H_2})^T} \quad ..(44)$$

Sensors for H_2S or S_2 based on β -alumina and Nasicon

Solid state sensors for H_2S or S_2 have also been developed using sodium β -alumina or Nasicon as the solid electrolyte and Na_2S as auxiliary electrode[65,66]. The advantages of the sodium ion conductors over CaF_2 are their higher melting point and higher electrical conductivity, which facilitate a more rapid cell response. A typical cell arrangement is

$$Au|\ (Ar'+H_2S'+H_2')|\ Na_2S\ \|\ \begin{matrix}\beta\text{-alumina}\\ \text{or}\\ \text{Nasicon}\end{matrix}\ \|\ Na_2S|\ (Ar''+H_2''+H_2S'')|\ Au \qquad (V)$$

Since pure Na_2S coexists with β -alumina or Nasicon and an $Ar+H_2+H_2S$ gas mixture at each electrode the cathode half cell reaction can be written as,

$$2Na^+\ (\text{in }\beta\text{-alumina or Nasicon}) + \frac{1}{2}\ S_2(g) + 2e^- \rightarrow Na_2S(s) \quad ..(45)$$

The corresponding anode half cell reaction is,

$$Na_2S(s) \;=\; 2Na^+_{(in\ \beta\text{-alumina or Nasicon})} + \frac{1}{2}\,S_2(g) + 2e^- \qquad ..(46)$$

The overall cell reaction is

$$\frac{1}{2}\,S_2(g)\ (\text{cathode}) \;=\; \frac{1}{2}\,S_2(g)\ (\text{anode}) \qquad ..(47)$$

and the cell emf is defined by Eq.(42) or (43). Since the present sensor is based on the cathode and anode half cell reactions, interpretation of the emf requires a knowledge of Na_2S activity at each electrode. It is convenient to have unit activity for Na_2S at both the electrodes. For this reason the application of this cell is ideally restricted to the thermodynamic stability field of solid Na_2S[65]. The main advantage of Nasicon over β -alumina is the three dimensional nature of its conductivity.

Sensor for H_2S or S_2 based on α-AgI

Cells incorporating AgI can be used as a sulphur probe by using Ag_2S as an auxiliary electrode[68,69]. The cell can be represented as

$$\text{Pt} \mid \text{Ag} \mid \text{AgI} \parallel Ag_2S \mid S_2 \mid \text{Pt} \qquad (\text{VI})$$

The sulphur pressure can be calculated from the equation

$$\ln P_{S_2} \;=\; (E - E^o)\,\frac{4F}{RT} + \ln P^o_{S_2} \qquad ..(48)$$

where E^o is the emf of the cell when liquid sulphur is present at the right hand electrode and $P^o_{S_2}$ is the equilibrium partial pressure of diatomic sulphur over pure liquid sulphur. The major disadvantage of this cell is the low melting point of AgI(827 K) and the onset of electronic conduction which restricts the operation of the cell to $T \leqslant 713$ K.

Sensor for H_2S or S_2 based on a two-phase sulphide-oxide electrolyte

In this case the solid electrolyte consists of a dispersion of CaS in a matrix of CSZ. The auxiliary electrode is embedded as a second phase in the solid electrolyte. This is the preferred configuration for the use of auxiliary electrodes. However limitations imposed by preparative techniques and materials compatibility do not always permit the realization of this ideal configuration in practice. The surface of the electrolyte is coated with a thin layer of CaS to prevent the formation of water vapour by reaction of H_2S with CaO or ZrO_2 present in the electrolyte The loss of emf resulting from the chemical potential gradient across the CaS, which is a mixed conductor, is minimized by reducing its thickness. The galvanic sensor based on the two-phase sulphide oxide electrolyte can be represented as[70)]

$$Pt(or\ Au)|\ Ar'+H_2' + H_2S'\|\ CaS + (CaO)ZrO_2\|\ Ar''+H_2'' + H_2S''|\ Pt\ (or\ Au) \quad \textbf{(VII)}$$

At each electrode CaS coexists with CaO dissolved in the zirconia electrolyte. The sulphur-oxygen exchange reaction

$$CaS(s) + \frac{1}{2}O_2(g) \rightleftarrows CaO_{(in\ ZrO_2)} + \frac{1}{2}S_2 \quad ..(49)$$

becomes important. There is only one value of P_{S_2}/P_{O_2} for this equilibrium at a given temperature. The composition of the electrolyte is thermodynamically stable only for this equilibrium ratio. The gas phase establishes the sulphur potential in the CaS coating applied to the external surface of the electrolyte and this is converted to an equivalent oxygen potential by virtue of reaction (49) at the CaS/electrolyte interface. The emf is given by Eq.(42). To evaluate the conduction domain for the biphasic electrolyte, the oxygen potential established at the electrodes during emf measurements in $Ar+H_2+H_2S$ mixtures must be evaluated using data for reaction (49). This requires the knowledge of activity of CaO in CSZ[70)] besides the Gibbs' free energies of formation of CaO and CaS available in the literature.

In atmospheres of low sulphur or oxygen potentials, the onset of electronic conductivity in CSZ matrix is a limiting factor. At high oxygen potential environments the CaS is readily oxidised. The range of condition for the applicability of the two-phase electrolyte for the sulphur potential measurements is discussed in reference (70).

Sensors for sulphur in high oxygen potential environments

At high P_{O_2} atmospheres sulphur is predominantly present as SO_x(x=2,3). Hence a solid state sensor for sulphur can be either based on a sulphate solid electrolyte or a nonsulphate material as the electrolyte with an auxiliary sulphate electrode.

Sensors for SO_x(x=2,3) based on sulphate electrolytes

The high temperature form of the alkali sulphates K_2SO_4[71,72], Na_2SO_4[73] and Li_2SO_4[74] have been successfully used as solid electrolytes to detect the oxides of sulphur. The following cell,

$$Pt|\ SO_3', O_2' \|\ A_2SO_4 \|\ O_2'', SO_3''\ |\ Pt \qquad \textbf{(VIII)}$$

where 'A' stands for Li, Na or K, has been used to detect sulphur oxides in air potentiometrically. The electrode reactions can be represented as

$$SO_3 + \frac{1}{2}O_2 + 2e^- \rightleftarrows SO_4^{2-} \qquad ..(50)$$

The resulting emf for the cell (VIII) is

$$E = \frac{RT}{2F} \ln \frac{(P'_{O_2})^{1/2} P'_{SO_3}}{(P''_{O_2})^{1/2} P''_{SO_3}} \qquad ..(51)$$

In oxidising atmospheres containing sulphur, SO_2 generally coexists with SO_3 at equilibrium

$$SO_3 \rightleftarrows SO_2 + \frac{1}{2}O_2 \qquad ..(52)$$

$$K = \frac{P_{O_2}^{1/2}\ P_{SO_2}}{P_{SO_3}} \qquad ..(53)$$

where K is the equilibrium constant. Equilibrium composition of the gas phase can be computed from material balance and thermodynamic data for reaction (52). Combining Eqs.(51) and (53) we get

$$E = \frac{RT}{2F} \ln \frac{P'_{O_2}\ P'_{SO_2}}{P''_{O_2}\ P''_{SO_2}} \qquad ..(54)$$

When the oxygen partial pressure is the same at each electrode Eq.(54) simplifies to

$$E = \frac{RT}{2F} \ln \frac{P'_{SO_2}}{P''_{SO_2}} \quad ..(55)$$

In atmospheric pollution monitoring, Eq.(55) can be used directly since P_{O_2} is the same at both the electrodes. In industrial atmospheres where the oxygen content is a variable such as stack gases of non-ferrous smelters, Eq.(54) must be used. Emf then responds to the product of partial pressures of SO_2 and O_2. To obtain P_{SO_2} a sensor based on sulphate solid electrolyte must be electronically coupled with a solid state sensor for oxygen.

Among the three alkali sulphates, Na_2SO_4-I appears to have the best combination of properties for use in sensors. The phase transformation to the high temperature structure is accompanied by large enthalpy and volume changes. The phase transformation temperature is 847K for Li_2SO_4, 857K for K_2SO_4 and 514K for Na_2SO_4. When the sulphates are cycled rapidly and repeatedly through the transformation temperature, microcracks develop in the solid electrolyte. Although the cracks heal under a small compressive stress when heated to ~ 1100K in some applications such in situ heat treatment may be inconvenient. Presence of interconnecting microcracks can lead to gas permeation and lowering of emf. Among alkali sulphates, the enthalpy of phase transformation is larger for Li_2SO_4 than for the other sulphates. Li_2SO_4 also exhibits higher reactivity to silica used in cell construction compared to the other alkali sulphates. At high partial pressures of SO_3 ($P_{SO_3} > 10^4$ Pa at 973K), K_2SO_4 will transform to a liquid rich in $K_2S_2O_7$, thus destroying the mechanical integrity of the sensor. In the pyrosulphate ion $[O_3S\text{-}O\text{-}SO_3]^{2-}$ the stability of the oxygen bridge diminishes with increasing polarization power of the cation. Since the Na^+ ion is more polarising than K^+ ion, $Na_2S_2O_7$ is less stable than $K_2S_2O_7$. Therefore the formation of $Na_2S_2O_7$ occurs only at a considerable high P_{SO_3} than required for $K_2S_2O_7$. From the point of view of conductivity, transformation enthalpy, phase stability and sinterability Na_2SO_4 appears to be the most promising material among sulphates.

Imanaka et al[75] have shown that the phase transformation in Na_2SO_4 can be suppressed by mixing with Li_2SO_4, $Y_2(SO_4)_3$ and SiO_2, without any adverse effect on ionic conductivity. Doped Na_2SO_4, can therefore withstand thermal cycling without the formation of microcracks.

Sensors for SO_3| SO_2 based on CaF_2| $CaSO_4$ couple

Since the alkali sulphates undergo a phase transformation on going from a low to a high temperature accompanied by large enthalpies and volume changes, Jacob et al[76] have tried CaF_2 solid electrolyte along with $CaSO_4$ as the auxiliary electrode for the measurement of SO_x. Calcium fluoride undergoes only a second order phase transformation to the high conducting phase and also has a very large ionic conduction domain[77].

Measurements on the solid state cell

$$Pt|\ O_2'+SO_2'+SO_3'|\ CaF_2+CaSO_4\|\ CaF_2|\ CaSO_4+CaF_2|\ SO_3''+SO_2''+O_2''|\ Pt \qquad \textbf{(IX)}$$

indicate that the emf is in agreement with that predicted by the Nernst equation when equilibrium is assumed in the gas phase near the electrodes[76]. Since the fluorine ions are the mobile species in CaF_2, the emf is given by

$$2FE = -[\mu'_{F_2} - \mu''_{F_2}] \qquad ..(56)$$

The fluorine potential, μ_{F_2}, at each electrode is established by the exchange reaction

$$CaF_2(s) + SO_3 + \tfrac{1}{2}O_2(g) = CaSO_4(s) + F_2(g) \qquad ..(57)$$

Since CaF_2 and $CaSO_4$ are present at unit activity and the μ_{SO_3} and μ_{O_2} are fixed,

$$\mu^o(CaF_2) + \mu'(SO_3) + \tfrac{1}{2}\mu'(O_2) = \mu^o(CaSO_4) + \mu'(F_2) \qquad ..(58)$$

$$\mu^{o}(CaF_2) + \mu''(SO_3) + \frac{1}{2}\ \mu''(O_2) = \mu^{o}(CaSO_4) + \mu''(F_2) \qquad ..(59)$$

Combining equations (56),(58) and (59) we get

$$2FE = -[\mu'(SO_3) + \frac{1}{2}\ \mu'(O_2) - \mu''(SO_3) - \frac{1}{2}\mu''(O_2)] \qquad ..(60)$$

Consequently

$$E = \frac{RT}{2F} \ln \frac{P''_{SO_3}\ P''^{\,1/2}_{O_2}}{P'_{SO_3}\ P'^{\,1/2}_{O_2}} \qquad ..(61)$$

Expressing P_{SO_3} in terms of P_{SO_2} and P_{O_2} the emf can be expressed as Eq.(54). The effect of moisture on the emf can be rationalized based on the equation

$$CaF_2 + H_2O + SO_3 = 2HF + CaSO_4 \qquad ..(62)$$

In most cases the test gases will have to be dehydrated before they enter the sensor. The response time of the cell varied from 8h at 850K to 2h at 1200K.

Sensors for SO_x(x=2,3) using β-Alumina| Na_2SO_4 and Nasicon| Na_2SO_4 couples

The main drawback of the SO_x sensor based on the CaF_2| $CaSO_4$ couple is its slow response time. To enhance the response time, fast sodium ion conductors - Nasicon and Li_2O stabilized β''-alumina have been used to monitor the partial pressures of SO_3 or SO_2 in the gas phase[78,79]. During emf measurement, a thin adherent layer of Na_2SO_4 was formed in situ on the surface of the electrolyte. Once a layer of Na_2SO_4 is formed, rate of reaction is slow. Since Na_2SO_4 is a Na^+ ion conductor, diffusion flux through the material will be limited by the trace concentration of electrons and holes. However, if the cell is cycled through the phase transformation temperature for Na_2SO_4, cracks develop in the protective layer, providing direct access for the gas phase to the electrolyte. Under such conditions, Na_2O from the electrolyte is continuously consumed, causing composition changes and mechanical deterioration of the solid electrolyte over long exposures. This problem is partly overcome by covering the electrolyte with a dense envelope of

Na_2SO_4 which acts both as an auxiliary electrode and a solid electrolyte[80,81].

The galvanic cell can be represented as[80,81]

$$\mathrm{Pt}|\ O' + SO_2' + SO_3'|\ Na_2SO_4 \| \begin{matrix}\beta\text{-alumina}\\ \text{or}\\ \text{Nasicon}\end{matrix} \|\ Na_2SO_4|\ SO_3''+SO_2''+O_2''|\ \mathrm{Pt} \qquad (\mathrm{X})$$

The Na_2SO_4 also acts as a protective coating preventing direct access of the gaseous SO_x species to the electrolyte. The chemical potentials of SO_x and O_2 are converted into an equivalent sodium potential. The electrode reactions are,

Anode reaction

$$Na_2SO_4 \rightarrow 2\,Na^+ + SO_3' + 1/2\ O_2' + 2e^- \qquad ..(63)$$

Cathode reaction

$$2Na^+ + SO_3'' + 1/2\ O_2'' + 2e^- \rightarrow Na_2SO_4 \qquad ..(64)$$

The Na_2SO_4 is at unit activity at both electrodes. The open circuit emf is same as expression (61) or (54) when equilibrium is established in the gas phase. With the use of catalytically active electrodes such as platinum equilibrium is generally attained in the gas phase at least in the neighbourhood of the electrodes at $T > 750$ K. The response times for sensors based on Nasicon and β-alumina with Na_2SO_4 auxiliary electrodes following a change in the gas composition for 99% response at 750K, 900K and 1100K are 0.5h, 0.06h and .01h respectively and are of the same magnitude as that obtained using pure Na_2SO_4-I as the solid electrolyte. The response times with pure Nasicon[82] are also of the same order of magnitude as for Na_2SO_4 - I at high temperatures. This suggests that the response time is not controlled by the bulk properties of the solid electrolyte. The electrode reactions involving Na_2SO_4, which is common to all these sensors is probably rate controlling.

The lowest partial pressure of SO_3 that can be measured by a sensor based on β-alumina and Nasicon with Na_2SO_4 auxiliary electrodes is determined by the decomposition of Na_2SO_4 according to the reaction

$$Na_2SO_4 \rightarrow SO_3(g) + Na_2O\ (\beta\text{-alumina or Nasicon}) \qquad ..(65)$$

Using standard thermodynamic data for Na_2SO_4, $SO_3(g)$ and Na_2O, and activity of Na_2O in Nasicon the lowest partial pressure limit is evaluated as $P_{SO_3} = 4.53 \times 10^{-3}$ Pa at 1000K.

Reference electrodes

Gas reference electrodes are very reliable and provide primary standards, but they are cumbersome to use in a practical detector. Although premixed gases have been used as reference electrodes in laboratories, solid reference systems are preferred in most industrial applications. A mixture of sulphate and oxide of the same metal at equilibrium will establish a unique partial pressure of SO_2, SO_3 and O_2 at any fixed temperature in a closed system. Jacob and Iyengar[83)] have shown that $Fe_2(SO_4)_3$ + Fe_2O_3can be used as a reference electrode when it is enclosed in a closed volume around the electrode. The oxide-sulphate mixture was kept out of physical contact with Na_2SO_4-I solid electrolyte to avoid complications arising from the formation of solid solutions. The equilibrium between the reference mixture and the platinum electrode was established via the gas phase. In this arrangement the oxygen partial pressure established by the dissociation of SO_3 is generally low. Gauthier et al[84)] provided a Knudsen orifice on the enclosure around the sulphate + oxide reference electrode and established an external partial pressure of oxygen (air) through this orifice. This design overlooks the outward flux of SO_x species through the orifice and consequent lack of equilibrium when partial pressures of SO_x species and orifice diameter are large. Gauthier[85)] et al have also studied the Ag/Ag^+ type reference electrode. Since Ag_2SO_4 is soluble in K_2SO_4 a solid solution in equilibrium with metallic silver can be adapted as a reference. The cell can be represented as

$$Ag \mid Ag_2SO_4 - K_2SO_4 \| K_2SO_4 \| SO_2, O_2 \mid Pt \qquad (XI)$$

and the emf of the cell is

$$E = E^o - \frac{RT}{2F} \ln a_{Ag_2SO_4} + \frac{RT}{2F} \ln (P_{SO_2})(P_{O_2}) \qquad ..(66)$$

The Ag/Ag^+ reference system works best when Ag is not exposed to an atmosphere containing oxygen. In the presence of oxygen, Ag_2O forms as a solid solution in K_2SO_4 leading to slow decay of emf. A major problem with metal-metal sulphate ($Ag-Ag_2SO_4$) reference electrode is that cationic interdiffusion between the sulphate in the reference and the alkali sulphate electrolyte establishes concentration gradients in the electrolyte which may give rise to unstable cell voltages. If a solid solution of Ag_2SO_4 and alkali sulphate is used as the solid electrolyte such that there is no gradient of Ag_2SO_4 between the reference and the solid electrolyte, interdiffusion problem is avoided. For successful use the solid solution should be a predominantly ionic conductor.

Worrell and Liu[86)] took advantage of a two-phase region in the Li_2SO_4 - Ag_2SO_4 phase diagram between 20 and 35 mole pct Ag_2SO_4[87)] and designed a solid electrolyte consisting of two phase. The advantage of the two phase system is that small composition gradients across the solid electrolyte does not introduce activity gradients and consequent ionic fluxes. Their cell exhibited exceptionally long-term chemical stability.

Analysis of non-equilibrium gas mixtures

Electrochemical studies on sulphur oxide sensors based on alkali sulphates have revealed that the active species is SO_3, not SO_2, and the use of Pt electrodes generally results in equilibrium between SO_2, SO_3 and O_2 at least in the vicinity of the triple contact between the electrolyte, metal and the gas phase at $T > 750K$. When equilibrium of the gas constituents prevails at the electrode, the sensor can be quite successfully used for measuring either SO_2 or SO_3 or both.

In combustion gas analysis, the sulphur oxide is mostly SO_2, although occasionally a small percentage of the highly corrosive SO_3 is also present In order to detect SO_3 selectively a metal electrode that would not catalyze the equilibration of SO_2 with SO_3 is essential. Gauthier et al[85)] have shown that non-equilibrium concentration of SO_3 in the gas can be detected using Au or Rh electrodes.

Summary

Doped-Na_2SO_4 and β-alumina and Nasicon solid electrolytes with Na_2SO_4 as auxiliary electrode are suitable for the detection of sulphur oxides over a large range of partial pressure of SO_x and temperature without the formation of microcracks when cycled through the phase transformation temperature of alkali sulphates. Non-equilibrium concentration of sulphur oxide can be detected using electrodes of low catalytic activity like gold and rhodium.

Sensors For CO_x (x=1,2)

Sensors based on alkali carbonates

Electrochemical cells using K_2CO_3[88] and Na_2CO_3[89] as solid electrolytes have been used for the measurement of the equilibrium partial pressure of CO_2 in CO_2+O_2 mixtures in the temperature range of 600 to 1000K. Alkali carbonates appear to be ionic conductors since Nernstian response is obtained with CO_2+O_2 mixtures. The electrochemical cell can be represented as,

$$\mathrm{Au}|\ CO_2'+O_2' \|\ A_2CO_3 \|\ O_2''+CO_2''|\mathrm{Au} \qquad \textbf{(XII)}$$

where A stands for Li, Na or K. The electrode reaction can be represented as,

$$A_2CO_3 \rightleftharpoons 2A^+ + CO_2 + 1/2\ O_2 + 2e^- \qquad ..(67)$$

and the corresponding cell emf is,

$$E = \frac{RT}{2F} \ln \frac{P'_{CO_2}\ P'^{1/2}_{O_2}}{P''_{CO_2}\ P''^{1/2}_{O_2}} \qquad ..(68)$$

By defining P'_{CO_2}, P'_{O_2} and P''_{O_2} the value of P''_{CO_2} can be calculated. If $P'_{O_2} = P''_{O_2}$ then the cell emf is,

$$E = \frac{RT}{2F} \ln \frac{P'_{CO_2}}{P''_{CO_2}} \qquad ..(69)$$

Solid reference electrodes such as mixtures of $CaCO_3$ + CaO may be used. The chemical reaction taking place in the electrode compartment is

$$CaCO_3 \rightleftarrows CO_2 + CaO_{(s)} \qquad ..(70)$$

$$\Delta G^o_{70} = -RT \ln P_{CO_2} \qquad ..(71)$$

where ΔG^o_{70} is the standard Gibbs' energy change for reaction (70). Since CO_2 is very stable at the cell operating temperature, unlike the SO_3 molecule, the oxygen partial pressure in the electrode compartment has to be fixed externally by supplying O_2 through a Knudsen orifice into an enclosure containing the solid reference mixture[88]. It has been found that the Nernstian response is only obtained for $P_{CO_2} > 10^2$Pa at ~1000K. At low oxygen potentials, where the gas phase consists primarily of CO and CO_2 molecules, the electrode reaction is[90],

$$Na_2CO_3 + CO \rightleftarrows 2Na^+ + 2CO_2 + 2e^- \qquad ..(72)$$

The emf at low oxygen potential will be proportional to $(P^2_{CO_2}/P_{CO})$ ratio.

Muruyama et al[91] have designed small sensors for CO_2 using Nasicon and β-alumina as the solid electrolyte and Na_2CO_3 as the auxiliary electrode. A simple cell of the type,

$$Au, CO_2{+}O_2 | Na_2CO_3 \| Na^+ \text{ ion conductor} \| O_2, Au \qquad (XIII)$$

has been found to respond rapidly to $CO_2 | O_2$ ratio at T > 700K. The anodic reaction can be represented as

$$Na_2CO_3 \rightarrow 2Na^+ + CO_2 + \frac{1}{2}O_2 + 2e^- \qquad ..(73)$$

and the cathodic reaction is

$$2Na^+ + \frac{1}{2}O_2 + 2e^- \rightarrow Na_2O_{(\text{solid electrolyte})} \qquad ..(74)$$

The sample gas is flown past both the electrodes. Therefore the oxygen partial pressures are equal at both the electrodes and the cell reaction may be represented as

$$Na_2CO_3 \rightarrow CO_2 + Na_2O_{(solid\ electrolyte)} \qquad ..(75)$$

consequently the cell emf is given as

$$E = -(\Delta G^o_{Na_2O} + \Delta G^o_{CO_2} - \Delta G^o_{Na_2CO_3})/2F - \frac{RT}{2F} \ln (a_{Na_2O} . P_{CO_2}) \qquad ..(76)$$

In order to use a reference oxygen potential different from that of the CO_2+O_2 mixture, the Na^+ ion conductor can be coupled with a stabilized zirconia oxygen sensor. For the cell,

$$Au,\ CO_2' + O_2' |\ Na_2CO_3 \|\ Na^+ \text{ ion conductor} \|\ YSZ |\ O_2'', Au \qquad (XIV)$$

the reference oxygen pressure can be established using either gas or condensed phase mixtures. If the reference electrode is maintained inside a stabilized zirconia tube, the oxygen partial pressure in the CO_2+O_2 mixture can also be measured by having an electrode attached to the outer surface of the zirconia tube. The emf of the cell is given by

$$E = -(\Delta G^o_{Na_2O} + \Delta G^o_{CO_2} - \Delta G^o_{Na_2CO_3})/ 2F - \frac{RT}{2F} \ln (a_{Na_2O}\ P_{CO_2} \left(\frac{P'_{O_2}}{P''_{O_2}} \right)^{1/2}) \qquad ..(77)$$

The activity of Na_2O in Na_2CO_3 in equilibrium with CO_2+O_2 mixtures is higher than the activity of Na_2O in Nasicon or alumina-rich β-alumina solid electrolyte. At temperatures in excess of 1000K partial decomposition of Na_2CO_3 to Na_2O dissolved in the solid electrolyte may be expected. If this reaction proceeds to a significant extent it will alter Na_2O activity in Na^+ ion conductor and consequently calibration of the cell. Also the nature of the atmosphere surrounding the junction between the two solid electrolytes is expected to influence the stability of emf. Ogata et al[92)] have used a combination of β-alumina and metal carbonate to design CO_2 gas sensors. However, electrode reactions in their cell are not well defined.

Sensors For NO_x(x=1,2)

The possibility of building a nitrogen oxide detector has been discussed by Gauthier et al[71] using a solid electrolyte obtained by melting a mixture of 1 mole pct silver chloride in barium nitrate at 873K and sintering at 673K for 24h. The cell may be represented as

$$Ag \| \ 1 \text{ mole \% AgCl in } Ba(NO_3)_2 \| \ O_2'' + NO_2'' \ | \ Pt \qquad (XV)$$

and the electrode reaction can be represented as

$$NO_2 + \frac{1}{2}O_2 + e^- \rightleftarrows NO_3^- \qquad ..(78)$$

The response time of the NO_x sensor varies from a few seconds to several minutes depending on the flow rate, the concentration range under study and the geometry of the apparatus. A number of laboratories are now working actively on NO_x sensors and new designs will be available in the near future.

Sensors for Alloying Elements

Silicon in hot metal

A solid state silicon sensor which incorporates magnesia-stabilized zirconia as a solid electrolyte and $ZrSiO_4$ as an auxiliary electrode has been designed by Iwase[93]. The cell can be represented as,

$$Mo | \ Mo + MoO_2 \| \ ZrO_2 \ (MgO) \| \ ZrO_2 + ZrSiO_4 | \ Si \text{ (in Fe)} | \ Fe \qquad (XVI)$$

At any given temperature, the activity of SiO_2 at the electrolyte/hot metal interface is fixed by the auxiliary electrode:

$$ZrSiO_4 = ZrO_2(s) + SiO_{2(\text{in } ZrSiO_4)} \qquad ..(79)$$

The oxygen potential, P_{O_2} (test), at the three phase interface of zirconia electrolyte + auxiliary electrode + liquid iron is established by the equilibrium reaction;

$$Si_{(\text{in iron})} + O_2(g) = SiO_{2(\text{in } ZrSiO_4)} \qquad ..(80)$$

or

$$ZrO_2(s) + Si_{(\text{in iron})} + O_2(g) = ZrSiO_4 \qquad ..(81)$$

Therefore

$$\log K = -\log h_{Si} - \log P_{O_2} \qquad ..(82)$$

where K is the equilibrium constant for reaction (81) and h_{Si} is the Henrian activity of silicon in liquid iron. From Eq .(82) it is clear that P_{O_2} correlates directly with h_{Si}.

If the n-type electronic conduction in magnesia-stabilized zirconia is negligible, then the emf developed can be given by the simple Nernst equation

$$E - E_t = \frac{RT}{4F} \ln \frac{P_{O_2 (ref)}}{P_{O_2(test)}} = \frac{RT}{4F} \ln \frac{P_{O_2(ref)}}{(Kh_{Si})^{-1}} \qquad ..(83)$$

where E_t is the thermal emf between Fe and Mo, and E is the cell voltage. If the n-type electronic conduction in the solid electrolyte is not negligible, open circuit emf is given by[94)]

$$E = \frac{RT}{F} \ln \frac{P_{O_2 (ref)}^{1/4} + P_n^{1/4}}{(Kh_{Si})^{-1/4} + P_n^{1/4}} + E_t \qquad ..(84)$$

where P_n is the oxygen partial pressure at which the ionic and n-type electronic conductivity are equal in the solid electrolyte.

Zinc in molten brass

At the casting temperature of molten brass, it is very difficult to control the zinc concentration because of its high affinity for oxygen to form ZnO and its high vapour pressure. The oxygen concentration of molten brass (70 wt pct Cu-30 wt pct Zn) at saturation with ZnO is of the order of 1 ppm. Hence in commercial practice brass is always saturated with ZnO. Hence the measurement of oxygen potential in brass with a solid electrolyte can be used to calculate zinc activity and thereby the zinc content in molten brass. Wilder and Walter[95)] have shown that the cell

$$Pt | Ni+NiO \| CSZ \| Zn_{(in\ molten\ brass)} + ZnO | Ta \qquad (XVII)$$

can be used to determine the zinc content of Cu-Zn alloys at 1268 K.

The cathode reaction can be represented as,

$$NiO_{(s)} + 2e^{-} = Ni_{(s)} + O^{2-}_{(electrolyte)} \qquad ..(85)$$

and the anodic reaction as,

$$Zn_{(in\ molten\ brass)} + O^{2-} = ZnO_{(s)} + 2e^{-} \qquad ..(86)$$

Thus the overall reaction is,

$$NiO_{(s)} + Zn_{(in\ molten\ brass)} = ZnO_{(s)} + Ni_{(s)} \qquad ..(87)$$

The cell emf, after correction for the junction emf between Pt and Ta, is

$$2FE = -(\Delta G_f^o\,(ZnO) - \Delta G_f^o\,(NiO) - RT\ln a_{Zn}) \qquad ..(88)$$

where ΔG_f^o (ZnO) and ΔG_f^o (NiO) are the standard molar free energy of formation of the respective oxides and a_{Zn} is the thermodynamic activity of Zn in molten brass. The activity of Zn can be related to its concentration using the known activity coefficient for Zn.

Cells of this type may also be used to determine the nickel content in cupro-nickels, aluminium in aluminium bronzes, chromium in stainless steels and other alloy systems which are expected to be saturated with respect to a stable oxide in normal processing operations.

7. ERRORS AND CALIBRATION

Possible Errors in EMF Measurements

Temperature gradients

In a cell with a solid electrolyte two temperature gradients should be taken into account. One is the temperature gradient across the electrolyte, and the other is a temperature difference at the interface itself when the contact area between the metal and the electrolyte is large. If there is a temperature gradient across the solid electrolyte, a part of the emf developed in the cell will be due to the thermo emf. If there is a temperature difference at the interface of the electrode/electrolyte itself, there will be a transfer of mobile species along the gradient resulting in a mixed potential. In the case of metal/metal oxide reference mixtures the equilibrium oxygen pressure also depends on the temperature.

Interference with a gas phase

In a galvanic sensor, if the two electrode compartments are not separated, there may be transport of the active species from the electrode with the higher chemical potential to the electrode with a lower potential. When a single condensed phase electrode is used, without any buffering reaction, the gas phase might supply or remove the active species thereby altering its chemical potential.

Permeability of the solid electrolyte

When there is a significant difference in the chemical potential of the active species at the two electrodes, effects of mass transfer through the solid electrolyte become important. The transfer of the active species through the electrolyte can be caused by an internal current due to electronic conduction, Knudsen diffusion of the active species through pores, chemical diffusion through grain boundaries or molecular diffusion through micro-cracks. The flux of the active species through the solid electrolyte can result in polarization of the electrodes and lowering of the emf. Physical permeability can be minimized through the use of dense material. The semipermeability flux cannot be avoided since it arises from the intrinsic electrochemical properties. The best method to minimize the effects of semipermeability flux is to use reference electrodes that establish a chemical potential of the active species close to that of the working electrode. Design of electrodes to dissipate semipermeability flux has been discussed earlier.

Reaction products between electrode materials and solid electrolytes

If an inter oxide compound, ABO_3, is formed between the material AO used in an electrode compartment and the solid electrolyte BO_2, with pure oxygen ion conduction, the cell can be expressed as,

$$P'_{O_2}\ Pt|\ A{+}AO\|\ \underset{P^{I}_{O_2}}{ABO_3}\|\ \underset{P^{II}_{O_2}}{BO_2}(t_{O^{2-}}=1.0)\|\ Pt,\ P''_{O_2} \qquad \text{(XVIII)}$$

ABO_3 may be mixed conductor. Thus the open circuit emf value will be different from the theoretical value deduced from the Nernst equation.

Induced emfs

The furnace winding can induce emf in the leads. This can be avoided by using non inductively wound furnaces or by shielding the cell with an earthed metallic sheet.

Calibration

Test for correct functioning

A number of tests can be performed to ensure that a solid state sensor is functioning properly.

a) The cell assembly should be checked for gas tightness by evacuating it at high temperatures and ensuring that no gas leakage occurs between the electrode compartments or between the cell and the ambient atmosphere.

b) At a fixed temperature with the same chemical potentials of the active species at both the electrodes, the open circuit emf should ideally be zero. Departure from this is usually indicative of a thermal gradient across the electrolyte.

c) The open circuit emf is measured as a function of temperature for a fixed gas composition at the electrodes. The emf, plotted as a function of absolute temperature, should be a straight line with slope equal to the theoretical value defined by the Nernst equation. In complex gas mixtures such as $SO_3+SO_2+O_2$, the chemical potentials vary non-linearly with temperature resulting in nonlinear emf variation for certain gas ratios.

d) Keeping the reference gas composition constant at a fixed temperature, the test gas composition is altered by a fixed amount using an electrochemical pump. The resulting gas composition determined by the sensor should match the new value calculated from the pumping current.

e) Cell reversibility can be checked by coulometric titration. A small quantity of current is passed through the sensor in either direction. The cell emf should return to its original value before titration. This ensures that the electrodes are not easily polarized.

f) The reproducibility of the emfs should be checked by temperature cycling.

g) The emf should be insensitive to small variations in the flow rate of gases through the cell.

SUMMARY

Solid state galvanic sensors have undergone significant technical development since their inception in 1921 by Weissbut and Ruka. Different types of galvanic sensors have been developed since then and it appears that galvanic sensors will find wide ranging applications within laboratories as well as in industries. However, several technical and scientific problems still remain. A better understanding of fundamental processes at electrodes and interfaces is needed. Interaction between different scientific disciplines viz., Metallurgy, Materials Science, Solid State Electrochemistry and Instrumentation is imperative to overcome these. Miniaturization of galvanic cells and their integration into microelectronic devices also needs to be addressed.

REFERENCES

1. Dietz, H., Solid State Ionics 6, 175 (1982).
2. Mizusaki, J., Yamauchi, S. and Fueki, K., Denki Kagaku Oyobi Kogyo Butsuri Kagaku, 50, 7 (1982).
3. Goto, K.S., Anal. Chem. Symp. Ser. 17, 338 (1983).
4. Goto, K.S., Nippon Kinzoku Gakkai Kaiho, 23, 978 (1984).
5. Williams, D.E., McGeehin, P., Electrochemistry, 9, 246 (1984).
6. Maskell, W.C., Steele, B.C.H., J.Appl.Electrochem. 16, 475 (1986).
7. Weppner, W., Sensors and Actuators, 12, 107 (1987).
8. Wagner, C., Z. Physik Chem. B, 21, 25 (1933).
9. Patterson, J.W., J. of Electrochemical Society, 118, 7, 1033 (1971).
10. Patterson, J.W.,in "The Physics of Electronic Ceramics", L.L. Hench and D.B. Dove, Editors, Marcel Dekker Inc., N.Y. (1971).
11. Akila, R., Jacob, K.T. and Shukla, A.K., Bull.Mater.Sci. 8, 453 (1986).
12. Fouletier, Fabry, P. and Kleitz, M., J. Electrochem. Soc., 123, 204 (1976).
13. Jacob, K.T., Iwase, M. and Waseda, Y., J. Appl.Electrochem., 12, 55 (1982).
14. Jacob, K.T., Hajra, J.P. and Iwase, M, Arch. Eisenhutenwes., 55, 421 (1984).
15. Worrell, W.L. and Isokoe, J.L., "Fast Ion Transport in Solid State Batteries and Devices", Ed. Van Gool W., North Holland Publishing Company, Amsterdam, p.513 (1973).
16. Tretyakov, Y.D., Maun, A., J. Electrochem. Soc., 116, 331 (1969).
17. Shores, D.A. and Rapp, R.A., J.Electrochem.Soc., 118, 1107 (1971).
18. Heyne, L. and den Engelsen, D., J. Electrochem.Soc., 124, 5, 727 (1977).
19. Winnubst, A.J.A., Scharenborg, A.H.A. and Burggraff, A.J., J.Appl.Electrochem., 15, 139 (1985).
20. Fouletier, J., Seinera, H. and Kleitz, M., J.Appl.Electrochem., 4, 305 (1974).
21. Scaife, P.H., Swinkels, D.A.J. and Richards, S.R., High Temp. Sci., 8, 31 (1976).
22. Anderson, J.E., Graves, Y.B., J.Appl.Electrochem., 12, 335 (1982).
23. Caneiro, A., Bommat, M. and Fouletier, J., J.Electrochem.Soc., 123, 204 (1976).
24. Ramasesha, S.K. and Jacob, K.T., J.Appl.Electrochem., in press.

25. Jacob, K.T. and Ramasesha, S.K., Solid State Ionics, in press.

26. Weissbart, J. and Ruka, R., Rev.Sci.Instrum. 32, 593 (1961).

27. Fouletier, J., Mantel, E. and Kleitz, M., Solid State Ionics 6, 1, (1982).

28. Meas, Y., Fouletier, J., Passelaigue, D. and Kleitz. M., J.Chim.Phys; 75, 826 (1978).

29. Fouletier, J., Sienera, H. and Kleitz, M., J.Appl.Electrochem., 5, 177 (1975).

30. Etsell, T.H., Flengas, S.N., Met.Trans., 3, 27 (1972).

31. Tuller, H.L. and Nowick, A.S., J.Electrochem.Soc., 122, 255 (1975).

32. Fouletier, J., Henault, M., Solid State Ionics, 9/10, 1277 (1983).

33. Ramanarayanan, T.A., Narula, M.L. and Worrell, W.L., J.Electrochem. Soc.,126, 1360 (1979).

34. Choudhury, N.S., J.Electrochem.Soc. 120 (12), 1663 (1973).

35. Jacob, K.T., Kale, G.M., Ramachandran, A. and Shukla, A.K., High Temperature Materials and Processes, 7, 141 (1986).

36. Pluschkell, W. and Yoshikoshi, H., Arch Eisenhuttwes, 41(8), 715 (1970).

37. Dompas, J. and Van Melle, J., J. Inst.Metals, 98, 304 (1970).

38. Scholes, W.A., J. Engg.Min. 175(10), 82 (1974).

39. Hens, L.J., in "Copper Metallurgy, Practice and Theory", Institution of Mining and Metallurgy, London, pp.377-425.

40. Nagata, K. and Goto, K.S., Solid State Ionics, 9/10, 1249 (1983).

41. Oguchi, Y., Suzuku, T., Yano, T. and Emi, T., Solid State Ionics 3/4, 641 (1981).

42. Janke, D.,Solid State Ionics, 3/4, 599 (1981).

43. Goto, K.S., Steel Making Proc. 61, 492 (1978).

44. Thomas, Y.D., Proc.Nat.Open Hearth Steel Making Conf; 59, 420 (1976).

45. Matsushita, Y. and Goto, K., Trans. ISIJ, 6, 131 (1966).

46. Fischer, W.A., Ackermaun, W., Arch. Eisenhuttwes, 36, 643 (1965).

47. Etsell, T.H. and Flungas, S.N., J.Electrochem.Soc. 119, 198 (1972).

48. Dietz, H., Maecker, W. and Jahuke, H., Advances in Electrochem. and Elechemical Engg. 10, Eds. H. Gerischer and C.W. Tobias 1, (1977).

49. Iwase, M. and Mori, T., Can.Met.Quart. 19, 285 (1983).

50. Vissers, D.R., McKee, J.M., Bartholme, L.G. and Nelson, P.A., Nucl. Tech., 21, 75 (1974).

51. Hamann, E., Manger, H. and Steinke, L., Tech. Paper SAE; 7, 770401 (1977).

52. Eddy, D.S., IEEE Trans. Veh.Tech. VT23, 125 (1974).

53. Camp, J. and Rachu, T., Tech. Paper SAE, 750369 (1975).

54. Spilski, R.A. and Creps, W.D., SAE Trans., 750371 (1975).

55. Saji, K., Kondo, H., Takahashi, H., Takeuchi, T. and Igarashi, I., Proc. Int.Conf. Solid State Sensors and Actuators, Philadelphia, 11-14 June (New York, IEEE) 336 (1985).

56. Soejima, S. and Mase , S. SAE Paper 850378 (1985).

57. Iwase, M. and Jacob, K.T., Proc.Symp. on High Temperature Materials Chem., Eds. D.D. Cubicciotti and D.L. Hildenbrand, Vol.82-1. The Electrochem.Soc.Inc., Pennington, N. J , 431 (1982).

58. Rao, C.N.R. and Pisharody, K.P.R., Prog.Solid State Chem. 10, 207 (1976).

59. Worrell, W.L., Tare, V.B. and Bruni, F.J., Proc. of the Third International Symp. on High Temperature Technology, Asilomar, CA, 503 (1967).

60. Nagata, K. and Goto, K.S., Met.Trans. 5, 899 (1974).

61. Nakamura, H. and Gunji, K., Nippon Kinzoku Gakkai-shi 42, 635 (1978).

62. Ono, K., Ishihara, K. and Moriyama, J., ibid. 44, 185 (1980).

63. Sato, M., Imanaka, N., Adachi, G. and Shiokawa, J., Mat.Res.Bull. 16, 215 (1981).

64. Jacob, K.T., Bhogeswara Rao, D. and Nelson, H.G., J.Electrochem.Soc., 125, No.5, 758 (1978).

65. Jacob, K.T., Iwase, M. and Waseda, Y., Advanced Ceramic Materials 4, No.3, 264 (1986).

66. Jacob, K.T. and Waseda, Y., Unpublished work.

67. Wagner, C., J. Chem.Phys., 21, 1819 (1953).

68. Sato, M., "Research Techniques for High Pressure and High Temperature", Ed. G.C. Ulmer; Springer-Verlag, N.Y., 43 (1971).

69. Schneeberg, E.P., Econ.Goel, 68, 507 (1973).

70. Jacob, K.T., Iwase, M. and Waseda, Y., J.Appl.Electrochem. 12, 55 (1982).

71. Gauthier, M. and Chamberland, A., J. Electrochem.Soc., 124, 1579 (1977).

72. Gauthier, M., Chamberland, A., Belanger, A. and Poirier, M., ibid, 124 1584 (1977).

73. Jacob, K.T. and Bhogeswara Rao, D., J. Electrochem.Soc., 126, 1842 (1979).

74. Liu, Q.G. and Worrell, W.L., Chemical Metallurgy - a tribute to Carl Wagner, Ed.Nev.A.Gocken, The Metallurgical Soc. AIME, p.43 (1981).

75. Imanaka, N., Yamaguchi, Y., Adachi, G. and Shioka, J., J. Electrochem. Soc. 133, 1026 (1986).

76. Jacob, K.T., Iwase,M., Waseda, Y., Solid State Ionics 23, 245 (1987).
77. Ono, K., Fujimura, T. and Moriyama, J., Trans.Jpn.Inst. Metals 20, 543 (1979).
78. Maruyama, T., Saito, Y., Matsumoto, Y. and Yano, Y., Solid State Ionics, 17, 281 (1985).
79. Itoh, M., Sugimoto, E. and Kozuka, Z., Trans.Jpn.Inst.Met. 25, 504 (1984).
80. Akila, R. and Jacob, K.T., J.Appl. Electrochem. 18, 245 (1988).
81. Akila, R. and Jacob, K.T., Sensors and Actuators 13, (1988).
82. Saito,Y., Maruyama, T., Matsumoto, Y., Kobayashi, K. and Yano, Y., Solid State Ionics 14, 273 (1987).
83. Jacob, K.T. and Iyengar, G.N.K., Met.Trans. B., 17B, 487 (1986).
84. Gauthier, M. and Bale, C.W., Met.Trans., 14B, 117 (1983).
85. Gauthier, M., Bellemare, R. and A. Belanger, J.Electrochem.Soc. 128, 371, (1981).
86. Worrell, W.L. and Liu, Q.G., J.Electrochem. Soc., 168, 355 (1984).
87. Qye, H.A., Acta Chem.Scan., 18, 361 (1964).
88. Cote, R. and Bale, C.W. and Gauthier, M., J.Electrochem.Soc. 131, 63 (1984).
89. Jacob, K.T. and Akila, R., Unpublished work.
90. Maruyama, T., Xu-Yun YHe and Saito, Y., Solid State Ionics 23, 113 (1987).
91. Maruyama, T., Sasaki, S. and Saito, Y., Solid State Ionics 23, 107 (1987).
92. Ogata, T., Fujitsu, S., Miyayama, M., Koumoto, K. and Yanagida, H., J.Materials Science letters 5, 285 (1986).
93. Iwase, M., Scandinavian J. Metallurgy 17, 50 (1988).
94. Schmalzried, H., Z. Elektrochem. 66, 572 (1962).
95. Wilder, T.C. and Galin, W.E., Trans. TMS AIME 245, 1287 (1969).

SOLID STATE PRIMARY AND SECONDARY BATTERIES

Roger G. Linford

School of Chemistry, Leicester Polytechnic
P.O. Box 143, Leicester, LE1 9BH
UNITED KINGDOM

ABSTRACT

An overview is presented of solid battery systems in which high conductivity solid ionic conductors are used as electrolytes and in some cases as matrix materials within electrode compartments. The battery systems are classified in terms of the type of ionic conductor that is used rather than by anode material, to facilitate cross reference with other chapters within this volume. Fuel cells and sodium sulphur batteries are omitted since they are covered in other chapters.

1. INTRODUCTION

Solid state batteries offer some attractive advantages over their liquid counterparts. They are inherently robust, spill-proof and usually non-corrosive. They are compact and well suited to miniaturization, and the ability of the components to stay in position can do away with the need for separators, membranes and/or diaphragms, which add to the mass and volume of liquid systems without producing concomitant increases in energy or power. Thin film configurations can be used, especially for polymer systems[1)] and sometimes these can be formed into unusual shapes that fit them for specialised applications. They can often perform adequately over wide temperature ranges, which is highly desirable for military and other uses. Their predominant disadvantages, with respect to liquid systems, are poor interfacial contact, slow mass diffusion and, for certain systems, formation of a highly resistive cell reaction product in a location within the battery where it impedes the cell reaction. The difficulty of overcoming such drawbacks accounts for the fact that few commercial solid state battery systems are available despite twenty years of intensive research and development work in this field.

Before commencing a detailed description of systems based on a number of types of solid ionic conductor, it is useful to clarify what is meant by some of the terms and concepts used, and to point to some of the attributes and difficulties that arise from the use of these materials.

Historically, the term battery denoted a power source consisting of a bank of interconnected electrochemical cells; an example is the lead-acid accumulator used in cars. In the field of solid ionic conductors, a range of impedance, polarization and other cells are used to characterise materials behaviour. To prevent confusion with these cells present usage, which will be employed here, favours the use of the word battery for a galvanic power source.

A power source is a transducer in which chemical energy from an electron - producing cell reaction is converted to electrical work performed on an external load such as a motor, heater or lamp. In a primary battery, the cell reactants are not replenished or recreated and when they are depleted, the battery is discarded. In a fuel cell, the cell reactants are gaseous or liquid and can be continually replenished. Such devices are considered in another chapter. Certain commercial systems such as the zinc-air battery[2], in which one electrode is fixed in amount and the other is unrestricted, represent an intermediate case.

In a secondary (or rechargeable) battery, after completion of the galvanic discharge process, an electrolyte recharge process is carried out, in which an external voltage and current source is connected to the battery so as to reverse the cell reaction and reform the cell reactants at the appropriate electrodes. The discharge-recharge sequence is called a cycle, and typical batteries can cope with several hundred cycles before intolerable degradation of performance occurs. Because reversal of the cell reaction may be imperfect, the depth of discharge that can be attained on each cycle depends on the nature of the cell reaction, the materials used and the way in which the battery is constructed.

A battery consiste of an anode at which oxidation takes place, separated from a cathode where a complementary reduction process occurs, by an electrolyte in which ions, but not electrons or electron holes, are free to move. The anode is negative during galvanic discharge since it is the pole or plate of the battery from which electrons flow externally, and the cathode is positive. For a secondary battery during recharge (or for an electrolyte cell such as that used for the electrolysis of brine[3]) in the chemical industry), the negative and positive electrodes remain in the same locations on the battery (conventionally drawn on the left and right respectively). The positive electrode is now the anode and the negative electrode is the cathode because the cell reaction is reversed, as are the directions of ionic flow through the electrolyte.

Liquid electrolytes are usually effective electronic insulators but solid ionic conductors in some cases are mixed conductors, permitting passage of both ionic and electronic species. Such mixed conductors are suitable for electrodes in which both types of conduction must occur, but are impractical as solid electrolytes because the cell reaction is free to proceed irrespective of whether the battery is connected to an external load or not. A battery utilizing such material would self-discharge on standing, leading to a short shelf-life[4,5)].

Many solid state battery systems utilize a cell reaction in which cations are inserted into an ion-insertion material which acts as the active cathode component. Ion-insertion compounds are usually intercalation materials; a common example is TiS_2 which can be converted during galvanic discharge to Li_xTiS_2, $0\leq x\leq 1$. Ion insertion causes volume expansion, and contraction occurs during recharge. Such active cathode material therefore needs to be embedded in a flexible and compliant matrix, for example a polymer electrolyte, if interfacial contact is not to be lost during battery use. This type of cathode material is often a poor electronic conductor when fully intercalated and carbon particles or some other form of electron carrier must, of necessity, also be incorporated into the cathode compartments. It is possible to use intercalation materials for both anode and cathode, as in the so-called "rocking chair cells".

A further handicap to the use of solid electrolytes is that, in contrast with liquid systems, both anions and cations can have non-zero transport numbers. This can lead to the formation of cell reaction product at an undesirable location, e.g. the anode-electrolyte interface, leading to lower current density and/or deteriorating voltage as the discharge process proceeds. In addition, loss of battery capacity on cycling can occur in secondary batteries.

A solid state battery will be defined in this chapter as one in which the anode, electrolyte and cathode are all mechanically solid. The term "mechanical" is chosen to exclude semi-solid pastes such as those used in "dry" cells[2)], but to include batteries incorporating polymer electrolytes. Polymer batteries are frequently operated at a temperature that exceeds both the glass transition temperature, Tg, of the amorphous region of the polymer and the melting point, T_m, of any crystalline polymer regions that may be present. The polymer might then be thought of as molten, at least in the sense that it behaves similarly to an ionic melt from the theoretical point of view[6)], but its tendency to flow is sufficiently slight for it to behave mechanically as a plastically deformable solid.

The above definition excludes coverage of both the sodium sulphur battery (covered elsewhere in this volume) and the related Zebra battery[7)] because, although the electrolyte is the solid ionic conductor β-alumina, the electrodes are molten at the operating temperatures. Batteries in which Nasicon or similar materials are substituted for β-alumina will also be omitted from this chapter.

A number of other devices, incorporating solid state materials, that behave as batteries are not comprehensively covered, on the basis that the production of electrochemical power is not their prime purpose. Examples include electrochromic cells and state-of-charge detectors[8)].

Despite the use of the term "high conductivity" in the title of this volume, for a solid electrolyte the relevant performance parameter is ionic conductance rather than conductivity. Even materials such as LiI in which the conductivity is as low as 10^{-7} S cm^{-1} can be used in a thin film configuration that can yield a conductance of 10^{-4} S, corresponding to current densities in the range 10-100 μA cm^{-2} for typical battery voltages of 0.4-4 V and cross-sectional areas of ~ 4 cm^2.

In summary, to compensate for the significant advantages over liquid systems that were set out at the beginning of this section, solid state battery systems are associated with the following problems:

i) multi-ion conduction, particularly in polymer batteries;

ii) electronic conduction, for example in compacted powder copper batteries;

iii) insufficient deformability to accommodate electrode dimensional changes, for example in glass batteries;

iv) low conductivities at low temperatures, restricting the available temperature range. Some systems embodying high conductivity solid materials such as Ag_4RbI_5, are beset by compensating disadvantages, in this case chemical instability and low cell voltage;

v) propensity to moisture attack, leading to variable performance parameters, especially for divalent polymer electrolyte systems.

Despite these difficulties, the goal of achieving a range of viable solid state battery systems remains very desirable. The strong international research and development interest is well displayed by the large number of reviews published this decade[9-36)], which show the steady progress that has been necessary to move towards some of the optimistic predictions that characterise many of the earlier overviews[37-45)].

2. POLYMER BATTERIES

2.1 Lithium Polymer Batteries

Solid state batteries in this class are nearing the commercialization stage. They have provided the focus for major coordinated research efforts in the European Economic Community (the Anglo-Danish project), between France and Canada (the ACEP project) and more recently in the USA and in Japan. Since the completion of the Anglo-Danish project, work has continued at Harwell UK[46)], and recent progress on the Harwell battery and on the not-dissimilar Canadian-French system is described in a forthcoming review by Hooper (Harwell) and Gauthier (Hydro-Quebec)[47)].

A major goal has been to develop a traction battery for electric cars[48], capable of sustaining modern day cruising speeds of ∿60 mph over a range of ∿250 miles before recharge. These figures are the minimum necessary to provide even a moderately plausible alternative to the gasoline-fuelled internal combustion engine for out-of-town driving. Within urban areas, the inherent characteristics of electrically powered vehicles, namely quietness, freedom from pollution and responsive acceleration without the need for mechanical gear changing whether manual or automatic, have been appreciated from the beginning of the century, when small delivery trucks powered by lead-acid accumulators began to appear.

The overriding limitation of the lead-acid accumulator is not its awkward construction, involving the corrosive liquid electrolyte sulphuric acid; technical advances have provided spill-proof sealed systems in which the electrolyte is trapped in a mat of fine glass fibres or similar material. Rather, it is the low theoretical mass and volume efficiencies that result from use of the Pb/PbO_2 system. It is possible to approach a calculated mass energy density of ∿200 Wh kg^{-1} (if cell separators, electrode grids, battery containers etc. are neglected) but in practice it is difficult to exceed 40-80 Wh kg^{-1}. Consequently, the combined energy reservoir-motor-transmission unit is substantially heavier and less powerful than for a corresponding internal combustion engine. The lithium polymer battery has a theoretical mass energy density that approaches 800 Wh kg^{-1} and an achievable performance of 425 Wh kg^{-1} (540 Wh h^{-1}) at 2.4 V has been anticipated[49]. A large battery consisting of 30 parallel connected units, each containing 20 bipolar connected cells, was shown to have a capacity of 30 Ah at 48 V and 120-140 ^oC, corresponding to an energy capacity of 1440 Wh[50].

The concept of the electric car has aroused some scepticism within the scientific and technical community but it is agreed that a lightweight rechargeable battery is a desirable, potentially low cost, marketable alternative to costly disposable primary batteries. The lithium polymer battery has some important advantages over rival, non solid-state systems. Pollution and hazard problems at both the

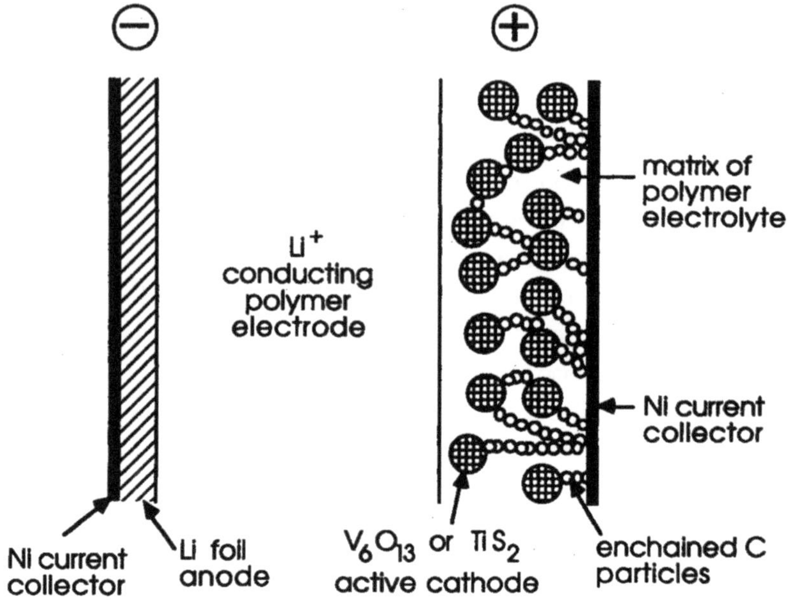

Figure 1. Galvanic Lithium Polymer Battery

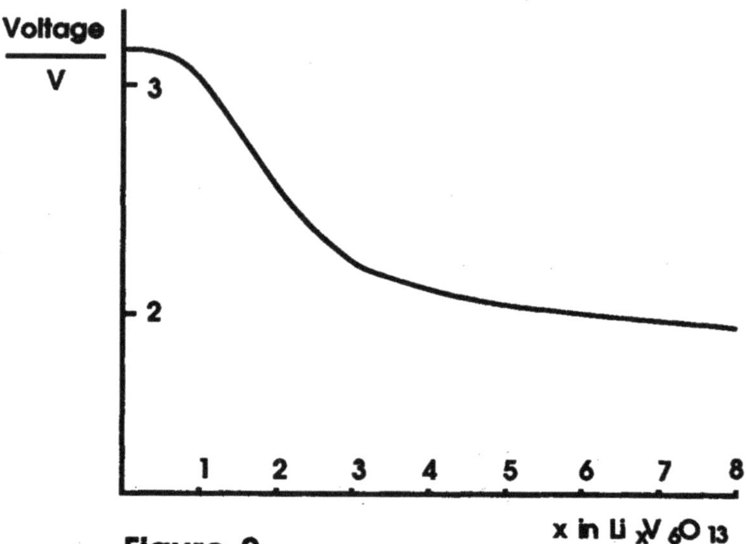

Figure 2.
Voltage as a function of amount of intercalation

manufacturing and the disposal stage are less severe than for the environmentally suspect NiCd and Pb-acid systems. Lithium is an intensely electropositive metal and its cell reactions therefore produce high voltage, and yet it has the lowest relative molar mass of any metal, so that large quantities of electrons are produced from a very low anode mass. It is half as dense as water, however, so that its advantages are not so outstanding from the point of view of volumetric efficiency.

The main features of a lithium polymer battery can be seen in figure 1 and are as follows:

i) It is a secondary battery system, capable of many cycles although there have been some problems of loss of capacity on recycling[51)] and gain of crystallinity on storage[52)]. It was initially a moderately high temperature system designed to operate at $\sim 100\ ^{o}C$ for reasons of electrolyte behaviour discussed in iv) below, but is now targeted to room temperature operation.

ii) The galvanic cell reaction on discharge is the insertion of lithium into a suitable intercalation material, and the recharge reaction is the removal of lithium from the intercalate followed by replating on the negative plate. Suitable intercalation materials are metal oxides and sulphides such as V_6O_{13} (Harwell) or TiS_2 (ACEP). Recently the use of V_3O_8 as an alternative material has been suggested[53)]. The voltage varies with the degree of intercalation as shown schematically in figure 2. The galvanic half cell reactions for complete discharge in the case of V_6O_{13} are

anode: $Li \rightarrow Li^+ + e^-$

cathode: $V_6O_{13} + 8Li^+ + 8e^- \rightarrow Li_8V_6O_{13}$

For TiS_2, the maximum value of x is 1[54,55)].

iii) The use of an intercalation material that undergoes dimensional changes on charge and discharge necessitates a flexible and deformable matrix within the cathode compartment, and the poor electronic conductivity of the fully intercalated material requires the addition of carbon. The matrix is usually provided by the ionically conducting polymer electrolyte, typical proportions for the cathode layer being 50% electrolyte, 45% active cathode material (intercalation compound) and 5% carbon mass, a small amount of surfactant being added to aid wetting and thus promote interfacial contact[56)].

iv) The anode usually takes the form of a thin Li foil, and in general, compatibility problems between Li metal and the polymer electrolyte do not arise. Battery systems based on Li/Li_3N anodes have also been examined[57)]. Both the anode and the electrolyte are moisture sensitive, and fabrication requires a dry box, dry room or other environment controlled to <1% relative humidity. The current collectors are usually Ni foil.

v) The electrolyte was originally a "solution" of a lithium salt of a soft anion such as perchlorate, $LiClO_4$, or triflate (trifluoromethanesulphonate) $LiCF_3SO_3$, in poly(ethylene oxide), PEO[58)]. This should be referred to by its more systematic name poly(oxyethylene), although the abbreviation PEO is specifically approved by IUPAC[59)] and it is typical to express the electrolyte composition in terms of a $[O]:[M^+]$ ratio, n, in the form $PEO_n:LiX$ or $PEO:(LiX)_{1/n}$. The repeat unit of the polymer "solvent" is $\{CH_2\text{-}CH_2\text{-}O\}$.

There are some major problems with the electrolyte in its original form. Both the perchlorate and the triflate are semi-crystalline polymers and consist of an amorphous phase region in which are embedded spherulitic, semi-crystalline phases of pure PEO and/or a low n, high melting complex of PEO and lithium salt. The relative properties of the amorphous region and the crystalline phases depend to some extent on the method of preparation and the thermal history. The concentration of salt dissolved in the amorphous region depends not only on the original amounts of reactants used, but also on the amounts of polymer and/or salt locked up in the crystalline phases. This is important because the conductivity occurs only in the amorphous region, and the value of σ is a strong function of the salt concentration within that region, being higher at high values of n, i.e. low salt concentrations. The conductivity does not reach 10^{-6} S cm^{-1} until the temperature is increased above the melting point of pure PEO ∿65 °C. For the perchlorate system above n = 7, no high melting complex is present[60] and the electrolyte is all in the high conductivity amorphous form at an operating temperature of ∿100 °C. For the triflate system, small amounts of high melting complex appear to be present[61] even at high n. The crystalline phases adversely affect the deformability as well as the conductivity, although they do provide mechanical rigidity and resistance to creep or flow.

Clearly, a viable room temperature system depends on maximising the amorphous nature of the electrolyte. In addition, conductivity is enhanced if the glass transition temperature of the polymer/salt solid solution is as low as possible[62]. A large number of alternative polymers have been examined in the quest for a technologically viable material[63-69], but in many cases low Tg and an avoidance of crystallinity has been achieved at the expense of poor mechanical properties. Recently, however, it appears that the introduction of a carefully controlled degree of irradiation-induced cross-linking, and/or the selection of appropriate anions, has enabled conductivities of the order of 10^{-4} S cm^{-1} to be obtained at room temperature in polymer electrolytes suitable for battery use. Chemical cross-linking leading to network polymers also shows great promise[70].

A further disadvantage of polymer electrolytes is that the measured anionic and cationic transport numbers are about the same[71]. The technological effect of this is uncertain; the cycling behaviour of the cells provides no evidence for involvement of mobile anions in the cell reaction, and the positive electrode is not an anion source. The positions that the anions take up within the amorphous polymer are not known at present, and experiments designed to locate them are handicapped by the relatively high anion mobility. This makes it difficult to employ a strategy in which scientific understanding of the behaviour of a material leads to consciously designed modifications of structure and/or composition in order to yield improved technological performance.

A direction in which great interest is currently being shown is to utilize conducting polymers[72] such as polypyrrole as the electroactive cathode component. Recently a composite cathode consisting of globules of polypyrrole dispersed in a $PEO_n:LiClO_4$ matrix has been examined; it shows high coulombic and energy efficiency but charge retention is poor because of a tendency to self-discharge[73].

In summary, the 1990's are likely to see the introduction of Li anode/Li polymer electrolyte/intercalation cathode solid state battery systems for a wide range of applications. They will operate at room temperature at ~3 V, >1 mA cm^{-2} and be of large cross-sectional area. They will use very small particle size material in the cathode region, to minimise film thickness. It is likely that techniques from the printing industry such as screen and ink jet printing will replace the doctor blade techniques used for the production of prototype thin film batteries.

2.2 Other Polymer Batteries

Two main alternatives to Li have been considered; Na and divalent metals. As early as 1981, Chiang[74] proposed an all-polymeric solid state battery in which the electrolyte was PEO:NaI and the electrodes were doped polyacetylene. The anode was sodium doped, i.e. $[Na_y{}^+(CH^{y-})]_x$ and the cathode was triiodide-doped, i.e. $[(CH^{y+})(I_3{}^-)_y]_x$. The current collectors were graphite paste spread on Pt plates. An open circuit voltage in the range 2.8-3.5 V could be obtained and although the long term cycling performance was poor, short-term charge-discharge curves revealed a power density of 10-20 W kg^{-1} and short-circuit current densities as high as 1 mA cm^{-2} could be obtained. A somewhat different Na/polyacetylene system, although using β"Na-Al_2O_3, has also been reported[75]. A monovalent alternative to sodium, namely silver, has also been used in conjunction with polyacetylene in a solid state battery system[76].

Later studies from Denmark[77] showed the feasibility of a Na/PEO: NaI/MoO_3 rechargeable battery which could be recycled at 15 μA cm^{-2}. The operating temperature range was between the melting point of crystalline PEO, 65 °C, and the sodium anode, 98 °C. Since adequate electrolyte performance at that time could only be achieved at ~100-120 °C, and Na-based solid state batteries cannot be run above 98 °C, it would be useful to re-examine an Na/polymer electrolyte system based on an amorphous polymer that is highly conducting at room temperature.

Divalent batteries based on Mg, Ca and Sr anodes have also been proposed[78,79]. The extremely low transport numbers of these "hard" cations, together with the presence of moisture in the battery components, has led to suggestions that the cell reaction in fact involves anion and/or proton conduction, especially for the magnesium system. This is countered by evidence of increasing Mg content in the cathode as discharge proceeds[79]. The high anionic conductivity in divalent polymer electrolytes, which makes them promising candidates for certain sensor applications, does not appear yet to have been exploited in power sources.

3. COMPACTED POWDER BATTERIES

Various configurations of small button cells, in which three layers of powder are pressed together to form a small disc, have been explored for low current density, ambient temperature, predominantly primary battery applications. Although other anodes have been considered, attention has been mainly focussed initially on Ag and subsequently on Cu systems. This is because AgI and CuI both have high temperature α-phase structures in which, unusually for the solid state, conductivity is comparable with that of dilute salt solutions, i.e. approaching 1 S cm^{-1}. As reviewed elsewhere in this volume, room temperature conductivity is enhanced above the value, $\sim 10^{-6}$ S cm^{-1}, for the natural γ room temperature phase by the addition of structural modifiers such as organic or inorganic iodides. A paradigm material is Ag_4RbI_5, and it and its many "descendants" such as $Cu_4RbCl_3I_2$ ($\sigma_{25\,^oC}$ = 0.47 S cm^{-1}, the highest room temperature conductivity for a solid state material) have formed the electrolyte, and also an electrode compartment matrix, in many solid state batteries. Organic-substituted alternatives based on Owens discovery[80] of the electrolyte $Ag_{13}[NMe_4]_2I_{15}$ have recently been reviewed by Linford[9].

In all compacted powder battery systems, a non-trivial problem is to minimise interparticle ionic impedance. Additional difficulties include

i) the thermodynamic instability of certain materials[81]), e.g. Ag_4RbI_5 below 27 °C;

ii) the deleterious effect of moisture;

iii) in copper systems, the discharge-induced structural modifications of the electrolyte[82]) that could account for anomalously low voltages and long induction periods.

3.1 Silver Batteries

Silver batteries typically produce voltages of ~0.8 V at room temperature and can sustain current densities of the order of ~100 μA cm^{-2}. The cost of both the anode metal and the electrolyte salts is a major factor restricting their wider application.

Results for early silver-based battery systems were first reported twenty years ago[83-91]). The electrolytes were Ag_3SI (which is in fact a mixed conductor), Ag_4RbI_5 or thin film ASgI, and in one case[91]), a reversible battery was proposed.

Throughout the 1970's and 1980's there has been considerable activity, especially in finding better cathode materials to replace the corrosive and volatile elemental iodine used in the early systems. Typical cathode materials are summarized in table 1. The majority of early cells used Ag_4RbI_5 as an electrolyte, sometimes in a colloidal form[117]), but a range of alternatives have been developed. These include Ag_4MI_5 where M is ammonium[100,102,118]) or potassium[93,118]); organic halides[96,97,104,119,120]); AgI with inorganic salts such as Ag_3SBr[103]), 4AgI-Ag_2X where X is chromate[121]), tungstate[101,121]) or molybdate[121]). Another major issue that has been addressed is that of long term storage life. Early predictions of a 10 year life[122,123]) were checked after five[94,105]) and ten[124]) years and the life span of alternative cathodes has also been examined[106]).

A recent battery system, difficult to classify in that a silver glass is powdered and subsequently compacted into a disk, has been described by Akridge[125,126]). It is in prototype production.

3.2 Copper Batteries

These systems have excited considerable interest because copper,

Table 1: Cathode Materials for Compacted Powder Ag Batteries

Cathode Material	Reference	Year
Elemental I_2	87	1969
	92	1973
	93	1976
	94	1978
Iodine Complexes		
I_2-perylene	95	1973
	96	1976
	97	1976
	98	1976
I_2-pyrene	99	1978
I_2-graphite	100	1985
I_2-phenothiazine	101	1986
Polyhalides		
KI_3	102	1978
$AuBr_{2.4}$	103	1973
$[NR_4]I_x$	91	1969
	104	1970
	105	1970
	106	1984
	107	1987
Ferrocenium Polyiodide	108	1983
	109	1983
Metals		
Pt, Pd	110	1971
	111	1980
Au	112	1981
Others		
$(CH)_x$	113	1985
WO_3	114	1974
Se	115	1970
Te	115	1970
	116	1975
	111	1980

like silver, is relatively free from environmental drawbacks, and can easily be handled with conventional battery technology; it is also a substantially cheaper material. As discussed earlier however, low voltage and long induction periods are two manifestations of an uncertain cell chemistry. Their chemistry is based on Cu^+, and the tendency of cuprous ions to disproportionate to copper metal and cupric ions leads to a third drawback, the likelihood of significant electronic conduction[127)].

Copper solid state batteries have recently been reviewed[128-131)]. Their development has benefitted from many of the lessons learnt from the closely analogous Ag^+ systems. Thus, it became relatively easy to focus on one particular chosen electrolyte, $Cu_{16}Rb_4I_{7-x}Cl_{13+x}$, although variants have also been used, as shown in table 2.

Table 2: Electrolyte materials for Cu cells

Material		Reference	Year
1.	$Cu_{16}Rb_4I_{7-x}Cl_{13+x}$	132	1982
		133	1983
		134	1984
		135	1986
		136	1986
2.	$Cu_{16}Rb_4I_6Cl_4$	137	1987
3.	$Cu_4RbI_2Cl_3$	138	1983
4.	CuI-sulphonium iodides	139	1978
5.	CuBr-N-methylhexamethylene-tetramine bromide	140	1979
6.	CuI-CuCl	141	1986

Some early primary batteries used organic polyiodides or I_2-perylene[131,139)] as cathode materials, and with Ag systems, metals[141)] and other alternatives such as Cu_2Se/Se couples[132,134)] have also been employed. The dominant cathode by far, however, is the intercalation compound TiS_2, because it permits the formation of a rechargeable system.

The Cu/TiS_2 intercalation, together with the less successful Ag/TiS_2 process, was first demonstrated by Scrosati in 1978[142)]. He and co-workers also developed[140)] a rechargeable concentration cell based on

$$Cu_{0.4}TiS_2/\text{electrolyte 5, table 2}/Cu_{0.8}TiS_2$$

and the Cu/TiS_2 system has been extensively studied and

developed[133,135,136,137]. Problems of voltage stability over long periods have been addressed[135] and the effect of moisture stability of the electrolyte on battery performance has also been considered[138]. Rechargeable systems based on the alternative cathode material, copper thiomolybdate, have also been reported[143].

3.3 Other Battery Systems

An interesting sodium-based, room temperature battery Na/Nasicon/MoO_2Cl_2 has recently been described[144]. Some compacted powder Li systems have also been considered, for example Li/Li_3N-LiI-$Ni(CO)_4$/PbI_2-$Ni(CO)_4$[145], and other Li_3N-based batteries[146]. A very curious Li-based system, albeit one that works at 500 °C rather than room temperature, has been described by West[147]. It is initially of the form Au/Li_2TiO_3/Au, but the gold layers act as current collectors, the anode and cathode being the decomposition products Li_2CO_3 and TiO_2 respectively.

An alternative monovalent conducting species is H^+, and a recent development is the battery TiNi:Misch metal/Sb_2O_5,$4H_2O$/γ-MnO_2 or β-PbO_2, where the anode is an alloy that absorbs hydrogen[148]. A much earlier attempt to construct a solid state proton galvanic cell utilized a quite different electrolyte, the potassium salt of a square-planar platinium complex[149]. Divalent ions such as Mg[150,151] and Ca[152,153] have also been used, although in the latter case, the mobile species is F^- rather than Ca^{2+}. The same is also true for the Pb/PbF_2/BiF_3-Br cell[154].

4. THIN FILM AND GLASS ELECTROLYTE CELLS

The electrolyte in these cells can be formed from a cast layer of polymer or glass, or by vapour or electro-deposition. Polymeric batteries have already been discussed above and their fabrication has been described[56].

Some glassy electrolyte batteries, involving compacted powder technology, have been considered in the previous section[125,126] and the performance of Li/2.5LiI:$Li_4P_2S_7$/TiS_2 batteries has recently been described in further detail[154]. Other battery systems involving

Li_2S-based vitreous electrolytes have recently been reviewed[155]. These differ from the later Eveready battery system in various ways. For example, the anode can be LiAl in place of Li[156]; B_2S_3 can replace P_2S_5 in the electrolyte composition[157]; oxide rather than sulphide glasses can be used[111,158]; and Li can be replaced by Na[157] or Ag[159]. Thin film Ag^+-conducting batteries based on AgI[93,112] or Ag_4RbI_5[102,114,160] have also been described. Chemical and electrolytic stability problems in vitreous electrolyte batteries have recently been discussed[161].

An alternative approach to thin film, solid state batteries is to be seen in such systems as the Pb/$PbCl_2$-AgCl/Ag battery prepared by vacuum deposition[160] and the reversible Pb/$PbCl_2$-BF_3/Bi battery[162].

5. THE Li-I_2 HEART PACEMAKER BATTERY

As long ago as 1970, lithium batteries of the form Li/LiI/I_2 were suggested[163]. A major problem with such systems is the very poor room temperature conductivity of LiI, 10^{-7} S cm^{-1}, and a further difficulty was caused by the high volatility of the corrosive I_2 cathode. The latter handicap could be overcome by the replacement of I_2 by AgI[164].

At almost the same time, a very curious but conspicuously successful battery system was described by Greatbatch et al.[165]. The problem of I_2 volatility was addressed by the use of a proprietary iodine-containing cathode and the conductance of the electrolyte was optimised by the use of an extremely thin-layer of LiI. The latter was formed *in situ*, and the cell contained no electrolyte at the assembly stage! This battery system is often considered as the major, indeed almost the only, success in the commercialization of solid state battery systems.

The proprietary cathode is a charge transfer complex of I_2 with poly(2-vinylpyridine), P2VP, and the characteristics of the battery have been reviewed by Owens[16,166], the leading worker associated with this system.

A range of alternative cathode formulations have been tried, including I_2-benzidine[167] which is prone to moisture attack, I_2-polyacrylonitrile[168] which gives higher cell capacity and I-doped polyacetylene[169]. Long term reliability, which is of vital importance for implantable prosthetics such as heart pacemaker batteries, has been assessed[170,171]. Alternative uses such as watch batteries for this power source have been described[172] and a D-cell battery configuration of 14 mAh capacity, capable of delivering 10 mA at room temperature and 500 mA at 130 °C has been claimed[172].

Recently, evidence has been presented[173] that the effective electrolyte in the battery is in fact a liquid decomposition product of P2VP and this regrettably means that the leading solid-state primary battery system is in fact not a true solid state system.

6. HIGH TEMPERATURE SULPHATE BATTERIES

Mixed sulphates act as conductors of both mono- and di-valent cations, and this property has been exploited in battery systems such as $Mg/Li_{1.76}Mg_{0.12}SO_4/MnO_2$[174]. A power density of 400 W kg^{-1} could be obtained at 745 °C. For some mixed sulphates, conductivities of 10^{-3} S cm^{-1} were obtained at room temperature, and systems involving Ag anodes could in fact operate at room temperature without the need for precautions to remove moisture.

Further work by Lunden and his group on Mg[174], Ag[174], Li[175] and Zn[176] batteries has recently been reviewed[177].

7. ELECTROLYTE MODIFICATION FOR BATTERY USE

Electrolyte properties can be modified by the addition of such materials as plasticisers or fillers. For example, rubber particles can be added to polymer electrolytes derived from PEO which, together with a cross-linking process, help to impede the crystallization that could otherwise reduce the conductivity[52]. Bentonite has been used in Cu and Mg batteries[150], and a special procedure has been suggested to modify the form of Ag_4RbI_5 used in thin film cells[117].

A particular class of electrolyte modified by non-conducting filler is the so-called "composite electrolyte" in which the conducting properties are significantly enhanced. The reasons for this increased conductivity are under active debate, the suggestion of some form of space-charge effect having recently been challenged by an alternative explanation based on mechanical deformation[178]. The performance of batteries incorporating the LiI/Al_2O_3 electrolyte developed by Liang has been described[179]. For a Li/LiI(Al_2O_3)/TiS_2 cell, 48% of cell capacity could be utilized over a 4.4 year period at an average current density of 2.6 μA cm^{-2}, and energy densities of the order of 1 Wh cm^{-3} have been anticipated[179]. Developments in composite electrolytes have recently been reviewed [180-183].

8. CONCLUSIONS

At present, there are few commercially available solid state battery systems. The most mature and best known existing example, the Li-I_2 heart pacemaker battery, does not in fact fall within the strict definition given in the introduction. Within a short time, however, the commercial availability of several other systems, especially Li polymer batteries perhaps together with Li/LiI:$Li_4P_2S_7$/TiS_2 vitreous batteries and certain Cu-based batteries, can be anticipated with reasonable confidence.

9. REFERENCES

1. Linford, R.G., editor, "Electrochemical Science and Technology of Polymers, 1" (Elsevier Applied Science, London), 1987.
2. Vincent, C.A., Bonino, F., Lazzari, M. and Scrosati, B., "Modern Batteries" (Edward Arnold, London), 1984.
3. Smith, P.J., "Perfluorinated Ionomer Membranes for Use in the Production of Chlorine and Caustic Soda", ref. 1, pp. 293-334.
4. Linford, R.G. and Hackwood, S., "Physical Techniques for the Characterisation of Solid Electrolytes", Chemical Reviews, <u>81</u>, 395-423.
5. Latham, R.J. and Linford R.G., "Ionic and Electronic Transport", ref. 1, pp.1-21.
6. Ratner, M.A., "Ion Transport, Free Volumes, Dynamic Percolation and Ion Interactions in Polymer Electrolytes", Brit. Polymer J. (1988), in press.
7. Coetzer, J., "Sodium-Iron Chloride Secondary Cells. An Interpretation of the Reaction Mechanism", Solid State Ionics (1988), in press.
8. Takahashi, T. and Kozawa, A., editors, "Applications of Solid Electrolytes" (JEC Press, Cleveland), 1980.

9. Linford, R.G., "Applications of Solid State Ionics for Batteries", Solid State Ionics, 1988, in press.

10. Munshi, M.Z.A., and Owens, B.B., "Solid State Batteries", Proc. Electrochem. Soc. 87-3, 116-126 (1987).

11. Chowdari, B.V.R. and Radhakrishna, S., editors, "Materials for Solid State Batteries" (World Scientific Publishing, Singapore), 1986.

12. Sequeira, C.A.C. and Hooper, A., editors, "Solid State Batteries" (NATO ASI Ser. E 101: Martinus Nijhoff, Dordrecht), 1985.

13. Hooper, A., "Solid-State Rechargeable Batteries", ref. 12, pp. 399-410.

14. Sequeira, C.A.C., "The d.c. Methods of Cell Characterization: Parts I and II", ref. 12, pp. 219-260.

15. Steele, B.C.H., "Thermodynamic Aspects of Solid State Battery Performance", ref. 12, pp. 93-107.

16. Owens, B.B., "Solid State Primary Batteries", ref. 12, pp. 389-398.

17. Kudo, T., "Recent Topics in Solid State Ionics in Japan", New Mater, New Processes 3, 84-86 (1985).

18. Owens, B.B. and Scrosati, B., "Solid State Batteries", Proc. Int. CODATA Conf., pp. 323-326 (1984).

19. Hooper, A. and Tolfield, B.C., "All-Solid-State-Batteries", J. Power Sources 11(1-2), 33-41 (1984).

20. Sekido, S., "Solid State Micro Power Sources", Solid State Ionics 9-10, 777-781 (1983).

21. Scrosati, B., Padula, A. and Lazzari, M., "Recent Developments in Electrochemical Energy, V", Chem. Ind. (Milan) 65(6), 415-422 (1983). (In Italian. Chemical Abstracts 99(18): 143024r)

22. Gabano, J-P., editor, "Lithium Batteries" (Academic, London) 1983.

23. Shahi, K., Wagner, J.B. and Owens, B.B., "Solid Electrolyte Lithium Cells", in ref. 22, pp. 407-448.

24. Liang, C.C., "Solid Electrolytes and Solid State Batteries", Chemtech 13(5), 303-305 (1983).

25. Velasco, G., Schnell, J.P., Nouailles, N., Croset, M., "Advanced Microelectronics Technology for Solid-State Ionic Devices: Sensors Displays, Capacitors and Batteries", Solid State Protonic Conduct. 1 Fuel Cells Sens., Dom-Fr. Workshop "Solid State Mater. Low Medium Temp. Fuel Cells Monit., Spec. Emphasis Proton Conduct.", 27-40 (1982). Edited by Jensen, J., Kleitz, M. CA 98(24): 208069k.

26. Huggins, R.A., "Solid Electrolytes", in Proceedings of the 1980 International Symposium on Electrochemical Industry, edited Landau, V., Yeager, E. and Kortan, D. (Plenum, New York), pp. 195-213 (1982).

27. Lin, Z., "Solid State Ionics and its Energy Related Applications", Ziran Zazhi 5(8), 593-596 (1982). (In Chinese, Chemical Abstracts 98(4): 19368w.)

28. Jensen, J., "Project on Electrochemical Energy Storage Systems", NTIS reports DE82-900845, NP-2900845, order no. N82-29770, pp. 1-83 (1981). (In Danish. Chemical Abstracts 98(10): 75368h.)

29. Li, M., "Solid Electrolyte Batteries", Wuli 10(12), 705-711 (1981). (In Chinese. Chemical Abstracts 97(6): 41376y.)

30. Owen, J.R., "Developments in Solid State Batteries", Power Sources 8, 77-89 (1981).

31. Kennedy, J.H., "Primary Batteries - Solid Electrolytes", in Comprehensive Treatise on Electrochemistry, edited Bockris, J.O'M., Conway, B.E. and Yeager, E. (Plenum, New York) 3, 279-302 (1981).

32. Skarstad, P.M. and Owens, B.B., "Application of Solid Electrolytes in Ambient Temperature Lithium Batteries", Proc. Electrochem. Soc. (Proc. Workshop Lithium Nonaqueous Battery Electrochem.) 80-7, 276-292 (1980).

33. Chandra, S. and Agrawal, R.C., "Solid-State Battery - Prospects and Limitations", in Golden Jubilee Commen. Vol. Natl. Acad. Sci. India, ed. Srivastava, U.S. and Naya, P., pp. 429-462 (1980).

34. Murphy, D.W., Broadhead, J. and Steele, B.C.H., editors, "Materials for Advanced Batteries" (Plenum, New York), 1980.

35. Dickinson, T., "Electrochemical Power Sources, Primary and Secondary Batteries, Room Temperature Cells with Solid Electrolytes", IEE Energy Ser. (Electrochem. Power Sources) 1, 464-481 (1980).

36. Collognes, R. and Colomban, P., "Solid-Electrolyte Storage Batteries", Recherche 11(108), 201-202 (1980). (In French. Chemical Abstracts 92(12): 101277a.)

37. Lazzari, M. and Scrosati, B., "A Review of Silver, Copper and Lithium Solid-State Power Sources", J. Power Sources 1(14), 333-358 (1977).

38. Dell, R.M., "The Application of Solid State Ionics to Batteries", NATO Advanced Study Int. Ser. C. C25, 387-410 (1976).

39. Feuillade, G., "Solid Electrolyte Electrochemical Generators", Rev. Gen. Electr. 84(6), 485-490 (1975.

40. Liang, C.C., "Solid State Batteries", Appl. Solid State Sci. 4, 95-135 (1974).

41. Liang. C.C., "Solid State Battery Systems", in Fast Ion Transport Solids, ed. Van Gool, W. (N-Holland, Amsterdam), pp. 19-31 (1973).

42. Gibson, J.G., "Solid State Batteries", J. Appl. Electrochem. 3(3), 251 (1973).

43. Scrosati, B., "Recent Advances in Lithium Solid State Batteries", J. Appl. Electrochem. 2(3), 231-238 (1972).

44. Liang, C.C., "Storage and Discharge Characteristics of a Lithium Solid State Battery System", in Proceedings of the Inter-Society Conference on Energy Conversion Engineering (Society of Automotive Eng., New York), pp. 673-676 (1971).

45. Hull, M.N., "Recent Progress in the Development of Solid Electrolyte Batteries", Energy Conversion 10(4), 215-226 (1970).

46. Hooper, A., "Solid-State Rechargeable Lithium Batteries", Chem. Ind. (London) (6), 198-201 (1986).

47. Hooper, A. and Gauthier, M., in "Electrochemical Science and Technology of Polymers 2", ed. Linford, R.G. (Elsevier Applied Science, London), 1988, in press.

48. Hooper, A., Lundsgaard, J.S. and Owen, J.R., "The Fabrication and Testing of Solid State Cells", Comm. Eur. Communities Report EUR 8660, Energy Conservation, Energy Storage, Advanced Batteries and Fuel Cells, 53-67 (1984).

49. Tolfield, B.C., Dell, R.M. and Jensen, J., "Rechargeable Solid-State Lithium Batteries", Comm. Eur. Communities Report EUR 9236, Energy Conservation Ind. 2, 198-209 (1984).

50. Hooper, A. and Tolfield, B.C., "Solid State Rechargeable Lithium Batteries", Prog. Batteries Sol. Cells 5, 64-68 (1984).

51. Hooper, A., North, J.M., Oliver, M. and Tolfield, B.C., "Solid State Rechargeable Lithium Batteries", Proc. 31st Power Sources Symposium, 120-124 (1984).

52. Le Mehaute, A., Crepy, G., Marcellin, G., Hamaide, T. and Guyot, A., "Polymer Electrolytes. The Route Towards Solid Polymer Electrolytes with Stable Electrochemical Performance upon Long Term Storage", Polymer Bull. 14(3-4), 233-237 (1985).

53. Bonino, F., Ottaviani, M., Seluaggi, A. and Scrosati, B., "A Polymeric Electrolyte Rechargeable Battery", Solid State Ionics (1988), in press.

54. Gauthier, M., Fauteux, D., Vassort, G., Belanger, A. and Duval, M. "Behaviour of Polymer Electrolyte Batteries at 80-100 °C and near Room Temperature", J. Power Sources 14(1), 23-26 (1985).

55. Gauthier, M., Fauteux, D., Vassort, G., Belanger, A., Duval, M., Ricoux, P., Chabagno, J.M., Muller, D., Rigoud, P. Armand, M.B. and Deroo, D., "Assessment of Polymer-Electrolyte Batteries for EV and Ambient Temperature Applications", J. Electrochem. Soc., 132(6), 1333-1345 (1985).

56. Hooper, A. and North, J., "The Fabrication and Performance of all Solid State, Polymer Based Rechargeable Lithium Cells", Solid State Ionics 9-10, 1161-1166 (1983).

57. Knutz, B. and Skaarup, S., "Cycling of Lithium Nitride, Titanium Disulfide Solid State Cells", Solid State Ionics 9-10, 371-374

(1983).

58. Armand, M., "Polymer Solid Electrolytes - An Overview, Solid State Ionics, 9-10(II), 745-754 (1983).

59. Linford, R.G., "Polymer Structures and Nomenclature" in Information Sources in Polymers and Plastics, ed. Adkins, R.T. (Butterworths, London), 1988, in press.

60. Neat, R.J., Hooper, A., Glasse, M.D. and Linford, R.G., "A Structural Model for the Interpretation of Composition Dependent Conductivity in Polymeric Solid Electrolytes" in Transport-Structure Relations in Fast Ion and Mixed Conductors (Risø National Laboratories, Roskilde, Denmark), pp.341-346 (1985).

61. Neat, R., Glasse, M., Linford, R. and Hooper, A., "Thermal History and Polymer Electrolyte Structure: Implications for Solid State Battery Design, Solid State Ionics 18&19, 1088-1092 (1986).

62. Owen, J.R., "Ion Conducting Polymers", ref. 1, pp. 45-66 (1987).

63. Vincent, C.A., "Ion Conducting Polymers", Prog. Solid State Chem. 17, 145-261 (1987).

64. Mrha, J., Novak, P. and Riha, J., "Conducting Polymers in Electrochemical Power Sources. III. Polymers as Solid Electrolytes and their Application in Secondary Cells", Chem. Listy 80(3), 274-285 (1986). (In Czech. Chemical Abstracts 105(2): 14037r.)

65. Marcoussis, F., "Solid Polymer Electrolytes", NTIS Report, order no. PB85-233856/GAR, 112 pp (1985). (In French. Chemical Abstracts 104(26): 227705y).

66. Le Mehaute, A., "Ambient Temperature Polymer Solid State Batteries", ref. 12, pp. 527-530.

67. Scrosati, B., "Electrode Kinetics and Electrochemical Cells", in Polymer Electrolyte Reviews, ed. MacCallum, J.R. and Vincent, C.A. (Elsevier Applied Science, London), 315-345 (1987).

68. Cameron, G.C., Ingram, M.D., MacCallum, J.R. and Vincent, C.A., "Proceedings of the 1st International Symposium on Polymer Electrolytes, St. Andrews" (1987).

69. Bonino, F. and Scrosati, B., "Polymeric Electrolytes", ref. 11, 41-52 (1986).

70. Watanabe, M., Nagano, S., Sanui, K. and Ogata, N., "Structure-Conductivity Relationship in Polymer Electrolytes formed by Network Polymers from Poly[dimethylsiloxane-g-poly(ethylene oxide)], J. Power Sources 20(3-4), 327-332 (1987).

71. Sorensen, O.T. and Jacobsen, T., "Conductivity, Charge Transfer and Transport Number - an A.C. Investigation of LiSCN-PEO", Electrochimica Acta 27(12), 1671-1675 (1982).

72. MacDiarmid, A.G. and Maxfield, M., "Organic Polymers as Electroactive Materials", ref. 1, 67-102.

73. Novak, P., Inganeaes, O, and Bjorklund, R., "Composite Polymer Positive Electrodes in Solid State Lithium Secondary Batteries, J. Electrochem. Soc. 134(6), 1341-1345 (1987).

74. Chiang, C. K., "An All Polymeric Solid State Battery", Polymer 22(11), 1454-1456 (1981).

75. Beniere, F., Boils, D., Canepa, H., Franco, J., Le Corre, A., Louboutin, J.P., "Polyacetylene Solid-State Batteries", J. Electrochem. Soc. 139(2), 2100-2108 (1985).

76. Beniere, F., Boils, D., Canepa, H., Franco, J., Le Corre, A. and Louboutin, J.P., "Solid State Batteries with Conducting Polymers", J. Phys. Colloq. (C3, Conf. Int. Phys. Chim. Polym. Conduct. 1982), 567-572 (1983).

77. West, K., Zachau-Christiansen, B., Jacobsen, T. and Atlung, S., "A Rechargeable All-Solid-State Sodium Cell with Polymer Electrolyte", J. Electrochem. Soc. 132(12), 3061-3062 (1985).

78. Cherng, J.Y., Munshi, M.Z.A., Owens, B.B. and Smytl, W.H., "Applications of Multivalent Ionic Conductors to Polymeric Electrolyte Batteries", Solid State Ionics, 1988, in press.

79. Patrick, A., Glasse, M., Latham, R. and Linford, R., "Novel Solid State Polymeric Batteries", Solid State Ionics 18&19, 1063-1067, (1986).

80. Owens, B.B., "A New Class of High Conductivity Solid Electrolytes: Tetraalkylammonium Iodide-Silver Iodide Double Salts", J. Electrochem. Soc., 117(12), 1536-1539 (1970).

81. Owens, B.B., "Applications of Halogenide Solid Electrolytes", in Solid Electrolytes, ed. Geller, S., Top. Appl. Phys. 21, 67-104 (1977).

82. Linford, R.G., Hall, P.G., Johnson, C. and Hasnain, S.S., "Changes in Electrolyte Structure in Solid State Cells During Discharge: Extended X-ray Absorption Fine Structure Study", Solid State Ionics 14, 199-203 (1984).

83. Hull, M.N., "Solid Electrolyte Batteries", Proc. Annu. Power Sources Conf. 22, 106-109 (1968).

84. Argue, G.R., Owens, B.B. and Groce, I.J., "Solid State Batteries", Proc. Annu. Power Sources Conf. 22, 103-105 (1968)

85. Scrosati, B., Boddy, P.J. and De Rossi, M., "Solid-State Battery with $RbAg_4I_5$ as Electrolyte", Ric. Sci. 38(12), 1366-1368 (1968).

86. Foley, R.T., "Solid-Electrolyte Galvanic Cells", J. Electrochem. Soc. 116(1), 161-170 (1969).

87. Owens, B., Argue, G.R., Groce, I.J. and Hermo, L.D., "Stability of the Solid State Cell $Ag/Ag_3SI/I_2$", J. Electrochem. Soc. 116(2) 312-313 (1969).

88. Vouros, P. and Masters, J.I., "Thin-Film Silver Halide Electrolyte Cells", J. Electrochem. Soc. 116(6), 880-884 (1969).

89. Liang, C.C. and Bro, P., "High-Voltage Solid State Battery Systems. I. Design Considerations", J. Electrochem. Soc. 116(9), 1322-1323 (1969).

90. Liang, C.C., Epstein, J. and Boyle, G.H., "High Voltage Solid State Battery Systems. II. Fabrication of Thin Film Cells", J. Electrochem. Soc. 116(10), 1452-1454 (1969).

91. De Rossi, M., Pistoia, G. and Scrosati, B., "Reversible Solid-State Battery with Rubidium Silver Pentaiodide as Electrolyte", J. Electrochem. Soc. 116(12), 1642-1646 (1969).

92. Nissen, D.A. and Carlsten, R.W., "Solid Electrolyte Thermal Battery", NTIS Report SC-RR-720386, 37 pp (1973): CA 79(24): 142210y.

93. Mohabey, V.K. and Chandra, S., "Solid State Battery using a Superionic Solid Film of KAg_4I_5", Proc. Nucl, Phys. Solid State Phys. Symp. 19C, 47-48 (1976).

94. Warburton D.L., Bis, R.F. and Owens, B.B., "Five Year Storage Tests of Solid State Silver/Iodide Batteries", Proc. 28th Power Sources Symp., 106-109 (1978).

95. Scrosati, B., "Electrochemical Properties of $RbAg_4I_5$ Solid Electrolytes. III. Chargeable Cells", J. Electrochem. Soc. 120(1), 78-80 (1973).

96. Linford R.G., Pollock, J.M. and Randell, C.F., "Serch for New Silver-Ion Solid Electrolytes for use in Batteries", Nature 259, 656-657 (1976).

97. Linford, R.G., Pollock, J.M. and Randell, C.F., "New Silver Iodide -bis- Sulphonium Diiodide Solid Electrolytes for Solid State Battery Systems", in "Power Sources 6", ed. Collins, D.H. (Academic Press, London), 511-526 (1977).

98. Chao, P-C, Lu, W-J, and Tzeng, J-S., "Studye of the $Ag/RbAg_4I_5/I_2$-Perylene Solid State Battery", Ts'ai Liao K'o Hsueh 8(4), 127-149 (1976).

99. Baranek, B.I., Faivovich, R.D., Perissinotti, L. and Franco, J.I., "Solid State Cell using Silver Iodide as Electrolyte", An Assoc. Quim. Argent. 66(1), 47-56 (1978).

100. Agrawal, S.S., Tomar, M.S. amd Srivastave, A.P., "Electrochemical Cell with Solid $NH_4Ag_4I_5$ Electrolyte", Bull. Electrochem. 1(5), 4901-4904 (1985).

101. Sunthanthiraraj, S.A., "Polarization and Discharge Characteristics of an All-Solid-State Battery Fabricated with $Ag_6I_4WO_4$ Solid Electrolyte", in ref. 11, 423-428 (1986).

102. Chandra, S., Agrawal, R.C. and Pandey, R.K., "A Thin Film Solid State Battery using Superionic $NH_4Ag_4I_5$", Natl. Acad. Sci. Lett. (India) 1(13), 112-114 (1978).

103. Kennedy, J.K., Chen, F. and Miles, R.C., "Solid-State Reserve

Battery", J. Electrochem. Soc. 120(2), 171-182 (1973).

104. Smyth, D.M., Tomkins, C.H. Jr. and Ross, S.D., "Solid Electrolyte Batteries with Modified AgI Electrolyte", Proc. 24th Power Sources Symp., 24-26 (1970).

105. Owens, B.B., Untereker, D.F. and Skarstad, P.M., "Low Temperature Degradation in the Solid State Battery $Ag/RbAg_4I_5/I_2$" in Fast Ion Transport Solids, ed. Vashiskta, P., Mundy, J.N. and Shenoy, G.K. (Elsevier North Holland, Amsterdam), 105-107 (1979).

106. Ernst, D.W., De Vries, L.E., De Graba, M.E. and Warburton, D.L., "Performance of the $Ag/RbAg_4I_5/Me_4NI_x$ Battery after Ten Years Storage", Proc. 31st Power Sources Symp., 271-276 (1984).

107. Mikhailova, A.M., Ukshe, E.A., "Electrochemical Circuits with Solid Electrolytes in the Silver-Complex Iodide Electrode System", Elektrokhimiya 23(5), 685-688 (1987). (In Russian. Chemical Abstracts CA 107(8): 51939n.)

108. Antonin, V., "Processes in $Ag-I_2$ Electrochemical Cell Electrodes" Telsa Electron 16(4), 119-122 (1983): CA 101(12): 100127u.

109. Antonin, V., "Primary Galvanic Cell with Solid Electrolyte", Tesla Electron. 16(2), 47-51 (1983): CA 101(6): 45382w.

110. Mrgudich, J.N. and Louzos, D.V., "Semidry and Solid Electrolyte Batteries", in "Primary Battery 1", ed. Heise, G.W. (Wiley, New York), 373-432 (1971).

111. Arora, M.R. and Giani, E., "Thin Films of Solid Electrolyte $Ag_{19}I_{15}P_2O_7$. II. Thin Film Electrolyte Cells", Thin Solid Films 71(1), 103-110 (1980).

112. Phadke, S.D., "Solid Electrolyte Thin Film Battery", Indian J. Phys. 55A(4), 331-334 (1981).

113. Beniere, F., Boils, D., Canepa, H., Franco, J., Le Corre, A. and Louboutin, J.P., "Polyacetylene Solid-State Batteries", J. Electrochem. Soc., 139(2), 2100-8 (1985).

114. Green, M. and Richman, D., "Solid State Electrochromic Cell $RbAg_4I_5/WO_3$ System", Thin Solid Films 24(2), S45-S46 (1974).

115. Takahashi, T. and Yamamoto, O., "Solid Ionics-Solid Electrolyte Cells", J. Electrochem. Soc. 117(1), 1-5 (1970).

116. Ivanov, V.E., Danilov, A.V. and Aleskovskii, V.B., "Discharge Characteristics of $Ag/Ag_4RbI_5/Te$ Cells in the Temperature Range from -50 to +150 ^oC", Elektrokhimiya 11(9), 1418-21 (1975).

117. Bodnaruk, L.I., Danilov, A.V., Kulinkovich, V.E. and Aleskovskii, V.S., "Thin Film Galvanic Cell Containing a $RbAg_4I_5$ Solid Electrolyte", Electrokhimiya 11(9), 1325-7 (1975).

118. Owens, B.B., "Solid Electrolyte Batteries", Advan. Electrochem. Electrochem. Eng. 8, 1-62 (1971).

119. Linford, R.G., Pollock, J.M. and Randell, C.F., "Solid State

Batteries Containing Silver Sulphonium Iodide Electrolyte", Nature 256, 398-9 (1975).

120. Mikhailova, A.M. and Ukshe, E.A., "Electrochemical Circuits with Solid Electrolytes in the Silver-Complex Iodine Electrode System", Elektrokhimiya 23(5), 685-688 (1987). (In Russian. Chemical Abstracts CA 107(8): 51939n.)

121.. Chiodelli. G., Magristris, A. and Schiraldi, A., "Solid Electrolyte Cells", Electrochim. Acta 19(10), 655-656 (1974).

122. Oxley, J.E. and Owens, B.B., "Solid State Batteries", in Power Sources 3, ed. Collins D.H. (Oriel Press, Newcastle upon Tyne). 535-549 (1971).

123. Owens, B.B., Sprouse, J.S. and Warburton, D.L., "Storage Life of the Solid-State Battery $Ag/RbAg_4I_5/I_2$", Power Sources Symp. Proc. 25, 8-11 (1972).

124. Owens, B.B., Patel, B.K., Skarstad, P.M. and Warburton, D.L., "Performance of $Ag/RbAg_4I_5/I_2$ Solid Electrolyte Batteries after 10 Years Storage", Solid State Ionics 9-10, 1241-1245 (1983).

125. Akridge, J.R. and Vourlis, H., "Solid State Batteries using Vitreous Solid Electrolytes", Solid State Ionics 18-19, 1082-1087 (1986).

126. Akridge, J.R. and Vourlis, H., "Li/TiS_2 High Temperature Solid State Batteries using Vitreous Phosphorus Chalcogenide Electrolytes", Solid State Ionics (1988), in press.

127. Jelfs, A.M.P., Johnson, C., Latham, R.J. and Linford, R.G., "Self-discharge in Solid State Cells", J. Power Sources 13, 115-122 (1984).

128. Yamamoto, O., Takeda, Y. and Kanno, R., "Solid Electrolyte Cells with Copper Ion Conductors", in ref. 11, 275-285 (1986).

129. Yamamoto, O., "Solid Copper Conductors" in ref. 11, 263-273 (1986).

130. Kukoz, F.I. and Skalozubov, D.M., "Electrochemical Systems with a Copper-Conducting Solid Electrolyte", Izv. Sev-Kavk. Nauchn. Tsentra. Vyssh. Shk., Tetch. Nauki 4, 10-13 (1984). (Chemical Abstracts 105(14): 122756h.)

131. Linford, R.G. and Hackwood, G., "CuI Electrolytes and their use in Miniature Battery Systems", Chem. and Ind., 523-526 (1980).

132. Takahashi, T., Kuwabara, K., Miura, M. and Nakanishi, M., "Solid Electrolyte Cell using CuCl-CuI-RbI", J. A pp. Electrochem. 12(2), 213-218 (1982).

133. Kanno, R., Takeda, Y., Imura, M. and Yamamoto, O., "Rechargeable Solid-Electrolyte Cells with a Cu^+ Conductor of $Rb_4Cu_{16}I_{7-6}Cl_{3+6}$ and a TiS_2 Cathode", J. Appl. Electrochem. 12(6), 681-685 (1982).

134. Kuwabara, K., Yoshizak, H. and Sugiyama, K., "Single and Multi-

layer Solid-Electrolyte Cells with a Cu^+ Conductor", J. Appl. Electrochem. 14(3), 373-377 (1984).

135. Kuwabara, K. and Sugiyama, K., "An Investigation of the Solid-State Cell $Cu/Rb_4Cu_{16}I_7Cl_{13}/Cu_xTiS_2$", J. Appl. Electrochem. 16(1), 23-27 (1986).

136. Kanno, R., Takeda, Y., Oda, Y., Ikeda, H. and Yamamoto, O., "Rechargeable Solid Electrolyte Cells with $Rb_4Cu_{16}I_{7-x}Cl_{13-x}$", Solid State Ionics 18-19, 1068-72 (1986).

137. Sotomura, T., Kambara, T. and Sekido, S., "Copper Solid State Secondary Battery", Prog. Batteries Sol. Cells 6, 25-29 (1987).

138. Turkovic, A., Grzeta-Plenkovic, B. and Etlinger, B., "Stability of the $RbCu_4Cl_3I_2$ Cell", Nuovo Cimento Soc. Ital. Fis. D. 20(6), 1977-1981 (1983).

139. Dahm, R.H., Hackwood, S., Linford, R.G. and Pollock, J.M., "Cu^+ Solid State Battery Systems Involving Sulphonium Iodide Electrolytes", Nature, 272, 522-524 (1978).

140. Scrosati, B., Voso, M.A. and Lazzari, M., "A Rechargeable Copper Solid-State Cell Based on Intercalation Electrodes", J. Electrochem. Soc. 126(4), 699-700 (1979).

141. Yu, W., Zhu, B. and Zheng, Q., "Thin Film Batteries with bi-Metal Electrodes and Solid State Electrolytes", in ref. 11, 481-485 (1986).

142. Bottini, A., Lazzari, M., Razzini, G., Rivolta, B., De Felici, G. Voso, M.A. and Scrosati, B., "Intercalation of Ag and Cu in TiS_2 from Solid State Cells", J. Electroanal. Chem. Interfacial Electrochem. 96(2), 165-173 (1978).

143. Sekido, S. and Tonomura, T., "Secondary Solid State Batteries", Jpn. Kokai Tokkyo Koho, Patent No. 87 47973 020387 A2 (1987).

144. Dennstedt, W., "Sodium-Molybdenum Dioxide Dichloride Solid State Batteries for Room Temperature Use", Dechema-Monegi 109, 421-429 (1987). CA. 107(22): 202048k.

145. Nagai, R., Wada, S. and Kawakami, A., "Solid Electrolyte Batteries using Li_3N-LiI", Prog. Batteries Sol. Cells 5, 69-72 (1984).

146. Bittihn, R., "Self Discharge of Li_3N-based All Solid State Cells" Solid State Ionics 8(1), 83-88 (1983).

147. Aceves, J.M. and West, A.R., "Electrochemical Decomposition of Lithium Silicate and Lithium Titanate in Solid-State Thermal Cells", J. Chem. Soc. Faraday Trans. 1 78(9), 2599-2608 (1982).

148. Yamamoto, O., Takeda, Y., Kanno, R. and Tone, K., "Solid Electrolyte Cell with a Solid Proton Conductor $Sb_2O_5.4H_2O$", ref. 11, 475-479 (1986).

149. Lecrone, F.N. and Perlstein, J.H., "Proton Injection in the Highly Conducting Mixed-Valence Square Planar Complex

$K_{1.64}Pt(C2/4)_2.xH_2O$. New Solid-State Proton Galvanic Cell", J. Chem. Soc. Chem. Commun. (2), 75-76 (1972).

150. Wang, D., Yu, W., Yuan, W. and Zhu, B., "The Application of Bentonite in Solid State Ionic Batteries", in ref. 11, 461-466 (1986).

151. Johnson, C., Latham, R.J. and Linford, R.G., "Room Temperature $Mg-I_2$ Solid State Cells", Solid State Ionics, 7, 331-334 (1982).

152. Koriath, M. and Braeutigaam, R., "Galvanic Cells with Doped CaF_2 as a Solid Electrolyte", report BMFT-FB-T-79-03, NTIS (1979). (Chemical Abstracts CA 93(6): 50468z.)

153. Baukal, W., Kuhn, W. and Voss, E., "High Temperature Rechargeable Cells with Doped CaF_2 as a Solid Electrolyte", Argonne Natl. Lab. Report, ANL-76-8, Proc. Symp. Workshop Adv. Battery Res. Des. B252-262 (1976).

154. Akridge, J.R. and Vourlis, H., "Solid-State Batteries", Prog. Batteries Sol. Cells 6, 81-88 (1987).

155. Gabano, J.P., "Applications of Glasses in All Solid-State Batteries. A Review", NATO ASI Ser 92, 457-480 (1985).

156. Malugani, J.P., Fahys, B., Mercier, R., Robert, G., Duchange, J.P., Baudry, S., Broussely, M. and Gabano, J.P., "New Glassy Li^+ Conductors and their use in Batteries", Solid State Ionics 9-10, 659-665 (1983).

157. Levasseur, A., Delmas, C., "Study of New Solid Electrolytes and Associated Cathodes. Study and Design of New All-Solid Batteries", Comm. Eur. Communities, 1405, 67 pp. (1984). CA 102(14): 116564r.

158. Di Pietro, B. and Scrosati, B., "A Lithium Solid State Cell Based on $Li_{3.75}Si_{0.25}O_4$ Electrolyte", J. Power Sources 2(4), 387-388 (1978).

159. Minami, T., "Solid State Batteries of Ag^+ Conducting Glasses", ref. 11, 169-179 (1986).

160. Mogil'nikskii, A.A., Presnov, V.A. and Kostyuk, N.P., "Solid-State Thin-Film Batteries for Microelectronics" in Issled. Obl. Khim. Ionnykh Rasplavov Tverd. Elektrolitov, 18-30 (1985). (Chemical Abstracts 103(24): 198507g.)

161. Tuller, H.L. and Barsoum, M.W., "Glass Solid Electrolytes: Past, Present and Near Future - the Year 2004", J. Non-Cryst. Solids 73(1-3), 331-350 (1985).

162. Danto, Y., Poujade, G., Pistre, J.D., Lucat, C. and Salardenne, J., "$Pb/PbF_2-BiF_3/Bi$ Thin Solid Film Reversible Galvanic Cell", Thin Solid Films 55(3), 347-354 (1978).

163. Schneider, A.A., Moser, J.R., Webb, T.H.E. and Desmond, J.E., "New High-Energy Density Solid Electrolyte Cell with a Lithium

Anode", Power Sources Symp. Proc. 24, 27-430 (1970).

164. Liang, C.C., "Self-Discharge Mechanism of the Li/LiI/AgI solid Electrolyte Cell", J. Electrochem. Soc. 118(6), 894-895 (1971).

165. Greatbatch, W., Lee, J.H., Mathias, W., Eldridge, M., Moser, J.R. and Schneider, A.A., "Solid State Lithium Battery. A New Improved Chemical Power Source for Implantable Cardiac Pacemakers", IEEE Trans. Bio-Med. Eng. 18(5), 317-324 (1971).

166. Owens, B.B., "The Role of Solid Electrolytes in Lithium Pacemaker Batteries", Solid State Ionics 3-4, 273-275 (1981).

167. Scrosati, B. and Torroni, M., "Lithium-Iodine Solid State Battery", Electrochim. Acta 18(2), 225-226 (1973).

168. Wiaux, J.P. and Bannehr, R., "Performance of Polyacrylonitrile-based Cathodes in Lithium-Iodine Solid State Batteries", J. Power Sources 19(1), 337-43 (1987).

169. Scrosati, B. and Owens, B.B., "Solid-State Lithium/Polyacetylene Battery", Solid State Ionics 23(4), 275-278 (1987).

170. Kraus, F.E. and Schneider, A.A., "The Lithium-Iodine Cell: A High Reliability Long-Life Solid Electrolyte System", Power Sources Symp. Proc. 27, 144-147 (1976).

171. Brennen, K.R., Fester, K.E., Owens, B.B. and Untereker, D.F., A Capacity Rating System for Cardiac Pacemaker Batteries", J. Power Sources 5(1), 25-34 (1980).

172. Schneider, A.A., Harney, D.E. and Harney, M.J., "The Lithium-Iodine Cell for Medical and Commercial Applications", J. Power Sources 5(1), 15-23 (1980).

173. Phipps, J.B., Hayes, T.G., Skarstad, P.M. and Untereker, D.F., "In-situ Formation of a Solid/Liquid Composite Electrolyte in Lithium-Iodine Batteries", Solid State Ionics 18-19, 1073-7 (1976).

174. Heed, B., Lunden, A. and Schroeder, K., "Electrochemical Cells with Sulphate-Based Solid Electrolytes", Proc. 8th Int. Symp. Reactivity Solids, ed. Wood, J., Lindqvist, O. and Helgesson, C. (Plenum, New York), 713-718 (1977).

175. Aronsson, R., Heed, B., Jansson, B., Lunden, A., Nilsson, L., Schroeder, K., Sjoeblom, C.A., Thomaas, J.O. and Tolfield, B.C., in Fast Ion Transport in Solids, ed. Vashiskta, P., Mundy, J.N. and Shenoy, G.K. (Elsevier, North Hollad, Amsterdam), 471-474 (1979).

176. Heed, B. and Lunden, A., "High Temperature Batteries with a Solid Sulfate Electrolyte", J. Power Sources 10(1), 103-104 (1983).

177. Lunden, A., "Sulfate Based Solid Electrolytes and their Applications in Batteries, Sensors and Heat Storage", in ref. 11, 149-1060 (1986).

178. Dudney, N.J., "Enhanced Ionic Conduction in Composite Electrolytes", Solid State Ionics (1988), in press.

179. Rea, J.R., Barnette, L.H., Liang, C.C. and Joshi, A.V., "A New High Energy D ensity Solid State Battery System", Proc. Electrochem. Soc. 80-4, 245-53 (1980).

180. Khandkar, A.C., "Fast Ion Transport in Composite Electrolytes" Diss. Abstra. Int. B. 1986, 46(8), 2768 (1985): Univ. Microfilms Int. DA 8522946.

181. Poulsen, F.W., "Composite Electrolytes", in Transport-Structure Relations in Fast Ions and Mixed Conductors, ed. Poulsen, F.W. (Risø Nat. Lab., Roskilde, Denmark), 67-78 (1985).

182. Poulsen, F.W., "The Effect of Different Internal Surfaces in Composite Lithium Electrolytes", J. Power Sources 20(3-4), 317-25 (1987).

183. Maier, J., "Composite Electrolytes", Mater Chem. Phys. 17(5), 485-498 (1987).

HIGH ENERGY BATTERIES WITH SOLID SODIUM ION CONDUCTIVE ELECTROLYTES

Wilfried Fischer

Brown,Boveri & Cie. AG
D-6900 Heidelberg, Germany

ABSTRACT

Three high energy/high power rechargeable battery types with sodium ion conducting solid electrolytes are presently under development : (1) Na/beta" alumina/S-batteries, (2) Na/glass/S-batteries and (3) Na/beta"-alumina/MCl_x-batteries.

Type (1) batteries are most advanced. Many batteries with about three times the energy to weight ratio than that of lead acid batteries have been successfully subjected to bench tests, EV-propulsion tests and load leveling tests. Results of these tests will be presented. Further development is directed towards increasing life and reducing cost.

Despite type (2) batteries do have the same performance potential than type (1) batteries, technical problems turned out to be more severe. Development concentrates on solving electrolyte and cell issues.

A relatively fast progress in type (3) battery development (MCl_x = $FeCl_2$ and $NiCl_2$) could be achieved by using knowledge accumulated with type (1) batteries. Potential advantages and disadvantages compared to type (1) batteries will be discussed. A few not yet optimized batteries have been used to propel EV's.

1. INTRODUCTION

The performance of conventional secondary electrochemical storage devices such as the lead acid battery (G. Planté 1859) and the nickel (or iron)/cadmium battery (W. Junger, Th. Edison, ca. 1900) have been considered to be sufficient for at least most of the applications during many decades. The situation has changed for the following reasons :

- Concerns about finite mineral oil resources have grown within the last twenty years. Hence, alternatives for gasoline operated automobiles are required.
- Air pollution and noise problems within our urban areas must be resolved. Substitution of combustion engine operated cars by electric vehicles could be an important contribution.
- Dependency from foreign primary energy resources could be reduced by using battery storage devices instead of gas turbines for generation of electric peak power.

These new needs caused search for rechargeable, less expensive batteries with a higher energy to weight and volume ratio than conventional batteries[1, 2]. With these aspects in mind, Kummer and Weber at Ford Motor Company[3] and independently Levine and Brown at Dow Chemical Company[4] tried to realize a Na/S cell in the early sixties. The most difficult problem was to find a solid electrolyte with a high sodium ion conductivity and stability against the agressive molten reaction materials. Finally, the Dow group decided to use glass[4] and the Ford group to use beta alumina[3].

Glasses of certain compositions were well known as sodium ion conductors, however, conductivity is low and chemical resistance against liquid sodium is poor. Consequently, development was directed towards improving these properties and assembling cells with thousands of glass capillaries.

In contrast to glass, beta alumina was not known to be an ionic conductor before Na/S cell development was commenced. First (1916) it was thought to be an alumina crystal type[5]. Sodium to be an essential constituent was established in 1931[6]. Speculations whether sodium ions are mobile were made in 1956[7], however, experimental evidence was gained not earlier than 1967[8]. This was the starting point for the development of Na/S cells at Ford[3]. Since that time Na/beta alumina/S cell and battery development has been persued by groups in Canada, France, Germany, Great Britain, Japan, the Peoples

Republique of China and the United States.

A review on the state of development of Na/S cells and batteries both with beta alumina and glass electrolytes will be given below. Cells and batteries in which the sulphur electrode is replaced by a $NiCl_2$ or $FeCl_2$ cathode (in order to reduce the operation temperature) will be included. Development of these systems was started in the late 1970's[9).

The major emphasis of this report is to review the recent literature in this field. For those interested in an earlier state of development reference is made to reviews on sodium ion conducting electrolytes[10-14) and sodium sulphur batteries[15, 16) published some years ago. Baisc questions will be discussed only to an extent necessary to understand recent results.

2. NA/BETA ALUMINA/S CELLS

Amongst the cells with sodium ion conducting solid electrolytes most effort was devoted towards developing Na/beta alumina/S cells. Hence, the state of development of this type of cell (and battery) is by far more advanced than that of other systems to be discussed in chapters 3 and 4.

2.1 Cell Function And Configuration

Fig. 1 shows a schematic of a fully charged Na/S cell. Major components are molten sodium (anodic reactand), molten sulphur (cathodic reactand), the beta alumina electrolyte membrane separating the two reactands, and the cell casing serving as current collector and container for the reactands.

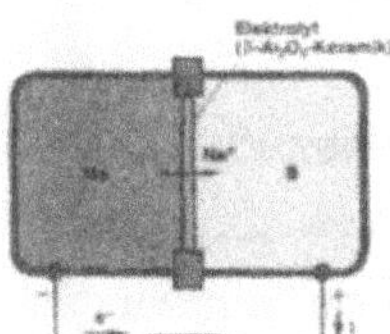

Fig. 1 : Schematic of a Na/S cell[17)

Sodium ions migrate through the electrolyte membrane during discharge and react with sulphur. The overall reaction is

$$2\ Na + xS \rightarrow Na_2S_x\ .$$

In order to maintain contact between the solid electrolyte and the electrode materials, Na, S and Na_2S_x must be kept above their melting points. Na_2S_4 has the highest melting point (285°C), see fig. 2. Therefore, the operation temperature must be > 285°C.

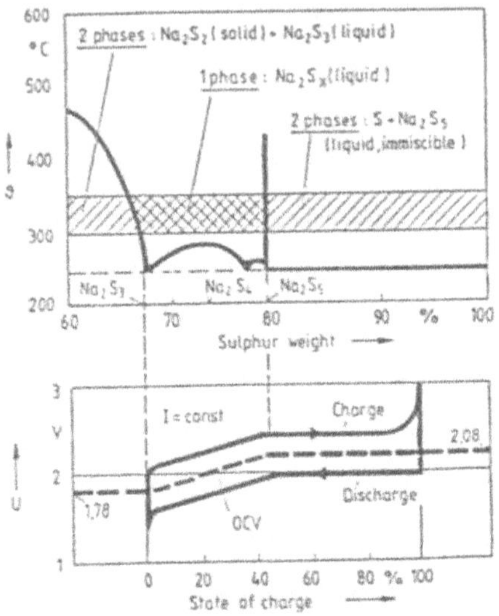

Fig. 2 : Simplified Na - S phase diagram[18] and charge/discharge curve of a Na/S cell at 350°C[15]

Two immiscible liquids are formed during discharge from 0 to about 60% discharge (see fig. 2). The EMF is constant within this two phase region. Further discharge results in formation of a single phase melt. The composition of this phase changes from $Na_2S_{5.2}$ to $Na_2S_{2.7}$ and the EMF declines from 2.076 V to 1.78 V when discharged to approximately 107%. Cells can be overdischarged, see fig. 2. In this region liquid Na_2S_3 and solid Na_2S_2 are coexisting and the EMF remains at 1.78 V.

The reactions during discharge are principally reversible. Only two restrictions have to be mentioned : Kinetics are slow at a depth of discharge > 100% and recharging a cell to pure sulphur is not possible. Effects related to this question will be discussed in paragraph 2.4.

Practical cells have a shape different from that shown in fig. 1. In spite of cells with disk shaped electrolytes have been built for test purposes[19, 20], cylindrical cells with tubular electrolytes are preferred from all developers. Reasons are that cylindrical cells are more tolerant to mechanical stresses and that hermetic seals are easier to realize. Multitubular cell designs have been considered[21, 22], but are not practically used.

Tubular Na/S cells can be subdivided into sodium core cells and sulphur core cells. Two typical cells are shown in fig. 3.

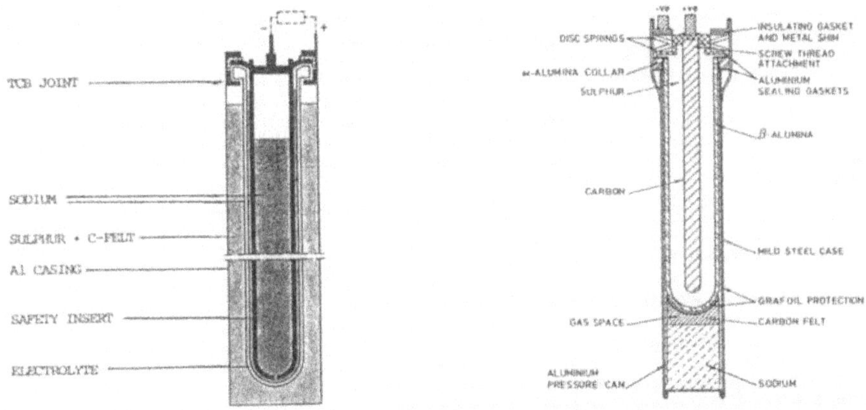

Fig. 3 : Sodium core cell[17, 23] (left)
and sulphur core cell[24] (right)

The major advantage of sulphur core cells is that the positive central current collector can be provided easier with a corrosion protection layer than the casing of a sodium core cell. At the other hand, it has a lower energy to weight ratio. Table 1 shows that developers have come to different conclusions with regard to the importance of these arguments. Fig. 4 shows components of a sodium core cell[23].

Table 1 : Capacity and aimed application of Na/beta alumina/S cells from different developers

Developer	Type	Design*	Capacity (Ah)	Application**	Ref.
BBC, FRG and Powerplex, Canada	A04	Na	42	EV + LL	17), 23)
Beta Research, UK (earlier Brit. Rail)	B30	S	100	EV	24)
CSPL, UK	NaS 7	S	150	EV	25), 26), 27)
	PB	Na	10	EV	
	XPB	Na	35	LL	
Ford, USA***	EVEM-1	Na	40	EV	28), 29)
	MARK II	Na	130	LL	
Hitachi, Jap.	-	Na	600	LL	30)
Yuasa + NGK Sparkplug, Jap.	-	Na	300	LL	31), 32)

* Na = sodium core cell, S = sulphur core cell

** EV = electric vehicle, LL = load leveling

*** activities terminated

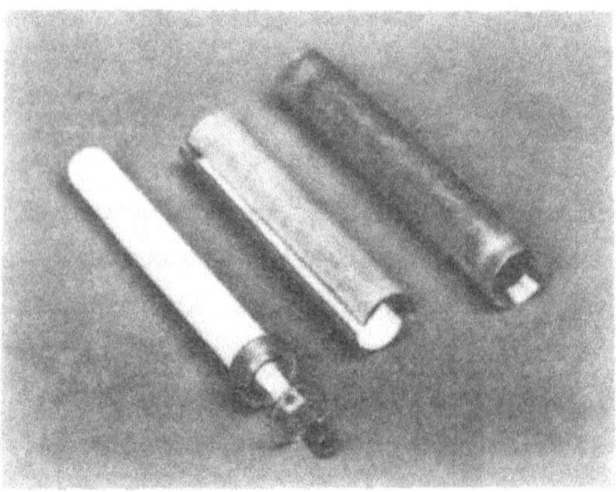

Fig. 4 : Sodium core cell components from left to right : solid electrolyte tube filled with Na and sealed, prefabricated sulphur electrode half shells, cell casing

Dimensions of cells are dictated by the kind of application (discharge time, available space), by manufacturability and by the strategy in which way to interconnect cells within a battery. A detailed discussion of these aspects follows later. Table 1 gives an overview of the various cell sizes (10 - 600 Ah) used from the different groups. Generally, electric vehicle cells are smaller in size (diameter) than load leveling cells. The difference reflects the difference in discharge time (2 - 3 hours for electric vehicle application, 4 - 8 hours for load leveling application). Large diameter cells cannot be discharged as fast as small diameter cells, since they contain more reactands per unit electrolyte surface. At the other hand, these considerations can be counterbalanced by aspects such as to use a large number of parallel strings for battery reliability reasons (see 5.1).

2.2 The Solid Electrolyte

2.2.1 Crystal structure and conductivity

Beta alumina, since its discovery as a fast sodium ion conductor[3], is the only ceramic electrolyte for Na/S cells. Attempts to replace beta alumina by Nasicon - a three dimensional sodium ion conductor with the composition $Na_{1+x}\ Zr_2\ P_{3-x}\ Si_x\ O_{12}$ ($0 \leq x \leq 3$) and a high conductivity[33, 34] - were not successful. The material turned out to be not stable against sodium at 350°C[35].

The unique properties of sodium beta alumina originate from its crystal structure. Several crystal forms do exist. Since β"-alumina has a higher conductivity, all developers use materials with a high β"-alumina content. The crystal structure of undoped alumina is shown in fig. 5. The material has the chemical composition

$$(1+x)\ Na_2O \cdot 11\ Al_2O_3\ .$$

Dopants such as lithia, magnesia or a mixture of these oxides in concentrations of several percent are added (a) to stabilize the β"-alumina structure and (b) to allow for additional x Na_2O and to increase the conductivity[37].

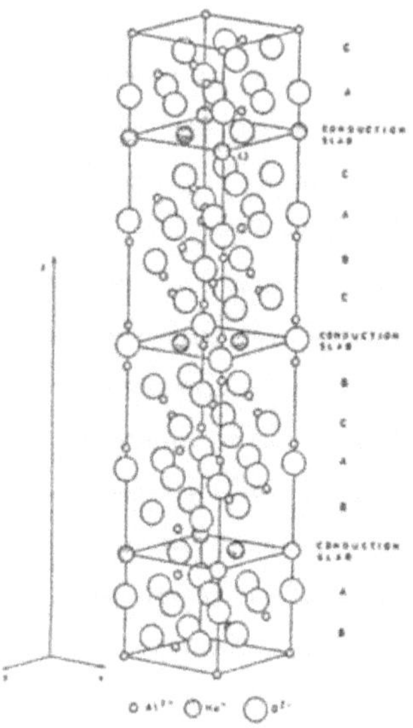

Fig. 5 : Perspective drawing of the idealized structure of β"-alumina[15)]

Beta aluminas conduct only in the conduction plane, not in the direction of the c-axis. The single crystal conductivity in the direction of the conduction plane is 1 ohm^{-1} cm^{-1} for β"-alumina and 0.3 ohm^{-1} cm^{-1} for β-alumina[38)]. Beta alumina ceramic tubes are polycristalline. The conductivity is smaller than that of single crystals since (a) microcristallites (direction of the conduction planes) are randomly distributed forcing current to make detours and (b) grain boundaries add to resistance. A typical cross section of beta alumina showing directions and dimensions of micro cristallites is shown in fig. 6.

Fig. 6 : Typical microstructure of beta alumina

2.2.2 Degradation

For most of the applications a battery life of at least five years and a cycle life of at least 500 cycles is required. Electrolyte properties should remain unchanged during this time.

It has been shown that a charge of several thousand Ah/cm^2 can be passed through electrolytes at current densities as high as 1 A/cm^2 in Na/Na cells[39]. This is by a factor ten more than needed. Nevertheless - partly because of more severe conditions in Na/S cells - degradation is observed. Two degradation types can be distinguished : (a) Progressive increase of electrolyte and cell resistance. This effect will be discussed in the next paragraph. (b) Development of an internal short after a number of successful cycles caused by a sudden electrolyte rupture.

The latter effect has been intensively discussed in the literature. Several failure mechanisms have been blamed as a possible cause. The most important mechanisms are :

- o Microcracks at the anodic beta alumina surface are filled with sodium. Charging results in sodium formation $Na^+ + e^- \rightarrow Na$ at the tip of the microcrack and to crack propagation into the electrolyte as the result of pressure caused by sodium flow out of the crack. While the first two-dimensional calculations led to a general understanding of the phenomenon but not to a quantitative agreement with the critical current density[*40], later three-dimensional calculations achieved full accordance with tests[41]. It can be concluded that the mechanism described is one of the major failure causes, especially since local current densities may be much higher than the average current density for three reasons : (a) poor sodium wetting of the electrolyte[42] leading to current concentration at the boarder line of wetted and non wetted areas[43], (b) inhomogeneous graphite felt distribution within the sulphur

* No electrolyte degradation below, but very fast degradation above critical current density

electrode compartment[44] and (c) variation in Na_2S_x composition because of gravitational effects[45].

- o Partial reduction of the electrolyte by interaction with sodium gives rise to an electronic/ionic transport number gradient across the electrolyte and to progressive deposition of sodium metal[46]. Discussion is still going on whether or not this failure mode does practically occur.
- o Intergranular attack of the electrolyte from the sulphur side leads to cracks and spallation[47]. Causes for this effect are not yet understood.

Evidence for the above mentioned failure types has in many cases been gained from electrolytes that have been operated in Na/S cells for hundreds or thousands of cycles[47]. It may be concluded (a) that an electrolyte life sufficient for practical applications can be achieved and (b) further improvements (more appropriate composition and manufacturing procedure) to be necessary for avoiding early failures.

2.2.3 Manufacturing methods

The most important requirements that determine the practical applicability of beta alumina for Na/S batteries are high conductivity, long life, high strength, tight geometrical tolerances and low cost. These quality figures depend on a number of material properties such as composition, impurity level, grain size distribution, β"-/β-alumina ratio, intergranular phases, macroscopic defects etc. They are highly dependent on the production process. Furthermore, a strong correlation exists between production parameters and the raw material used. Hence, both production procedure and raw materials are important for electrolyte quality. Various electrolyte compositions and production processes (table 2) have been used.

Advanced production routes compared to standard methods in table 2 have been investigated. The investigations include sol gel routes for better mixing, seeding with β"-alumina cristallites in order to accelerate transformation into β"-alumina, forming by extrusion or slipcasting etc. Furthermore, in process control procedures are in development. Improved production and quality control methods will certainly lead to electrolytes with an even higher quality.

Table 2 : Standard beta alumina production processes

Company	BBC	Beta Research	Ceramatec	CSPL/GE
Type of β-alumina	Mg-stabilized β"	Li-stabilized β"	Li-stabilized β"	Li (+ Mg)-stabilized β"
a) Powder Preparation	o Calcine MgO + Al_2O_3 → $MgAl_2O_4$		o Calcine $LiNO_3$ + Al_2O_3 → $LiAl_5O_8$	
	o Mix Al_2O_3 + Na_2CO_3+$MgAl_2O_4$	o Make slurry from Al_2O_3 + NaOH + LiOH	o Mix Al_2O_3 + Na_2CO_3+$LiAl_5O_8$	o blend raw materials
	o Calcine + grind + make slurry		o Calcine + grind + make slurry	o calcine and mill
	o Spray dry	o Spray dry	o Spray dry	
b) Forming	o Half automatic isostatic dry bag pressing	o wet bag isostatic pressing	o wet bag isostatic pressing	o make suspension, charge particles (e.g. vibro-milling in amyl alcohol)
				o electrophoretic deposition
				o remove from mandrel + evaporate volatile constituents
c) Sintering	o Gas furnace double peak batch sintering in MgO crucibles	o Electric furnace double peak batch sintering in MgO crucibles	o bisquing*) remove volatile constituents	o rapid unencapsulated zone sintering
			o electric furnace batch sintering in Pt-lined Al_2O_3-crucibles	o annealing
d) References	48)	5)	49)	50) - 52)

*) pre-temperature-treatment at about 600°C to remove binders

2.3 The Sodium Electrode

Both performance and safety of Na/S cells are essentially depending on fabrication parameters and on the design of the sodium electrode.

2.3.1 Performance

The complete electrolyte surface must be kept in contact with liquid sodium during all states of charge/discharge. Methods used are (a) Na feed by gravity from a reservoir, (b) capillary action and (c) inert gas pressure. The methods are simple, nevertheless difficulties have been encountered.

A first problem is poor wetting of sodium on beta alumina and on other construction materials (wicks). Investigations in several laboratories show temperatures of 300°C and more to be necessary to achieve contact angles ≤ 90°C[42, 53]. The most common way to attain sodium wetting on sodium compartment materials is to operate cells with small current at temperatures ≥ 300°C for some time. Decline of cell resistance from a higher initial to a lower final resistance indicates wetting to have occured.

A second more serious problem is sodium/electrolyte interfaces to exhibit considerable polarisation during discharge under certain conditions. Cell resistance has been observed to be higher in this case on discharge than on charge. The effect has been designated as asymmetry effect. Typical asymmetric voltage current curves are shown in fig. 7[54].

The cause for asymmetric cell behaviour has shown to be calcium impurities for instance leached out from the electrolyte and oxidized at the electrolyte sodium interface[54-56]. The most straight forward way to avoid asymmetry and reduced power as a consequence, is to reduce calcium in the electrolyte raw materials to a level lower than about 50 ppm[54, 55]. Earlier assumptions that a high concentration of lithium[15] or water pick up at the electrolyte[57] would cause asymmetry seem to be of minor importance.

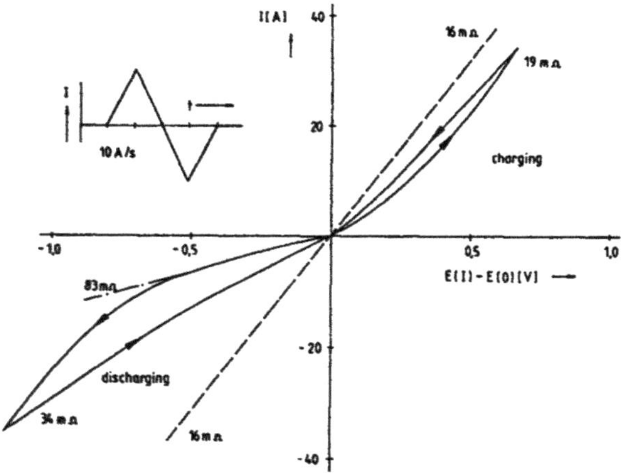

Fig. 7 : Polarisation curve of a asymmetric (——) and symmetric (- - - -) NaS cell

2.3.2 Cell safety

One of the most important features of the sodium electrode is to minimize the amount of sodium in direct contact with the electrolyte. Generation of heat by chemical reaction in case of electrolyte rupture is limited in this way. Two practical solutions have been applied : (a) Utilization of a safety insert with a sodium flow restriction hole and (b) keeping sodium within a wick.

Two cells equipped with sodium electrodes of the first type are shown in fig. 3. The sodium compartment of both cells is divided into a small volume (annulus) in contact with the electrolyte and into a larger volume serving as sodium reservoir. The subcompartments are connected by a small hole allowing for a sodium flow high enough for fast discharge but restricting sodium flow and chemical reaction in case of electrolyte rupture. Flow of sodium during discharge is maintained either by gas pressure or by capillary action.

The sodium compartment of cells of the second type is simply filled with a wick[58]. Temperature rise after electrolyte rupture is higher than that of cells of the first type, but still within limits necessary to keep cells leaktight.

A generally accepted safety test is to deliberately fracture the electrolyte tube by overcharging the cell to overvoltages of about 30 V. The temperature rise after such an event is an indicator for cell safety. Fig. 8 represents the result of a safety test with

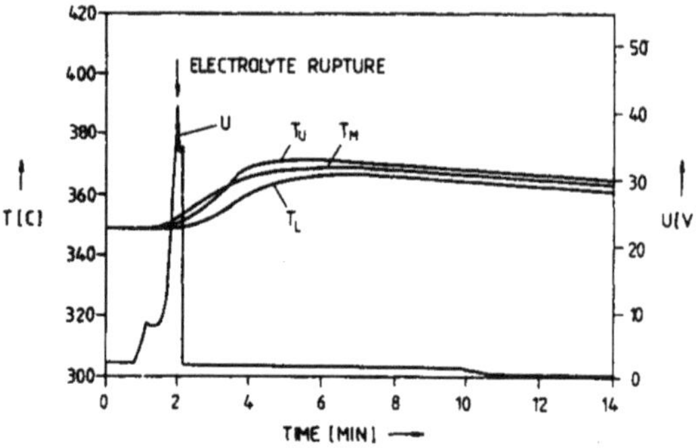

Fig. 8 : Temperature at the outside of the cell case T_o and cell voltage V versus time during a safety test of a Na/S cell (fig. 3, left). Charging current 10 A

a cell shown at the left hand side of fig. 3. The diagram reveals ΔT to remain low ($\sim$ 30°C). Sodium electrode design changes have resulted in reducing the initially high ΔT values (several 100°C) to these low ΔT values[24, 59]. Safety optimized cells remain leaktight after electrolyte rupture.

2.4 The Sulphur Electrode

As outlined in paragraph 2.1 and as a consequence of the sodium sulphur phase diagram (fig. 2), operation of Na/S cells is limited to the range $x > 2.7$ (of Na_2S_x) and $T > 285°C$[*)]. The sulphur electrode has partly to be operated in the one phase region ($2.7 \leq x \leq 5$) and partly in the two phase region ($x \geq 5$). Major areas of development were (a) to find appropriate electronically conducting materials at which the electrochemical reactions can take place and (b) to achieve a high rechargeability.

(a) Since both sulphur and sodium polysulfides are electronic insulators, a corrosion resistant material with a high electronic

*) Overdischarge to $x < 2.7$ leading to formation of solid Na_2S_2 is normally avoided, however, principally possible.

conductivity and a high surface area is required. Carbon or graphite fibers in form of a felt or a mat extending over the whole sulphur compartment have turned out to fulfill these requirements. The sulphur and/or the polysulfides are soaked into the pores of these fibrous materials. The electrochemical reactions take place at the phase boundary graphite/sulphur. They are distributed over the whole sulphur electrode if high resistive felts (resistivity higher than that of the polysulfide melt) are utilized. This follows from theory[60-63] and from experiments. Consequently, rechargeability as well as cell resistance are high.

(b) Sulphur electrodes with a low resistive felt, however, cannot be charged into the two phase region. This can qualitatively be understood as follows : The charging current I when entering the sulphur electrode from the cathode current collector is subdivided into I_P and I_C (I_P = current through polysulfide, I_C = current through carbon/graphite fibers). I_P/I_C is determined by the ratio of the corresponding conductivities. Most of the current is carried by the fibers if their conductivity is high. As a consequence, the electrochemical reactions (oxidation of polysulfide anions to sulphur) are taking place only in a thin layer close to the electrolyte surface. An insulating sulphur layer at the electrolyte surface blocks further transport of ions and terminates the charging process. Sulphur utilization is limited to approximately 40% of the theoretical value in this way. This consequence of the phase diagram (fig. 2) turned out to be a major development problem.

Two methods have been used to attain a high rechargeability even in case of a low sulphur electrode resistance : (a) additives to sulphur and (b) structured electrodes e.g. electrodes with more than one porous material as wick for sulphur and polysulfides.

A number of additives has been used : Se and J_2[64], $NaC(CN)_3$ and other thiocyanates and rhodanides[59] and even more succesfully tetracyanoethylene (C_6N_4)[65]. In spite of the high rechargeability achieved with C_6N_4 some disadvantages such as toxidity, gas pressure above Na_2S_x[59] and increased corrosion of the aluminum positive current collector[66] forced developers to look for other solutions such as structured sulphur electrodes.

Structured electrodes are composed of a high conductive carbon or graphite felt filling most of the sulphur compartment and of a thin porous layer of another material adjacent to the electrolyte. The latter material may be (a) a felt with a lower conductivity or (b) a material with a wetting behaviour different from that of the carbon or graphite felt in the main part of the sulphur compartment.

The higher rechargeability in case (a) can be explained by shifting the electrochemical reaction into the vicinity of the boundary between the two felts[67]. Even better results are achieved if powders such as Cr_2S_3, TiO_2, alpha-Al_2O_3 or beta alumina are used (case (b)).Alumina powders have advantages with regard to durability, availability, low resistance and cost[59].

2.5 The Cathode Current Collector

The positive current collector is either the electronically conducting pole in a sulphur core cell (fig. 3 right) or the metallic cell casing of a sodium core cell (fig. 3 left). It is exposed to molten sulphur or sodium polysulfide at 300 - 350°C during the whole life of the battery. Corrosion may have several undesirables consequences : (a) corrosion consumes sulphur and reduces capacity, (b) leaks develop at the cell casing and (c) corrosion products migrate to and deposit at the sulphur electrode/electrolyte interface[68]. Two consequences may result from effect (c) : (d) deposits change the properties of the structured sulphur electrode front layer and lead to a lower rechargeability and/or (e) generate an uneven current distribution with electrolyte rupture as the final consequence.

The requirements onto cathode current collector materials against the highly agressive polysulfide melt are not easy to fulfill. Corrosion tests without and with current have shown, that electronically conducting materials may be divided into four groups[39, 64, 68, 70] : (a) Materials at which no corrosion is observed. Examples are graphite or refractory compounds such as AlN, SiC, B_4C, TiC; (b) metals which develop dense sulfide layers preventing further corrosion. Aluminum is such a meterial; (c) materials such as Cr, Mo, nickel base and cobalt base alloys exhibiting only slow corrosion and (d) fast corroding metals, for instance copper.

No single material possesses all the properties required for a cathode current collector. Type (a) materials have too low a mechanical strength and conductivity. Aluminum - as the most important type (b) material - is not suited because the corrosion product Al_2S_3 is non conductive. Type (c) materials are too expensive. Therefore, composits of at least two materials are used for practical cathode current collectors. The following solutions have hitherto been applied :

(a) Some companies developing sodium core cells use chromized steel as cathode current collector[30, 71-73]. Chromizing is a powder pack technique leading to a chromium rich coating on the steel casing. The cell casing is filled with a powder mixture of Cr, an activator (e.g. NH_4Cl) and Al_2O_3 as filler. The pack is heated to about 1000°C in an inert atmosphere, at which chromium chloride is formed and Cr deposition on and diffusion into the substrate occurs. Mild steel can be used as substrate, however, care must be taken to avoid coating defects since fast corrosion of mild steel would result. Cells with those kind of casings have been operated up to several 1000 cycles.

(b) In order to avoid the relatively high costs related to chromizing, Cr-plating on steel casings with subsequent heat treatment was tried[64]. Laboratory tests exhibited good corrosion resistance, however, in-cell life tests have not yet been performed. Alternatively, Cr-plated steel casings without heat treatment have been used to build cells and batteries and have been cycled up to three years[28, 75].

(c) Several thousand charge/discharge cycles have been attained with sodium core stainless steel casing cells, in which an adhesive of phenol aldehide resin and carbon powder was used to fix a 50 µm thick graphite foil at the inner surface of the cell casing[76].

(d) An excellent solution for sulphur core cells is to isostatically press a graphite foil onto a nichrome coated Al current collector pole. Cells of this type could be operated for more than 1000 cycles[77).

(e) Aluminum cell casings coated with superalloys[78) or plated with chromium[79) have been used to operate cells up to more than thousand cycles.

These examples demonstrate suppression of corrosion to be possible to an extent required for traction batteries. Improved fabrication techniques for corrosion portection layers and/or better composits will be necessary for batteries with life times of about or more than ten years required for load leveling applications.

2.6 Cell Performance

The most important properties characterizing cell performance are energy and power per weight and volume and the stability of these properties.

The theoretical*) energy to weight ratio based on a full discharge from pure sulphur to $Na_2S_{2.7}$ is 760 Wh/kg. Table 3 gives an overview onto experimentally achieved energy and power to weight ratios**).

The table shows practical energy to weight ratios to be approximately 6 times as high as those of lead acid cells (20-40 Wh/kg) and higher than most of the advanced competing cells. Practical energy to weight ratios of Na/S cells cover the range 18 - 25% of the theoretical value. W/kg values are higher than Wh/kg-values for most of the cells in table 3. This means fast discharge (< 1 hr) to be possible or - in case of slow discharge - efficiency to be high.

Performance degradation (life) is the next serious issue. Continuous and abrupt performance degradation is observed.

---

*) Only the weight of the reactands Na and S is taken into account

**) The weight of all cell components including electrolyte, cell casing etc. is taken into account

Table 3 : Energy and power to weight ratios for NaS cells achieved from different developers. Additional information including references : see table 1

Developer	Cell Type	Wh/kg a)	W/kg b)
BBC and Powerplex	A04	190	290
Beta-Research	BR 30	156	120
CSPL	PB	140	195
	XPB	200 c)	240 c)
Ford	EVEM-1	140	175
	MARK II	140	56
Hitachi	600 Ah	150 c)	-
Yuasa/NGK	150 Ah	235 c)	-

a) 2 hrs discharge to Na_2S_3

b) at 2/3 EMF

c) calculated

Continuous degradation is caused by effects such as impurities migrating into the electrolyte leading to electrolyte resistivity rise, corrosion leading to sulphur consumption (see 2.5), corrosion products migrating into the front layer of the structured sulphur electrode leading to rechargeability problems (see 2.5) etc. The practical consequences are a capacity decline and/or a resistance rise of one to several % per hundred cycles depending on cell design. The achievements made with regard to performance decline do almost allow to operate traction batteries during the required life of 500 to 1000 cycles, but are insufficient for load leveling batteries with a necessary life of several thousand cycles.

Abrupt failures are mainly correlated to ceramic components (electrolyte, seal) rupture and to complete loss of cell voltage. Causes are ceramic components rupture as a consequence of thermal stress, of inhomogeneities within the electrolyte (flaws, inhomogenious composition or microstructure) leading to current related degradation (compare 2.2.2) or of uneven current distribution as a consequence of inhomogenious electrodes (incomplete Na-wetting, inhomogenious graphite felt distribution, inhomogenious distribution of corrosion products). The overall results are abrupt cell failures after a certain number of cycles. The failure distribution is usually presented in terms of a Weibull plot as shown in fig. 9.

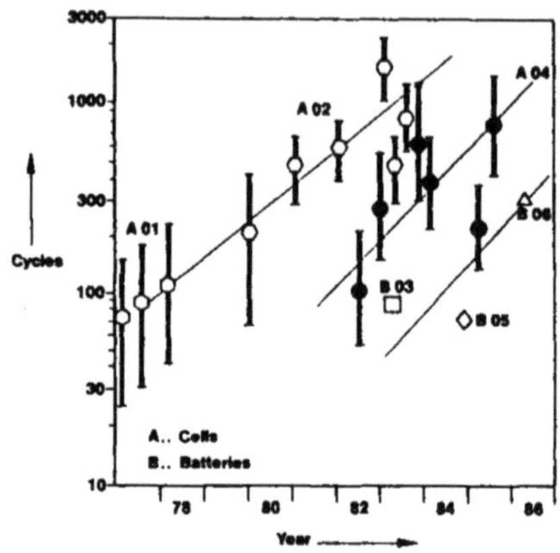

Fig. 9 : Cycle life of Na/S cells and batteries. Data points and scatter of cells A01, A02 and A04 were obtained with groups of 10-50 cells. Bars contain 68% of the cells. Data points for batteries B03, B05 and B06 were obtained with 1 battery for each point [23]

The state of development can be characterized by a "mean time to faulure" of more than thousand cycles. Further development goals are directed towards tightening the distribution (achieving a higher slope of the Weibull curves) and attaining a higher mean time to failure.

3. NA/GLASS/CELLS

The most obvious difference between Na/S cells with ceramic electrolyte and with glass electrolyte is that the single beta alumina tube is replaced by thousands of glass capillaries. This becomes necessary because of the several orders of magnitude lower conductivity of glass compared to beta alumina. The design of a cell with glass capillary electrolytes is shown in fig. 10. The principal function is

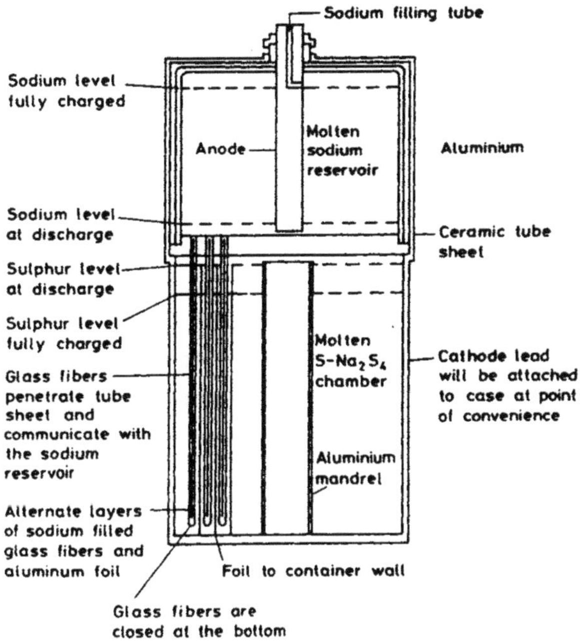

Fig. 10 : Schematic of a Na/glass/S cell[80)]

the same as that of cells shown in fig. 3. Cell performance is similar to that of cells with beta alumina electrolyte (table 3)[81)]. Cell reliability - according to results at Dow Chemical - has turned out to be a more serious problem than for beta alumina based cells[82)]. Problem areas are the seal of the glass capillaries to the sulphur compartment header and degradation of the capillaries. Development has been terminated at Dow Chemical. A new attempt has recently been commenced at Argonne National Laboratory[83)]. A quarterny glass (42 mole% Na_2O, 8 Al_2O_3, 5 ZrO_2, 45 SiO_2) is used as electrolyte instead of B_2O_3 based composition.
The conductivity is considerably higher than for the glasses used at Dow Chemical. Too few results are available to judge whether the problems mentioned above can be overcome in this way.

4. NA/BETA ALUMINA/MCl_x CELLS

Several attempts have been made to substitute sulphur in Na/S cells by metal chlorides dissolved in or in contact with $AlCl_3$ in order to reduce the operation temperature or to expand the operation temperature range. Lower operation temperatures are a consequence of the low melting point (153°C) of $AlCl_3$. The cell reactions are

$$x\ Na + MCl_x + x\ AlCl_3 \longleftrightarrow x\ Na\ AlCl_4 + M.$$

Three cell types have been investigated : MCl_x = $SbCl_3$, $NiCl_2$ and $FeCl_2$. Open circuit voltages at 250°C are 3.7, 2.59 and 2.35 V, respectively. The theoretical energy to weight ratios are > 700 Wh/kg. This means theoretical performance characteristiques to be as promising as those of Na/S cells. Experiments have been performed with cells of a design similar to that as Na/S cells, i.e. beta alumina tubes have been used as electrolytes in sodium core and cathode core cells.

Experimental work on the $Na/SbCl_3$ cell with $SbCl_3$ dissolved in the acidic $NaAlCl_4$ melt showed that corrosion problems in spite of the low operation temperature (210°C) are not easy to overcome and that the practical energy density - because of the limited solubility of $SbCl_3$ in $NaAlCl_4$ - is low in comparison to Na/S[84, 85]. The program was terminated for these reasons.

Promising results have been obtained with $Na/NiCl_2$ and $Na/FeCl_2$ cells (Zebra cells), first described by Coetzer[86]. The positive electrodes are prepared in the discharged state by sintering mixtures of MCl_2 and M powder. The electrodes become porous during the first charge by transferring MCl_2 into NaCl which dissolves in $AlCl_3$. $NiCl_2$ and $FeCl_2$ - in contrast to $SbCl_3$ - do not dissolve in $AlCl_3$. Dissolution is avoided by keeping the operation temperature well below the NaCl-MCl_2 eutectic point. This limits the operation temperature range to temperatures below about 500°C for $Na/NiCl_2$ cells and 300°C for $Na/FeCl_2$ cells[87-89].

Cells are not yet optimized for special applications. Performance goals for traction cells are an energy to weight ratio of 140 Wh/kg at a 3 hrs discharge rate and a power to weight ratio of 150 W/kg at 2/3 EMF[90]. This is slightly less than the values achieved for Na/S cells (compare table 3), but still high in comparison to other systems. Durability seems to be good. No or only minor performance degradation has been observed up to 500 fast cycles [88]. Cells are claimed to be very safe because of the solid positive reactands.

5. BATTERIES

5.1 Battery Design

The design of a solid electrolyte high temperature battery differs from that of a conventional battery for at least two reasons :

- o The energy for heat up of the battery amounts to 30 - 50% of the electric energy. Energy economy requires to keep the battery at operation temperature during the major part of the battery life and to keep the thermal losses small. This is achieved by means of a vacuum thermal enclosure in case of a traction battery. In addition, a cooling system is needed for waste heat removal during high power operation.
- o Power and voltage requirements for all major applications as well as safety and redundancy considerations lead to batteries with a large number of relatively small cells. For these reasons even a small traction battery with an energy content of approximately 30 kWh does contain several parallel cell strings.

Fig. 11 shows the principal design features of a high temperature solid electrolyte battery and fig. 12 is a photo of a Na/S traction battery, designed as mentioned above.

Major efforts were necessary to develop traction battery enclosures with a low thermal conductivity, cooling systems which guarantee an even temperature distribution and cell interconnection strategies minimizing the impact of cell failures onto battery performance.

The development of atmospheric pressure bearing vacuum thermal enclosures has led to effective thermal conductivities in the order $2 \cdot 10^{-3}$ W/m·K. This is achieved by utilization of glass fibers plus an opacifier powder in between the two walls of the thermal insulation[91, 92] or by a multilayer glass fiber board interleaved with Al foil

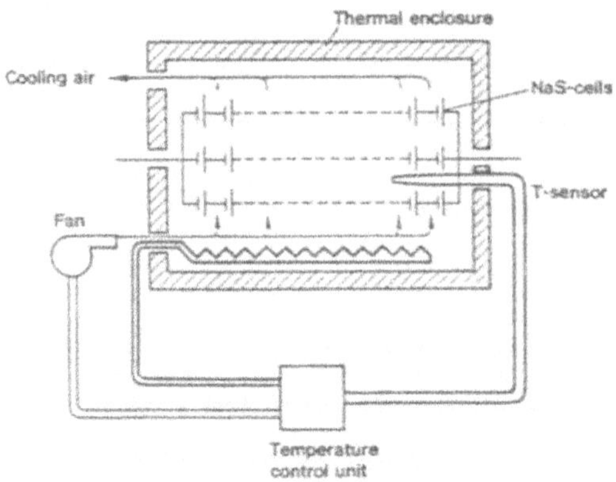

Fig. 11: Design features of a high temperature solid electrolyte battery

Fig. 12: Pre-prototype Na/S traction battery including all thermal management components schematically shown in fig. 11

radiation barriers[93, 94]. The thermal conductivity is equivalent to losses of about 100 W for a traction battery. A conventional thermal insulation is sufficient for stationary batteries, where space is not so important.

The sustained power of a battery is determined by waste heat removal [79]. Forced air cooling is used by most of the developers. A comparison between direct cooling and indirect cooling via heat exchanger shows that both methods do result in temperature gradients within the battery not higher than 30°C[95]. Both methods and fluid cooling methods have been used.

Two adverse effects are important with regard to the impact of cell failures onto battery performance degradation : (a) A high ohmic failing cell in a battery with n not crossconnected strings results in a relative capacity loss of 1/n. A second cell failure will most probably result in an additional capacity loss or an overall loss of 2/n. (b) The probability for the second cell failure to cause a second capacity loss can be drastically reduced by crossconnecting the strings (making substrings) after a few cells. The disadvantage of this solution is that state of charge imbalance between substrings with a different number of failed cells will be severe.

The compromise adopted from most groups is to make longer (> 7 cells) substrings. Two other solutions should be mentioned : (a) The battery consists of many small cells allowing for a high n and making the capacity loss 1/n small (CSPL solution[93,96]). (b) The battery does not contain crossconnectors. Instead looping elements are parallel connected to cells. Activation of the looping element is initiated (shorting the cell) after cell failure.

An overview on battery design solution realized up to now is given in table 4.

5.2 Battery Properties

5.2.1 Performance

Battery performance, as far as energy and power to weight ratios for traction batteries and efficiency for load leveling batteries are concerned has been steadily improved over the years.

Traction battery performance at BBC has been increased from the first 96 cell battery with 10 Wh/kg - 5 W/kg in 1978[23] to 85 Wh/kg - 120 W/kg for the most recent 360 cell battery[79]. The Beta Research program can be characterized by 71 Wh/kg - 60 W/kg for a 76 cell Na/S battery in 1983[102] and 88 Wh/kg - 65 W/kg for a 66 cell Na/$NiCl_2$ battery in 1987[88, 90]. Similar development steps have been made at Ford: Performance data for a 52 cell Na/S battery fabricated in 1981 were 68 Wh/kg - 22 W/kg[103] and design values for a 480 cell Na/S battery (1983) were 92 Wh/kg - 97 W/kg[28]. Energy to weight ratios refer to a 2 - 3 hours discharge time and power to weight ratios to 2/3 EMF. Whilst performance of the early batteries was lower, performance of recently fabricated batteries is generally higher than that of lead

Table 4 : Battery Design Features

General Battery Characterisation				Thermal management		Cell Interconnection Method			
Company	Type	Design	Application	Kind of Insulation	Cooling system	Number of Cells	Cells per String	Crossconnection after ... cells	References
BBC + Powerplex	Na/S	B06	EV	glass fibers + vacuum	forced air, direct	224	56	8	23)
		B11	EV, LL*	glass fibers + vacuum	forced air, direct	360	45, 60, 90	2, 6, 8, 90**	79), 97)
Beta Research	Na/S	-	EV	conventional	none	176	22	11	88), 90)
	$Na/NiCl_2$	-	EV***	glass fibers + vacuum	none	66	33	?	
	$Na/FeCl_2$	-	EV***	glass fibers + vacuum	liquid circulation	152 (76)	38 (38)	?	88), 90)
CSPL	Na/S	for Bedford CF van	EV	glass fiber boards + Al interleaves	none	1328	96	4	25), 93), 96)
Ford (FACC)	Na/S	100 kWh	LL	conventional	none	512****	8	8****	28), 98)
		for EVEM-1	EV	glass fibers + vacuum	forced air + recuperator	480	120	15	29)
Yuasa + NGK Sparkplug	Na/S	80 kWh	LL	conventional	none	224	28	7	99)
		100 kWh	LL	conventional	none	280	28	7	31)

* 6 B11 batteries to be used for a load leveling experiment 100), 101)

** no interconnections at a few 180 V B11 batteries, looping element parallel to each cell

*** cells not yet optimized for EV-application

**** fuse in series to each cell

acid batteries. Target values for Na/S traction batteries are about 140 Wh/kg - 200 W/kg.

A less spectacular, but nonetheless important progress was made with regard to energy efficiency of load leveling batteries. The efficiency (at a 8 hrs discharge/8 hrs charge rate) for batteries fabricated earlier than 1984 was in all cases < 80%[28] whilst the efficiency of batteries fabricated after 1984 is > 80%[31] and higher than that of competing advanced batteries[104].

5.2.2 Safety

As far as safety is concerned accelerated tests to simulate vibration during a more than 100 000 km electric vehicle operation have shown that Na/S and Na/MCl_2 cells, battery components and complete batteries do not develop any malfunction during these tests[90, 93, 79]. Furthermore, BBC Na/S batteries have been exposed to squeeze and crash tests, external shorts and gasoline fire, immersion into cold water and failure propagation tests. Fig. 13 shows a squeezed battery. No safety issues were discovered during and after these tests.

Fig. 13 : Squeezed sodium sulphur battery

As a result, these batteries are permitted from the German technical supervision institution (TÜV) to be used for EV-operation on public streets[105].

5.2.3 Life

Two major statements describe the present status of battery life:

- The maximum battery life has been improved to several hundred cycles[23]. Goals are 500 - 1000 cycles for traction batteries and 1500 - 3000 cycles for load leveling batteries. The 512 (later 384) 150 Ah cell load leveling battery at Ford has attained 675 cycles during 32 months of operation[29]. Only 3 failed cells were recorded during a 500 cycle life test of a 120 10 Ah cell traction module[25, 106]. A low energy efficiency decline (1% per 100 cycles) was observed with three batteries, each containing sixty 150 Ah cells during a 460/640/760 cycles operation time at Yuasa[107]. A larger Yuasa battery (280 300 Ah cells) is now in operation for 140 cycles[31, 104].
- Early battery failures have been observed in a number of cases : Ground faults caused test termination for several batteries in different laboratories. The Ford 480 cell traction battery could not be operated because of shorts within the battery[108]. Several cells had to be replaced in the 76 cell Beta Research traction battery prior to operation[102].

In conclusion : The maximum cycle life of Na/S batteries of several hundred cycles is already close to EV-requirements, however, reliability has to be improved.

5.3 Application Tests

Evidence that Na/S, $Na/NiCl_2$ and $Na/FeCl_2$ batteries can be used to propel electric vehicles (EV's) was gained already during the early stages of development (CSPL 1972[109]; Yuasa 1977[32]; British Rail 1980[102]; BBC 1981/83[89]). An acceptable drive performance characterized by a range > 150 km, a maximum speed > 100 km/h and an acceleration to 50 km/h in < 15 sec has been achieved only recently with Chrysler Minivans[113], BMW passenger cars and VW passenger cars[114]. The EV's were equipped with 360 cell Na/S batteries from BBC and Powerplex. Four Na/S battery powered cars ranked at places 1, 3, 5 and 6 in an EV race[115]. Two of these EV's are shown in fig. 14. Batteries

Fig. 14 : Two Na/S battery propelled passenger cars

shaped for volume production will most probably have a shape different from that of present batteries. A Na/S battery tailored to the propulsion of a Bedford (GE) van is in development at CSPL [93, 96].

Utilization of batteries for load leveling or peak shaving in electric grids (charging in off-peak periods and discharging during high demand periods) would allow to substitute oil for peak power generation by more abundant primary energy sources and to generate peak power more efficiently and economically [115, 116].

A first Na/S load leveling battery with an initial energy of 100 kWh was tested at Ford, USA from 1982 - 1985 [29] (see 5.2.3) and several Na/S batteries with a power/energy in between 1 kW/8 kWh and 12 kW/100 kWh were operated within the Japanese Moonlight Project since 1985. The energy efficiency of these batteries is higher than that of competing advanced batteries [31, 104, 107]. A test with a 120 kWh Na/S battery composed of six traction batteries was commenced recently [100, 101]. Development should come to an end in the mid 1990's, since an increasing demand in load leveling batteries is to be expected at this time [117].

6. CONCLUSIONS

Solid electrolytes have always been considered to be a scientifically interesting new class of materials. At least one of those electrolytes, beta alumina, will most probably become technically important for two reasons :

o The principal problems of Na/beta alumina/S- and to a lesser degree of Na/beta alumina/MCl_2-cells and batteries are essentially solved. High energy and power densities have been achieved. Risks to overcome the remaining technical issues (fabrication technique, life) have become minor. Raw materials are abundant and inexpensive, the price should become low as soon as volume production starts.

o Applications such as propulsion of EV's and load leveling in electric grids will not only become economically feasible, but highly desirable for environmental reasons because of flexibility in primary energy utilization.

7. REFERENCES

1) Cairns, E.J. and Steunenberg, R.K., in "Progress in High Temperature Physics and Chemistry", C.A. Rouse edit., Pergamon Press Oxford and New York 5, 63 (1973)

2) Fischer, W., in "Handbook of Thermodynamic and Transport Properties of Alkali Metals", R.W. Ohse edit., Blackwell Sci. Publ., 181 (1985)

3) Kummer, J.T. and Weber, N., Proc. SAE Congr., Paper 670 179, 1-6, (1967)

4) Levine, C.A., Proc. 10th IECEC, Newark, Del., Aug. 18 - 21 (1975) p. 621

5) Rankin, G.A. and Merwin, H.E., J. Am. Chem. Soc. 38, 568 (1916)

6) Bragg, W.L., Gottfried, C. and West, J., J. Krist. 77, 255 (1931)

7) Saalfield, H., Naturwiss. 43, 420 (1956)

8) Yao, Y.F. and Kummer, J.T., J. Inorg. Nucl. Chem. 29, 2453 (1967)

9) Coetzer, J., J. Power Sources 18, 377 (1986)

10) Kummer, J.T., in "Progress in Solid State Chemistry" 7, 141 (1972)

11) May, C.J., J. Power Sources 3, 1 (1978)

12) Dell, R.M. and Moseley, P.T., J. Power Sources 6, 143 (1981)

13) Stevens, R. and Binner, J.G.P., J. Mat. Sci. 19, 695 (1984)

14) Ansell, R.O., J. Material Sci. 21, 365 (1986)

15) Sudworth, J.L. and Tilley, A.R., "The Sodium Sulfur Battery", Chapman and Hall, London and New York (1985)

16) Tischer, R.P., "The Sulfur Electrode", Academic Press, New York (1983)

17) Angelis, J.W., Birnbreier, H. and Haase, H., Elektrische Bahnen, 85, 252 (1987)

18) Gupta, N.K. and Tischer, R.A., J. Electrochem. Soc. 119, 1033 (1972)

19) Sudworth, J.L., Hames, M.D. et al., Power Sources, J.M. Collins edit., Oriel Press, 4 (1973)

20) Silverman, H.P., EPRI Final Report EM-226, Project 127 (1976)

21) Knoedler, R. and Krapf, H., J. Power Sources 10, 263 (1983)

22) Chatterji, D., EPRI-Report EM-266, Project 128-3 (1976)

23) Fischer, W., Birnbreier, H. and Benninger, G.N., Proc. 21st IECEC, San Diego, Aug. 25 - 29 (1986) p. 1041

24) Hames, M.D. and Sudworth, J.L., Proc. of the Institution of Electrical Engineers 126, 1157 (1979)

25) Molyneux, J., Sands, G. et al., Proc. 22nd IECEC, Philadelphia, Pa., Aug. 10 - 14 (1987) p. 975

26) Auxer, W., Proc. 8th Battery and Electrochem. Contractors Conf., Vienna, Va., Nov. 16 - 19 (1987) p. 59

27) Mulcahey, T.P., Tummillo, A.F. et al., 8th Battery and Electrochem. Contractors Conf., Vienna, Va., Nov. 16 - 19 (1987) p. 85

28) Gates, A.B., Proc. 18th IECEC, Los Angeles, Ca., Aug. 21 - 26 (1983) p. 1690

29) Minck, R.W., Proc. DOE/EPRI Beta Battery Workshop VI, Snowbird, Utah, May 19 - 23 (1985) p. 4-13

30) Mori, S., Ogino, M. et al, Abstract No. 64, 170th Electrochem. Soc. Meeting, San Diego, Ca., Oct. 19 - 24 (1986) p. 94

31) Kita, A., Nomura, E. et al., 22nd IECEC, Philadelphia, Pa., Aug. 10 - 14 (1987) p. 1066

32) Fuchida, K., Okada, K. et al., 17th IECEC, Los Angeles, Ca., Aug. 8 - 11 (1982) p. 591

33) Hong, H.Y.-P., Material Sci., Bull. 11, 173 (1976)

34) Von Alpen, V. et al., Final Report BMFT - FB 83-095, May (1983)

35) Tennenhouse, G.J. and Crosbie, G.M., J. Ceram. Soc. 67, 204 (1984)

36) Linn, Z., in : "High Conductivity Solid Ionic Conductors : Recent Trends and Applications", T. Takahashi edit., World Scientific Publishing Co., Singapore (1988)

37) Harbach, F., Solid State Ionics 13, 53 (1984)

38) Bates, J.B., Wang, J.C. and Dudney, J.C., Physics Today, July, 46 (1982)

39) Fischer, W., Haar, W. et al., J. Power Sources 3, 299 (1978)

40) Armstrong, R.D., Dickinson, T. and Turner, J., Electrochim. Acta 19, 187 (1974)

41) Virkar, A.V. and Viswanathan, L., J. Material Sci. 18, 1202 (1983)

42) Viswanathan, L. and Virkar, A.V., J. Material Sci. 17, 753 (1982)

43) Virkar, A.V., Viswanathan, L. and Biswas, D.R., J. Mat. Sci. 15, 302 (1980)

44) Breiter, M.W., in :"Trends in Electrochemistry", J. O'M Bockris edit., Plenum Publishing Co., New York (1977) p. 9-11

45) Kawamoto, H. and Wada, M., DOE/EPRI Beta Battery Workshop VI, Snowbird, Utah, May 19 - 23 (1985) p. 9-11

46) De Jonghe, L.C., EPRI-Report EM-3631-SR, DOE/EPRI Beta Battery Workshop V, Washington D.C., Oct 18 - 20 (1983) p. 5-85

47) De Jonghe, L.C., EPRI-Report EM-4151, July (1985)

48) Haar, W., DOE/EPRI Beta Battery Workshop VI, Snowbird, Utah, May 19 - 23 (1985) p. 5-6

49) Gordon, R.S., Mc Entire, B.J. et al., in : "Processing of New Crystalline Ceramics", H. Ralmour, R.F. Davis and T.M. Hare eds., Plenum, New York, 405 (1978)

50) Powers, R.W. and Mitoff, S.P., in "Solid Electrolytes", P. Hagenmüller and W. van Gool eds., Academic Press, New York, 123 (1978)

51) Tan, S.R. and May, G.J., Sci. Ceram. 9, 103 (1977)

52) Heavens, S.N., DOE/EPRI Beta Battery Workshop VI, Snowbird, Utah, May 19 - 23 (1985) p. 8-12

53) Taylor, J.W. and Ford, S.D., "Solid Metal - Liquid Metal Interaction Studies, Part II, Contact Angle Relationships for Sodium on Solids", Harwell-Report AERE MR 1729 (1955)

54) Lehnert, G. and Hartmann, B., Ext. Abstr. of the Electrochem. Soc. Meeting, Detroit, Mich., Oct. 17 - 21 (1982) p. 539

55) Demott, D.S., J. Electrochem. Soc. 10, 2312 (1980)

56) Breiter, M.W., Choudhury, N.S. and Hall, E.L., Solid State Ionics 14, 225 (1984)

57) Breiter, M.W. and Dunn, B., Electrochim. Acta 26, 1247 (1981)

58) Iwabuchi, S., Takashima, K. and Okada, K., EPRI-Report EM 3631-SR, DOE/EPRI Beta Battery Workshop V, Washington D.C., Oct. 18 - 20 (1983) p. 6-181

59) Mennicke, S., Dechema Monographien, VCH Verlagsgesellschaft 109, 409 (1987)

60) Euler, J. and Nonnenmacher, W., Electrochim. Acta 2, 268 (1960)

61) Gibson, J.G., J. Appl. Electrochim. 4, 125 (1974)

62) Kao, Y.K. and Wayner, P.C., J. Electrochem. Soc. 123, 632 (1976)

63) Kawamoto, H., Denki Kagaku 52, 589 (1984)

64) Bauer, R., Haar, W. et al., J. Power Sources 1, 109 (1976/77)

65) Weddigen, G. and Fischer, W., Chem.-Ing. Techn. 49, 345 (1977)

66) Demott, D.S., DOE/EPRI NaS Battery Workshop IV, Heidelberg, Oct. 6 - 8 (1980)

67) Breiter, M.W. and Dunn, B., J. Appl. Electrochem. 9, 291 (1979)

68) Battles, J.E. and Mrazek, F.C., EPRI-Report EM-4341, Research Project 1198-15, Dec. (1985)

69) Hartmann, B., J. Power Sources 3, 227 (1978)

70) Knoedler, R., Ext. Abstracts, Electrochem. Soc., Cincinnati, Ohio, May 6 - 11 (1984)

71) Park, D.S. and Chatterji, D., Thin Solid Films 83, 429 (1981)

72) Iwabuchi, S., Progress in Batteries and Solar Cells 5, 310 (1984)

73) Stackpool, F.M., Ceramatec Special Report 8601, DOE/EPRI Beta Battery Workshop VI, Snowbird, Utah, May 19 - 23 (1985) p. 7-8

74) Wicker, A., Deplanches, G. and Saisse, H., Thin Solid Films 83, 437 (1981)

75) Smaga, J.H. and Battles, J.E., Ext. Abstr., 7th Battery and Electrochem. Contractors Conf., Crystal City, Va., Nov. 18 - 21 (1985) p. 40

76) Fischer, W., Dechema Monographien 92, 103 (1982)

77) Wright, M.L. and Tilley, A.R., 16th IECEC, Atlanta, Ga., Aug. 9 - 14 (1981) p. 841

78) Knoedler, R., Ceramatec Special Report 8601, DOE/EPRI Beta Battery Workshop VI, Snowbird, Utah, May 19 - 23 (1985) p. 7-15

79) Fischer, W., Birnbreier, H. and Benninger, G.N., Proc. 8th Intern. Electric Vehicle Symp., Washington D.C., Oct. 20 - 22 (1986) p. 324

80) Levine, C.A. in "The Sulfur Electrode", P. Tischer edit., Chapter 8, Academic Press, New York (1983)

81) Levine, C.A., Dean, D. and Li Wu, M., 2nd Ann. Battery and Electrochem. Techn. Conf., Arlington, Va., June 5 - 7, (1987)

82) Pierini, P., Ext. Abstracts, 7th Battery and Electrochem. Contractors Conf., Crystal City, Va., Nov. 18 - 21 (1985) p. 101

83) Bloom, I., Nelson P. et al., Electrochem. Soc. Meeting, San Diego, Ca., Oct. 19 - 24 (1986) p. 110

84) Werth, J., Klein, I. and Wylie, R., Ext. Abstract No. 242, Electrochem. Soc. Fall Meeting, Dallas, Tx., Oct. 5 - 10 (1975)

85) Chreitzberg, A.M. et al., J. Power Sources 3, 201 (1978)

86) Bones, R.J., Coetzer, J. et al., Ext. Abstr., 170th Soc. Meeting, San Diego, Oct. 19 - 24 (1986) p. 1122

87) Bones, R.J., Teagle, S.D. et al., Abstract 553, 171st Electrochem. Soc. Meeting, Philadelphia, Pa., May (1987)

88) Sudworth, J.L., Galloway, R.C. and Demott, D.S., UPAC Conf., Sofia, Bulgaria, July (1987)

89) Tilley, A.R. and Wedlake, R.J., Electric Vehicle Developments 6, 119 (1987)

90) Tilley, A.R. and Bull, R.N., Paper 879327, Proc. 22nd IECEC, Philadelphia, Pa., Aug. 10 - 14 (1987)

91) Reiss, H. and Ziegenbein, B., Proc. 8th Int. Thermal Conductivity Conf., Rapid City, Oct. 3 - 5 (1983)

92) Reiss, H., Spektrum der Wissenschaft, Nov. (1985), p. 112

93) Mangan, M.F., Ext. Abstracts, 8th Battery and Electrochem. Contractors Conf., Vienna, Va., Nov. 16 - 19 (1987) p. 52

94) Chilenskas, A.A., Biwer, R.L. and De Luca, W.H., Proc. 8th Intern. Electric Vehicle Symp., Washington D.C., Oct. 20 - 22 (1986) p. 338

95) Ziegenbein, B. and Reiss, H., BBC-Nachrichten 66, 348 (1984)

96) Auxer, W., Proc. 8th Intern. Electric Vehicle Symp., Washington D.C., Oct. 20 - 22 (1986) p. 318

97) Birnbreier, H., Fischer, W. and Benninger, G.N., 170th Meeting of the Electrochem. Soc., San Diego, Ca., Oct. 19 - 24 (1986) p. 100

98) Bridges, D.W. and Haskins, H.J., Proc. 17th IECEC, Los Angeles, Ca., Aug. 8 - 12 (1982) p. 563

99) Kita, A., Takashima, K. et al., Ext. Abstracts, IAE Intern. Symp. on Fuel Cell and Advanced Battery, Tokyo, Jan. 21 - 22 (1986) p. 135

100) Freund, H., Tagungsbericht, VDE-Fachtagung "Hochenergiebatterien in Verkehr und Energieversorgung", Stuttgart, Sept. 29 - 30 (1987) p. 147

101) Seele, W., Tagungsbericht, VDE-Fachtagung "Hochenergiebatterien in Verkehr und Energieversorgung", Stuttgart, Sept. 29 - 30 (1987) p. 163

102) Sudworth, J.L., EPRI-Report EM 3671-SR, Proc. DOE/EPRI Beta Battery Workshop V, Washington D.C., Oct. 18 - 20 (1983) p. 7-1

103) Harlow, R.A. and Mc Clanahan, M.L., Proc. 30th Power Sources Symp., Atlantic City, N.J., June 7 - 12 (1982) p. 51

104) Itoh, N., Hiramatsu, T. et al., Int. Conf. "Batteries for Utility Energy Storage", Berlin, Nov. 9 - 11 (1987) p. 267

105) Birnbreier, H., Int. Conf. "Batteries for Utility Energy Storage", Berlin, Nov. 9 - 11 (1987) p. 251

106) Mc Namee, M., 8th Battery and Electrochem. Contractors Conf., Vienna, Va., Nov. 16 - 18 (1987) p. 41

107) Yuasa/NGK Sparkplug-Report to NEDO, Aug. (1986)

108) Electric Vehicle Progress 8, No. 17, Sept. 1 (1986)

109) Churchman, A.T., J.R. Soc. Arts (1975)

110) Fischer, W., Jahresbericht, Enzyklopädie Naturwissenschaft und Technik, verlag moderne industrie, Landsberg (1982)

111) Reschke, R. and Fischer, W., 7th Intern. Electric Vehicle Symp., Paris-Versailles, June 26 - 29 (1984) p. 119

112) Angelis, J. and Hughson, E.A., Proc. 8th Int. EV-Symp., Washington D.C., Oct. 20 - 22 (1986) p. 194

113) BBC-Druckschrift Nr. DHB 1221-87 D, 15.6. (1987)

114) Electric Propulsion No. 3, 2, 6 (1987)

115) Seele, W., Adler, T. and Gels, H.B., A techno-economical comparison of battery storage systems for peak shaving in distribution networks, BMFT research report T 84-141 (1984)

116) Seele, W., International Conf. "Batteries for Utility Energy Storage", Berlin, Nov. 9 - 11 (1987) p. 443

117) Klein, K.W., 8th Battery and Electrochem. Contractors Conf., Vienna, Va., Nov. 16 - 19 (1987), Ext. Abstr. p. 7

SOLID OXIDE FUEL CELLS

Jack T. Brown

Director, Electrochemical Projects
Westinghouse Research and Development Center
Pittsburgh, Pennsylvania
U.S.A

ABSTRACT

Solid oxide fuel cells offer the potential for high efficiency, low cost electric power plants for many applications. The fuel cell generator itself is an all solid state reactor operating at about 1000°C, which produces direct current electricity from the conversion of the energy in gaseous fuel by an electrochemical reaction with the oxygen in air.

System studies have shown 50-60 percent efficiency can be obtained in multi-megawatt all electric power plants in which the high quality exhaust heat is used to produce electricity in a bottoming cycle. Alternatively the exhaust heat can be used in industrial and commercial co-generation systems for space cooling or raising high pressure steam. Space based and military applications for electric power plants have been proposed with high power per unit weight and volume designs which use hydrogen fuel and pure oxygen.

Truly significant accomplishments have been made in the 1980's in the technological development of the special materials and processes for producing them for the various cell designs and in the conception of device designs. These advancements were due to the fundamental work of the several decades earlier. In 1986 through 1988 experimental test units in the several hundred to several thousand watt sizes have been designed, fabricated, shipped, and tested for thousands of hours by user organizations in the United States and Japan.

The extent of the commercial market for Solid Oxide fuel cells for various applications, vis-a-vis alternate electrical generation options, will depend on what system capital cost can be achieved.

1. INTRODUCTION

A number of literature references are available on the development of solid oxide fuel cells (SOFC's) since the pioneering work of Bauer and Preis[1] fifty years ago. It is not the purpose of this chapter to summarize the many past activities which have led up to the worldwide renewed and increased interest in SOFC's beginning about 1980. Instead this chapter will offer a synopsis of the technology position as it appears to be in the latter half of the 1980's, from reports and seminar communications. It will also attempt to predict the relatively near term events (over the next five to eight years) which will take place as significant efforts to find a commercial market niche proceed.

A brief explanation of the operating principle of the solid oxide fuel cell is offered below and is illustrated in the crossection of the three layered schematic diagram of Figure 1.

When an external load is applied to the cell, the following occurs: oxygen from air is reduced at the porous air electrode/electrolyte interface to produce oxygen ions. Since the oxygen ion vacancies that exist in the yttria-stabilized zirconia solid electrolyte provide high mobility at 1000°C to the oxygen ions, these ions readily migrate through the solid electrolyte to the fuel electrode. At specific reaction sites where the fuel gas, fuel electrode and electrolyte coexist, the oxygen ions react with the fuel, H_2 or $CO + H_2$, for example, to produce H_2O or $CO_2 + H_2O$ and release electrons to the external load. The number of oxygen ions reacting at the fuel electrode equals the number of oxygen ions entering the electrolyte, so that the electrical neutrality of the electrolyte is maintained. This reaction continues as long as fuel and air are supplied.

The open circuit voltage of the SOFC is dependent upon the temperature and the ratio of the oxygen partial pressure in the fuel and air streams and is given by the equation:

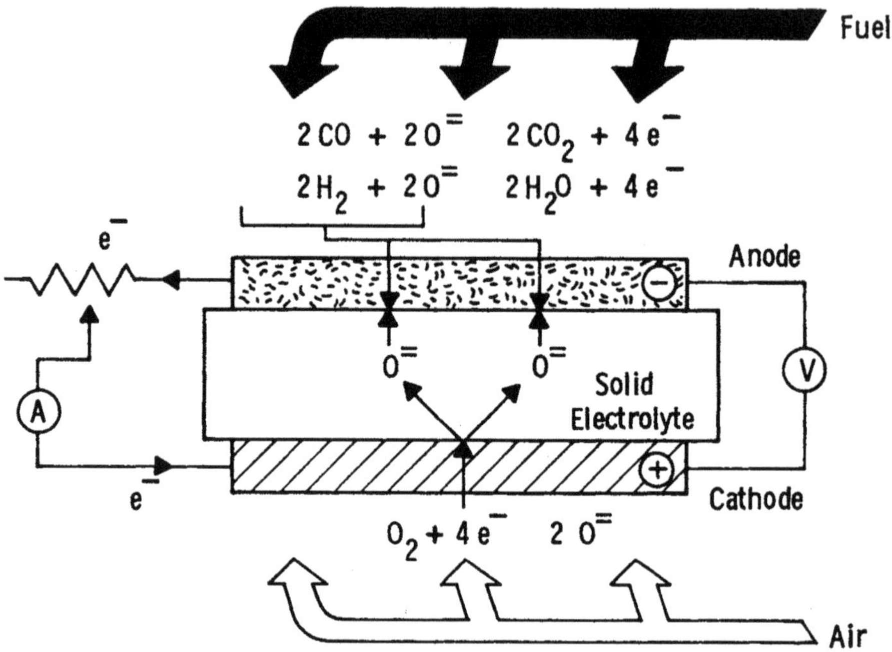

Figure 1. Schematic diagram of a solid oxide fuel cell, indicating how oxidation of the fuel generates electric current to the external load.

$$E = \frac{RT}{4F} \ln \left[\frac{P_{O_2}(\text{airstream})}{P_{O_2}(\text{fuelstream})} \right]$$

where:
E = the open circuit voltage
R = the universal gas constant
T = the absolute temperature
F = Faraday's constant
P_{O_2} = oxygen partial pressure

This equation has the form of an oxygen concentration cell. At about 10^{-17} atmosphere of oxygen in the fuel stream (and 0.21 atmosphere oxygen on the air side), the theoretical open cell voltage at 1300K is about 1 volt.

Over the past thirty years researchers have conceived various structural embodiments of this principle, but it is fair to say that the success with relatively long life tubular designs has sparked the recent renewed interest by a number of organizations worldwide. It can also be stated that the success in the identification of various compatible materials for the components of the cell has been key to the renewed interest. Some research groups have put forward the hypothesis that with these materials one could obtain conceptually, at least, greatly increased power per unit weight and volume of the generator than that achieved with tubular designs. The possible fault with this logic however is that the materials must be both chemically and mechanically compatible with the cell/module design, and while it has been proven they are compatible with the tubular design, it is yet to be determined that they are also mechanically compatible with the so-called monolithic/planar designs.

Table 1 summarizes the organizations presently known to be working on the various SOFC designs.

The author suspects that in 1988 the effort (full-time personnel) at Westinghouse exceeds the sum of all the other efforts cited above, and that the approximate sum of all annual R&D budgets for SOFC research and development is \$15-\$18,000,000.

The balance of the chapter will discuss the state of the technology for the various programs.

2. TECHNOLOGY FOR SOFC GENERATORS

2.1 Materials

After decades of research looking for compatible materials for SOFC generators, there is general agreement on the general compositions of materials for each of the four components of the basic cell. These are shown below:

Electrolyte: Yttria Stabilized Zirconia (dense)

$$(ZrO_2)_{0.92} \cdot (Y_2O_3)_{0.08}$$

Table 1.

Tubular SOFC Designs

1. Westinghouse (USA)
2. Electrotechnical Laboratory/M.I.T.I. (Japan)
3. Dornier (Germany)

Monolithic/Planar SOFC Designs

4. Argonne Laboratory/D.O.E. (USA)
5. Garrett (USA)
6. United Technologies/Ceramatech (USA)
7. Z-Tek (USA)
8. Standard Oil Co. (USA)
9. Hitachi (Japan)
10. Toa Neuro Kogyo (Japan)
11. CEC (Europe)

Air Electrode(Cathode): Strontium Modified Lanthanum Maganite

(~ 40% porous)

$$La_{0.9}\,Sr_{0.1}\,(MnO_3)$$

Fuel Electrode(Anode): Nickel Zirconia Cermet (~ 40% porous)

$$Ni\,(ZrO_2),\ (\sim 60\ \text{wt. \% Ni})$$

Interconnection: Magnesium Modified Lanthanum Chromite (dense)

$$La_{0.95}\,Mg_{0.05}\,(CrO_3)$$

There are detailed differences in the exact composition of these materials from that shown that are small but quite important, and have to do with how to prepare the materials for fabrication into the various component shapes used. These are however proprietary processing variables.

Compatibility means, as stated earlier, both chemically and mechanically stable when in direct contact with each other. For example the requirements are; minimum interdiffusion of the various elements up to the processing temperatures, which are above the operating temperature, resistance to oxidation and reduction in the particular atmosphere to which they are subjected, closely matching expansion characteristics, stability in both high and low partial pressure of oxygen (interconnection), acceptable conductivity.

Figure 2 shows the resistivity as a function of temperature for the various SOFC materials mentioned. This information is needed for cell design to minimize ohmic losses balanced against diffusional losses in the porous electrodes. It points out the basis for the need for very thin electrolyte and interconnection (the higher resistivity materials) and the hoped for improvement in cathode material conductivity (order of magnitude higher resistivity than the anode material).

2.2 Processes

The various structural embodiments of the SOFC dictate to a large extent the processes used for constructing the various components of the cell and module. For example the sequence required by the cell design for constructing the various components affects the choice of processing, i.e. a decreasing processing temperature for the various components from beginning to end of cell construction so that latter processes do not destroy the structures created by earlier processes. Also the need for maintaining controlled porosity in the electrodes (and in some cases the substrate), and high density (no porosity) in the electrolyte and interconnection, also strongly influence the processes chosen for constructing the cells. Basically well-known advanced ceramic processing techniques are used such as powder extrusion, high temperature pressing and sintering of powders, filter press agglomeration and laying down powder films on substrates, i.e., tape casting.

A unique process for fabricating very thin gas tight, uniformly coated layers over porous substrates was introduced by Westinghouse in

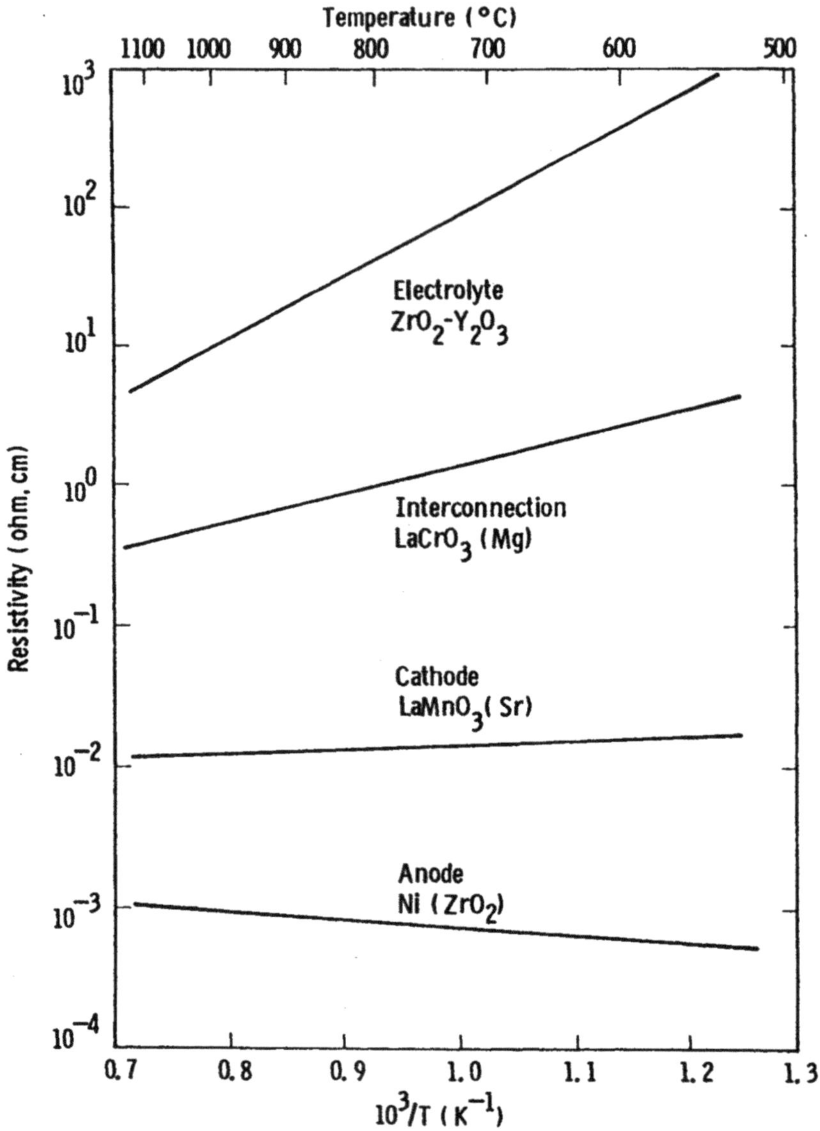

Figure 2. Resistivity - temperature behaviour of SOFC materials.

the 1970's, and was called electrochemical vapor deposition.[2)] The principle of the EVD process is shown in Figure 3. It takes place in

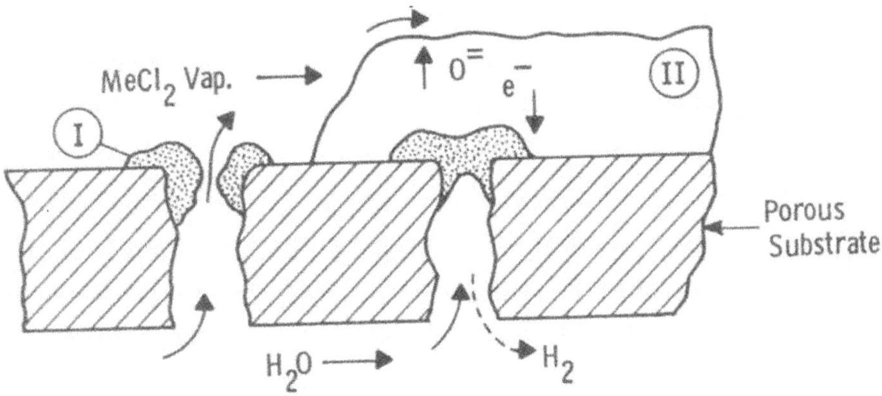

Phase I — Pore Closure by CVD

$$MeCl_2 + H_2O \rightarrow MeO + 2HCl$$

Phase II — Scale Growth by EVD

$$MeCl_2 + O^{=} \rightarrow MeO + Cl_2 + 2e^{-}$$

$$H_2O + 2e^{-} \rightarrow H_2 + O^{=}$$

Figure 3. Principle of electrochemical vapor deposition.

two phases. The first is the chemical vapor deposition for closing of the pores in the substrate, and the second is the uniform growth of the oxide material due to both ion conductivity and a very small amount of electronic conductivity in the film.

2.3 Cell and Module Configurations

2.3.1 Tubular Designs. Three designs are presently being pursued by the three organizations mentioned earlier. These can be referred to as series connected cells on a tubular substrate support, (Electrotechnical Laboratory), self-supporting series connecting cells, (Dornier Systems), and single cell on a support tube, (Westinghouse). The multiple series connected cells on a support tube was originated by Westinghouse but abandoned due principally to the difficulty of interconnecting the ends of individual tubes in an oxiding atmosphere to construct a module. Self supporting series connected cells, originally conceived by Brown-Boveri, are now used only by Dornier for high temperature water vapor electrolysis modules, not fuel cells, and will not be considered in this chapter. The single cell on a support tube design of Westinghouse will be described in some depth as many thousands of such cells have been built and tested and in addition experimental modules, generators, and power systems have been constructed from them.[3] Figure 4 is a perspective sketch showing the five components of the single tabular cell. The circumferences enclosed by the electrolyte and the interconnection provides the dense, gas impervious layer which separates the air on the inside of the tube from the fuel on the outside. It should be appreciated that during cell operation both of these materials are subjected simultaneously to an oxidizing and reducing atmosphere on opposite sides of their surfaces.

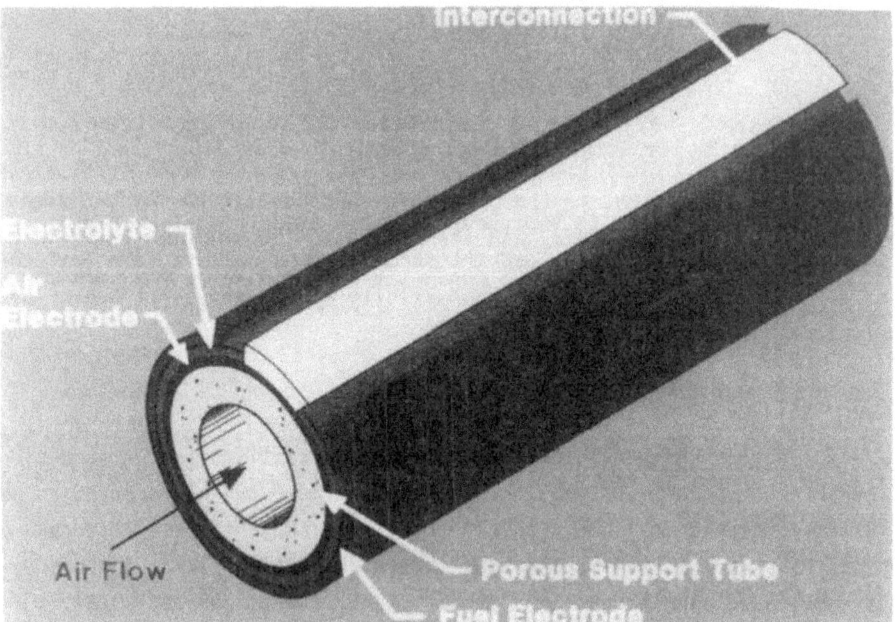

Figure 4. Sketch of a single solid oxide fuel cell.

The gaps on each side of the interconnection at the ends of the fuel electrode provides the insulation between the positive (interconnection/air electrode) and the negative (fuel electrode) terminals of the cell. Diffusion of reactants and products can readily occur in the porous cermet fuel electrode and porous support tube and oxidic air electrode. Figure 5 is a photograph of an actual cell illustrating that it is closed at one end, and the actual cell structure does not extend over the entire length of the tube. Actual size is approximately 110 square centimeters of projected surface area, with a outside diameter of about 1.5 cm. and a length of about 30 cm.

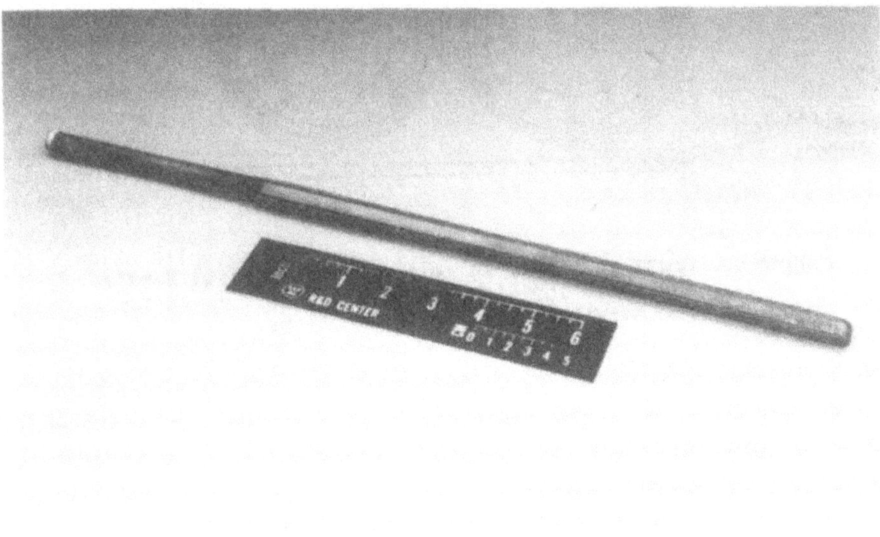

Figure 5. Single solid oxide fuel cell.

Figure 6 shows how cells are bundled together in both series and parallel to provide higher power in a module. A special material "nickel felt" consisting of nickel metal fibers sinter bonded to each other and to the nickel metal in the fuel electrode provides a soft, mechanically compliant, low electrical resistance connection between the cells.

Actual operation of the cells in a generator is illustrated in Figure 7, showing an air injection tube is inserted into the fuel cell tube to provide co-flow of oxidant and fuel. Excess air over that

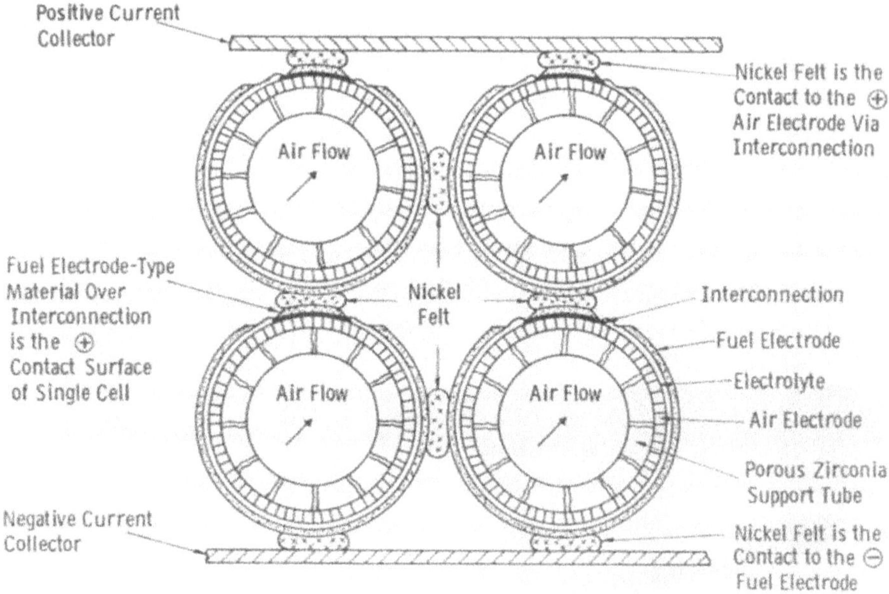

Figure 6. Cross sections, to cell axes showing solid oxide fuel cell individual layers, as well as series and parallel electrical conections.

needed for the electrochemical oxidation of the fuel is needed for thermal control. The oxygen depleted air and depleted fuel gas combine above the open end of the cell to chemically burn the residual fuel content of the exhaust gases.

A photograph of 24 cells bundled together with the first insulation layer surrounding them on two sides is shown in Figure 8a. The sketch, Figure 8b shows how they are connected, three in parallel and eight in series. The top nickel fiber bus has not been assembled yet in Figure 8a.

The generalized design concept for a generator module is illustrated by the perspective drawing of Figure 9. This shows the cell bundle bottom end resting on a sole plate above the fuel plenum, the cell top ends extending through a porous tube sheet and the air injector tubes hanging from their tube sheet. The volume between the tube sheets is the combustion chamber and heat exchanger which preheats

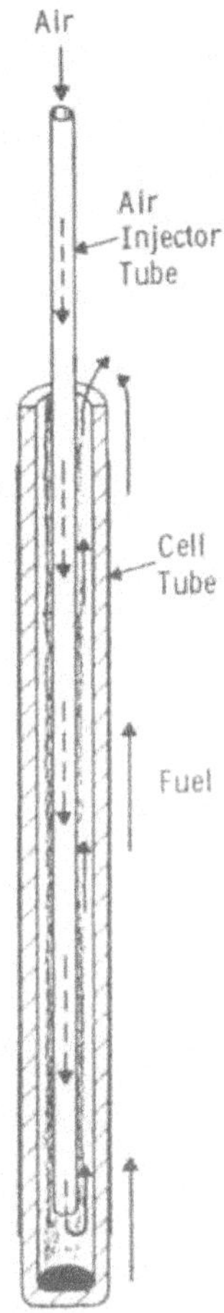

Figure 7. Schematic of cell tube showing both the open end and the closed end. Air or oxygen is introduced through an injector tube into the cell near the closed end.

the incoming air before it gets into the fuel cells. Details of this modular design are proprietary.

2.3.2 Monolithic/Planar Designs. The monolithic/planar design approaches appear to offer the possibility of providing a solid oxide fuel cell module with perhaps 10 to 100 times greater power per unit weight and volume than the tubular designs. By far the greatest effort in this direction has been performed by the Argonne National Laboratory

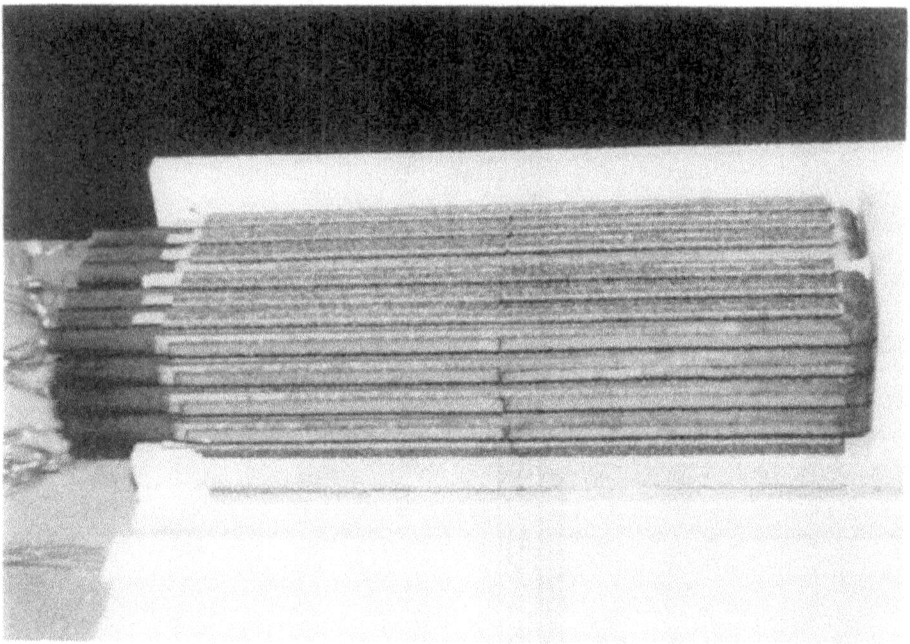

Figure 8a. View of 24 assembled cells.

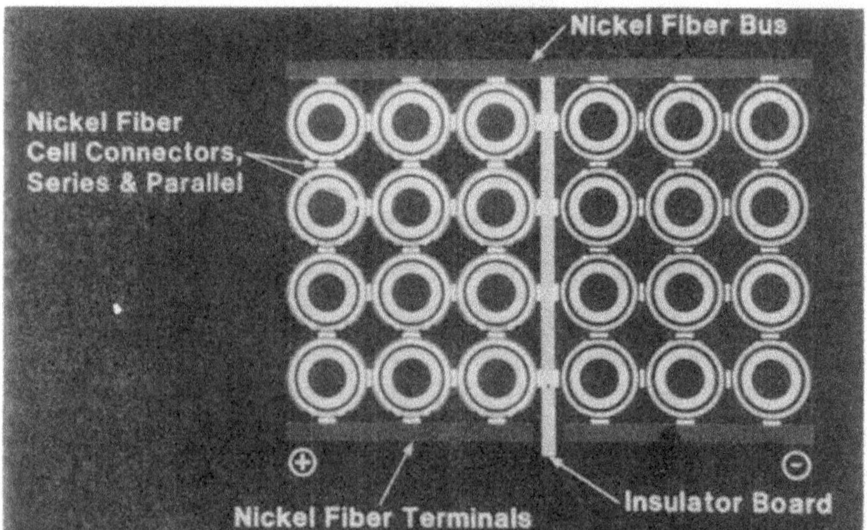

Figure 8b. Schematic arrangement of interconnecting cells.

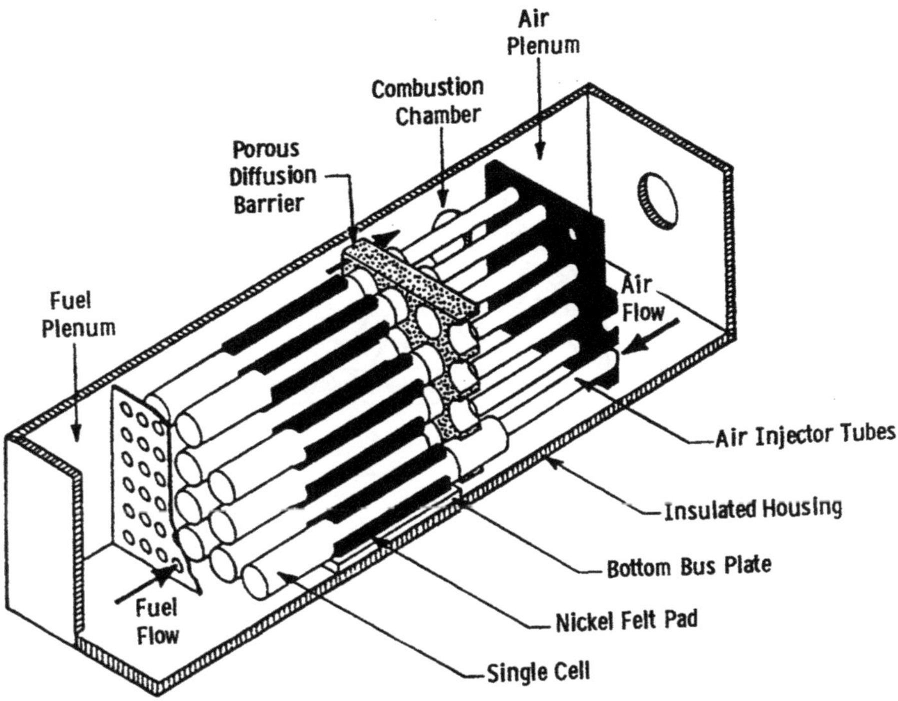

Figure 9. Generator Module Design Concept (Westinghouse).

who first proposed the concept shown in crossection of Figure 10.[4]) This conceptual design requires advances in the manufacturing of ceramics based on tape-casting technology. Two triple layer sub-assemblies shown on the left of Figure 10 are prepared by sequentially tape casting the anode and cathode material and layering them to the tape cast electrolyte or interconnecting material. Next, the anode/cathode/electrolyte layered material is corrugated. Finally, after drying the sub-assemblies are stacked and co-fired to produce the monolithic structure indicated in Figure 10, and designated the MOD-1 configuration. Clearly, due to the 180 degree counterflow arrangement of fuel and air flow and the small (several millimeter) dimensions of the channels, the manifolding of the reactants presents a very difficult design problem. A different fabrication procedure has been suggested by Garrett Corporation to alleviate this problem, but it is not known whether those efforts have been successful or not.

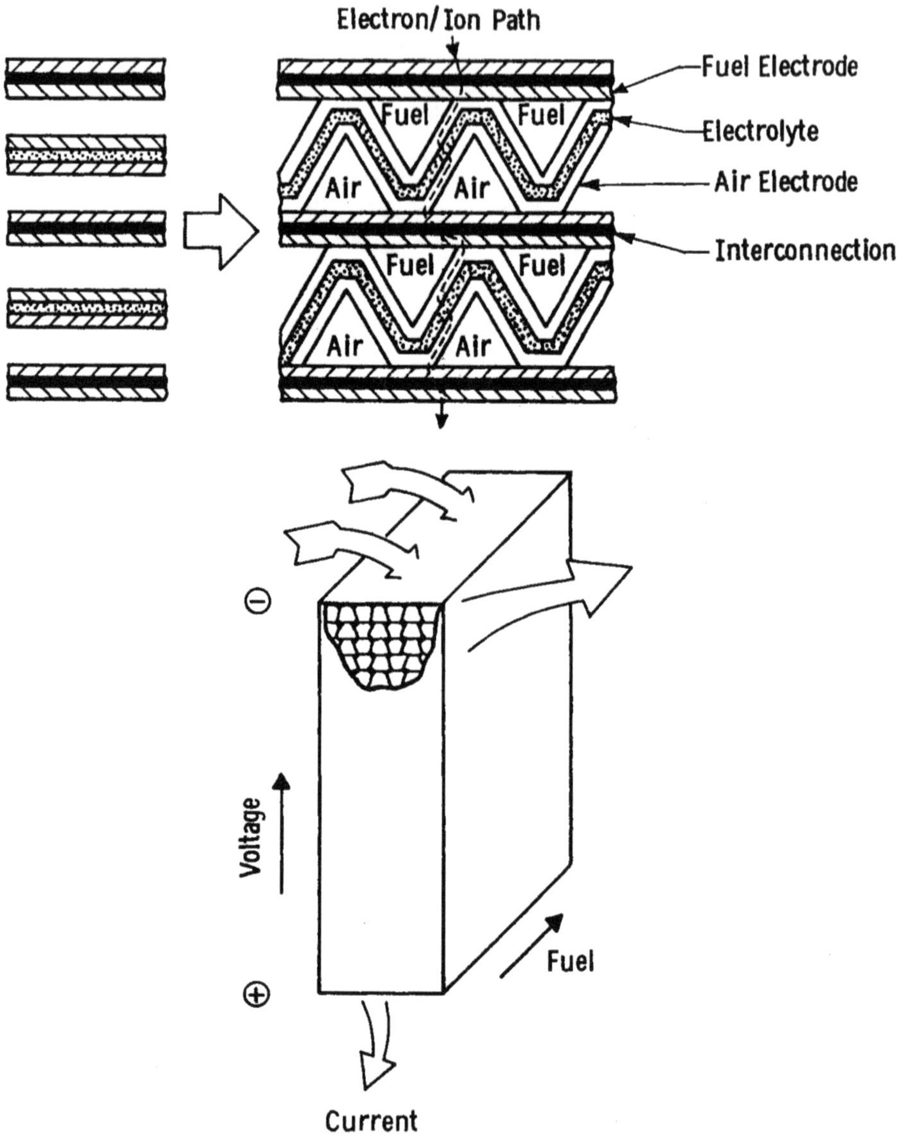

Figure 10. Argonne Multi-Channel Corrugated Design, (MOD 1 configuration).

Argonne has tried an alternative arrangement they have called the MOD-0 configuration. This permits the tape-cast triple layer subassemblies to be combined in a cross-flow arrangement shown in Figure 11 (except that alternate fuel and air layers are turned 90° to that

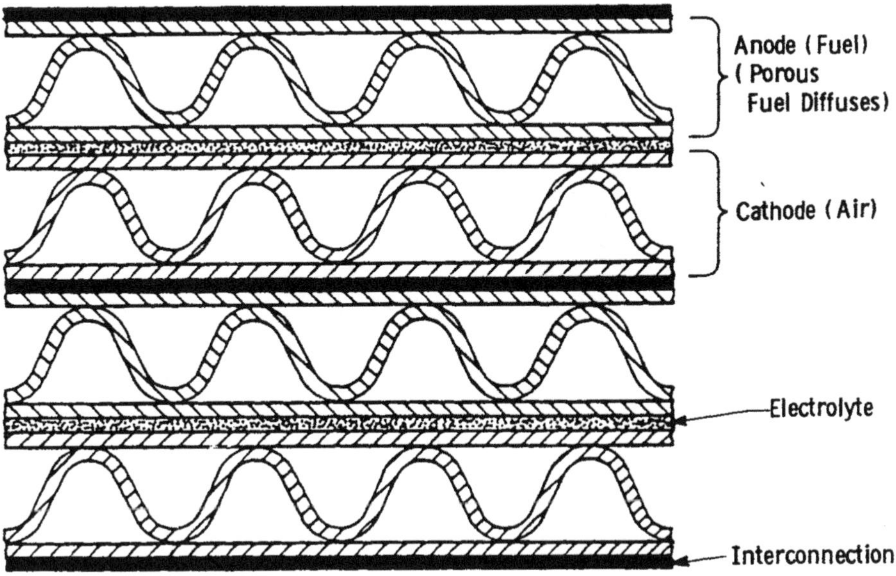

Figure 11. Argonne Corrugated Monolithic Reactor, (MOD-0 configuration).

shown in the figure). The air and fuel channels are formed by corrugated sections of porous cathode and anode material. In this configuration only half of the reactant channels are in immediate contact with the electrode/electrolyte region. It has been rationalized that diffusion of the reactants can take place through the porous electrodes and allow a sufficient fraction of reactant to be utilized. Figure 12 shows how the MOD-0 module can be manifolded for reactant input and product output.

Argonne has indicated the specific power of the MOD-0 is projected to be about 40% that of the MOD-1 configuration. No performance data is known to exist on the MOD-1. Some performance results have been reported on the MOD-0 which will be related in the performance section.

Materials used are basically the same as those mentioned earlier for SOFC's. However special additives are made to the tapes to enable all materials to co-sinter into the monolith. Especially difficult in this respect is the magnesium modified lanthanum chromite interconnection. Long time tests will be required to determine if diffusion of the sintering aids has an effect on performance.

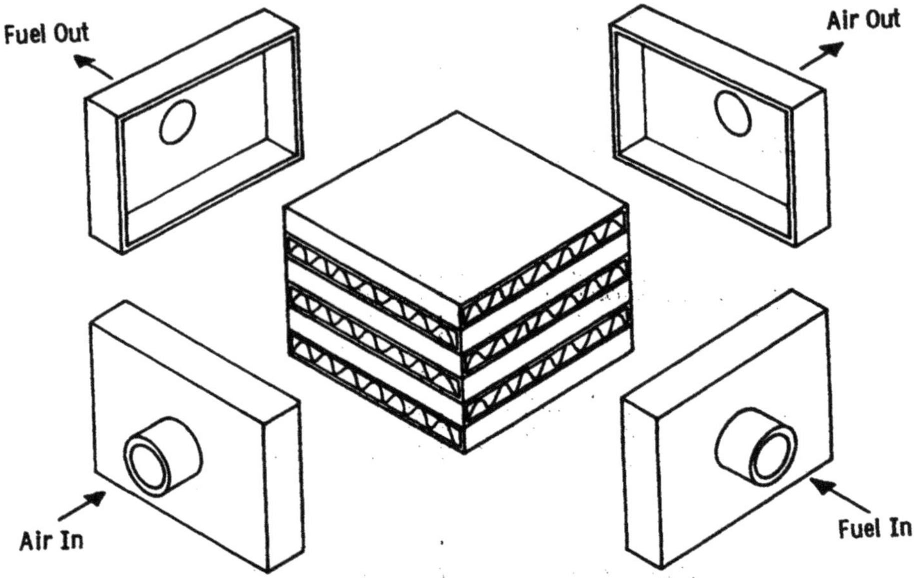

Figure 12. Manifold arrangement for Argone MOD-0 configuration.

A honeycomb monolith structure has been suggested by B. C. H. Steele[5)] as a configuration that could provide a power density of 1 MW/m^3 based on its conceptual design surface to volume ratio about 10 cm^2/cm^3. This configuration is illustrated in Figure 13. Although ceramic honeycomb structures are made by the millions for automobile catalytic converters, the inventor states critical technical issues are whether processing can be developed to put the electrode materials on the electrolyte honeycomb and whether they will stay there during thermal cycling. Also the conductivity of the cathode material is too low to be accommodated in this structure and a new material would have to be developed.

No known performance data exists for this configuration.

A recent planar design that looks very much like a phosphoric acid fuel cell stack has been suggested by the UTC/Ceramatec consortium organization mentioned earlier. Two three layered sub-assemblies as shown in Figure 14 would be assembled by stacking. This would provide for a cross-flow arrangement of fuel and oxidants. Critical technical issues appear to be materials and processes for edge sealing and

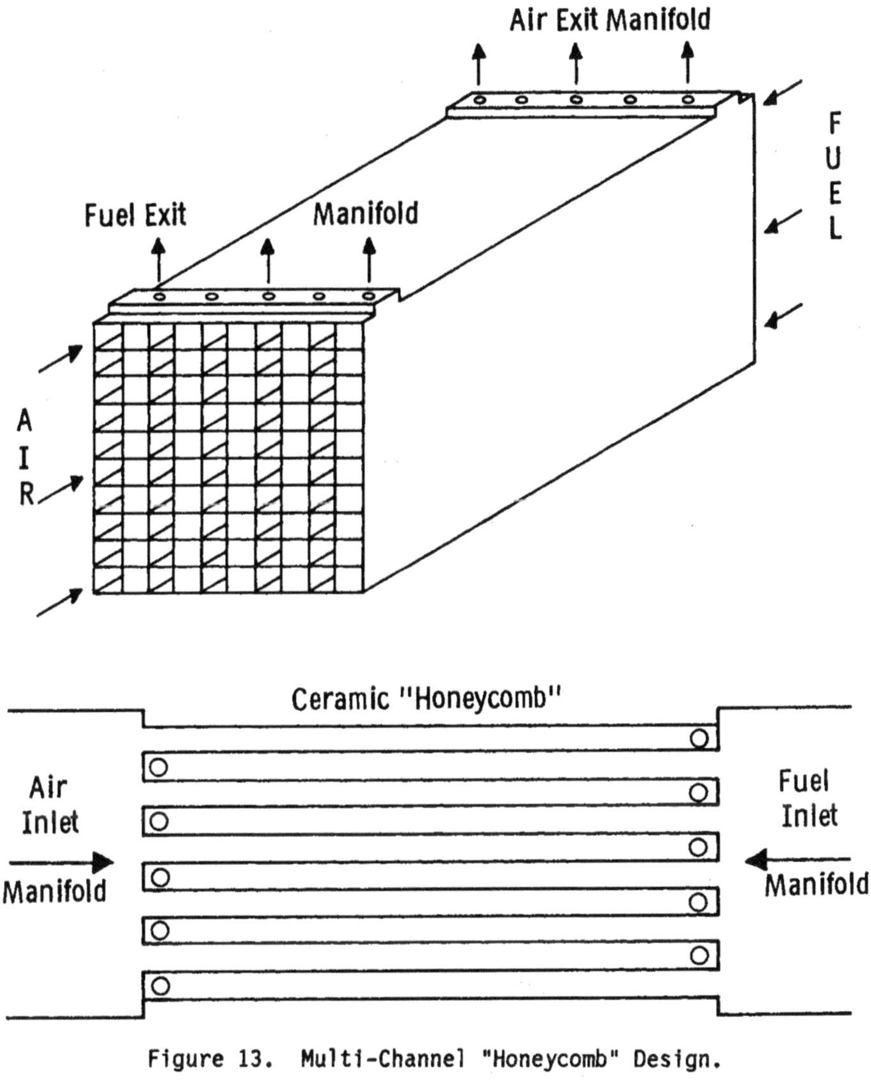

Figure 13. Multi-Channel "Honeycomb" Design.

providing a structural design which is sufficiently mechanically compliant during thermal cycling to prevent cracking.

No known performance data is available for this configuration.

ZTEK has been developing a stack constructed of thin zirconia electrolyte plates with applied electrodes on each side.[6)] These are stated to be designed to be about 5 cm. diameter plates with metallic

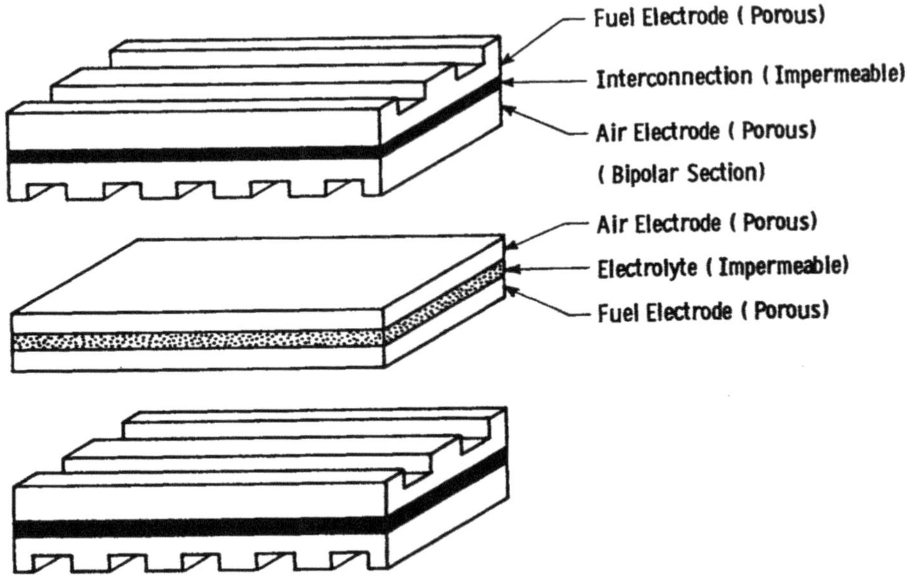

Figure 14. Recent design for SOFC Planar Configuration.

interconnector plates interspersed as bipolar type interconnectors. Internal gas passages entrance and exhaust are provided by grooving the interconnector plates which are pressed against the electrolyte plates and the stack held together with tie bolts. Some preliminary data has been presented on this structure which will be discussed briefly in the performance section.

2.4 Cell and Module Performance

Fuel cell performance data are usually presented as voltage-current graphs. However in order to make fair comparisons between various systems detailed information about how the V-I information was obtained is required. For example fuel composition, fuel utilization (i.e. the percentage of the high heating value of the fuel introduced to the cell that has been electrochemically converted to electricity and heat), oxidant composition and utilization all can dramatically affect the data presented. If the fuel is passed over the reacting cell surfaces such that only 5-10 percent of the heating value is electrochemically reacted the voltage will be higher at any current than when the fuel is 85-90% electrochemically reacted.

Because of the above it has not been possible to obtain good comparison data on even all types of solid oxide fuel cells, much-the-less on various types of fuel cells. Nevertheless data that is available will be presented for the various solid oxide type fuel cells.

2.4.1 Tubular. Figure 15 shows the V-I performance data on a single Westinghouse tubular state-of-the-art solid oxide fuel cell over the temperature range of 700°C to 1000°C at constant fuel and oxidant utilization. Also shown is the difference in performance between air and oxygen at 1000°C. This is the type data that must be available for valid comparisons between systems to be made. It is interesting to note that lower temperatures extrapolate to higher open circuit values as thermodynamics dictate should occur, but at the higher temperatures, higher voltages are obtained at the higher current densities due to higher electrolyte conductivity and lower cell ohmic resistance and polarization losses, also as expected. At the typical design point for use of this cell in a generator, namely 250 ma/cm^2, one obtains about 640 mV, giving a design power density of 0.16 $watts/cm^2$.

The actual power vs current curves for this cell at various temperatures is shown in Figure 16. This data is derived from the V-I curves. Peak power for this cell under these reactant conditions at 1100°C is seen to be about 26 watts at 47 amps (0.55 volts and 350 ma/cm^2) or approximately 0.193 $watts/cm^2$.

Efficiency of the cell, i.e. the ratio of the dc power output compared to the power possible from the high heating value of the fuel flow rate input, can also be calculated. Figure 17 shows this graph for a different single cell under the defined conditions shown. This somewhat shorter cell is rated at 16 watts (250ma/cm^2), where its efficiency with air is about 45 percent. Over 50 percent efficiency can be obtained at lower power outputs, and at peak power (about 140 percent of rated power) the efficiency drops off to 31 percent. This type of performance actually permits generator design options. Where

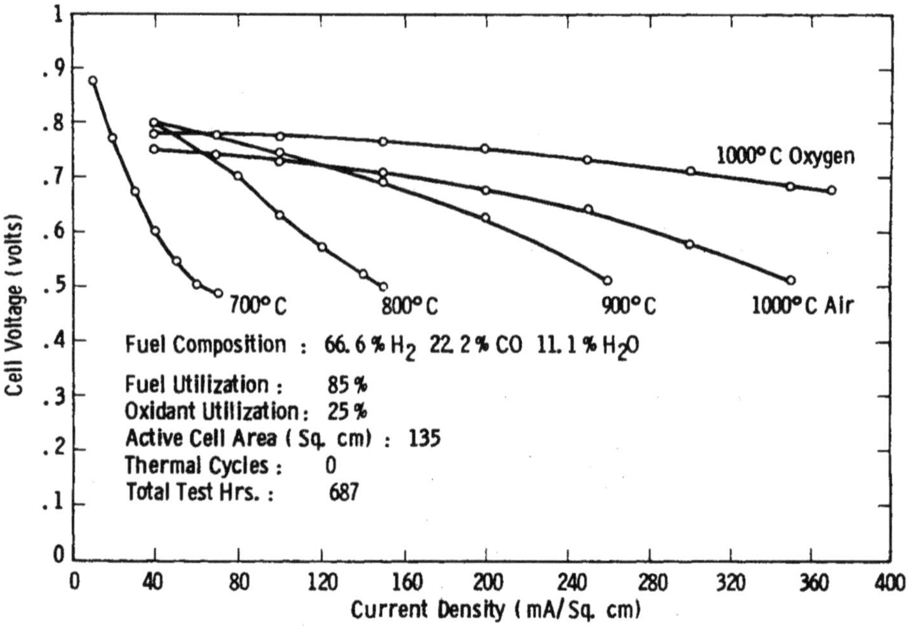

Figure 15. V-I curve for state-of-the art SOFC.

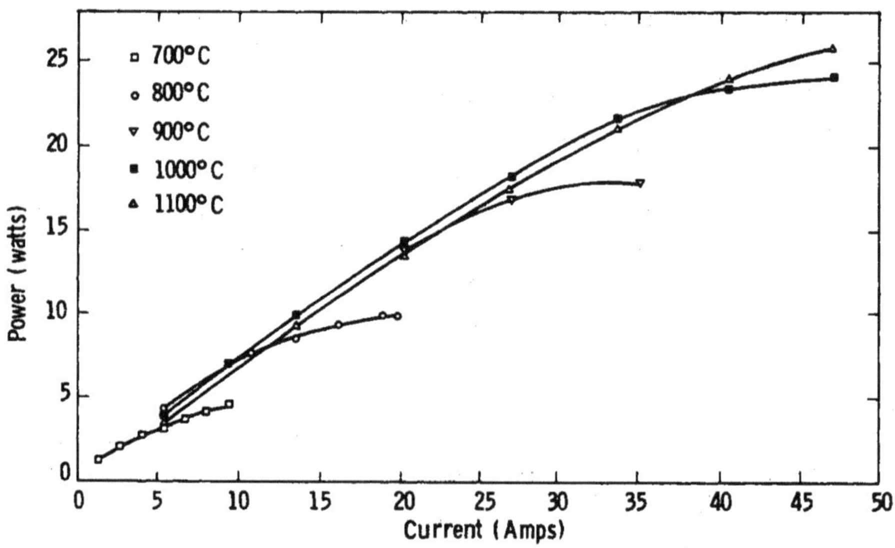

Figure 16. Peak Power Curves for Test Cell

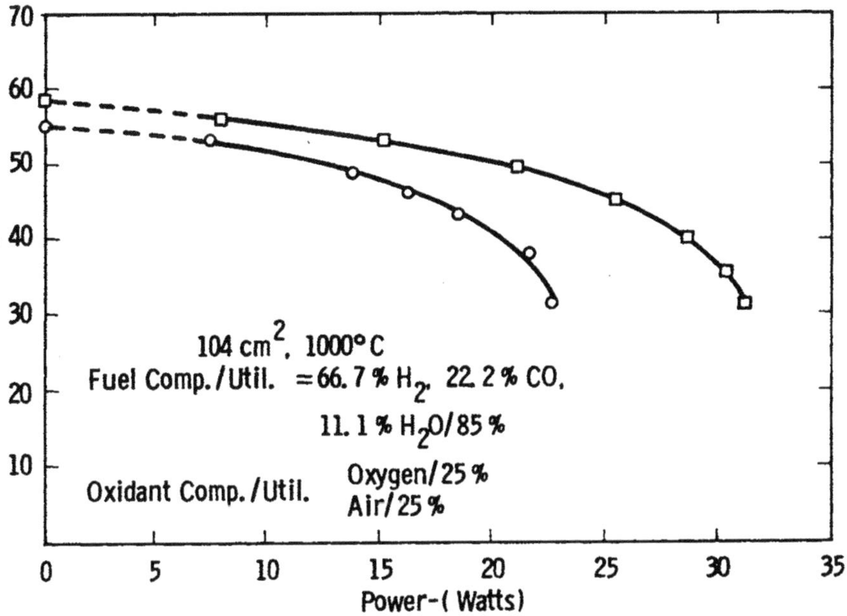

Figure 17. Efficiency vs. power of state-of-the-art SOFC's (Methane Equivalent).

fuel cost is high units can be designed for high efficiency. Where capital costs are most important, designs can favor higher cell power output conditions.

Life of a fuel cell is a major issue to be investigated. Figure 18 shows the stable voltage performance at the designed current density obtained from a three cells in series test that ran for over one year. At the end of this 9000 hour test the cells were examined for morphological changes and interdiffusion of various elements between the components. Scanning electron microscopy and electron beam microprobe analysis found neither of the above.

Single cell performance data is of course basic, but stack data or as tubular cell arrays are more appropriately called "bundles" data are also needed for system design. Figure 19 shows the performance of an early 24 cell bundle module. These cells were arrayed as eight in series and three in parallel. Eighty amperes was approximately 250 ma/cm^2. A voltage of about 5 volts at 85 percent utilization gives

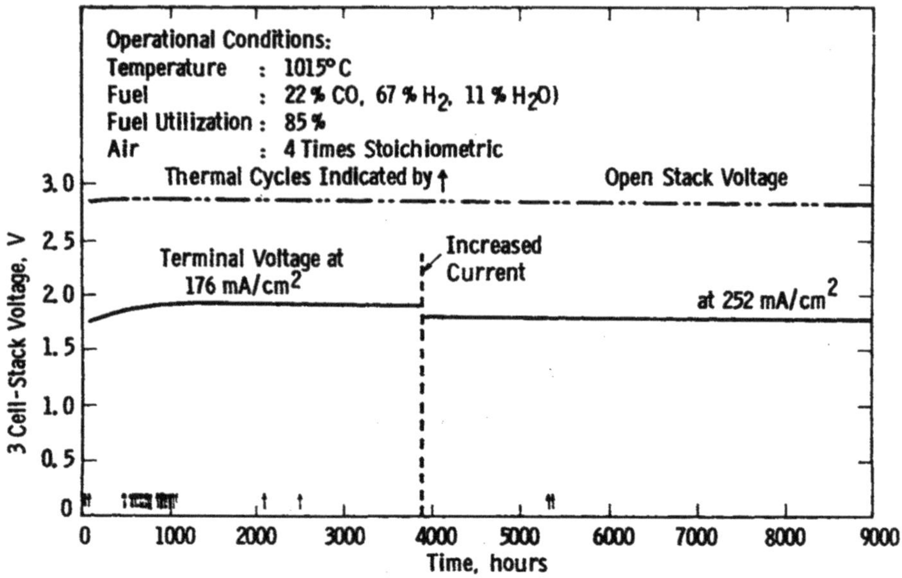

Figure 18. Voltage-time curve for three-cell stack.

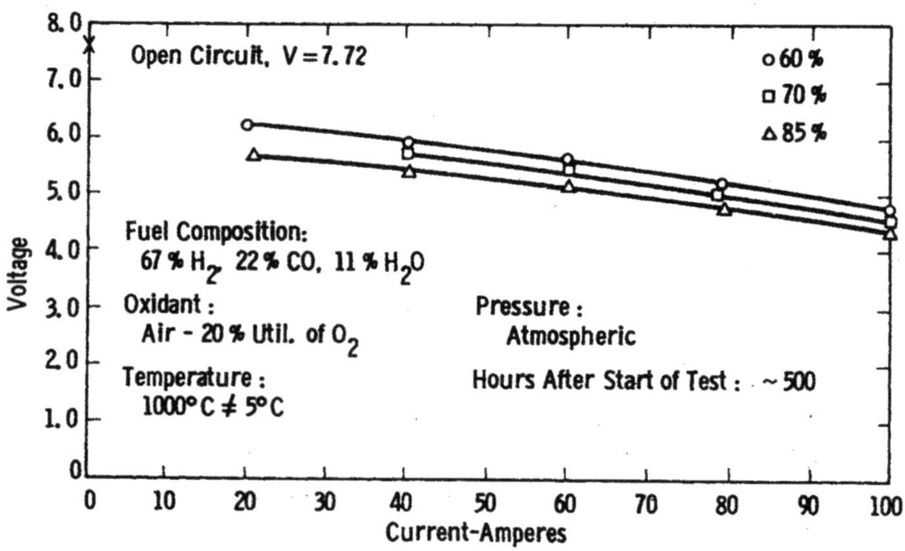

Figure 19. V-I performance at constant fuel utilization of a 24 cell SOFC test module.

performance results very close to those expected from single cell tests. This figure also illustrates that at the lower fuel utilization higher voltage is obtained, as was pointed out in the earlier discussion should be expected. The efficiency of this module test as a function of power output is also shown in Figure 20. Here again results similar to single cell test are obtained at rated power (16 watts/cell x 24 cells = 384 watts) and 85 percent fuel utilization, i.e. about 45 percent efficiency. Notice the overall lower efficiency also shown for the lower fuel utilizations. This is of course the reason generators are designed for higher utilization, in general. An exception could be for a cogeneration application where the need is for a high heat to electricity ratio.

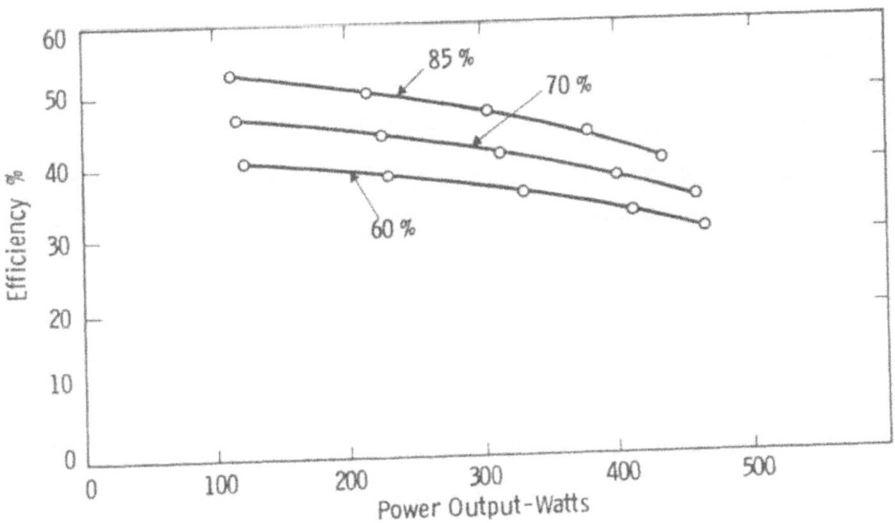

Figure 20. Efficiency vs. power at constant fuel utilization of a 24 cell SOFC test module.

Larger modules have also been built and tested. Figure 21 shows 324 cells being aligned prior to final assembly of a 5 kW test generator. These cells were interconnected three in parallel and 108 in series. A photograph of the test loop with the generator module in the center is shown in Figure 22. The fuel and air monitoring system are on the left, and the load bank and data acquisition equipment is on the right.

Figure 21. Alignment of 324 cell SOFC test module.

The general arrangement of the internal components of the module is shown in the perspective drawing of Figure 23. The fuel inlet into a manifold under the cell bundle, the electrical terminals on the bottom, the air inlet manifold holding the air injector tubes, the depleted fuel combustion chamber, and the path for the combustion products around the air manifold and out the top, all surrounded by thermal insulation, is diagramed.

A power curve at various currents, about 100 hours into test, is shown in Figure 24. Eighty amperes is about 250 ma/cm^2 and the efficiency at this point is just under 45 percent, again correlating very well with single cell tests.

2.4.2 Monolith. Data on monolith configuration solid oxide fuel cells available from Argonne National Laboratories will be discussed in this section.

Figure 22. 5 kW Test Loop.

Performance of initial cell tests of a thin electrolyte film with electrodes laminated to it is shown in Figure 25. The cell was about 2 cm in diameter and was set up so that fuel impinged upon one side of the cell disk and air on the other. About 600 mv was obtained at 50 ma/cm^2 using hydrogen as the fuel. The voltage remained stable for about 75 hours and then fell off rather abruptly. No information on the utilization of the fuel was given but it is suspected that is was low in order to obtain the highest voltage possible, which is understandable for exploratory testing. A polarization plot for improved cells of this type is shown in Figure 26. Again the utilization of the hydrogen fuel was undoubtedly very low.

Results on an early four cell actual monolith structure are presented in Figure 27. The cells were arranged in the so-called Mod-0,

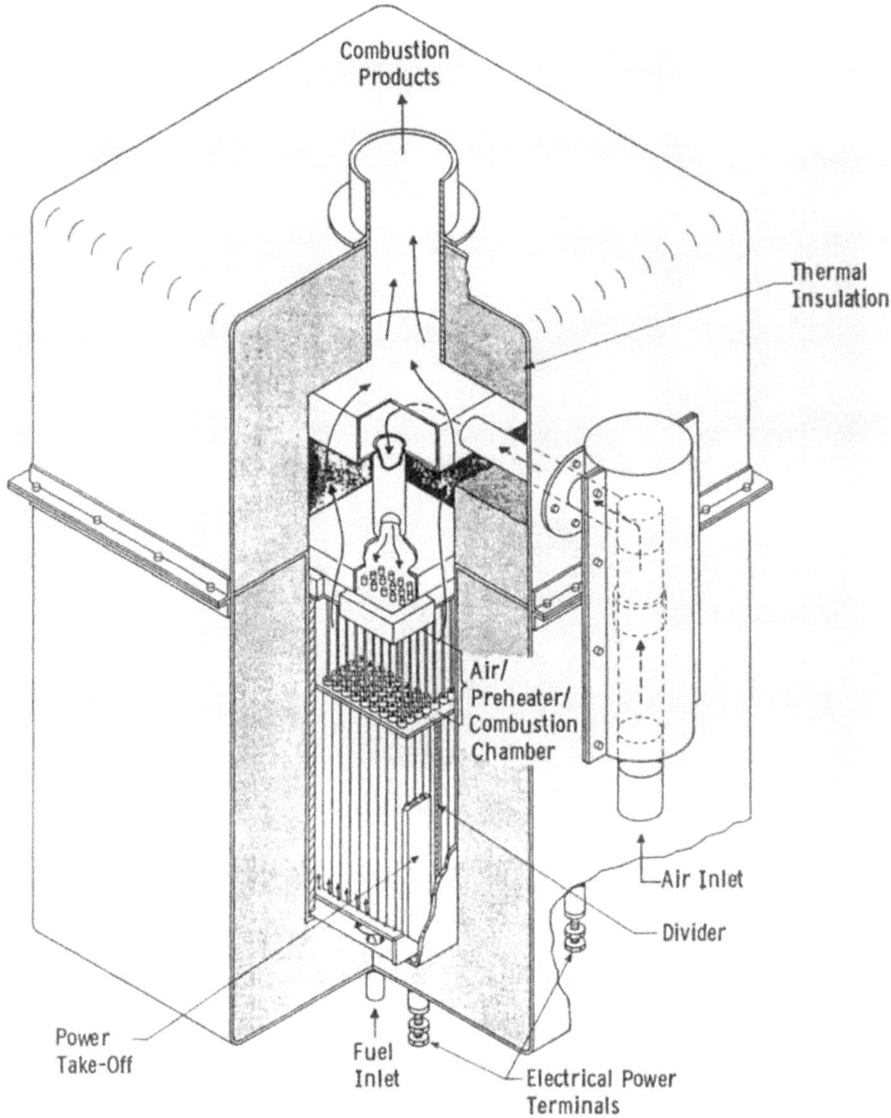

Figure 23. Cut-away view of the Westinghouse design SOFC generator module.

i.e. crossflow of fuel and oxidant (lower power density) design. The cell layers were about 10 cm on a side and 1-2 mm thick. Relative voltage stability for over 600 hours at about 500 mV per cell was achieved. Current density was not stated. Fuel utilization was not

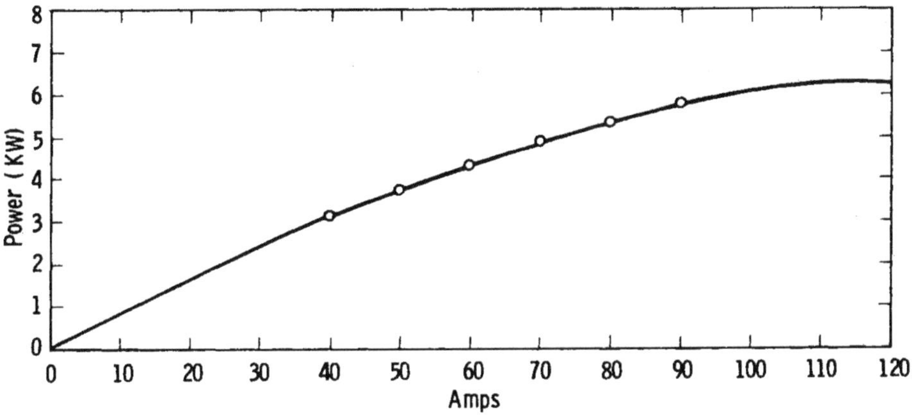

Typical Data 100 Hours into Test II

Fuel Composition	$H_2 = 66.3\%$; $CO = 22.7\%$; $H_2 = 11.0\%$		
Fuel Utilization	84.5 %	Generator Current	79.96 Amps
Generator Voltage	65.8 Volts	Generator Power	5,261 Watts

Generator Efficiency $\left[\frac{\text{Electric Power Out}}{\text{HHV } CH_4 \text{ Eqiv. In}}\right] = \frac{5,261}{11,731} = 44.8\%$

Figure 24. Power curve of 5 kW test module.

stated. The cell stack unit was known to have a number of cracks in the various components. Even with small cracks stability can be achieved at low current density when the fuel and air are sent rapidly, i.e. at low utilization, through the stack.

2.4.3 Planar. ZTEK has stated that a single cell and two cell planar stack has demonstrated 0.12 watts/cm^2 area power density, and further improvement can be anticipated. Fuel and oxidant composition and utilization have not been reported.

A five cell stack has been thermally cycled at 100°C/minute. No performance data is reported before and after, but the ability to thermally cycle the stack is attributed to the highly conductive metallic interconnectors.

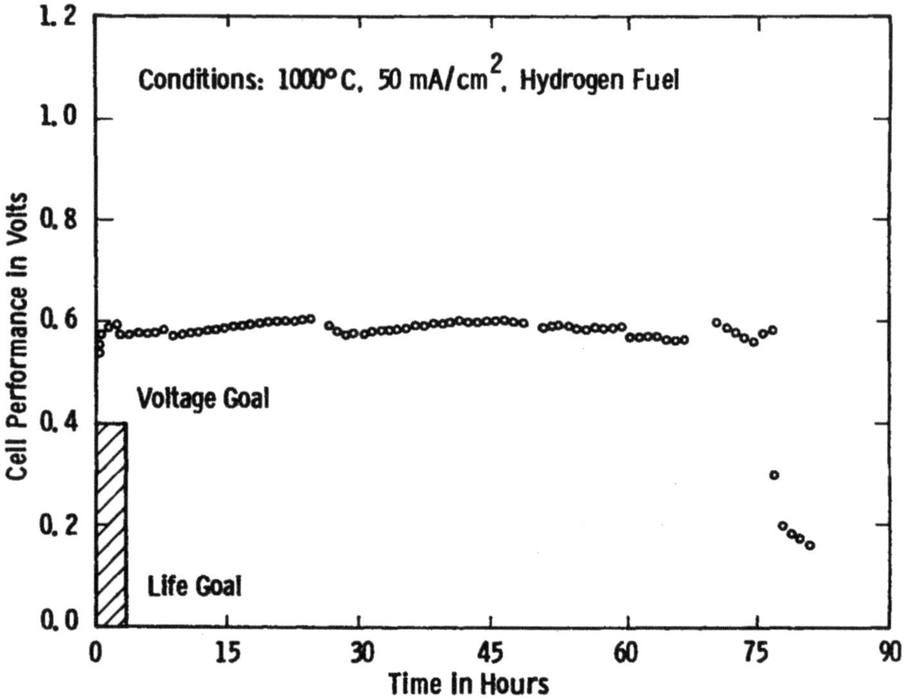

Figure 25. single cell monolith performance test.

3. DEVELOPMENTAL ACTIVITY

3.1 Materials Research

Current materials used for SOFC's have been subjected to more than twenty years of research and development and a great deal of operating data has been obtained. Actually processes for producing these materials are known and of reasonable cost. However one always can envision an important improvement in design could occur if new materials were available. The $LaSrMnO_3$ cathode material, as has been previously shown, has a higher than desired resistivity. It is interesting to note that world wide activity on similar type materials for "warm superconductors" could give clues to obtaining a higher conductivity cathode material for SOFC's.

It has also been known that an electrolyte material that could operate in the 600 to 800°C range as well as stabilized zirconia does at 1000°C would have advantages for system design. The temptation to

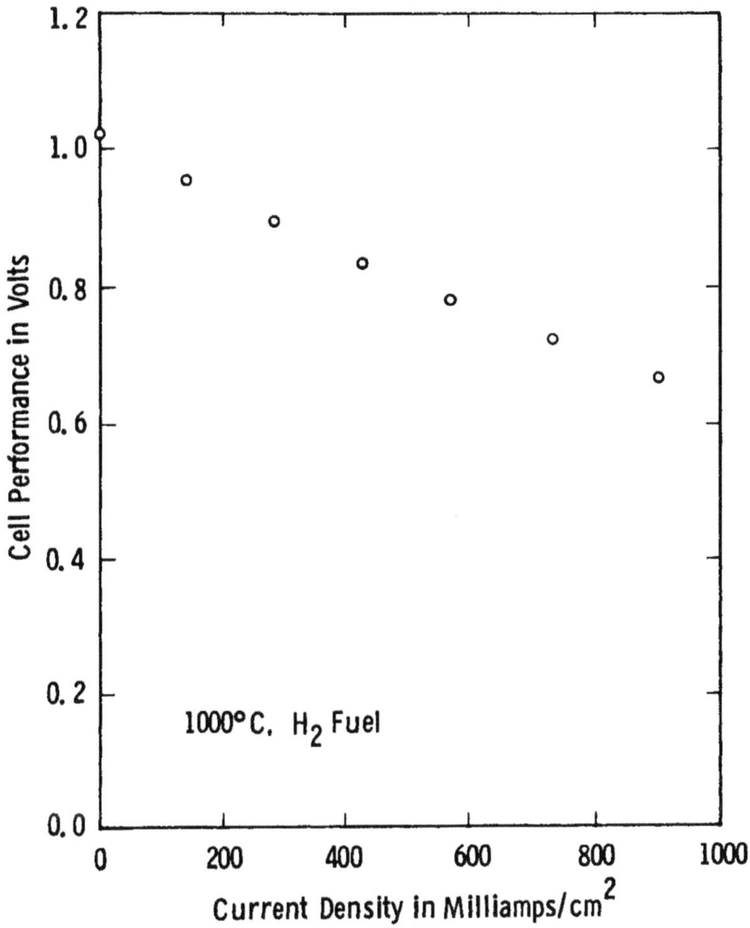

Figure 26. Single cell monolith polarization plot.

look for such materials has been pursued by many materials scientists but no new electrolytes with all required properties has yet been found after many years of good but frustrating research.

Studies on the effects of impurities in fuels are being carried on. Principally sulphur interaction with SOFC materials is a concern and although SOFC's operating at 1000°C do have greater tolerance to sulphur impurity than other types of fuel cells, desulfurization of fuel to less than 10 ppm is likely to be a requirement. Operation at

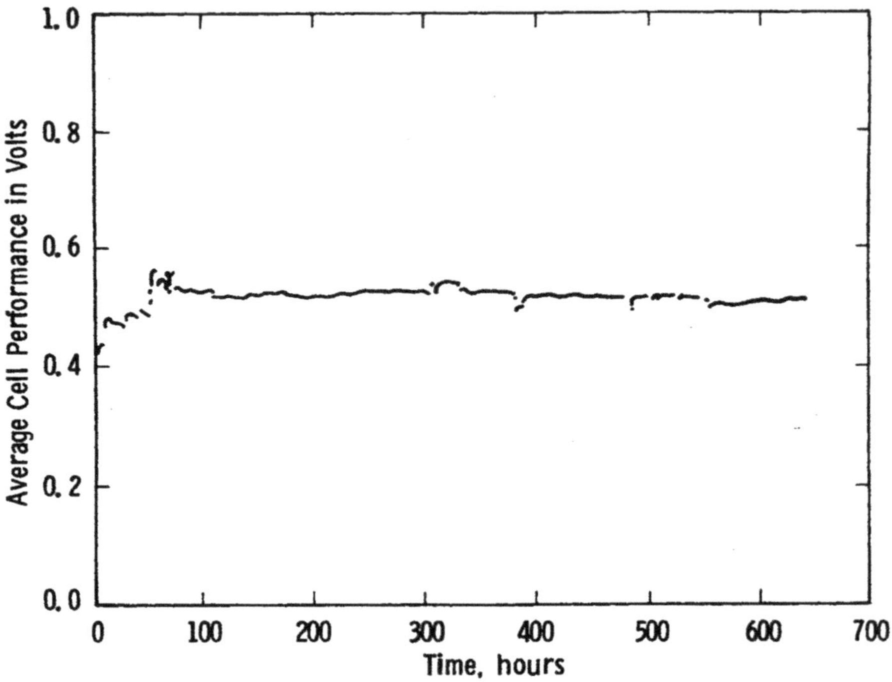

Figure 27. Four cell monolith array test (Mod-0).

1000°C precludes the formation of sulphides of nickel when sulphur is present at less than 100 ppm in the fuel, but adsorption on reaction sites can still take place. Sulphur will desorb when eliminated from the fuel. Alkali metals in coal-derived gas could have a deliterious effect on SOFC materials. Studies still need to be performed.

Modification of materials to allow co-sintering of the various component films for monolithic designs is a major thrust. Testing of these structures to show no harmful effects from the introduction of sintering aids, still needs to be done.

3.2 Cell and Stack Design Modeling

Three dimensional mathematical models are being developed for the various configurations which help understand the complex interaction of current variation, heat generation and transfer, and stresses developed. This is absolutely necessary in ceramic structures that are susceptible to cracking during thermal disturbances. In the tubular configuration increasing the cell size up to perhaps several hundred watts per cell is a significant ongoing developmental effort. Models to understand cell behavior when the design is changed are being developed.

3.3 In-Situ Reforming

Of major interest for SOFC's is that the conditions of operation are appropriate for the reforming of natural gas directly on the cell surface. Water for the reaction:

$$CH_4 + H_2O \rightarrow CO + 3H_2$$

is formed on the anode of the cells. Also the SOFC can react both CO and H_2 in any mixture. This leads to the possibility of systems being developed which do not need separate reformers, which could lead to simplification, efficiency improvement, and significant cost reduction. Test modules are being operated at Westinghouse to demonstrate this capability of SOFC's. A concern is that the above reforming reaction is endothermic and inducing local thermal disturbances on cells could have a deleterious effect on the cell bundle, if not adequately accommodated by proper thermal design.

4. CONCLUSIONS

Solid oxide fuel cell systems are expected to have all of the usual advantages mentioned for all types of fuel cell systems vis-a-vis traditional energy systems. Usually mentioned are the potential for being:

- Clean
- Efficient
- Quiet
- Reliable
- Modular
- Simple Thermal Management
- Versatile

In addition however the solid oxide fuel cell will have significant advantages over other fuel cell systems. Operating at the high temperature of 1000°C it shows the promise of attaining higher overall system efficiency. Furthermore SOFC's produce higher quality exhaust heat at greater than 600°C which can be a valuable energy source for cogeneration systems or a bottoming unit for all electric power plants, which will increase plant efficiency.

The high temperature of operation of the SOFC makes catalysts unnecessary and leads to simplification of fuel processing and simplified system design. Equally important, the use of a solid state electrolyte eliminates material corrosion and electrolyte loss maintenance. There are, of course, no constraints on carbon monoxide in the fuel, and the solid oxide fuel cell is also tolerant to sulfur impurities in the fuel. This tolerance to impurities reduces fuel clean-up cost and eliminates restrictive operating procedures. Thermal control of the system is by cooling with process air mitigating the need for separate liquid or liquid/vapor cooling loops. Units can be packaged so they are thermally self-sustaining at power levels that range from a few kilowatts to hundreds of megawatts. The SOFC can perform across a wide power level range, from peak power to about 10 percent of peak power. Further it can follow load changes at a rate adequate for virtually all residential, commercial and industrial applications. Maintenance costs should be low and there are no acid or lubricant leaks.

With all these attractive features and the added benefits of the ruggedness of the technology and the low cost of materials, the SOFC systems, when fully commercialized, should serve a wide range of power and heat applications, such as:

- Industrial cogeneration
- Electric power production
- Commercial/residential total energy systems
- Space power

Successful commercialization in the mid-1990's will depend on a combination of timely developmental program progress and good experience with mechanized process equipment which can produce cells rapidly and at low cost.

5. REFERENCES

1. Bauer, E., Preis, H., Z. Electrochem, 43, 727, 1937.

2. Feduska, W., Isenberg, A.O., "High Temperature Solid Oxide Fuel Cell-Technical Status," J. Power Sources, 10, 89, 1983.

3. Brown, J. T., "Solid Oxide Fuel Cells; Technology and Applications," Energy Technology Conference, Wash. D.C. March, 1986.

4. Fee, D. C., et. al., "Monolithic Fuel Cell Development," Fuel Cell Seminar abs. 1986, Tuscon USA, 40, October 1986.

5. Steele, B. C. H., Science of Ceramics, 12, 697, 1984.

6. Hsu, M. S., "Zirconia Fuel Cell Power System Planar Stack Development," Fuel Cell Seminar abs. 1986, Tuscon, USA, 84, October 1986.

7. Brown, J. T., "Solid Oxide Fuel Cell Technology," IEEE Transactions on Energy Conversion, 3, No. 2, 193, June 1988.

6. ACKNOWLEDGEMENT

Much of the data presented on the Westinghouse tubular solid oxide fuel cell was obtained from work performed on cost share contracts with the Department of Energy and the Gas Research Institute. The support of these agencies is greatly acknowledged.

Future Expectations for the Application of Solid Ionic Conductors

Robert A. Huggins

Dept. of Materials Science and Engineering
Stanford University
Stanford, CA 94305 USA

Introduction

As evidenced by the other articles in this book, there is a great deal of interest in solid ionic conductors within the research community. It seems quite reasonable to ask where this field is going. Is it going to remain only a playground for the practitioners of science, or is it going to be nourished by many applications, in either the scientific or the technological realms?

This article will focus upon the question of applications, and will emphasize the wide range of possible opportunities that have become visible from a somewhat historical point of view. Except for a few comments at the end, the temptation to inject value judgements, to put them in rank order, and to make specific prophecies will be deliberately avoided. Because of the breadth of this topic area, there will also be no attempt to be comprehensive, or to be entirely up-to-date in discussing the latest trends and outlook.

Because of the great economical incentives for the development of effective ways of converting and storing energy, much of the attention that has been given to solid electrolytes and mixed ionic-electronic conductors has been either directly or indirectly related to their potential use in batteries or fuel cells. There are, however, many other types of applications of materials with highly mobile ionic species, both scientific and technological. Because battery and fuel cell - related matters are included elsewhere in this book, they will be, for the most part, not included here.

Practical utilization of solid ionic conductors has a long history, dating from the end of the last century, when doped zirconia was used as a broad spectrum light source(1-4), known as the "Nernst Glower", although the mechanism of its operation was not understood until much later(5).

Almost the same material was used(6) by Baur and Preis in 1937 as an oxide ion - conducting membrane in a fuel cell operated at 1000°C. This approach is still being pursued at the present time. Similar materials are now being employed in large number as oxygen sensors in the exhaust systems of automobiles, as well as other combustion systems, and for the the measurement and control of the composition of gas streams and liquid metals.

An early unusual application of a solid electrolyte made use of the change in ionic conductivity accompanying a phase transformation in silver mercury iodide at 50°C to make a fire detector, resulting in a Dutch patent in the late 1930's. Timing devices based upon solid state galvanic cells were used in the rockets developed near the end of World War II(7).

Solid electrolytes have also played an important role in the development of the understanding of point defects in solids and of many aspects of fundamental materials science over many years. They have both been subjects of investigation themselves, as well as being used as tools to investigate other materials and phenomena. There has been a long history of the use of solids as electrolytes in galvanic cells for the determination of thermodynamic data(8-15).

The following discussion of applicational areas is organized in terms of the type of configuration and conditions in which the material might be employed.

Applications of Solid Electrolytes Between Chemically Similar or Inert Electrodes

It has been found that the ionic conductivity of materials such as Li_5AlO_4 and Li_5GaO_4 changes by several orders of magnitude with variation of the moisture content in air. This has lead to the proposal that such materials might be usefully employed as humidity sensors at high temperatures(16).

Galvanic cells can be used as ionic or atomic pumps by imposing an electrical potential difference across them, causing a charge flux to occur. In such applications, it is important to employ electrodes that do not introduce serious impedances, and that the potential difference that is applied does not exceed the decomposition limit of the electrolyte itself. Oxygen can be readily pumped into or out of closed or flowing gas systems using doped zirconia electrolytes at temperatures above 600oC. Discussion of this application can be found in several places(17).

It has also been shown that one can electrolyze the CO_2 in human breath to reclaim oxygen, and thus produce a life-support device for use in prolonged space of undersea missions(18).

A similar scheme can be used to pump oxygen or other species into or out of liquids, including molten metals. It should also be possible to remove oxygen from other liquids as well.

At very high temperatures some solid electrolytes become mixed conductors, and can carry both ionic and electronic fluxes. Zirconia can be used as a heating element for very high temperature furnaces in oxidizing environments, for its melting point is over 2500 oC. This material is one of the leading candidates for use in electrodes in magnetohydrodynamic generators(19).

A novel type of thermoelectric device, based upon the use of beta alumina as a sodium - conducting solid electrolyte, has been described(20-22). In this case, the force causing the transport of the sodium ions through the solid electrolyte is due to the pressure difference between two heat reservoirs. Work can be extracted from the electronic charge flux in a coupled external circuit.

Passive Applications of Mixed Conductors

Mixed conductors, in which both ionic and electronic species can have appreciable fluxes, have some unique properties, leading to several types of possible applications.

Palladium has the electronic properties of a metal, and yet readily passes hydrogen. It is especially useful as a permselective hydrogen filter because it is not subject to oxidation in the atmospheres normally encountered in that use. A number of other metals, several with the BCC crystal structure, also have very high chemical diffusion rates for hydrogen. These materials exhibit many of the same characteristics as electronically insulating fast ionic conductors.

Some mixed conductors, such as certain insertion compounds of graphite(23,24) and some oxide bronzes, have high values of electronic conductivity, and have been considered as alternatives to metallic conductors in specific chemical environments in which normal metals are not stable. Other electronically conducting insertion compounds have also been found to be interesting chemical reagent reservoirs(25-27).

In addition, it is now recognized that a number of the important catalysts are mixed conductors. Insertion compounds of graphite are of interest for such purposes(28,29). It appears that some of the useful hydrogenation catalysts are materials which have relatively rapid transport of hydrogen through their lattices, and it has also been known for some time that some of the best oxidation catalysts for certain reactions, including variable valence metal oxides such as bismuth molybdate, behave as though the heterogeneous reaction involves lattice oxygen, rather than only species adsorbed on the surface(30).

Passive Applications of Solid Electrolytes Between Chemically Different Electrodes

Following the important early work of Carl Wagner(31-33), there have been numerous examples of the use of galvanic cells with solid electrolytes as chemical potential sensors.

These often involve the use of doped zirconia for the measurement of differences in oxygen activity or pressure. Doped thoria has been used in cases involving lower oxygen pressures, where zirconia becomes electronically conducting. Materials based upon bismuth oxide can be employed if the oxygen activity is maintained at sufficiently high values.

There have been many examples of the use of oxygen sensors for the measurement of the composition of gases(17). Reference electrodes are generally air, pure oxygen, or two-phase metal/oxide solid mixtures, such as Ni/NiO, Fe/FeO, Mo,MoO_2, or Cr/Cr_2O_3. A thin layer of porous platinum is typically used as the electronic conductor at the three-phase contact between electrolyte, gas, and electronic carrier.

The most common application of oxygen sensors is the measurement of the composition of the exhaust gases from internal combustion engines. Such a device, called the "Lambda Sond", was first introduced commercially by Volvo in its 1977 cars. They are now installed in many manufacturer's automobiles(34-36). Such a sensor is especially well suited to the determination of the stoichiometric ratio during combustion, due to the sudden change of about one volt in the measured output when the gas composition changes from reducing to oxidizing. Similar sensors have been used to monitor the gas composition in industrial combustion processes, metal heat treating atmospheres, and the analysis of the output of blast furnaces and the basic oxygen furnace.

Oxygen - conducting solid electrolytes have also been employed as sensors to evaluate the oxygen activity in liquid metals. Examples are molten Cu, Sn, Pb, Ag, Na, and various of their alloys. This is being used commercially in the production of copper alloys(17,37). There has also been a considerable amount of work on the much more difficult, but very important, problem of the use of oxygen sensors in liquid steels, typically at about 1600 °C(17,38). A number of papers were presented on this topic at the Tokyo Solid State Ionics conference(39).

Solid state cells can also be used to measure the chlorine activity in certain gases and liquids. For example, it has been demonstrated that the solid electrolyte $LiAlCl_4$ can be used for this purpose over a wide range of chlorine pressure(40). Cells using AgCl are being developed for use with minute integrated circuits for biological applications(41).

A lithium - conducting solid electrolyte has been used to determine the lithium content in molten metals, such as aluminum(42), and there are many examples in which materials of the beta alumina family have been utilized to obtain thermodynamic information in molten metal and metal/metaloid systems, primarily those involving sodium.

There have been commercial instruments available for a number of years that employ solid electrolytes or electrolyte/electrode combinations as "specific ion" sensors in aqueous solutions(43). The solid electrolyte LaF_3 is commonly used as the critical element in a sensor for fluorine, and a combination of silver sulfide (a mixed conductor) and various other phases has been used to determine the concentrations of a number of different ions.

Solid electrolytes have been used to determine thermodynamic data in a host of solid systems. The accuracy of the data obtained in this way is often considerably higher than that which can be readily achieved by the use of more conventional calorimetric techniques. A survey a few years ago(44) included data on the Gibbs free energy of formation in 85 binary oxide systems studied with solid electrolytes. Many ternary oxide phases have been investigated as well, and 53 examples are listed in (44), as is the free energy change involved in 31 cases of spinel formation.

The oxygen sensor technique, combined with coulometric titration, first introduced by Wagner(31-33) has been utilized to study the thermodynamic properties and compositional ranges of a good number of vario-stoichiometric oxide phases, and 52 were included in (44).

Fluorine - conducting solid electrolytes have also been used in galvanic cells between appropriately designed electrode combinations to evaluate thermodynamic parameters in binary and ternary oxide systems. 21 such systems are listed in (44).

Solid electrolyte cells have also often been used to obtain data on the thermodynamic properties of metal alloy systems. Experiments utilizing both oxide and fluoride solid electrolytes to study 40 solid solutions and 22 intermetallic compounds were listed in (37). Also cited were 13 examples of the use of these techniques to study liquid metal alloys. The influence of third elements upon the oxygen activity has been studied, and 28 ternary systems in which the oxygen activity has been measured are listed in (37).

While most of the work to date has involved metals and oxides, similar techniques utilizing anionically - conducting solid electrolytes have also been employed to study carbides, borides, phosphides, and sulfides(37).

Cation - conducting solid electrolytes have also been used to make similar measurements on solids(31-33). An early alkali metal example was the evaluation of the difference in potassium activity between two potassium oxide bronze samples using potassium beta alumina(45). There have been many subsequent applications.

Similar methods have been employed recently to study lithium systems using molten salt electrolytes instead of a solid electrolyte. The combination of coulometric titration and potential measurement techniques has been used over wide ranges of composition, including a number of intermediate phases in the Li - Sb(46), Bi(46), Al(47), Sn(48), Si(49), Cd(50), In(51), and Ga(52) binary systems.

This experimental method has also been applied to the study of several ternary lithium - transition metal - oxide systems(53,54), providing thermodynamic data on individual phases as well as elucidating the reaction mechanism that gives rise to the various potential plateaus observed when such materials are employed as electrodes in lithium battery systems.

Active Applications of Galvanic Cells with Solid Eelctrolytes and Chemically Different Electrodes

There have been many applications of solid electrolytes to control or modify the composition, and hence the properties, of materials which act as electrodes, or which are in contact with the electrodes. Some of these applications involve a combination of the imposition of a potential or a current or charge, along with a sensing function.

If a solid electrolyte has a very low electronic transference number, the amount of mass transported is proportional to the electric charge passed. This has been the basis of the use of such materials as coulometers and timers, and discussions of such uses can be found in a number of places(55,56). Many applications involve the deposition or "stripping" of the electroactive species from a chemically inert electrode. The potential across the galvanic cell undergoes a significant (e.g. 0,5 to 1.0 volt) step with the appearance or disappearance of this species.

There are a number of mixed conductors whose optical absorption coefficient in the visible range can be changed from essentially transparent to a dark blue-black by variation of the composition. Tungsten trioxide and the related "oxide bronzes" are good examples. The addition of hydrogen, lithium or sodium, for example, occurs by the insertion of both ions and electrons, the latter contributing to the optical absorption. In some cases, the rate of chemical diffusion of these guest species in the crystal structure is sufficiently fast and reversible that relatively rapid optical display devices can be constructed using such materials as electrodes in a galvanic cell(57,59).

While most of the developmental work to date has focussed upon the use of proton injection in WO_3 using aqueous electrolytes, there is the possibility of the discovery of materials and techniques such that all-solid-state devices of this type might be feasible. Work has been undertaken to use beta alumina as an electrolyte for this application(60) and efforts are underway to investigate lithium - based systems(61).

The electrical conductivity of some mixed conductors can be varied appreciably by changing the composition. This has lead to the development of variable resistors in which the mixed conductor is one electrode of a galvanic cell. The passage of current through the cell changes this property, which is measured by the use of separate electrodes.

Similarly, the electric potential across such a cell generally varies with the composition of the mixed conductor. This has been employed to make coulometers, integrators, analog memories, potential - sensitive timers, automatic sweep signal generators, and control systems(62,63). A number of such devices are on the market, primarily in Japan. One of the more useful mixed conductors is a silver selenide - silver phosphate solid solution, which has an almost linear relationship between composition and potential (0 to 150 mV) relative to silver(64).

Another property of a mixed conductor was employed in an electrically controllable and renewable oxygen getter based upon the oxygen affinity of reduced zirconia(65,66).

A galvanic cell can be used to control the activity of one of the species in a mixed conductor which has an appreciable vapor pressure. By measurement of the current through the cell as a function of potential under proper conditions, one can obtain information about species in the gas phase. This is the so-called "electrochemical Knudsen cell"(67).

One can use observations of the activity - dependence of the kinetics, as measured by the use of a galvanic cell, to determine the critical intermediates in a heterogeneous reaction(68), and measurements of the interfacial impedance in the present of various gases have also provided information about rate - controlling processes in such reactions(69,70).

Solid state cells have also been used in an active sense to greatly enhance the rate of both decomposition(71,72) and synthesis(73,74) reactions, with very large effects in some cases.

Another broad area of application of solid state ionic techniques involves the measurement of kinetic processes inside solids. Information on quantities such as the chemical diffusion coefficient, self diffusion coefficient, partial ionic, electronic and hole conductivities, parabolic tarnishing rate constant, and the mobilities of electrons and holes can be obtained in a number of cases. This approach often provides unusual precision with relative ease and rapidity. This area is too extensive to discuss here. General principles are reviewed elsewhere(75).

Applications Based Upon the Properties of the Solid Electrolyte/Electrode Interface

A number of simple experiments have been made to evaluate the amount of charge that can be stored at an ionically - blocking solid electrolyte / electrode interface(53,54). It is generally found that such interfaces store between 1 and 10 microfarads/cm^2 of interfacial area. This is considerably greater than can be achieved with conventional electronic capacitors. By the use of multiple layer or fine particle configurations, very large values of capacitance density can be obtained. Examples are 7 farads/gm of positive electrode material in a silver rubidium iodide/graphite device(43). and 80 farads/cm^3 of electrode volume in a beta alumina system(55,56). Because of these huge capacitance values, such devices can only be used at relatively low frequencies, for their time constant is the product of the capacitance and series resistance. The latter is typically dominated by the resistivity of the solid electrolyte.

Closing Comments

It should be obvious from this brief overview that there is a great wealth of potential applications available for solid ionic and mixed conductors in the future. There surely are many more that have not been included.

In most cases, applications will not involve solid electrolytes alone, but in configurations in which they will be coupled to appropriate electrodes. In some cases the electrodes may be simple porous metals. But in others, they will involve mixed conductors. In the author's opinion, an increasing fraction of the research and development community's attention will be directed toward the many potential applications of mixed conductors, either alone or in conjunction with solid ionic conductors.

Acknowledgement

The author gladly acknowledges with gratitude the important contributions by his students and associates to the ideas and results reported here. Much of the work in this area at Stanford University has been supported by the U.S. Department of Energy under Lawrence Berkeley Laboratory subcontracts 4519410 and 4536310.

References

1. W. Nernst, German Patent 104872 (1897)

2. W. Nernst, Z. Elektrochem. 6, 41 (1900)

3. W. Nernst and W. Wild, Z. Elektrochem. 7, 373 (1901)

4. H. Reynolds, Ph.D. Dissertation. Göttingen (1902)

5. C. Wagner, Naturwissenschaften 31, 265 (1943)

6. E. Baur and H. Preis, Z. Elektrochem. 43, 727 (1937)

7. F. Haber et al., Z. Anorg. Chem. 41, 407 (1904)

8. F. Haber et al. Z. Anorg. Chem. 51, 245 (1906)

9. F. Haber et al., Z. Anorg. Chem. 51, 289 (1906)

10. W. D. Treadwell et al., Z. Elektrochem. 22, 415 (1916)

11. N. W. Taylor, J. Amer. Chem, Soc. 45, 2865 (1923)

12. H. Reinhold, Z. Elektrochem. 40, 361 (1934)

13. W. D. Treadwell et al., Helv. Chim. Acta 19, 1255 (1936)

14. B. A. Rose, G. J. Davis and H. J. T. Ellingham, Disc. Faraday Soc. 4, 154 (1948)

15. E. Stuhlinger, personal communication (1978)

16. R. T. Johnston, Jr. and R. M. Biefeld, Mat. Res. Bull. 14, 537 (1979)

17. P. Jagannathan, et al., in "Solid Electrolytes and Their Applications", ed. by E. C. Subbarao, Plenum Press (1980), p. 201

18. J. Weissbart, W. Smart and T. Wydeven, Aerospace Medicine 40, 136 (1960)

19. S. K. Adams and R. E. W. Castleton, J. Amer. Ceram. Soc. 53, 117 (1970)

20. N. Weber, Energy Conversion 14, 1 (1974)

21. T. K. Hunt, N. Weber and T. Cole, "Proc. 10th IECEC Addendum" (1975), p. 231

22. N. Weber, in "Superionic Conductors", ed. by G. D. Mahan and W. L. Roth, Plenum Press (1976), p. 37

23. A. R. Ubbelohde, Carbon 14, 1 (1976)

24. J. E. Fisher, in "Molecular Metals", Plenum Press (1979), p. 281

25. J. M. Lalancette, G. Rollin and P. Dumas, Can. J. Chem. 50, 3058 (1972)

26. M. S. Whittingham and L. B. Ebert, in "Physics and Chemistry of Materials With Layered Structures", Vol. 6, ed. by F. Levy, D. Reidel, Dordrecht, Holland (1979), p. 533

27. H. Selig and L. B. Ebert, Adv. in Inorg. Chem. & Radiochem. **23**, 281 (1980)

28. M. A. M. Boersma, Cat. Rev.-Sci. Eng. **10**, 243 (1974)

29. M. A. M. Boersma, in "Advanced Materials in Catalysis", Academic Press (1977), p. 67

30. J. L. Callahan, R. K. Grasselli, E. C. Milberger and H. A. Strecker, Ind. Chem.Prod. Res. Devel. **9**, 134 (1970)

31. C. Wagner, J. Chem. Phys. **21**, 1819 (1953)

32. K. Kiukkola and C. Wagner, J. Electrochem. Soc. **104**, 308, 379 (1957)

33. C. Wagner, Proc. Int. Comm. Electrochem. Thermo. Kinetics (CITCE) **7**, 361 (1957)

34. "Lambda-Sond: Complete Emission Control?", Automotive Engineering **85**, 45 (1977)

35. G. T. Engh and S. Wallman, SAE Paper 77, 0295, Feb. 1977

36. E. Hamann, H. Manger and L. Steinke, SAE Paper 770401. Feb. 1977

37. M. S. Chandrasekharaiah, et al., in "Solid Electrolytes and Their Applications", ed. by E. C. Subbarao, Plenum Press (1980), p. 99

38. G. R. Fitterer, in "Metallurgical Chemistry", ed. by O. Kubaschewski, Inst. of Metals (1972), p. 589

39. Proceedings of the 3rd International Meeting on Solid Electrolytes - Solid State Ionics and Galvanic Cells, ed, by T. Takahashi, K. Fueki, B. B. Owens and C. A. Vincent,Solid State Ionics, Vol. 3/4

40. W. Weppner and R. A. Huggins, in "Fast Ion Transport in Solids", ed. by P. Vashishta et al., North-Holland (1979), p. 475

41. D. Harame, Ph. D. Dissertation, Stanford Univ. (1987)

42. J. D. Fray, British patent 2,025,626 (1980)

43. "Ion-Selective Electrodes", ed. by R. A. Durst, Nat. Bur. of Stand. Special Pub. 314 (1969)

44. S. Seetharaman and K. P. Abraham, in "Solid Electrolytes and Their Applications", ed. by E. C. Subbarao, Plenum Press (1980), p. 127

45. M. S. Whittingham and R. A. Huggins, in "Solid State Chemistry", ed. by R. S. Roth and S. J. Schneider, Nat. Bur. of Standards Spec. Pub. 364 (1972), p. 139

46. W. Weppner and R. A. Huggins, J. Electrochem. Soc. 125, 7 (1978)

47. C. J. Wen, B. A. Boukamp, R. A. Huggins and W. Weppner, J. Electrochem. Soc. 126, 2258 (1979)

48. C. J. Wen and R. A. Huggins, J. Electrochem. Soc. 128, 1181 (1981)

49. C. J. Wen and R. A. Huggins, J. Solid State Chem. 36, 2233 (1981)

50. C. J. Wen, Ph. D. Dissertation, Stanford Univ. (1980)

51. C. J. Wen and R. A. Huggins, Mat. Res. Bull. 15, 1225 (1980)

52. C. J. Wen and R. A. Huggins, J. Electrochem. Soc. 128, 1636 (1981)

53. N. A. Godshall, I. D. Raistrick and R. A. Huggins, Mat. Res. Bull. 15, 561 (1980)

54. N. A. Godshall, Ph. D. Dissertation, Stanford Univ. (1980)

55. J. H. Kennedy, in "Physics of Electrolytes", ed. by J. Hladik, Academic Press (1972), p. 931

56. M. Voinov, in "Electrode Processes in Solid State Ionics", ed. by M. Kleitz and J. Dupuy, D. Reidel Pub. Co., Dordrecht, Holland (1976), p. 431

57. I. F. Chang, in "Nonemissive Electrooptic Displays", ed. by A. R. Kmetz and F. K. von Willisen, Plenum Press (1976), p. 155

58. B. W. Faughnan and R. S. Crandall, Topics in Appl. Physics 40, 181 (1980)

59. T. Oi, in Annual Review of Materials Science, Vol 16, ed. by R. A. Huggins, Annual Reviews, Inc. (1986), p. 185

60. M. Green and K. S. Kang, Thin Solid Films 44, L19 (1977)

61. S-K. Joo, I. D. Raistrick and R. A. Huggins, Mat. Res. Bull. 20, 897 (1985)

62. T. Takahashi and O. Yamamoto, J. Appl. Electro. 3, 129 (1973)

63. T. Takahashi, in "Superionic Conductors", ed. by G. D. Mahan and W. L. Roth, Plenum Press (1976), p. 379

64. T. Takahashi and O. Yamamoto, Japanese patent 73 55396 (1973)

65. J. Fouletier and M. Kleitz, Vacuum 25, 307 (1975)

66. P. Fabry and M. Kleitz, in "Electrode Processes in Solid State Ionics", ed. by M. Kleitz and J. Dupuy, D. Reidel Pub. Co. Dordrecht, (1976), p. 331

67. R. J. Ratchford and H. Rickert, Z. Elektrochem. 66, 497 (1962)

68. H. Schmalzried, Prog. Solid State Chem. 13, 119 (1980)

69. V. B. Tare et al., in "Solid Electrolytes and Their Applications", ed. by E. C. Subbarao, Plenum Press (1980), p. 165

70. T. M. Gür, Ph. D. Dissertation, Stanford University (1976)

71. S. Pancharatnam, R. A. Huggins and D. M. Mason, J. Electrochem. Soc. 122, 869 (1975)

72. T. M. Gür and R. A. Huggins, J. Electrochem. Soc. 126, 1067 (1979)

73. T. M. Gür and R. A. Huggins, Solid State Ionics 5, 563 (1981)

74. T. M. Gür and R. A. Huggins, "Proc. Symp. on Electrocatalysis", ed. by W. E. O'Grady, P. N. Ross, Jr. and F. G. Will, Electrochem. Soc. (1982), p. 85

75. W. Weppner and R. A. Huggins, in "Annual Review of Materials Science", ed. by R. A. Huggins, Annual Reviews Inc. (1978), p. 269

www.ingramcontent.com/pod-product-compliance
Ingram Content Group UK Ltd.
Pitfield, Milton Keynes, MK11 3LW, UK
UKHW022109170926
8624IPUK00024B/361